Communication Technology

Today and Tomorrow

Second Edition

Communication Technology Today and Tomorrow

Second Edition

Mark Sanders
Virginia Polytechnic Institute

Glencoe
McGraw-Hill

New York, New York Columbus, Ohio Woodland Hills, California Peoria, Illinois

Contributing writer for Section III: John Bies

Consultant for Section VI: Robert Bloompott, Illinois Central College

Reviewers for the second edition:
Mike Bobbitt
Hominy High School
Hominy ,OK

Douglas W. Cotton
Lyon County Jr/Sr High School
Eddyville, KY

Dedication:
For Gail Marie, Nicole Grace, and Rachel Leigh

Glencoe/McGraw-Hill

A Division of The ***McGraw-Hill*** *Companies*

Send all inquiries to:
Glencoe/McGraw-Hill
3008 W. Willow Knolls Drive
Peoria, IL 61614

ISBN: 0-02-838759-7 (Student Text)

ISBN 0-02-838760-0 (Student Workbook)
ISBN 0-02-838761-9 (Teacher's Resource Binder)

Printed in the United States of America

4 5 6 7 8 9 10 071 06 05 04 03 02 01 00

TABLE OF CONTENTS

FEATURES (stories about the history, applications, and impact of technology)

Health & Safety

Technolinks

Technology's Impact on Your World

User's Guide to Technology

COMMUNICATION TECHNOLOGY TODAY . . .

How do people use communication technology today? They use it to help them communicate faster, farther, and more often than ever before. They use it to exchange information, not only with other people but also with machines and with animals. At work, people are using communication technology to become more creative and more productive. In school, they're using it to learn new things and explore their own interests and abilities. At home, people are using communication technology to stay informed, to be entertained, and to keep in touch with family and friends.

© 1987 Pixar

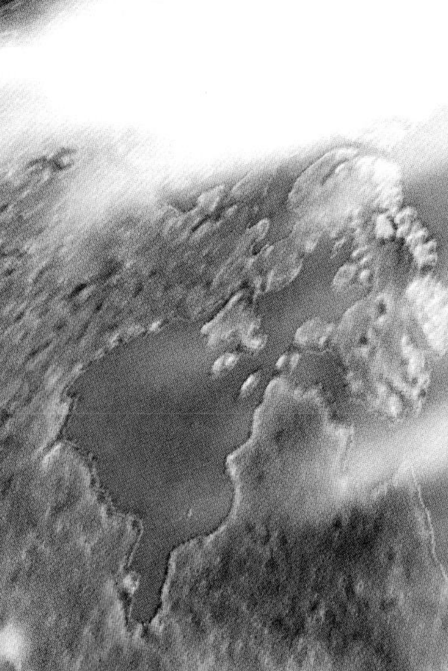

How will people use communication technology in the future? During the 1990s, we can expect improvements in existing devices as well as exciting new products. The technology for virtual reality continues to improve. The photo below shows a virtual reality game. Computers are taking on new forms and functions. For example, wearable computers are being used by some technicians and paramedics to diagnose problems and to communicate with headquarters (see bottom photo). An electronic desk blotter (see photo at far right) contains input and output devices that communicate with a separate computer unit via infrared rays. Other forms of wireless communication are becoming commonplace. Laptop computers can be linked to a cellular phone for wireless transmittal of data. One day soon, we may each have one personal number that will signal our telephone, pager, or any other communication device.

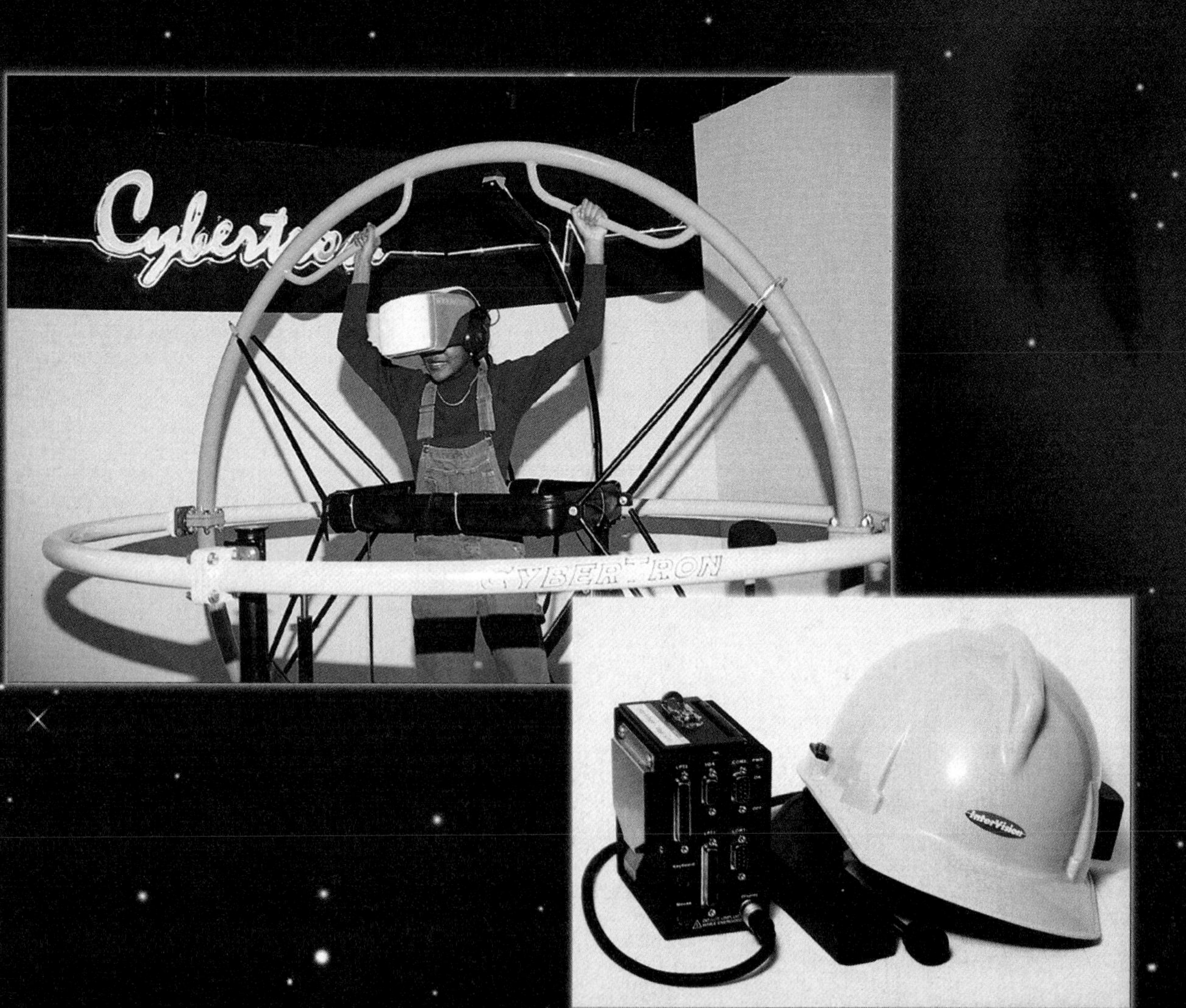

SECTION I

Introduction to Communication Technology

Chapter 1: Understanding Communication Systems
Chapter 2: The Changing Nature of Communication Technology
Chapter 3: The Impact of Communication Technology

If you ask ten of your friends what "technology" means, you will probably get ten different answers. The term is used so commonly today that it means different things to different people. When they hear "technology," some people think of computers. Others think of hammers and saws. Still others think of ideas.

Technology is more than just computers and tools. It's more than ideas. **Technology** is using knowledge, tools, and skills to solve problems. Technology suggests "doing."

The study of technology can be approached in many different ways. In technology education, we usually start by looking at technical processes. We are involved in the hands-on application of what we know about our technological world.

To make technology easier to study, we can divide its subject matter into three general areas: communication, production, and energy/power/transportation. Each of these areas may be subdivided. Production, for example, includes construction and manufacturing.

In this book, we will be focusing on the technologies that help us communicate. Communication is the sharing of information, thoughts, and ideas. You'll see that the different subject areas have been grouped into systems. The communication systems you will learn about in this book are

- Data Communication Systems
- Technical Design Systems
- Optic Systems
- Graphic Production Systems
- Audio and Video Systems

Why bother to call them "systems" instead of using terms such as "printing" or "television"? You will see as you read through this book that communication technology includes more than is suggested by the traditional names. For example, there is much more to reproducing graphic images today than putting ink on paper. In fact, many graphic production systems do not use ink. Therefore the term "printing" isn't broad enough. The same is true in each of the other areas. Communication technology is much more complex than it used to be!

CHAPTER 1

Understanding Communication Systems

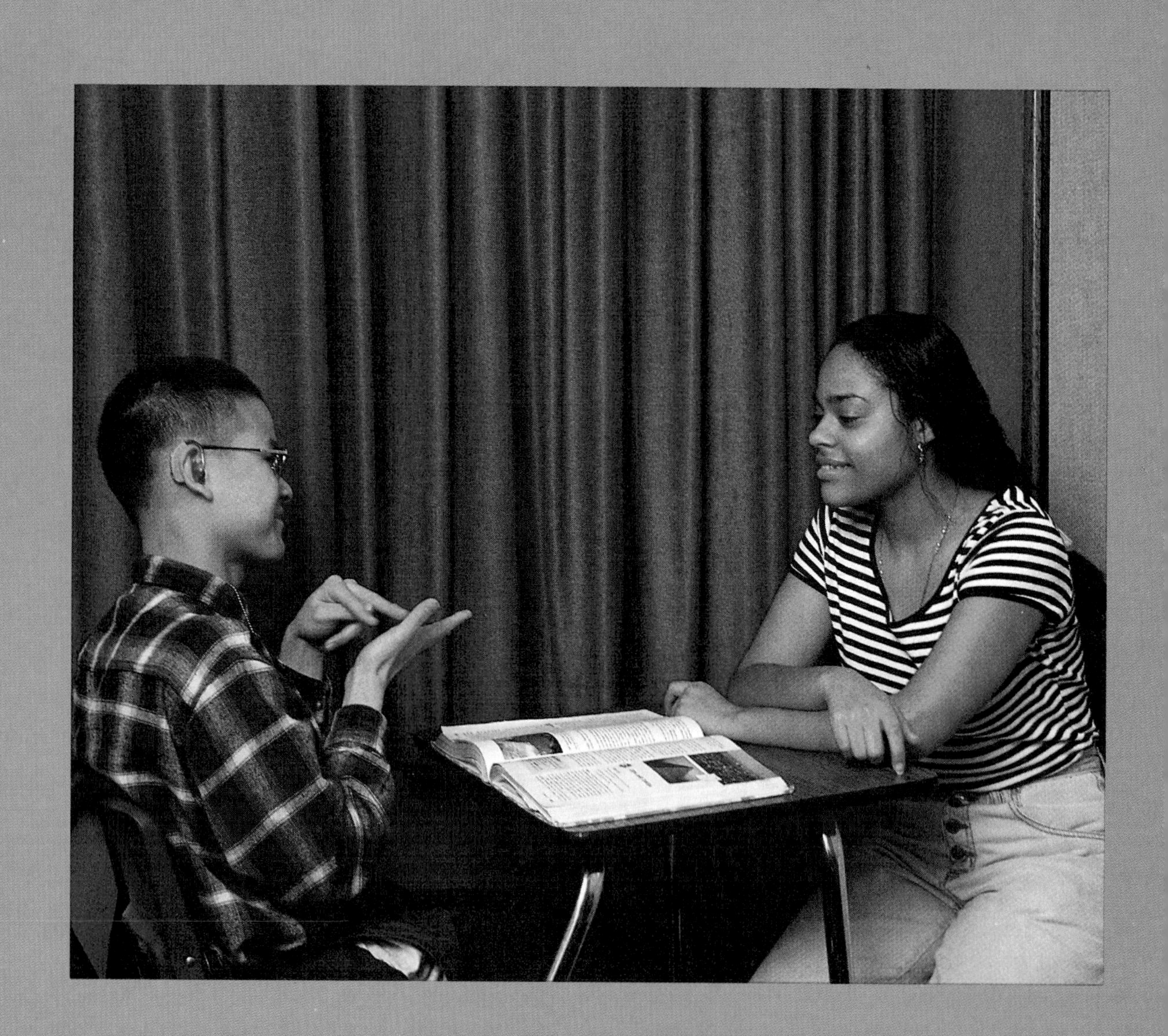

If technology is using knowledge, tools, and skills to solve problems, just what is **communication technology**? We might say it is using knowledge, tools, and skills to communicate.

Of course, this definition covers a lot of ground. We communicate in many different ways. The telephone clearly fits into our definition, but so do the hand signals used by a police officer directing traffic. You may not have thought of it that way, but the furnace thermostat in your home is also a communication device.

This chapter, then, will set the stage for the study of communication technology. It will also introduce you to each of the communication systems covered in the text: data communication, technical design, optics, graphic production, and audio and video systems.

Terms to Learn

communication technology
computer control systems
development
feedback
input
output
problem solving method
process
research
technical communication systems
telecommunication
universal systems model

As you read and study this chapter, you will find answers to questions such as:

- What is the universal systems model?
- What are the basic parts of a communication system model?
- What are the concepts that help us understand communication systems?
- Are humans the only ones who communicate?
- What are the basic problem-solving steps used in technology?

UNIVERSAL SYSTEMS MODEL

One way of studying technological systems is by examining the inputs, processes, and outputs of that system. Consider the automobile as a "technological system" for a moment. On a very basic level, you put gas in the tank (input), the engine burns the gas (process), and the car moves (output). However, there is really a lot more to the automobile than this. If we study the automobile in a broader sense, we must look at what it took to build it: people, money, energy, and so forth. Also, there's a lot more output from a car than just motion. What about pollution? How about the jobs created by the automobile industry? What about the changes in society that came about as a result of the automobile?

Every system includes input, process, and output. Therefore this way of describing systems is called the **universal systems model**. Fig. 1-1. Looking at each part of this universal systems model forces you to think about aspects of technology you may not have considered. Take inputs, for example. Regardless of the technology, its development and use require **input**: information, materials, energy, financial resources, and human effort. Can you name a technology that does not rely upon these inputs?

The **process** is what happens to the input. The process part of the model includes technical processes as well as the concepts and principles upon which the technology is based. An automobile engine burns gas. That process is possible because of the principles of internal combustion. The brakes rely upon hydraulic pressure principles, and so forth.

Most systems have many outputs. **Outputs** are the result of the process. Some are desirable; some are not. Movement of the car is a desirable output, but the pollution from the car's exhaust is not. Use of automobiles affects our environment, our economy, our entire society. When we consider outputs, we need to look at the impact these have on our world.

Finally, there is a feedback loop in the model. **Feedback** is something that happens as a result of the outputs. It has an impact on the overall system. Poor gas mileage in the 1970s led to the design of smaller engines that got more miles per gallon. In the preceding example, the input/process/output model was applied to a transportation system. It may just as easily be applied to a communication system. As you study various communication systems in this book, keep the universal systems model in mind. It will give you a broader picture of the technological system you are studying.

Fig. 1-1. The universal systems model describes systems in terms of input, process, output, and feedback.

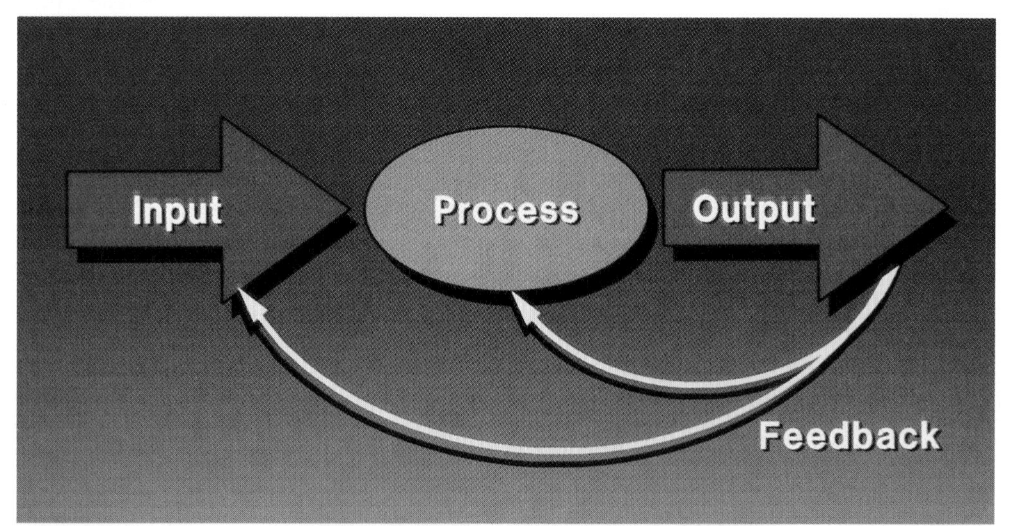

COMMUNICATION SYSTEM MODEL

There are many different communication technologies, but all can be thought of in terms of their basic parts. First, there must be a *message*. It could be in the form of a picture, a sound, or the written word, among others. Next, there must be a *sender* to "launch" the message. The message travels through a communication *channel* to the *receiver*.

These basic parts make up the communication system model. All communication technologies may be thought of in terms of the message, sender, receiver, and channel over which the message is sent. Fig. 1-2.

Something as complex as satellite communication may be simplified if you look at it in terms of this model. Consider a newscast from Japan. The program is the message. The sender is a transmitter and antenna. The communication channel is electromagnetic radiation travelling through our atmosphere (more about this in Section VI). The receiver is your television set. Actually the message is first received by an antenna and then relayed to your television set, but you get the idea: message, sender, channel, receiver. You should be able to fit all communication systems into this model.

COMMUNICATION CONCEPTS

When you study science, you zero in on the principles and concepts of science. Gravity, for example, is a scientific concept with which you are familiar. The concept of gravity helps explain the motion of planets and why powerful rockets are needed to launch spacecraft. It also explains how you can weigh 150 pounds on earth and 25 on the moon without going on a diet.

Fig. 1-2. Every communication system includes a message, sender, channel, and receiver. Most of them also include feedback.

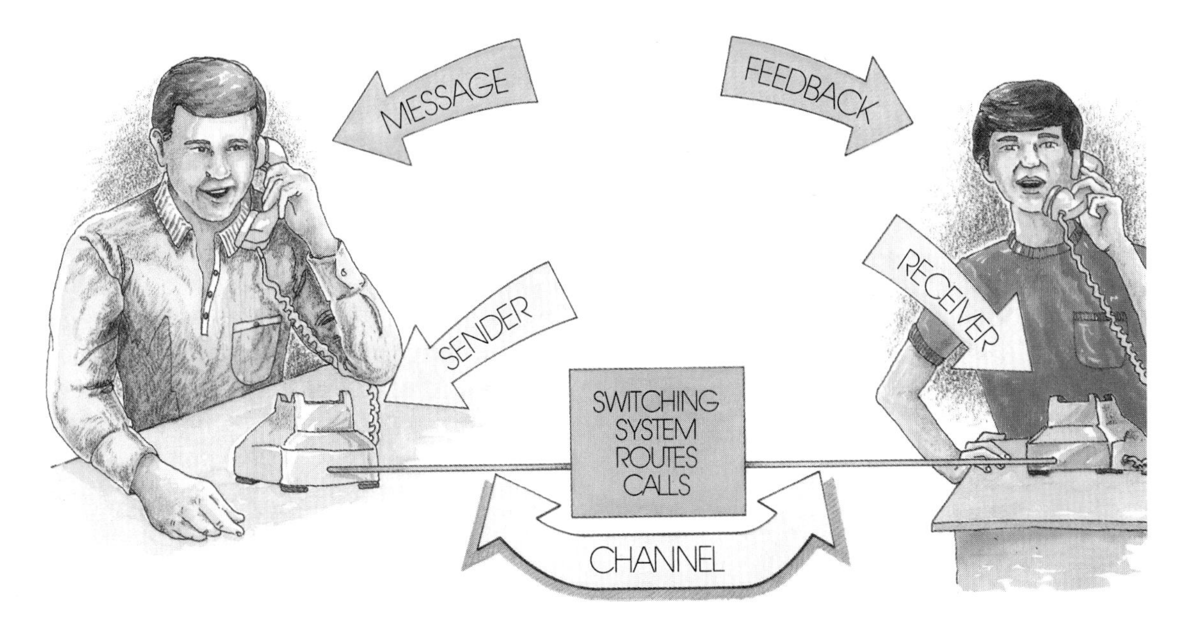

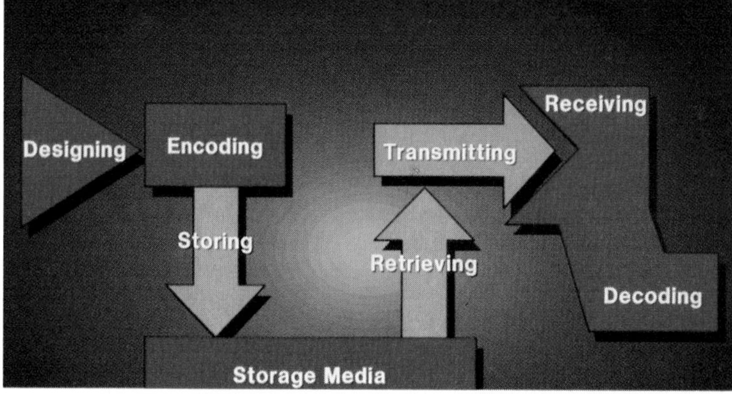

Fig. 1-3. Understanding these concepts will help you understand the communication process.

SCIENCE FACTS

Gravity

The English scientist Isaac Newton (1642-1727) discovered some basic principles of gravity. He found that the greater the mass (amount of matter), the greater the force of gravity. That's why you would weigh less on the moon; the moon has less mass than the earth. Newton also found that the force of gravity decreases over distance. That's why the moon doesn't fall to the earth. At that distance, the earth's gravity isn't strong enough to bring the moon closer, but it is strong enough to keep the moon in orbit around the earth.

Newton believed gravity was an unseen force that pulled objects toward one another. Albert Einstein (1879-1955) developed a different explanation as part of his general theory of relativity. Imagine a sheet stretched flat. If you put a tennis ball on it, the sheet will bend a little. If you put a basketball on it, the sheet will bend more. If the tennis ball is close enough, it will roll towards the basketball. According to Einstein, the universe behaves in a similar manner. The mass of an object actually curves the space-time around it. The greater the mass, the greater the curvature of space-time.

We might very well think of communication in terms of the basic communication concepts. One set of concepts that is very important includes designing, encoding, storing, retrieving, transmitting, receiving, and decoding. Fig. 1-3.

To help explain these concepts, consider a newspaper illustration produced by computer. A growing number of newspaper illustrations are being created this way.

First, an artist must design the illustration. We'll call it the "message," since messages can be drawings as well as words. The process begins in the artist's head and may be refined on paper or directly on a computer.

Once the message has been designed, the computer encodes it (turns it into code). One common way computers do this is with a "bit map." Think of the screen of the computer as a sheet of graph paper with thousands of squares. If you painted some of these tiny squares black and left the others white, you would create a patterned picture. That's the "bit-mapped" coding system. The computer simply encodes the picture this way.

For convenience, the message is saved on some sort of magnetic media, usually a floppy or hard disk. This is the storing concept. Likewise, the message may be "called up" from the storage disk. This is the concept of retrieving.

The message may then be transmitted over telephone lines or by satellite to newspapers across the country. Does that sound futuristic? It's not. Newspapers have been communicating news and illustrations "over the wire" for decades!

The message has been sent, but the communication is not completed. There must be a receiver. This particular message will be received as computer data. It will be in the form of "bit map" code, as described earlier. It must now be decoded into the graphic message that will appear in the newspaper. In this case, this is done by a computer output device such as a printer.

These basic concepts may be found in any communication system. Language, for example, may be broken down into these basic concepts. Of course, different parts of the brain take the place of the computer, but the concepts remain the same.

MODES OF COMMUNICATION

Perhaps the most common type of communication is one person talking to another. We can call this "human-to-human" communication. People aren't the only ones that communicate, though. Animals and machines also communicate.

Crossover occurs when communication takes place among these three modes. Humans can communicate with machines, for example. This is human-to-machine communication. It also can happen the other way around: machine-to-human communication.

Human Communication

Technology is involved when you talk to a friend and when you read this book. Think about it this way: What if there were no words? How would you communicate? In this sense words are tools we use. They are sounds that we've all agreed mean certain things. Likewise, the letters in our alphabet are shapes we've agreed to use to represent specific sounds.

Communication is more difficult for the visually or hearing impaired. However, it still takes place. American Sign Language (ASL) is a communication system developed for those who cannot hear. Fig. 1-4. Braille consists of different patterns of raised dots that represent the alphabet. This type of communication makes it possible for blind persons to read.

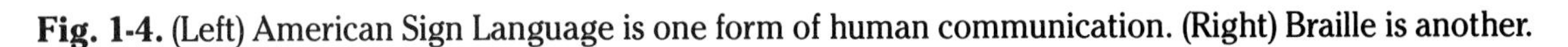

Fig. 1-4. (Left) American Sign Language is one form of human communication. (Right) Braille is another.

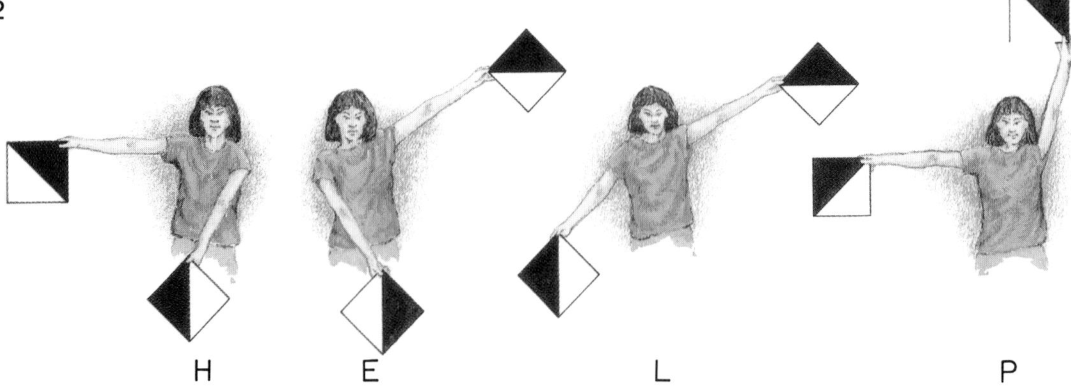

Fig. 1-5. In semaphone, the flags and their positions indicate different letters.

Language and alphabets are good examples of the application of knowledge to solve a problem. They are, in this sense, communication technologies. The problem they help to solve, of course, is human communication.

Language works well when the sender and receiver are standing next to each other, but what if they are separated by distance? Over the centuries, human beings have devised hundreds, perhaps thousands, of ways to communicate with those far away. **Telecommunication** is communicating over distance.

Early forms of telecommunication included signals tapped out on drums and the smoke signals of native Americans. Paul Revere's now famous "one if by land, two if by sea" lantern code communicated the route taken by the British during the American Revolution. Semaphore is a telecommunication system that uses two flags held in different positions to represent letters in the alphabet. Ships have used semaphore since 200 B.C. Fig. 1-5.

Police, firefighters, surveyors, athletes, and movie directors also use signals to communicate. Can you describe these and other communication systems used among humans?

Scientists have noted complex systems of communication within many different animal species. You don't have to be a scientist to realize the ants invading your picnic are well organized. We often refer to ants as an "army," partly because of their orderly behavior. Obviously, communication is going on among them.

It is well known that honey bees communicate with each other by "dancing." Fig. 1-6. After a honey bee locates a new source of food, it returns to the hive and dances in a way that communicates the location to the other bees. The basic step is a figure eight, with lots of buzzing and wing flapping!

Dolphins are considered by some scientists to be among the most intelligent of all animals. They communicate with one another using a series of high-pitched sounds. Their squeals may not be understandable to us, but scientists believe they are part of a complex language.

Fig. 1-6. A honey bee dances its message to other bees.

SCIENCE FACTS

Whale Songs

Among the most intriguing forms of animal communication are the songs of the whales. A whale song consists of clicks, moans, chirps, and whistles. The whales make these sounds by forcing air through valves and flaps located below their blowholes.

There is a pattern to each song, and some songs last up to thirty minutes. Different groups of whales sing different songs, and what's popular one year may be "out" the next.

Scientists aren't sure why whales sing. They may be trying to attract mates or warn away competitors or, perhaps, they simply enjoy singing!

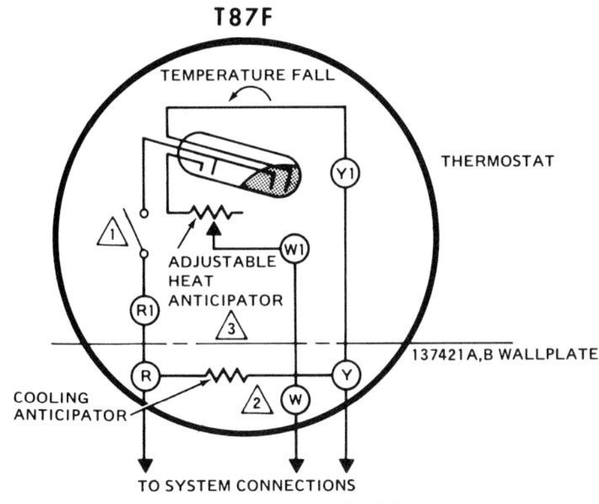

1. MODEL WITH POSITIVE OFF SWITCH OPENS THE THERMOSTAT CIRCUIT WHEN SET POINT DIAL IS MOVED TO THE OFF POSITION.
2. MAKE SYSTEM WIRING CONNECTIONS TO TERMINALS ON 137421A,B WALLPLATE.
3. R1, W1, Y1 TERMINALS ON THERMOSTAT ARE DIRECTLY CONNECTED TO R, W, Y TERMINALS ON WALLPLATE WHEN THERMOSTAT IS MOUNTED ON WALLPLATE.

3884A

Fig. 1-7. This is a schematic of a thermostat. How do you think the principle of feedback applies to this device?

Machine Communication

When you work at a computer, you are communicating with a machine. When the text you type appears on the monitor, the machine is communicating back to you. When you tell the computer to print out the text, the machine talks to the printer, another machine. These are examples of human to machine, machine to human, and machine to machine communication.

Machine communication is all around us. A thermostat probably controls the temperature of the room you are in. It does this in two stages. First, it senses the current temperature. (Using a machine to collect data is called "instrumentation.") If the room is too cold, the thermostat tells the heating system heat is needed. (When one machine directs another, it is called "control.") Fig. 1-7.

The most complex machine communication systems involve computers. **Computer control systems** collect input, process data, and then output controlling signals to other devices. Fig. 1-8. (A computer control system is a good example of the systems model—input/process/output—shown in Fig. 1-1.)

A typical computer control system collects data with one or more sensors. Sensors can detect such things as light, pressure, temperature, or sound. Computer control systems are now used routinely to control the environments of large buildings. They handle heating, cooling, lighting, and air-exchange systems.

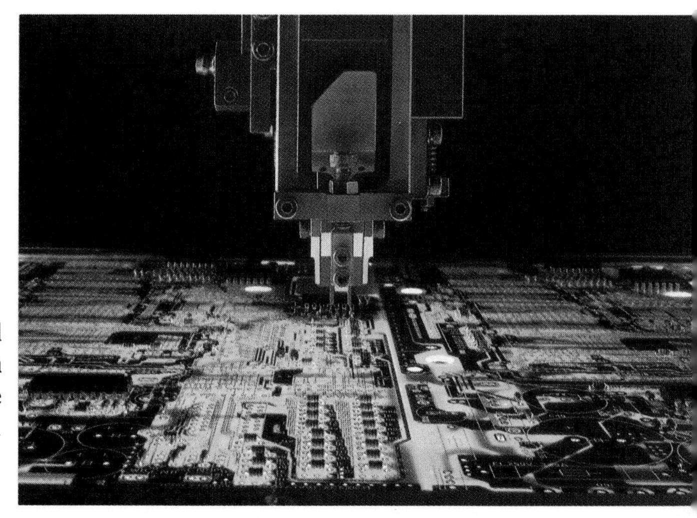

Fig. 1-8. A robot is one type of computer-controlled device. In many of today's robots, visual and touch sensors are important parts of the control system. The robot shown here is assembling a personal computer.

TYPES OF COMMUNICATION SYSTEMS

An amazing number of communication systems exist in the world around us. This book concentrates on **technical communication systems**. These are communication systems that depend upon specific tools and equipment.

Data Communication

Data communication (computer) systems play a very important role in modern communication. They have become part of all the other technical communication processes you will be studying.

In the data communication section, you'll learn the basic principles upon which computer systems are based. How computers gather data, how they process these data, and what they may do with the data after processing will all be discussed.

Technical Design

Engineers, architects, and industrial designers all require plans for their work. Sketches are fine for the idea stage, but a product or a building must be constructed from a plan. That plan must be accurately drawn.

Technical design systems are used to produce these drawings. In the past, technical design has been referred to as drafting, mechanical drawing, and engineering design. None of these titles suggests all that technical design systems actually accomplish.

The tools and equipment used range from T-square and drawing board to computers. Most technical work is now done using computer-aided design (CAD) systems.

Optics

Optic systems use light to transmit and record the message. The most common type of optic system is photography. Photographic optic systems focus light onto a recording medium, such as film. Light, a lens, and a way of recording the image are needed.

You can learn a lot about most optic systems by studying photography. The principles, materials, and equipment used mirror those in other optic systems. For this reason, the section on optic systems focuses on photography.

There is more to optic systems, though, than just photography. You will learn about the principles of light, lenses, and light-sensitive materials. You'll also study about fiber optics, lasers, and holograms. These are state-of-the-art optic systems.

Graphic Production

"Ink on paper" used to be the way graphic production systems were described. You probably know them simply as "printing." There is still a great amount of ink being put on paper, but many more ways of reproducing images now exist.

Sometimes the printing is not even done with ink. A photocopy machine, for example, fuses a powder to the paper to form the image. Often, the image is reproduced on a surface other than paper. Metal, plastic, and fabric are frequently used.

Audio and Video

The role of audio and video communication systems in our lives cannot be overstated. The average American watches six or seven hours of television a day! The radio has remained popular as well. Many homes have several radios and TV sets.

The section on audio and video systems deals primarily with radio and television. There is a great deal more to this type of communication than most people realize. While you may tune in to several different stations from time to time, there are millions of broadcast messages bouncing around in our atmosphere!

In this section, you'll learn how radios and televisions work, as well as telephones, record players, and tape players. The basic principles underlying broadcast communication will be discussed. You'll also learn about some interesting uses for audio and video systems.

Integrated Systems

As you study about communication systems, remember that the areas overlap. For instance, computers are used in each of the areas already discussed. The lenses that are described under optic systems show up in video cameras. Color theory is found in optic systems and in graphic production, and so on.

It is important to realize that no communication system exists all by itself. For example, telephone lines can be used to transmit computer data as well as human voices. The telephone lines may be optic fibers rather than copper wires. This *integration* of the systems will be discussed further in the next chapter.

As you read about the communication systems, keep thinking about the impact each has made on the world around you. More importantly, what impact have they made on your world? What would your day be like without computers, photographs, movies, printed material, radio, television, and telephones?

RESEARCH AND DEVELOPMENT

We hear a lot about "cutting-edge" technologies. New ways of doing things are always being found. These new techniques are discovered through the processes of research and development.

Research is the search for new knowledge. This new knowledge may or may not have an immediate use in our world. New knowledge is important for its own sake. We never know all the places knowledge will lead us.

Development, also called applied research, is done with the idea of solving a specific problem. It results in a product or method. The discovery of electricity was the result of research. The invention of the telephone, which used electricity, was development. Fig. 1-9.

Fig. 1-9. (Left) Ben Franklin's discovery of electricity added to our knowledge. (Above) Alexander Graham Bell used the knowledge of electricity to develop the telephone.

Problem Solving

Thomas Edison said, "Genius is one percent inspiration and 99 percent perspiration." In other words, using technology to find solutions is mostly hard work. The way in which researchers go about discovery and invention is often called the **problem solving method**. This method has a number of steps. Fig. 1-10.

1. *The problem is defined.* A clear description of the problem is given.
2. *Research is done.* Information relating to the problem is gathered and studied.
3. *Possible solutions are identified.* As many different solutions as possible are imagined. This step is often referred to as "brainstorming" or "ideation."
4. *Possible solutions are evaluated.* The list of solutions may be tested by trying them out. This often means building a model or "prototype."
5. *The problem may be revised.* When testing solutions, researchers often learn things that lead them to go back and change the definition of the problem. This feedback may result in a better solution.
6. *The best solution is discovered.* After steps 1-5 are repeated enough times, a solution to the problem is reached.

This problem solving method is used all of the time in the area of communication technology. Fiber optic engineers use it to invent new ways of sending messages. Graphic designers use it to develop eye-catching messages. It's basically the same process, whether it's used by an engineer or a designer.

As you study the different systems described in this book, you'll notice ways the problem solving method can be used. Whether you are taking photographs, producing a video, or printing a business card, there will be problems to be solved. Following the six steps will help you find the best solutions.

Fig. 1-10. (Right) Philo T. Farnsworth, one of the inventors of television, demonstrated this "receiving cabinet" to reporters in 1935. When developing new technologies, inventors use the problem solving method (shown below).

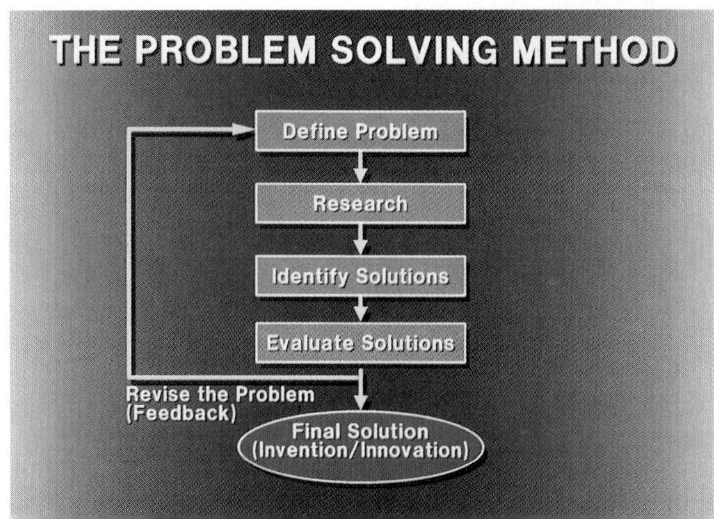

CAREERS IN COMMUNICATION TECHNOLOGY

A world of opportunity exists for those interested in communication careers. For instance, the printing industry claims to be the largest industry in the world! This is because there are more facilities in the world for printing than there are for any other industry. As you've just seen, though, graphic production is only one of the five general areas under the umbrella of communication technology. Careers are of all types: technical, creative, managerial, production, administrative, and so on.

A few of the many opportunities in communication technology are shown in Fig. 1-11. Specific careers are described at the end of each section in this book. For more detailed information, see the *Dictionary of Occupational Titles* in your school or local library.

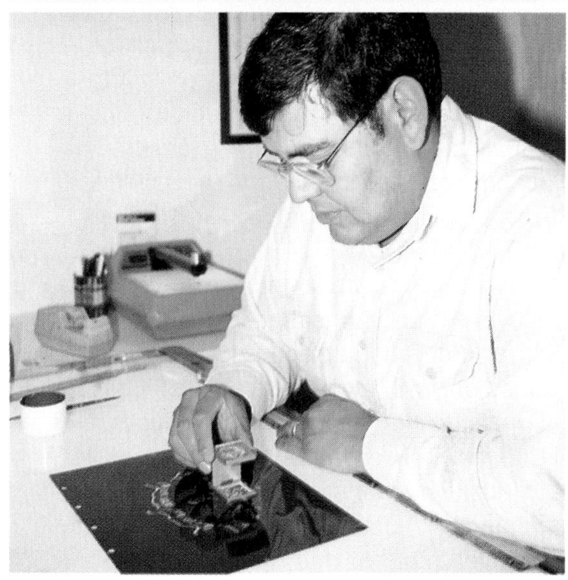

Fig. 1-11. Communication technology offers many career opportunities. Shown on this page are some of the ways people use communication technology on the job.

TECHNOLOGY AIDS THE DISABLED

When Alexander Graham Bell invented the telephone, he hoped it would help his mother and wife, who were deaf, hear. Although Bell's invention did not work out quite as he had planned, he did start something. Since that time many devices have been invented that do help disabled people communicate.

Have you ever noticed the words "Close-captioned for the hearing impaired" appear across the bottom of your TV screen? This means that with the use of a special device, deaf people can see words of dialogue flashed on the screen.

Technology has also come to the aid of the blind or visually impaired. An organization called Educational Tape Recording for the Blind uses volunteers to record textbooks for blind students. Other recordings, funded by the Library of Congress, bring "talking books" to those who cannot see well enough to read. Some radio stations are organized for the visually impaired and provide special programming just for them.

A "Brailler," which resembles a typewriter, creates characters in Braille. The machine has only six keys with different arrangements of dots and a space bar. More than one key is pressed at a time to create the raised dots of a Braille character. Computers are also available that translate print into Braille using a special printer. Other devices receive Braille input and then output the spoken word.

For those who are visually impaired but not blind, special closed-circuit TV systems magnify print up to 45 times using a zoom lens. The user then reads the TV screen.

For those with speech handicaps, two new devices, designed with the help of NASA, are now available. Many deaf people tend to speak with high-pitched voices. They cannot hear how they sound to others. To help them control their voice pitch, a device called Speech Teacher™ shows them their voice amplitude on a small monitor. Also on the monitor is an indicator of what the normal range should be. Users can then raise or lower the pitch as needed.

A second device, designed for people who cannot speak because of injury, cancer, or other diseases, is built into a denture. Some components are so small they can fit like fillings into a couple of artificial teeth. The devices produce tones that can then be shaped into words using the tongue, teeth, and lips.

People who are paralyzed or confined to wheelchairs may have a different set of difficulties in dealing with the world. They may not be able to speak or even move their arms.

A special sensor developed by a group called Volunteers for Medical Engineering, permits users to spell out words on a monitor. A scanner travels over the alphabet. Users blink when the right letter is found. Gradually, words are spelled out.

CHAPTER 1 REVIEW

Review Questions

1. Define communication technology.
2. What are the four parts of the universal systems model? Describe each.
3. Describe the basic parts of the communication system model.
4. Describe how the communication concepts of designing, encoding, storing, retrieving, transmitting, and receiving apply to a television show as it is produced, taped, and aired.
5. Describe the human, animal, and machine modes of communication.
6. Define telecommunication. Give examples.
7. List five types of technical communication systems.
8. How does research differ from development?
9. List the six basic steps involved in the problem solving method.
10. When should the problem solving method be used? Give examples.

Activities

1. Survey ten people to find out what the word "technology" means to them. Report your findings to the class.
2. Create a display that illustrates the three modes of communication.
3. Analyze a state-of-the-art communication device in terms of the models described in this chapter. Make a model or display and label the different parts as to sender, receiver, input, output, etc.
4. Research five different technology-related careers in the *Dictionary of Occupational Titles*. Write a paragraph describing each.
5. Apply the problem solving method to a specific communication problem in your school. Write a report that describes the steps you followed in search of your solution. Read your report to the class.

CHAPTER

2

The Changing Nature of Communication Technology

Benjamin Franklin once said that in this world the only two things that were certain were death and taxes. We can probably add "technological change" to his list. Technology is dynamic. This means it changes all the time—whether we like it or not. The changing nature of technology presents new opportunities, as well as new challenges.

In this chapter we will take a look at some of the changes that have been taking place in communication technology. Most of those changes have been influenced by the computer. Some trends have cut across all of the technical communication systems. These will also be discussed.

A historical time line will give you an idea of changes that have occurred in the past, and you'll get a peek at the future.

Terms to Learn

computerization
data
digitization
E-mail
integration
miniaturization
online database
printing on demand

As you read and study this chapter, you will find answers to questions such as:

- During what time period have most changes in communication technology taken place?
- What impact has the computer had on communication?
- What have recent developments brought to the different communication systems?

CONTRIBUTIONS OF THE PAST

The history of communication systems is rich with spectacular innovations and inventions. Many of them completely changed people's lives. Johannes Gutenberg's use of movable type in the middle of the fifteenth century made it possible to print books quickly. For the first time, books became available to *all* people, not just the elite. Alexander Graham Bell's telephone allowed two people to talk to each other without being face-to-face. Computers, too, have changed our lives in countless ways.

Placing the major technological developments in communication on a time line (pp. 34-35) lets you see how the whole field developed. It is interesting to see the relationships among the different technical systems (optic, graphic production, etc.). What do you think might appear on the time line in the next decade? What might appear in the twenty-first century?

THE INFORMATION AGE

Until the middle of the 1800s, the majority of people spent most of their waking hours growing and harvesting food. Farming was *the* way of life.

The Industrial Revolution of the mid-nineteenth century changed that. For the first time in history, more people began to work in industries than on farms. The Age of Industry had begun.

Now, we live in the Information Age. As our factories become more automated, fewer people are employed in the production of goods. At the same time, there has been an explosion of information. This information is often called **data**. New communication systems allow us to handle this information with relative ease. That's why many people refer to our time as the Information Age.

Fig. 2-1. The record players of the past have been replaced by compact disk players. How does this change reflect overall trends in communication technology?

CHANGES IN COMMUNICATION

If you look at the "big picture" of communication technology, you can spot some changes taking place. Fig. 2-1. These include computerization, miniaturization, digitization, and integration. Another change to be examined is the growing use of computers in the home.

Computerization

The single most significant change in communication technology is **computerization**. More and more, communication systems are relying upon the power of computers to make them work. The computers need not be the type that sit on a desktop. There are microprocessors (computing "chips") inside your telephone, television set, photocopier, radio, compact disc player, videodisc player, or camera. Fig. 2-2. Almost every important piece of communication equipment you can think of has some sort of computer inside, or soon will.

Computerization offers many benefits. Increased quality is one. If you compare the picture on a new color television set with one from twenty years ago, you will see it is much improved. The sound quality of a compact audio disc is much better than that of a vinyl phonograph record.

Reliability of communication devices has also been greatly improved. When computers were new, people blamed the computer when something went wrong. Now, people have begun to realize that computers are very reliable. Also, the parts last far longer than those used in the past. When problems do arise, they are usually the result of human error.

Quicker turn-around time is another benefit of computerization. In many communication systems, use of computers has eliminated some tasks. The result is quicker communication. Text and illustrations (graphics), for instance, may be sent around the world in the time it takes to make a phone call.

While we often think of computers as expensive devices, the fact is, they often save money. For instance, speeding up a communication process may mean fewer work hours and reduced wage expenses.

Fig. 2-2. Many cameras use computer chips to control focus and exposure.

COMMUNICATION TECHNOLOGY TIME LINE

Before 35,000 BC: Language is developed; smoke signals and drums are used for long distance communication.

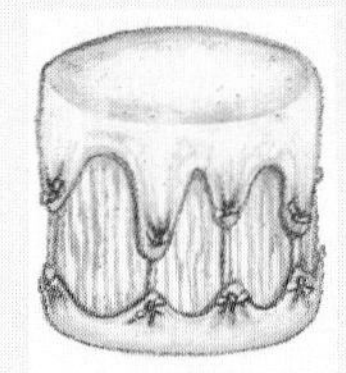

About 3000 BC: The abacus (perhaps the earliest computing machine) is developed.

About 1500 BC: Middle Eastern cultures develop the beginnings of the alphabet.

350 BC: Aristotle discovers the principle of the "camera obscura." Light entering a hole in one side of a darkened box will form an image on the opposite side.

AD 105: Paper is invented in China by Ts'ai Lun.

1822: Joseph Nicéphore Niépce, a French physicist, invents the photographic process called "heliography." Four years later, he produces the first permanent photograph.

1823: Charles Babbage, an Englishman, develops the "Difference Engine," a calculator that could handle algebraic equations.

1833: Charles Babbage develops the "Analytical Engine," a computer prototype that could be programmed with punched cards.

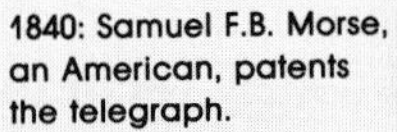

1840: Samuel F.B. Morse, an American, patents the telegraph.

1884: Ottmar Mergenthaler, a German-born mechanic, patents the Linotype, a machine that casts a complete line of relief type at a time.

1888: Americans Thomas A. Edison and William Dickson develop the kinetograph, the first motion picture camera.

1888: George Eastman, an American, develops the "Kodak" camera.

1895: Guglielmo Marconi, an Italian, develops the wireless telegraph, later to become the radio.

1930s: Audio tape recorders are invented by German engineers. The machine shown is a later model.

1944: IBM unveils the Harvard Mark I computer.

1947: John Bardeen, Walter Brattain, and William Shockley, all Americans, invent the transistor for Bell Telephone Laboratories.

1947: Dennis Gabor, a British engineer, invents holography.

Mid-1950s: Television networks begin recording on magnetic videotape.

1952: Sony Corporation of Japan markets the first pocket-sized transistor radio.

1962: *Telstar*, the first commercial satellite, is launched from the U.S. It allows telephone and television transmissions between the United States and Europe.

1970: Corning Glass Works, an American company, introduces the first commercial optical fiber.

1971: Theodore Hoff, an American, develops the first microprocessor (a computer on a chip) for the Intel Corporation.

Early 1970s: NV Philips, a Dutch firm, and Music Corporation of America (MCA) develop the laser videodisk.

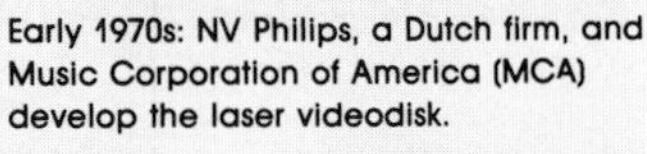

1976: Steve Wozniak and Steve Jobs invent the original Apple® microcomputer.

AD 868: The earliest known "book," *The Diamond Sutra*, is printed with woodblocks in China.

About 1000: Screen printing is developed in China and Japan.

1045: Movable type is invented by the Chinese printer Pi Sheng.

1440s: Johannes Gutenberg, a German, develops the first system of movable metal type known to the Western world.

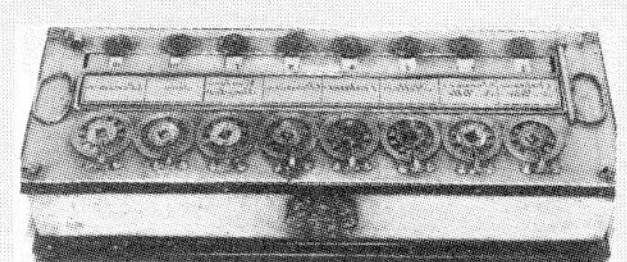

1600s: Frenchman Blaise Pascal develops calculator prototypes, which lead the way to the computer.

1811: Friedrich Koenig, a German, patents the first steam-powered relief printing press.

1864: James Clerk Maxwell, an Englishman, develops a theory of electromagnetism. This theory forms the basis for the invention of the radio.

1868: Christopher Sholes, an American, patents the first practical typewriter.

1876: Alexander Graham Bell, an American, patents the telephone.

1877: Thomas A. Edison patents the phonograph.

1898: Valdemar Poulsen, a Dane, invents the first audio recording device.

1904: Ira Rubel, an American, builds the first offset lithographic printing press.

1908: C.A. Holweg of Alsace-Lorraine patents the first flexographic printing press.

1912: German chemist Rudolph Fischer patents the first color photographic process.

1923: Vladimir K. Zworykin, a Russian-born physicist living in the U.S.A., develops the television camera and receiver.

1930: Vannevar Bush, an American electrical engineer, builds the first reliable analog computer.

1959: Xerox Corporation of the United States markets the first practical office copier, the Xerox 914.

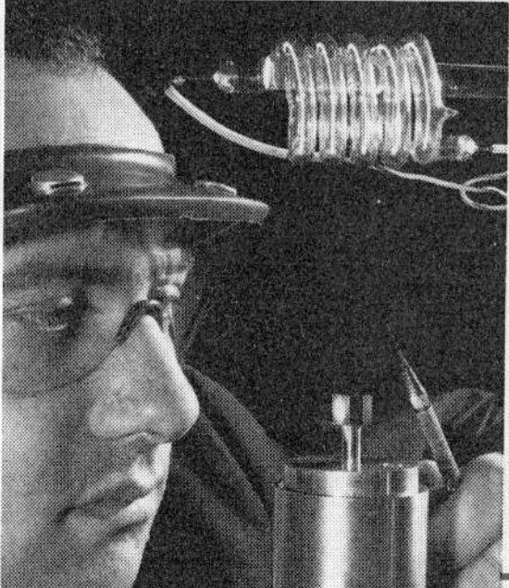

1960: Theodore Maiman, an American, develops the first laser at Hughes Research Laboratory.

1960: Laser holograms are developed by Emmett N. Leith.

1960: *Echo I*, the first satellite to receive/reflect radio signals back to the earth, is launched from the U.S.

1978: Music Corporation of America (MCA) markets the first 50 videodisk players in America. They cost $20,000 each.

1982: The Hattori-Seiko Company of Japan develops the flat video display screen using liquid crystal display (LCD) technology.

1983: The first compact audio disks (CDs) are introduced.

1984: Macintosh computer goes on the market. Though its sales trail behind those of the PCs, its innovative user interface–with icons, pointing devices, and a desktop metaphor–is eventually adapted by major computer hardware and software makers.

1990s: Wireless personal communication devices come into common use worldwide. Examples of such devices include pagers, cellular phones, cellular modems, and personal digital assistants.

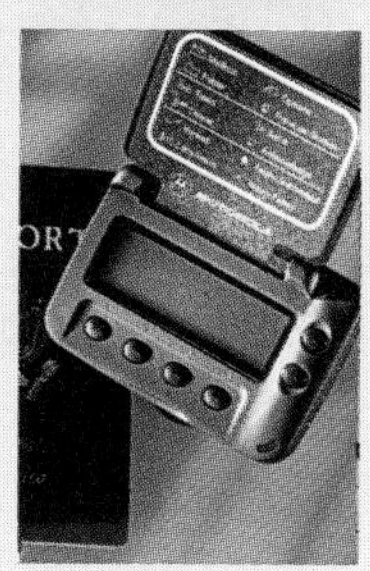

Miniaturization

Along with computerization comes miniaturization. **Miniaturization** means making things smaller. Computers used to be monstrous in size. Now they are small and getting smaller all the time. Electronic parts that used to fill a large room have been replaced with "boards" that hold computer chips. These boards fit easily into a desktop computer. Fig. 2-3.

Of course, everything that relies upon these chips gets smaller as well. When Thomas Edison invented the phonograph, do you think he imagined joggers would one day carry stereophonic sound in their shirt pockets? Now we even wear computers on our wrists.

Miniaturization will continue. As the devices get smaller, they will also get cheaper. The calculators of the early 1970s cost hundreds of dollars and took up most of a desktop. Today a pocket-size calculator may be purchased for several dollars.

SCIENCE FACTS

Optical Computers

Today's computer chips are small and fast, but the connections between the chips slow things down. Scientists are now working on an optical computer. It will use laser beams instead of electric wires. Laser beams can carry more information than wires. Also — unlike wires — the beams of light will not have to be kept separate. They can pass through one another without affecting the information they carry. That means an optical computer can be even smaller than today's electronic computers.

Fig. 2-3. (Left) The Harvard Mark I computer filled a large room. (Above) Today's home computers are more powerful and fit on a desktop.

Digitization

There are two types of wristwatches. One kind has hour and minute hands, and the other displays just numbers. Fig. 2-4. The watch that displays numbers is digital. The time is never in between numbers. It is measured in whole seconds.

The watch with hour and minute hands is an analog timekeeper. Since the hands are in constant movement, it's never really any exact time. Analog systems don't measure in separate chunks. They work continuously.

All of the communication systems discussed in this text (graphic production, optic, technical design, audio and video) have traditionally been analog systems. That, however, is changing. The systems are becoming digital.

The shift from analog to digital systems is called **digitization**. It is happening because computers are taking over communications. Since computers are digital systems, wherever they are used the data must become digital in nature.

Integration

Since data for print, audio, video, and photographic media are all becoming digital, the different systems can be linked to one another. Digital data may be transferred easily between systems. For example, a digital photograph can be passed to a digital printing system very easily. In fact, this is now common in the graphic communication industry. The result of this trend is **integration**, or combining, of communication systems.

A lot of integration takes place in business. Fig. 2-5. Fax machines routinely send and receive digital text and graphics over thousands of miles. Graphics and text are easily merged on a computer. It's more complex to bring in video and audio as well, but that's what interactive video does. It puts all of these together on the same computer screen.

Fig. 2-4. Digital watches display numbers. Analog watches show hands moving around a numbered dial.

In the 1970s, audio, video, print, and photographic media were thought of as being very different from one another. In the 1980s, as each of these media became digital, mixing took place. By the year 2000, there will be a lot of overlap. At the same time, computerization will be widely accepted in each of these areas. Communication will be, for the most part, based on digital systems.

Integration is having a growing impact on our lives. Such items as "picture" telephones, "talking" newspapers, voice-operated keyboards, and interactive television are all currently in development.

Computers in Every Home

How many homes in America have television sets? When black and white television was first introduced, sets were expensive. Not everyone could afford them. As the price came down, more and more people bought one. It also took a while for color television sets to make their way into our living rooms. Nearly every household in the United States now has at least one color television set. Many have several.

The same thing appears to be happening with computers. More and more computers are finding their way into people's homes. As this hap-

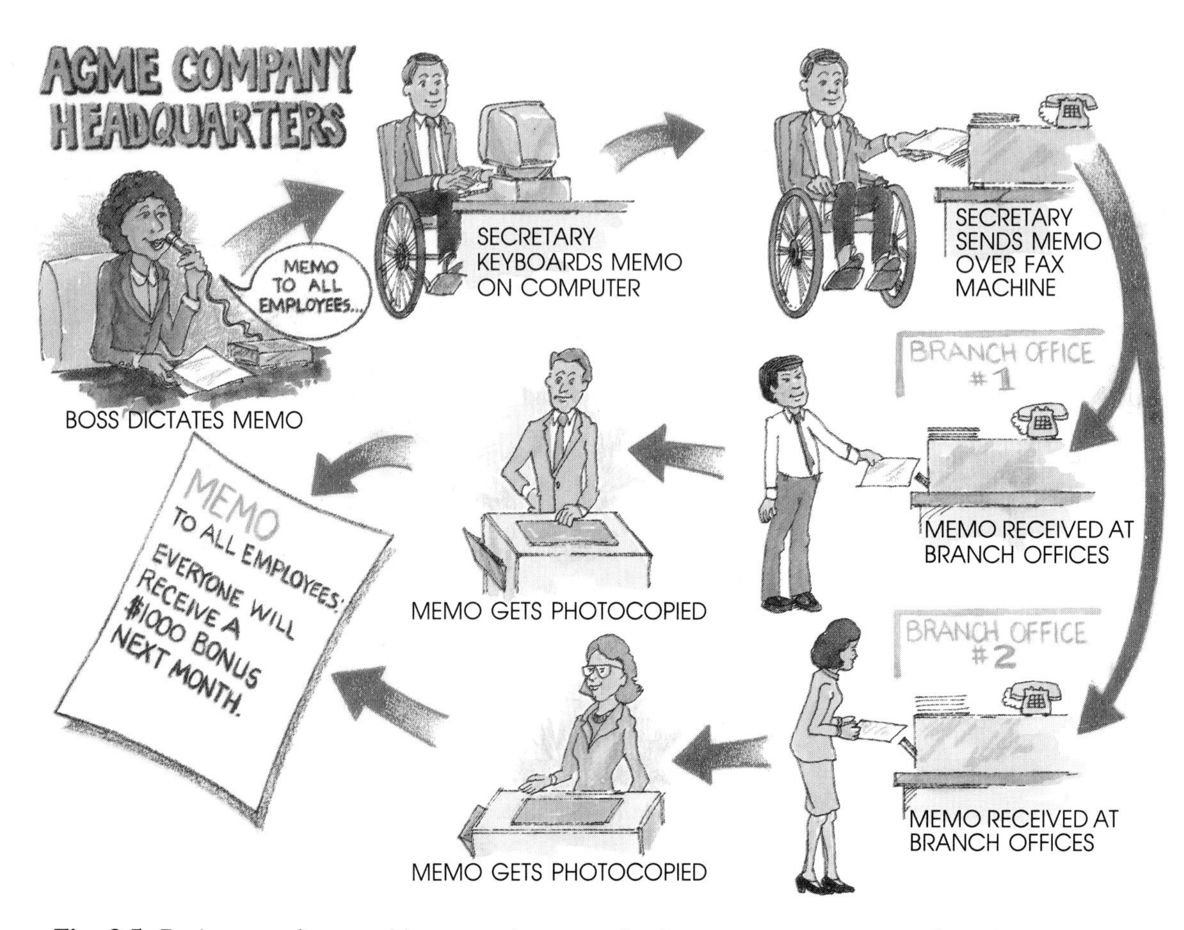

Fig. 2-5. Businesses often combine several communication systems to create and send a message.

pens, American businesses are finding more ways to communicate with consumers through the home computer.

Interactive television is one example. Some television stations have experimented with computers hooked up to viewers' TV sets. Viewers are asked questions. Then they answer using the computer. The answers are immediately received by the television station, which is linked to the TV by cable.

Using such a system, viewers could play along with a TV game show. They could also record a football game and watch it from different vantage points. Applications such as these are being developed for use by the television industry.

A computer in the home means more than interactive TV. There are a growing number of services already available. Home computers may be used to trade stocks on the stock market, check books out of a library, buy computer software, or find information on nearly any topic.

DIRECTIONS FOR THE FUTURE

What changes can we expect to see in communication technology in the near future? In this section we will examine trends that seem to be leading in definite directions.

Computers

Three important computer trends have to do with information storage, networks, and databases.

Storage Media

The very first computers did not have the ability to save information. When you unplugged them, all of the data vanished. The early computers used holes punched in cards to represent data. Computer users were often seen carrying their programming cards around in shoe boxes. Today, "floppy" and "hard" disks are used. Removable cartridges, also common, are hard disks enclosed in a plastic shell. They may be removed from the drive and transported like floppy disks. Optical disks have become a very good means of storing large amounts of data. These include compact audio disks (CDs) and CD-ROMs. Lasers are used both in the manufacture of optical disks and in reading the data from them. Fig. 2-6.

At the same time, the memory inside the computer has increased dramatically. Today, a desktop computer may have more than 1000 times as much memory as did the first desktop computers in the 1970s. This means computer programs can be much, much larger. Today's desktop computers are far more powerful than those that occupied large rooms several decades ago!

Fig. 2-6. Improvements in storage media have made it possible to store more data in more convenient ways. A CD-ROM, one type of optical disk, holds more than 400 times the amount of data stored on a typical floppy disk.

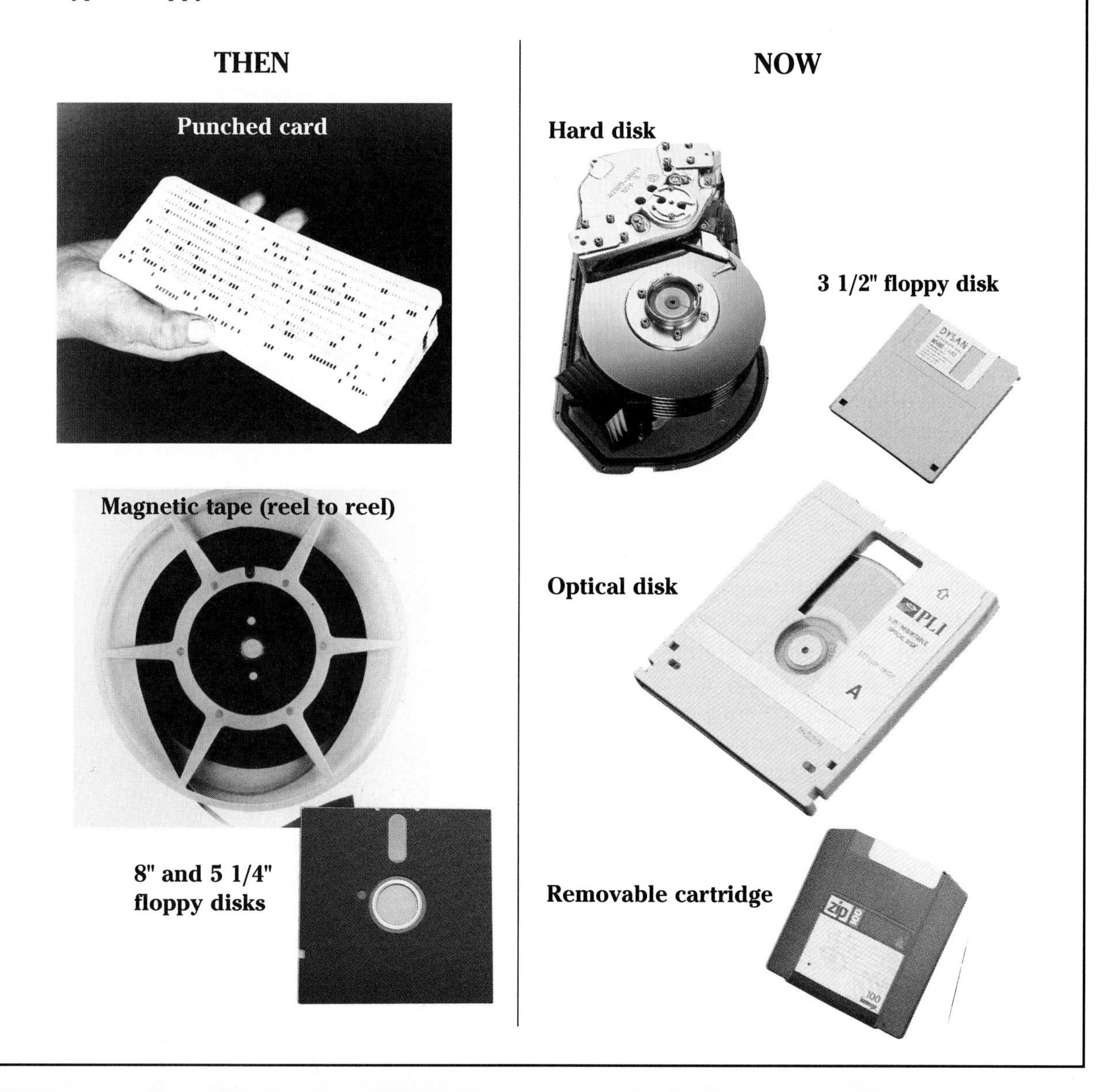

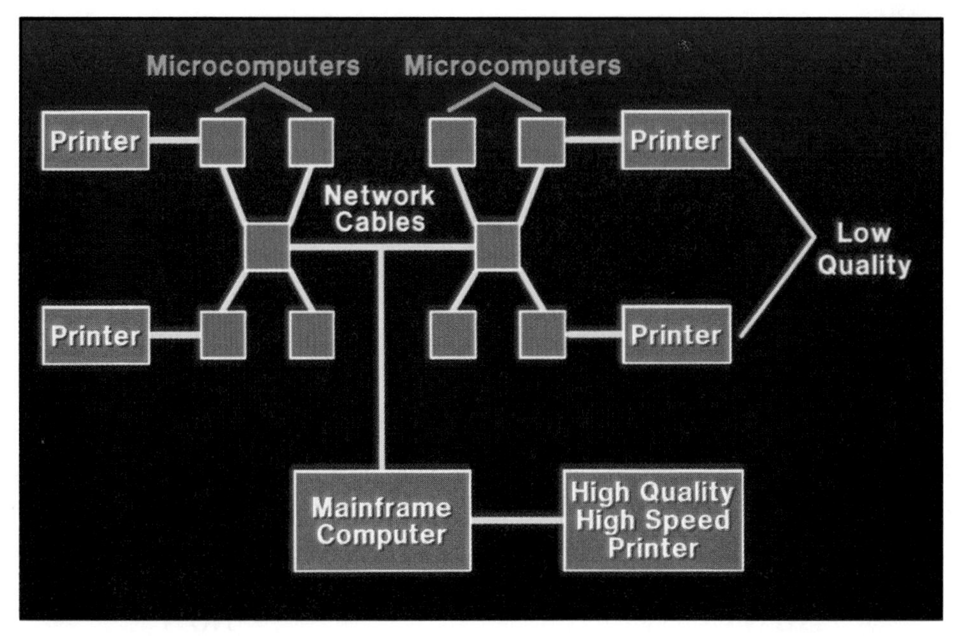

Fig. 2-7. A typical office network allows each computer to communicate with the others. Some networks include a mainframe computer or minicomputer. Others consist entirely of microcomputers.

Computer Networks

When you hook one computer up to another they can "talk" to each other — that is, one user can communicate with another. When you connect a number of computers together, as in an office, everyone can communicate through a computer network. Fig. 2-7.

Computers can also communicate easily to one another over telephone lines and by means of satellites. This means that it's possible to send data back and forth around the world via computer.

Sending messages through computer networks is called electronic mail, or **E-mail**. It is rapidly becoming a standard practice in offices. People are sending messages and documents to one another over the computer network instead of using the telephone. E-mail won't replace the telephone, but it does have advantages. For example, you can send one message to ten (or more) people at the same time. Imagine how much longer it would take to call those ten people on the telephone.

Online Databases

An **online database** is a package of information that may be stored and retrieved by computer. For example, you are familiar with the card catalog in your library. An online library catalog is one in which the information from the cards is stored in a computer system. The user types in the author's name, and the system lists all the books by that author. The same may be done for subject headings and titles. Fig. 2-8. What do you think might be some advantages of an online library catalog?

Online databases are of all different types. Using a database, you may book an airline reservation, locate a restaurant in Chicago, find out what the weather is like today in any major city, find information on different universities, and so on. On the job, scientists, engineers, teachers, doctors, artists, and others use various databases to gather information.

Fig. 2-8. On-line databases are replacing card catalogs in many libraries.

The Internet

The Internet is a worldwide computer network that is changing the way we communicate in the Information Age. The Internet is used for electronic mail, accessing large on-line databases of information, and transferring data files across the planet. Distance means nothing on the Internet, since information is sent and received the same way, whether it's going to the computer next door or to one on the other side of the earth.

The Internet is also used to send/receive audio and video data. This means it can be used like a "videophone" or for videoconferencing, with people at both ends of the connection seeing and hearing the other(s) on their computers.

This era will be remembered as the time when worldwide computer networks such as the Internet developed into an indispensable communication channel. Worldwide commerce is just beginning to flourish via electronic pathways. See Chapter 6 for more information about these networks.

Technical Design

The area of technical design lends itself to computerization. For one thing, most technical designs rely on geometric shapes. Computers have an easy time of creating and displaying geometric shapes.

Also, technical designers often redraw the same objects many times. House plans, for example, generally include sinks, lights, showers, electrical outlets, and so on. With a computer, symbols representing these objects may be called up from computer memory and used in the drawing. For these reasons, an increasing number of technical designs are being done with computers.

There is also a trend to link designing and production by means of computers. This is known as computer-aided manufacturing, or CAM. A related trend uses computers to control the entire manufacturing process. This is known as computer-integrated manufacturing, or CIM. Fig. 2-9. As production facilities become more and more automated, CAD, CAM, and CIM will play an increasingly important role.

Fig. 2-9. In a CIM system, all aspects of the manufacturing operation are linked by computers.

Optics

Cameras have come a long way in the last century. Again, the most significant change is a result of computers. Lenses have improved greatly because computers have been used to design them. The lenses produce better quality images, even when little light is available.

Another trend is toward fully automatic cameras. Microprocessing chips inside cameras are doing much of the work. These chips decide on the proper exposure and focus. This means the photographer may concentrate instead on setting up the photograph.

Digital storage of information is coming more slowly to photography than to the other communication systems. This is because a photograph contains a huge amount of data. Five hundred pages of text can easily be stored in the computer space it takes to store a simple black and white photograph. Photographic film, therefore, is still more convenient.

Graphic Production

The use of computers in graphic production systems is called electronic, or computer-aided publishing. Now that many of the computers used are relatively small, people also call this "desktop publishing." Computers are now involved in nearly every part of the publication process. Fig. 2-10.

The increased use of color is one change computers have made possible. It was rare to see color used in a newspaper in the 1970s. Now, it is rare *not* to see color printed in newspapers. Likewise, magazines and books are using much more color than ever before.

Color is also getting better. A 1988 reprinting of the award-winning children's book *Where the Wild Things Are* looks different from the original 1963 edition. That is because the color now looks the way the author, Maurice Sendak, wanted it to look. In 1963, that wasn't possible.

Another trend is **printing on demand**. Documents are being stored in computer files and printed as needed. This saves the cost of storing large quantities of printed material in warehouses. Since much of the cost of graphic production is for the paper on which the product is printed, printing on demand saves money in materials as well. Only the materials already "sold" are printed.

Fig. 2-10. Computers are used in many phases of graphic production. Shown here is a pagination system, which allows users to compose pages electronically.

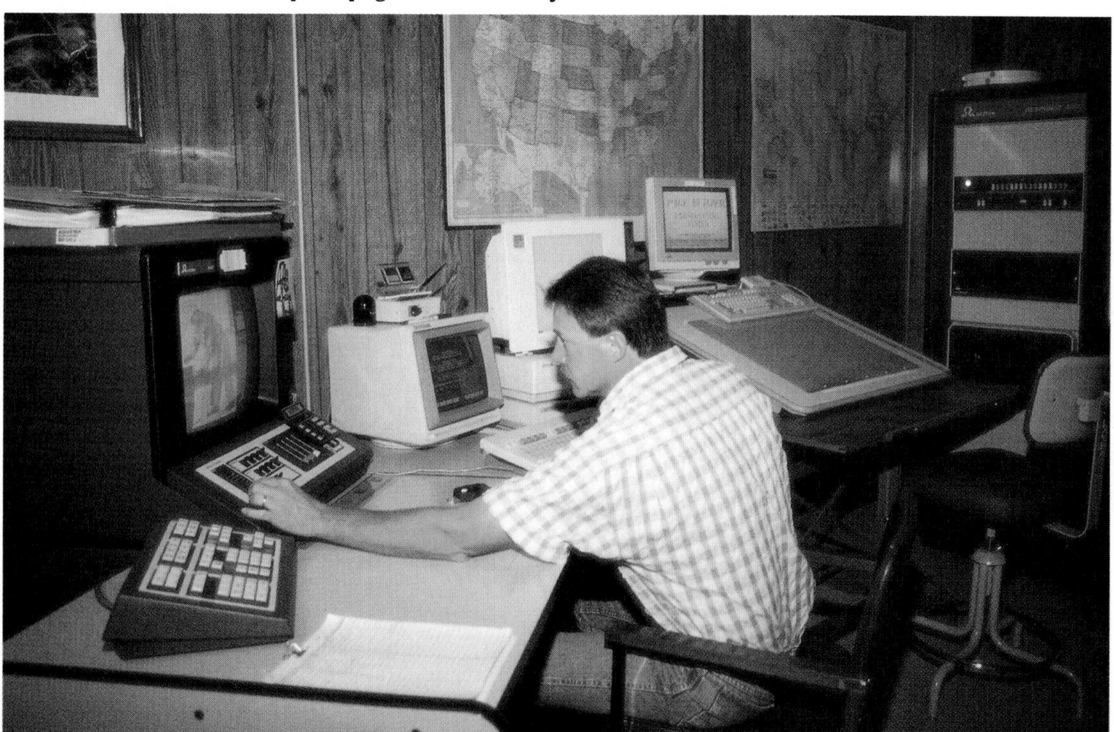

Computer design has shifted more control over the final product to the designer. This important trend saves time and expense since fewer corrections are necessary. It makes the designer's role more important than ever. Since better communication between the designer and printer takes place, better quality may also result.

Audio and Video

Compact audio discs, which provide crystal clear digital sound, have been one of the most successful consumer electronics products in history. Right behind them are videocassette recorders (VCRs). Both have brought quality and convenience to the entertainment area.

The trend in both audio and video is toward digital systems. Yesterday's television sets will be replaced by digital "entertainment systems." These new systems will receive digital (CD quality) audio, high quality digital video, and computer data.

The fact that these systems will be completely digital means they will be more like computers than television sets. This will make them much more flexible. They will have a direct link to computer networks, such as the Internet, to allow the user to access the incredible array of information stored and distributed by computer.

Unlike the entertainment systems of the past, the new systems will have two-way communication. They will ask for and respond to input from the user. In this way, we will increasingly be able to communicate with the message sender. This means our entertainment will be more interactive. For a price, we'll be able to access many types of interactive entertainment. The options will be vast compared to those that currently exist.

Voice data is just another form of audio that can also be transmitted and received digitally. Digital systems will increasingly carry our telephone conversations, with or without video of the person at the other end of the line.

SCIENCE FACTS

Satellite Orbits

Communication satellites travel at the same speed as the earth rotates. Thus, they remain above the same part of the earth at all times. The satellites therefore are said to move in *geostationary* orbits. Just three satellites in geostationary orbits 22,300 miles above the earth can relay information from any part of the globe to any other part.

To transmit the vast amount of digital data, the new systems will increasingly rely upon satellites, optical fiber, and coaxial cable. Digital satellite systems (DSS), introduced in 1994, were the first to capitalize on the new digital transmission technologies. DSS systems use an 18"-diameter satellite dish to receive digital video and audio signals relayed from satellites. They send these digital signals to a box that converts them into an analog signal that conventional television sets can display. They offer very high quality video and CD quality audio at prices that roughly compare with the cost of cable television.

Fig. 2-11. Digital Satellite Systems such as these offer a wide variety of high quality video and audio programs.

THE BLACKSBURG ELECTRONIC VILLAGE

Computers have long been an essential tool for faculty, staff, and students at Virginia Polytechnic Institute and State University. Virginia Tech, as it is known, has been recognized as having more personal computers than almost any other college campus in the world.

In 1987, Virginia Tech entered into a partnership with the town of Blacksburg and Bell Atlantic, the local telephone company. The partnership was named the Blacksburg Electronic Village (BEV).

The BEV was established to connect the citizens, town government, and businesses of Blacksburg with each other and the Internet. The required software was distributed at no cost, and student volunteers helped many citizens "get connected." For those without computers, free access was provided via computers across the campus and at the town library.

The following list describes some of the information and services easily accessible on the BEV.

- electronic mail and electronic mailing lists
- full Internet access
- thousands of "news groups"
- Chamber of Commerce guides about Blacksburg (to help citizens, visitors, and Internet explorers discover local history, commerce, and community activities)
- an on-line Village Mall that features information on almost 100 local businesses. At the Village Mall, users can print and use an on-line coupon, browse a take-out menu, order flowers, review the listings for a local radio station, send electronic mail to an accountant, compare local bank services, check out the new releases at the video store, electronically mail a letter to the editor of the local paper, etc.
- events calendars, updated weekly, that provide the schedules for arts and entertainment, public meetings, health-related events, senior citizen activities, and community activities
- guide to local organizations and clubs
- on-line newsletters, essays, and other information from local schools, much of which is posted by students
- medical tips and treatment from the medical database composed by a local doctor
- information provided by the town government. BEV users may read about recreation programs, send e-mail to the police chief to schedule vacation watch, use the fire department's information to teach their children fire prevention and safety, etc.
- educational opportunities in the area, including colleges and universities, day care and pre-school, private and public schools
- local weather forecasts
- schedules for the Blacksburg Transit (bus system)
- guide to restaurants, accommodations, and retail shopping
- guided walking tour of Blacksburg's historical sites
- information about other points of interest

The BEV is an ongoing experiment that has received worldwide recognition and interest. It allows the participants (town, university, local businesses, and citizens) to explore first-hand how a widespread network such as this might impact the community. The BEV could be a model for similar networks in other communities.

CHAPTER 2 REVIEW

Review Questions

1. What was probably the earliest computing machine?
2. Define computerization.
3. What has caused the trend toward miniaturization?
4. What are the differences between a digital watch and an analog watch?
5. How have computers brought about integration of communication systems?
6. Name at least three uses for computers in the home.
7. What is a computer network? How is it used?
8. What are online databases used for?
9. Why is digitization coming more slowly to photography?
10. What is the most successful consumer electronics product in history?

Activities

1. Find three examples of miniaturization in your home or school. Make a poster giving measurements between the old and new versions of the products.
2. Obtain a sheet of very fine graph paper. Place the paper over a black and white photograph. Lay them both on a light table or tape them to a sunny window. Using a #2 pencil, fill in squares that correspond to the dark areas of the photograph. Leave blank squares where the photograph is light. Show the results to the class. Explain how you have converted an analog image to a digital image.
3. Interview your classmates. Find out how many have computers in their homes. Make a list of the different applications for which they or their families use their computers. Make a chart of your findings.
4. Does your school library have an online database? If not, go to your local public library and see if it does. Conduct a data search to find out what books the library has on telecommunications. If the library doesn't have an online database, try to find a business in your town that does. Ask how the database is used. Report your findings to the class.
5. Write a report on one of the technologies shown on the time line on pp. 34-35.

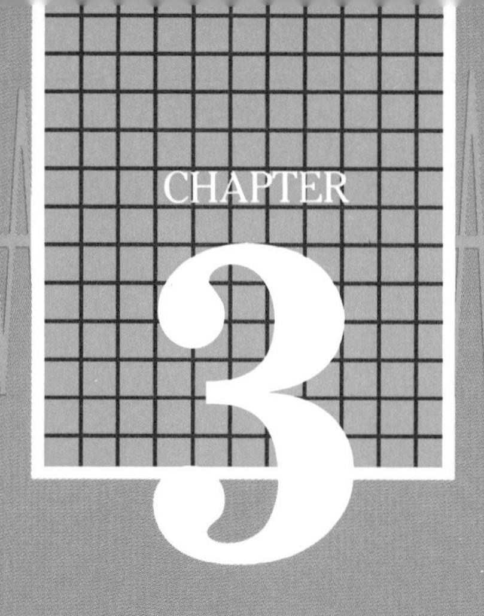

The Impact of Communication Technology

"High tech" communication devices are just the tip of the technological iceberg. The tip of the iceberg is the part that you see because it's above water. Fig. 3-1. Most of the iceberg is under water, where you can't see it. What you don't see can cause problems if you're not careful.

New technological devices often solve one set of problems, but at the same time create a new set. Of course, the impact need not be bad. Sometimes changes are good. Other times they are neither good nor bad. They are just different.

In this chapter, you will learn about some ways in which communication technology affects our world. Some of them might surprise you. For example, do you know how communication technology affects forests?

Terms to Learn

computer viruses
cultural
digital editing
economic
environmental
ethical
mass media
political
sampling
social
technology assessment

As you read and study this chapter, you will find answers to questions such as:

- How can technologies be evaluated so their impact is determined?
- What are some effects of communication systems?
- In what categories are evaluations made?

Fig. 3-1. Like the 90 percent of an iceberg that's underwater, the impact of a new technology is often not obvious.

TECHNOLOGY ASSESSMENT

To assess something means to evaluate it. When you study the effect of a new technical device, you are doing **technology assessment**. Technology assessment allows you to see the whole picture, not just a part of it.

Consider, for example, the invention of the telephone. For the first time, people could talk to their friends in town without having to leave their houses. This simple fact radically changed their lives. How would business be conducted today without telephones? How would the relationships with your friends change if you couldn't call them on the phone? How many people are employed by the telephone companies? How have telephones changed our country's foreign policy?

Technological change can have major impacts on our society. Some impacts are good, while others are harmful. To make intelligent choices about the way we respond to and use technology, we need to be informed about those impacts. Fig. 3-2.

Fig. 3-2. Small transmitters attached to a car enable someone, such as the police, to track the vehicle. What effect might these devices have on the privacy and anonymity of people who live in a free society?

Categories of Impact

There are so many different issues involved in assessment that it's hard to keep track of them. Putting them in categories can help. Listed below are six different categories that relate to communication technology.

- **Political** — Relating to the government.
- **Social** — Having to do with the ways in which communities of people live.
- **Economic** — Having to do with the economy.
- **Environmental** — Relating to our physical environment.
- **Cultural** — Having to do with the skills and arts developed during a given period.
- **Ethical** — Relating to matters of right and wrong.

In this chapter we will discuss different impacts and issues. They will be categorized into each of the areas described above. Keep in mind that areas may overlap. For example, economic issues may be related to political issues. Also, most technological changes create effects in more than one category.

Politics and the Media

Long ago someone wrote, "The pen is mightier than the sword." His point was that the written word could bring about more political and social change than violence or war.

The **mass media** (television, radio, newspapers, magazines, and books) have a real impact on our political system. Consider what happens during a large campaign. Each candidate employs one or more "media advisors." The advisors' job is to make their candidates look and sound good in the media. They rehearse speeches with the candidates and advise them on what to wear. They are concerned less with *what* their candidates say than with *how* they say it.

Media advisors often boil down complex campaign issues to very short, simple statements. These have come to be known as "sound bytes." Sound bytes are more likely to be picked up and reported by the media than a long, involved speech. They are also more likely to be remembered by the public.

POLITICAL IMPACT

If a technology has political effects, the effects relate to government. This includes the way our government works in the United States. It also includes how our government relates to other governments.

Fig. 3-3. Politicians use the mass media to inform and persuade the public. How might politics be different if there were no televisions, radios, or newspapers?

During the year and a half of campaigning, the media influenced the political process. Some candidates dropped out overnight after reports that appeared in the media. The popularity of the front-runners went up and down depending upon their media image.

Some would argue that the media selected the president of the United States in 1988, not the people. Others would disagree. Few would deny that mass communication plays an important role in the political process.

Satellite Communication

During the War of 1812, soldiers fought the Battle of New Orleans 15 days after the war was officially over. The news had simply not reached them. About 315 people were needlessly killed. Today, communication satellites allow us to send messages around the world almost instantly. Officials in our government routinely telephone officials in other governments. The calls travel via satellite. Television signals sent by satellite bring us news from other countries even while it is happening. Such rapid flow of information affects how governments relate to one another.

Communication satellites are also used to spy on other countries. Fig. 3-4. Mapping satellites photograph areas being watched for defense purposes. Any suspicious-looking construction or troop movements are discovered immediately. Potential wars can be halted before they begin.

SCIENCE FACTS

Instant Communication — Almost

When you speak into a telephone, your listener doesn't hear your words at that same moment. It takes time for the telephone signals to travel from one place to another. In most cases, the delay is just a tiny fraction of a second, so you don't notice it.

However, if you were to call someone very far away—in Australia, for example—you would find that the other person doesn't reply immediately after you stop talking. The gap in the conversation occurs because of the long distances the signals must travel. In the call to Australia, the signals would be beamed to a satellite, relayed to another satellite, and then sent back to the earth. The elapsed time might be as much as a second.

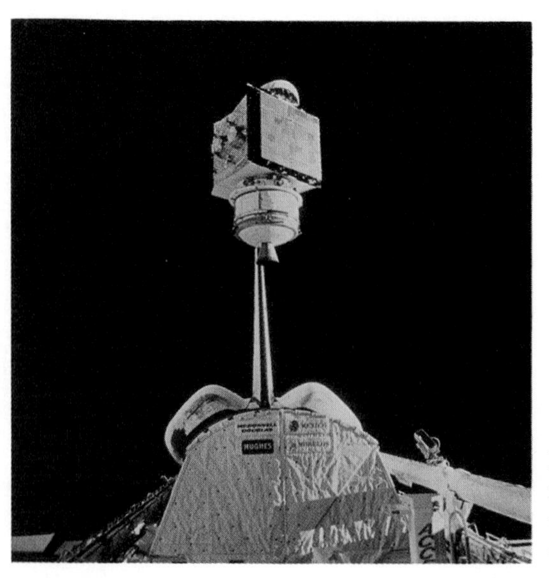

Fig. 3-4. The Space Shuttle has been used to place a number of commercial communications satellites, like the one shown here, into orbit. It has also been used to deploy military satellites.

SOCIAL IMPACT

How people live together has been deeply influenced by communication technology. The effects range from the power government has over us to the way we spend leisure time.

Public Opinion

Throughout 1995, O.J. Simpson's trial on murder charges took place in front of the entire world. Television, radio, and print media carried this story every day. It was nearly impossible to find a jury of peers who had not been influenced by the early media coverage. By the time the trial came to an end, almost everyone in the United States, and millions of others around the world, had become familiar with the details of the trial and had an opinion about it. Polls taken of the American public just before the verdict was announced showed most white Americans believed Simpson was guilty, while most black Americans felt he should be acquitted. The intense media coverage caused us all to learn a little more about each other and to rethink issues of race in America.

Brave New World

A half-century ago, Aldous Huxley wrote a book called *Brave New World*. In this book he told of a world in which the government had access to information about all of the people. This information gave the government power over its citizens.

Most people in the United States would agree our lives are not controlled by our government. However, government computers do in fact keep track of us, at least a little. For example, almost every person has a Social Security number. Originally, the purpose of the number was to help the government run the Social Security program. Today, this number is being used in connection with drivers' licenses, bank accounts, income taxes, and so forth. People in certain government offices can learn a great deal about a person simply by typing that person's Social Security number into a computer.

When we spend money, we leave a trail that may be followed. If we use a credit card or write a check, the credit card company or bank keeps a computer record of how much money we spent, and where we spent it.

When all of this information is combined, it provides an economic profile of a person. This profile is known as a credit rating. When we ask to borrow money from a bank, the bank checks this credit rating to see if we are a "good risk." The bank's decision helps to determine the kind of home we live in, the car we drive, and so on. Computers make such files as credit ratings easier to maintain and faster to access.

Fig. 3-5. Because of the intense media coverage of O.J. Simpson's trial, nearly everyone in the nation was aware of the trial's events and had an opinion about them.

Education and Training

Microcomputers made their way into classrooms in the 1980s. Now, teachers have found ways for computers to help them in their work. It's likely, for example, that your communication technology laboratory now uses computers to generate type. At one time this typesetting was all done by hand.

Educational videotapes have become a useful aid for many teachers. They provide information that before was not readily available for classroom use, such as public television programs and training videos. At home, television programs such as "Sesame Street" have made spelling and arithmetic fun for millions of preschool children. Educational programming is also available on television for adults.

Industry has made good use of new technologies for training workers. One popular device is interactive video. Videodiscs and computers are combined to deliver information. For example, pilots are taught to fly planes by watching a video screen. A computer reads how they use the "controls" and the picture changes accordingly. Some schools are experimenting with interactive video instruction as well. Fig. 3-6.

With increased use of such teaching aids, the classroom is changing. Although teachers will never be replaced by technology, their time may be freed up so they can give more individual attention to each student.

Fig. 3-6. This class is learning about science from an interactive videodisc program.

Information Storage and Retrieval

Communication technology is changing the way in which information is being stored and retrieved. For the past five centuries, all important information was stored in written or printed documents. The documents, usually books and magazines, were kept in libraries. Of course, that's still the case, but libraries are beginning to store information in other ways. Optical disks are being used for large amounts of data. An entire encyclopedia, for example, fits on one-fifth of an optical disk. Using a computer, you can instantly locate the information you want in the encyclopedia.

More importantly, you don't have to travel to the library to get the information. Since computer data can now be sent over telephone lines, you can access it from your home using a computer. You don't have to have a "hard" copy, such as a book. If you see something on your computer screen that you want to keep, you may print it out on your printer.

The same might be done with a catalog of items for sale. You select the item you want to buy and the system automatically withdraws the cost from your bank account and ships the purchase to you.

What will be the effect on society? Retrieving information in this way will make it easier for researchers to gather information. Major problems in science and medicine, for example, may be solved much more quickly than with traditional research methods.

On the other side of the coin, there will be far more information available to the average person. We will have to become more selective about what we choose to read or view. Otherwise, we may begin to suffer from "information overload."

Leisure Time

What did Americans do with their spare time before there was television? We read books, listened to the radio, talked to our families and friends, and entertained ourselves with hobbies. Now TV and computers have become the ways many people entertain themselves. Fig. 3-7. People don't go visiting as often as they did 100 years ago.

Fig. 3-7. Communication technology has changed the way people spend their leisure time. How much of your time is spent with friends? How much is spent watching television?

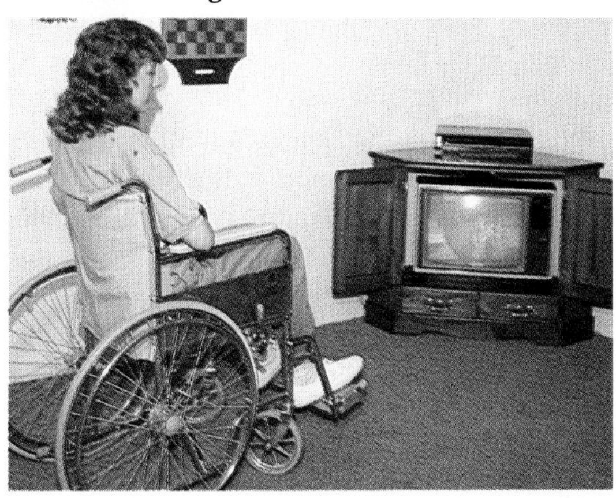

ECONOMIC IMPACT

Today, businesses rely on computers, high tech telephones, facsimile (fax) machines, and local area computer networks. These systems have a real effect on our economy.

The World Economy

With global communication comes a global economy. Our economy has always been linked to that of other countries. With new technologies, economic trends spread faster than ever before. Things that happen this morning in New York can affect business in Tokyo, thousands of miles away, a few hours later. For example, on October 19, 1987, the U.S. stock market dropped 508 points. This was a decline of more than 20% in one day! Japan has many investments in this country. The next day, the Tokyo stock exchange took a similar nosedive. Fig. 3-8.

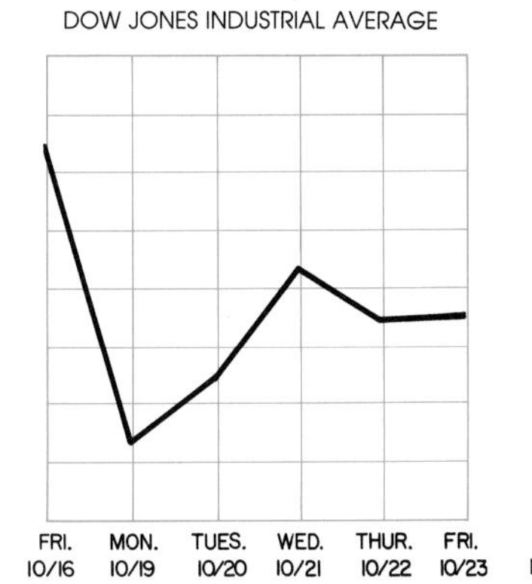

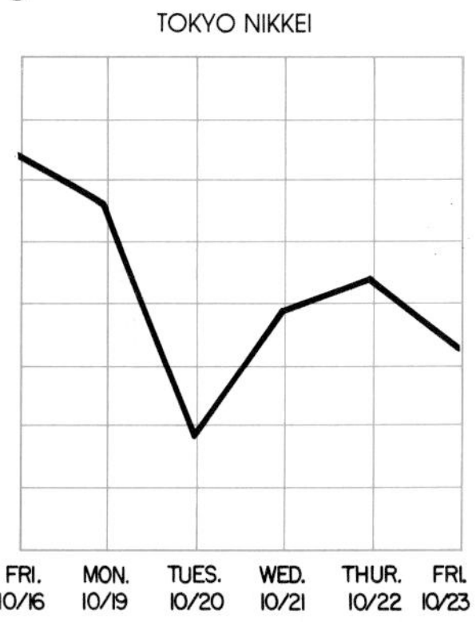

Fig. 3-8. An economic crisis in one part of the world quickly affects other countries. Notice how the changes in the Dow Jones averages are echoed the following day on the Tokyo Nikkei.

During the weeks that followed, the Americans and the Japanese watched the news for what was happening in both countries. Both stock markets seemed to go up and down together. Both markets were closed early on separate occasions to prevent panic selling. If the information had reached Tokyo days instead of hours later, would people have reacted so emotionally?

How would you classify this effect? Good? Bad? Neutral? In any case, economists must now figure into their plans the speed with which information travels.

From a Production to a Service Economy

Communication technologies are helping to move us from a production economy to a service economy. Instead of building things in factories for a living, more and more people are performing services. Teaching is one example.

The shift from production to service creates a new employment cycle. Technologies make factories more efficient. Some people are laid off. Large-scale layoffs present an immediate problem for the economy. To combat the rising numbers of unemployed, new education and training programs are established. The cycle is completed when graduates of these programs return to the workforce, often in service-related jobs.

Use of Credit

Many Americans live on credit. Many spend more than they earn. Unlike their grandparents, they seem unwilling or unable to save and wait for what they want. Do you think high-speed communications might be part of the reason people seem to want everything right away? What about TV advertising? Every day we watch while hundreds of commercials urge us to buy, buy, buy. Is TV affecting our spending habits?

ENVIRONMENTAL IMPACT

Communication tends to be a "clean" technology. Compared to "smokestack" industries, like steel production, communication industries are easier on our environment. However, this doesn't mean there are fewer environmental impacts involving communication technologies. A cleaner environment *is* an impact. Others have occurred as well.

The Use of Paper

When microcomputers arrived in the late 1970s, many people talked of the "paperless" office of the future. They imagined that information would stay in computer memory. We could read it on the computer monitor, so we wouldn't need paper. Filing cabinets would no longer be needed.

How wrong they were! More paper floats around the modern office now than ever before. Computers are a big part of the reason. Word processing software makes it much easier to create written documents, so there are more of them. They also tend to be longer than in the past.

The real culprit, though, is another communication device: the photocopy machine. Since it is now possible to make large numbers of paper copies in almost no time at all, we do just that. As a result, we use paper as if it grew on trees!

Of course, paper does grow on trees. When we harvest trees to make paper, we make an impact on the environment. Using energy to manufacture paper also has an impact. Bleaches and other chemicals used to make paper pollute our rivers and streams and even the air. Fig. 3-9. When we throw paper away, we create other problems, such as litter.

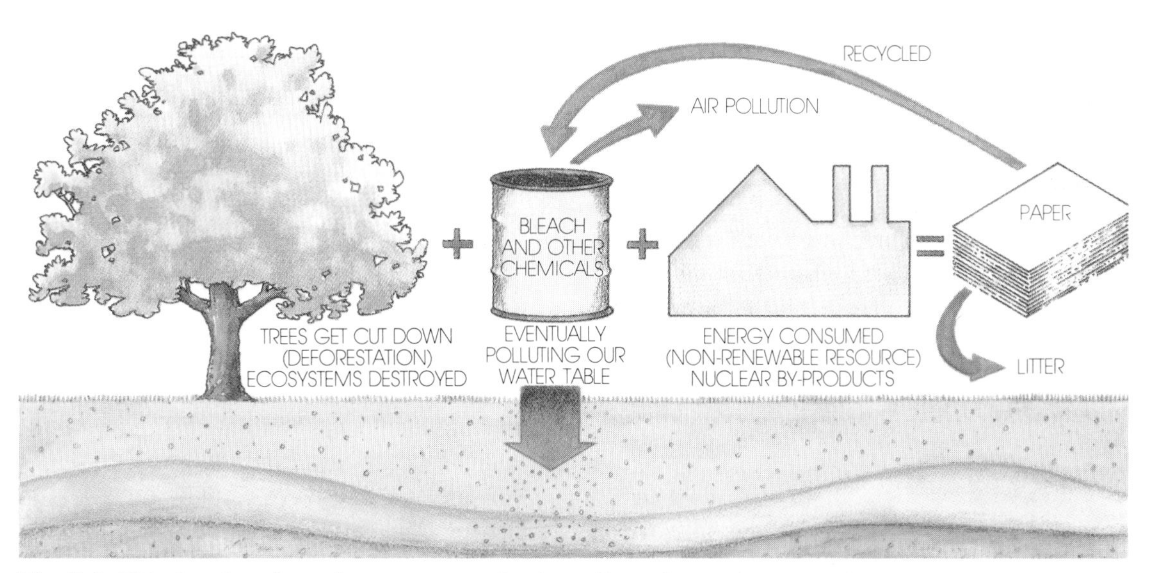

Fig. 3-9. This drawing shows how paper production affects the environment. Recycling paper conserves trees and reduces litter.

CULTURAL IMPACT

Five hundred years ago a society's culture was based on the spoken word. If people wanted a story, they listened to story tellers and minstrels. When books became available in the fifteenth century, people began to read stories. Eventually, new art forms, such as the novel, appeared. How has our culture changed now that we have television and other new technologies?

The Video Revolution

Many people believe television has made us a less literate society than we were a generation or two ago. (People who are literate are able to read and write well.) School test scores have declined. Could that be because we read less and watch television more?

Many television shows change scenes or images every couple of seconds. Something seems to be happening constantly. We have come to expect this constant action or we're bored. Many teachers report short attention spans among today's students. Is television to blame for this?

What about our attitudes toward violence? Have they changed because of television? When we see battles in bloody detail on the evening news each night, does it have an impact on our feelings about war?

Shopping has become a national pastime. Fig. 3-10. Instead of spending Sundays going to museums or visiting, many families wander around shopping malls. Do you think the constant advertising on television might have something to do with this?

Fig. 3-10. Americans enjoy shopping. What effect do you think advertising has had on our buying habits?

ETHICAL ISSUES

Ethical concerns have to do with matters of right and wrong. Things that are unethical may not necessarily be against the law. However, most people agree they are wrong. New developments in communication bring with them some ehtical problems that did not exist before.

Digital Editing

You read in Chapter 2 that computers are playing a growing role in each of the major communication technologies. Words, images, and sounds are being stored or transmitted digitally by computer.

One of the wonderful things about digital files is that they may be changed very easily. This process is called **digital editing**. While digital editing makes many tasks much simpler, it may also be used for questionable purposes.

For example, when commercial photographers work for magazines, they are paid for their photographs. Often, the photographer's name appears beside the photographs. The photographs may be copyrighted, just as books are copyrighted.

With digital editing systems, however, it is now possible for these photographs to be changed by someone other than the photographer. Fig. 3-11. The color of a fashion model's eyes may be changed. A background may be changed from a building to a beach. With a few keystrokes, images may be so altered you cannot even tell they came from photographs.

Now, the question is, what rights do the photographers have regarding their pictures? Is paying them enough? Photographers are used to seeing their names beside their photographs. What happens if the printer changes the photograph and the photographer doesn't like the change? Should the printer be allowed to make the change anyway? Who owns the copyright if many changes are made?

Musicians are also concerned about editing practices. Digital sound editing enables an audio technician to electronically capture a piece of music. This process is known as **sampling**. Sampling is so simple, in fact, that children's toys now have sampling capability. Once captured, this collection of notes and sounds may be edited and electronically placed in another composition by another recording artist. Now, who deserves the credit and payment for the edited piece? The original musician? The second recording artist whose name appears on the new composition? The audio technician?

Movie "colorization" fits in the same category. Early movies were all in black and white. While many of these classics are still enjoyed by movie buffs, more people seem to prefer color movies. As a result, companies have started to use computers to add color to the old black and white movies. Fig. 3-12. Should they be allowed to do so? Some people say yes; others disagree.

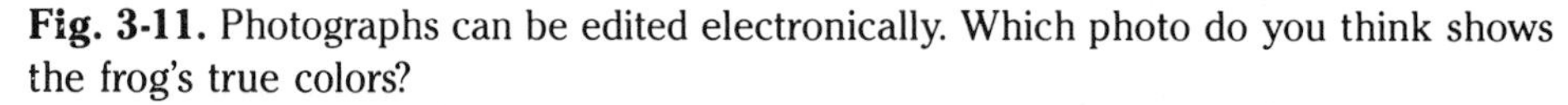

Fig. 3-11. Photographs can be edited electronically. Which photo do you think shows the frog's true colors?

Fig. 3-12. The film *Yankee Doodle Dandy* was shot in black and white. Here, the center portion of a frame has been colorized to show the difference between the original and colorized versions.

Data Security

Computer programmers know a great deal about software development and computer data storage. They often can change some programs and data. This allows them to customize the software to a specific application.

This ability to alter software creates a new problem known as data security. Fig. 3-13. When should a programmer be allowed to alter software developed by someone else? Who should be allowed access to data stored by computer? When should software be copy protected so it may not be easily duplicated? Should companies that purchase software allow employees to make a copy for home use? In 1984, Congress passed the Electronic Communication Privacy Act to answer some of these questions. They followed it with the Computer Fraud and Abuse Act in 1986. Two decades earlier, in 1968, Congress had passed the Wiretap Act to deal with a related ethical issue—the wiretapping of telephone conversations.

In the late 1980s, **computer viruses** became a major data security issue. A computer virus is a program hidden within another program. When that program is loaded into the computer, the virus copies itself onto floppy or hard disks that are run on that computer.

Sometimes these viruses are meant to be harmless. For example, some just take up space in computer memory and make the computer run more slowly. Others may display a graphic and/or a message on the monitor, surprising the computer user. The user usually has no idea where the virus came from. The programmer who created the virus has no idea which computers the virus will infect.

Some viruses have been written that destroy data in the computers they infect. They can completely wipe out large databases and destroy years of work! The problem has become increasingly serious. In the fall of 1988, the computers in the United States Department of Defense were infected by a virus written by the son of one of the employees. It was done as a prank, but it turned out to be a serious matter.

Some people have predicted that a virus could bring our country to its knees if it infected certain key databases. Others have suggested it is now easy to "fix" a presidential election, since the voting results are handled by a fairly small number of computers. All these legal and ethical questions must be answered.

Fig. 3-13. Data security has become a major issue in the computing world. Most attempts to "lock" software to keep it from being copied or altered have been unsuccessful.

HEALTH & SAFETY

MODERN "DEATH" RAYS

In early science fiction comics and movies, battles were fought with ray guns. Every Martian had one and would zap you in a minute if you weren't careful. Today, we all know that "death" rays are only make-believe, right? Well, maybe not.

Electromagnetic ("radio") waves, which are all around us in the atmosphere, are not as harmless as once believed. Many cases are on record of problems resulting from electromagnetic radiation. In fact, some people believe we are suffering from "electronic pollution."

In most cases, electromagnetic waves have only interfered with electronic devices. For example, when garage door openers first came on the market, the radio waves they emitted sometimes opened the neighbor's garage door, as well.

The Federal Communications Commission (FCC) takes care of most of these problems by regulating the use of radio waves. Radio stations, for example, must broadcast their programs at a specific wavelength specified by the FCC. This wavelength corresponds to a number on your radio dial.

However, the FCC can't control all electronic interference. In the late 1970s, American auto manufacturers put complex electronic devices in cars and other vehicles. They found out that some of these were subject to electromagnetic interference. In one instance a bus in Chicago came to an abrupt stop every time it tried to cross a certain bridge. It turned out that a signal from a radio transmitter on a nearby building reflected off the bridge and reached the bus. The signal caused the bus's electronic brakes to lock up.

Now, before new cars are allowed on the road, they are bombarded with a wide range of radio frequencies. If nothing goes wrong with a car's electronic systems during this test, the car passes. However, when car owners install their own CB radios or use cellular phones, they may still have difficulties.

Computers are another source of electronic pollution. Their microchips and other electronic components radiate signals that can interfere with other devices. In 1985, the FCC began regulating computer radiation. Now, computer cases are designed to contain these unwanted signals.

Sometimes, however, electromagnetic waves cause truly serious problems. During the Vietnam War, a radio signal set off an MX missile on a U.S. aircraft carrier, killing 134 crew members. In 1984, signals from a Voice of America radio transmitter interfered with the electronics in a West German fighter plane. The plane crashed. In Japan, electronic interference has caused industrial robots to maim or kill plant workers.

Evidence suggests also that electromagnetic radiation may present general health hazards to humans. In 1987, the New York Department of Health found that children exposed to electromagnetic radiation from power lines suffered from a higher rate of brain cancer and leukemia. A study done in Maryland showed electrical workers to have a higher than expected number of brain tumors.

Research continues on the effects of electromagnetic radiation. Our society could not function without electrical and electronic devices. We may need to develop shielding for these devices to protect ourselves.

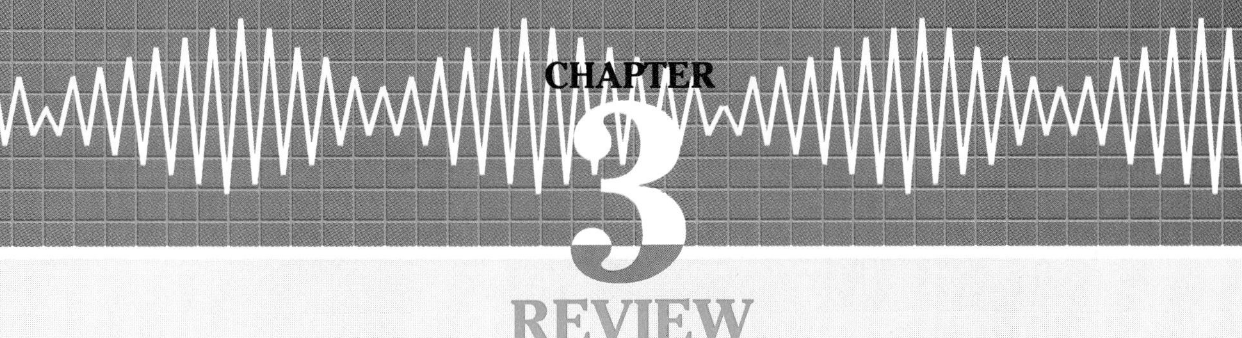

CHAPTER 3 REVIEW

Review Questions

1. What is the purpose of technology assessment?
2. What are the six different categories used in assessing technology impacts?
3. How do media advisors affect political campaigns?
4. Name ways in which a government could use computers to control its citizens.
5. Describe at least two social changes brought about by communication technologies.
6. How does a service economy differ from a production economy?
7. Name three ways in which TV has affected our culture.
8. Define *ethical.*
9. Why does digital editing raise ethical problems?
10. What is a computer virus?

Activities

1. The "picture phone" is a telephone that displays a video picture of the person spoken to. Make a technology assessment of this device. What effects do you think it will have on our lives? Be sure to consider each of the six categories listed on page 49.
2. Ask ten people to identify what they consider the most important communication device currently in use. Ask each person's reasons for the opinion. Compile your findings in a brief written report.
3. Read an article that describes an issue or impact relating to a communication device. Tell the class about the article. Give your own opinion of the situation.
4. Read Aldous Huxley's *Brave New World.* Report to the class on the differences and similarities between our own world and that described in the book.
5. Make a chart similar to the one shown in Fig. 3-14. For each column in the chart, identify a device used in that area and name some of its impacts.

Fig. 3-14.

	GRAPHIC PRODUCTION	TECHNICAL DESIGN	OPTICS
DEVICE			
Political Impact			
Social Impact			
Economic Impact			
Environmental Impact			
Cultural Impact			
Ethical Impact			

Careers

Communication Technology Teacher

Why did your communication teacher decide to teach? Why did he/she decide to teach communication technology? You might be surprised by the answers.

Teaching is among the most rewarding of all professions. Most teachers are people who like people. They like to help others learn new things. They often think education is the most important thing they can give to others.

Those who become technology teachers combine their interest in people with a need to know how things work. They often have interesting hobbies that relate to technology. For example, many communication teachers enjoy photography, printing, or video production in their spare time.

Most teachers have two or three months of "free" time in the summer. This gives them time to pursue their own interests. Others work a second job in the summer to earn extra money. While teachers aren't as highly paid as some other professions, they tend to lead comfortable lives and enjoy the respect of those in their community.

Education

A four-year college degree is required to teach communication technology. Most communication technology teachers have a degree in technology education. This degree prepares them to teach other courses in technology as well, such as production or power/energy courses.

There are about 250 colleges and universities in the United States that offer bachelors' degrees in technology education or a related field, such as industrial education. For more information, start by asking your teacher. Then contact the International Technology Education Association, 1914 Association Drive, Reston, Virginia, 22091.

Related Careers

Technology teacher (construction materials and processes, manufacturing, graphic communication, computer-aided design, photography, principles of technology, power and transportation, introduction to engineering, exploring technology, etc.)
Industrial trainer
Educational administrator

Other Careers in Communication Technology

There are many career opportunities in the field of communication technology. A good source for a complete listing is the *Dictionary of Occupational Titles*. Your library should have a copy. Some of the career opportunities include:

Audio engineer
Commercial photographer
Computer programmer
Customer service representative
Designer
Director (video or film)
Editor
Illustrator
Journalist
Marketing specialist
Production manager
Production worker (audio, video, printing)
Public relations specialist
Sales representative
Technical writer
Technician (audio, video, photography, printing)

Correlations

 *Language Arts*

Develop a crossword puzzle that has communication technology as its theme. See how many words from the lists of "Terms to Learn" you can use.

 *Science*

Select a communication device and conduct a study of its environmental impact. Possible choices might be the telephone, television, or audio tape players. Your study should be based upon research and scientific observation. Collect as much data as you can. Be sure to consider materials from which the device is made as well as effects like noise pollution. Then, prepare a report of your findings that includes the following sections:

a. History and background
b. Definition of the problem
c. List of possible effects
d. Data collected to study possible effects
e. Findings and results

If you can, work with your science teacher on this project.

 *Math*

One of the trends discussed in Chapter 2 was increasing memory in computers. One character (number or letter) requires eight bits (one byte) of computer memory. A kilobyte is about 1000 bytes. A megabyte is about one million bytes.

A typical 5¼-inch floppy disk holds 360 kilobytes of information. A typical 3½-inch floppy disk holds 1.4 megabytes of data. A typical optical disc holds 400 megabytes of information.

If you assume there are 1500 characters on a typical typewritten (or word processed) page, how many pages of text can each of the disks hold?

 *Social Studies*

The system of democracy we live under in the United States depends upon informed citizens. How does your local government communicate information about key issues and concerns to the people in your community?

Visit at least three different departments in your city hall. Interview a clerk. Ask what communication systems they use to "get the word out." What systems can be used by the citizens to communicate their opinions to those who govern? Report your findings to the class.

Basic Activities

Basic Activity #1: Interrelationship of Technologies

In the introduction to this section, you learned that technology can be divided into three general areas: communication, production, and energy/power/transportation. This does not mean that the areas are completely separated from each other. What happens in one area affects other areas, and one could not exist without the others. The activity described here will help you see how technologies are related to each other.

Materials and Equipment

poster board
markers
photos or illustrations cut from magazines

Procedure

1. Select a communication device you often use, such as a radio. How do other technologies contribute to your use of the device? Think of as many examples as you can.
2. Create a poster to illustrate these relationships. Make your poster colorful by using markers and photos or magazine illustrations. The example shown in Fig. I-1 illustrates some of the ways production and power/energy/transportation contribute to telephone communication.

Basic Activity #2: Job File

As you learned in Chapter 1, communication technology offers many career opportunities. There are careers for people of all abilities and interests. Now is the time to start learning about them. One source of information is this book. At the end of Sections I through VI, you will find descriptions of careers. You can also learn about careers from newspapers and magazines. In this activity you will start a "job file" that contains information about communications careers. As you finish each section of this book, add to the job file. By the end of the course, you will have learned valuable information about the types of jobs available.

Materials and Equipment

newspapers and magazines
3″ × 5″ index cards

Procedure

1. Check "help wanted" ads for jobs in communication. Your local newspaper is one place to look, but also go to the library and check other newspapers, especially those from large cities. Also look in trade-related magazines. These are magazines published for a certain industry, such as printing, computer hardware or software, photography, or drafting.

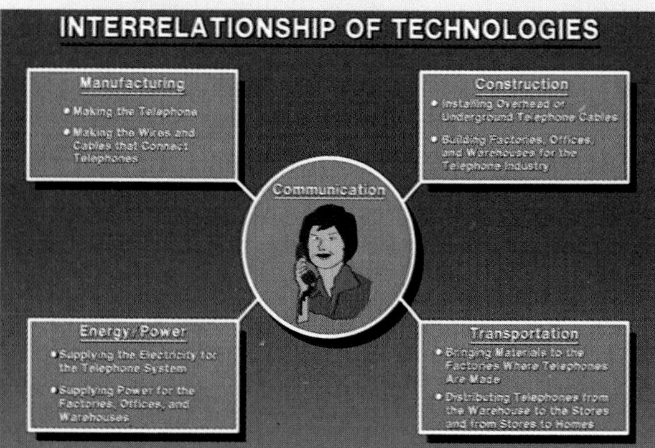

Fig. I-1.

2. For each job you find, prepare a 3″ × 5″ index card with the following information: name of the job, name of the company offering the job, location (city or state), education or experience required, and salary offered. See the example in Fig. I-2. (Some ads may not include all of this information.)
3. As your file grows, see if you can detect any patterns. For example, in which parts of the country are most of the publishing jobs located? How many jobs require a college education?

Basic Activity #3: Job Fair

At a job fair, people from various companies offer information about the jobs they have available. Your class can conduct its own version of a job fair. You and your classmates can share information you've learned about communications jobs and perhaps discover new career interests.

Materials and Equipment

Resources such as the *Dictionary of Occupational Titles*, trade magazines, etc.

Poster board, paint, and similar items for making displays

Procedure

1. Select a career which interests you—such as news announcer, disc jockey, graphic artist, photographer, or computer programmer—and research it. Find out what is done, what tools and equipment are used, what education is needed to prepare for the job, and so on.
2. Prepare a poster or other type of display to inform others about the job. Suppose, for example, you were a recruiter for a television station and you were looking for someone to be a reporter. How would you interest people in the job?

Fig. I-2.

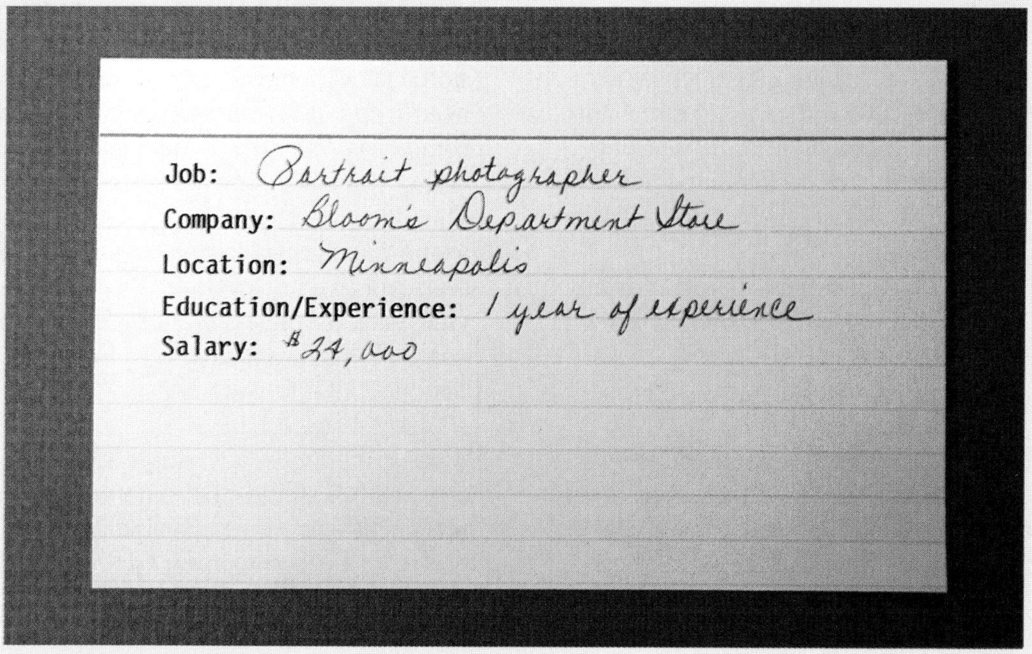

Job: Portrait photographer
Company: Bloom's Department Store
Location: Minneapolis
Education/Experience: 1 year of experience
Salary: $24,000

Basic Activity #4: Resumé

A resumé is a short description of one's education, job background, and interests. When you are job hunting, you will probably send your resumé to a number of companies. Some job ads ask that you include a resumé when responding to the ad.

This activity will give you practice in writing a resumé. Here are some guidelines to keep in mind.

- Be honest. Lying on your resumé can cost you your job later.
- Make your resumé brief, neat, and accurate. Check your spelling and grammar carefully.
- Learn as much as you can about the job. Then prepare your resumé to show what qualities and experience qualify you for that job.

Materials and Equipment

Typewriter or word processor
Paper
Dictionary

Procedure

1. Suppose you are applying for a "job" as a student in this class. Prepare a resumé that includes the following information:
   - Your educational background—schools attended, what years, any communications courses taken.
   - Your work experience—particularly any jobs in communications, such as writing for a school paper.
   - Your goals—what do you hope to learn from the course?
2. When you have finished, trade resumés with a classmate and discuss your reactions. What sort of impression do the resumés make? How could they be improved?

Intermediate Activities

Intermediate Activity #1: Job Interview

A job interview is an opportunity for an employer and a job seeker to learn more about each other. Most people feel a little nervous when going to a job interview; so it helps to be well prepared. This activity will give you practice in interviewing. You'll have a chance to act both as a job seeker and an employer.

Materials and Equipment

To help in evaluating the interview afterwards, you may want to tape-record or videotape it.

Procedure

1. Select an interview partner. (Your teacher may select one for you.)
2. With your partner, create an imaginary company where the interview will take place. For example, it might be an advertising agency, a "quick print" shop, a TV station, computer store, or telephone company. Both of you will need to do some research to learn what such companies do and how they are organized.
3. One of you will be the personnel manager. The other will be the job seeker. The personnel manager will be interviewing the job seeker for an entry-level job. Before the interview, each of you should make a list of the questions you want to ask.
4. Conduct the interview. You may find Figs. I-3 and I-4 helpful.
5. Afterwards, discuss your impressions. What do you think are the most important aspects of a job interview from an employer's standpoint and from the job seeker's?

Which person would you hire?	
Someone who . . . • is neat and well-groomed. • sits up straight; looks alert. • speaks clearly and audibly. • shows an interest in the company and what his/her role might be. • appears eager to learn and willing to work.	Someone who . . . • is dirty and unkempt. • slouches in the chair. • mumbles. • doesn't seem to know or care what the company does. • acts as though you owe him or her a job.

Fig. I-3.

The law protects people from having to answer personal questions that do not relate to their ability to perform a job.	
The interviewer may ask about . . . • your educational background. • your previous job experience. • your career goals.	The interviewer may *not* ask about . . . • your marital status. • whether or not you have children. • your age. • your weight or height. • your religious background. • your military or financial status.

Fig. I-4.

Fig. I-5.

Intermediate Activity #2: Forming a Company

Your teacher may lead your class in a student enterprise—a project in which the class forms a company and produces goods or offers services. Participating in a student enterprise will give you an understanding of how companies work. You'll learn how to operate tools and equipment. You'll also get practice in working with others to achieve a goal. Fig. I-5.

Procedure

1. Incorporate the company. Incorporation is a legal process that allows a group of people to form a corporation. In a corporation, people invest money in a company in return for part ownership. They share in the company's profits and have a voice in running the company. Your teacher will advise you of the steps required to form a corporation.
2. Finance the company. You will need money to run your company. Materials must be purchased. The school will have some tools and equipment, but you may need to rent or buy additional items. Also, some student companies pay their workers, so money is needed for wages. There may be additional costs, such as advertising.

   One way to finance a company is mentioned in Step 1—incorporation. People could buy shares in your company at $1.00 per share, for example. Later, when the products have been sold and you are ready to dissolve your company, you will distribute the profits to the shareholders. The shareholders might be the students in your class, or they might include people outside the class as well. Your teacher will guide you in financing the company.
3. Select a board of directors. The shareholders elect a board of directors to control the company. In a student enterprise, the entire class may be on the board of directors. The board elects one of its members to be chairperson. It also elects a president, a secretary who keeps a record of the board's actions and a list of the shareholder's names, and a treasurer who handles the company's money. The president, in turn, may select managers to be in charge of the company's various departments. For example, your company might have a research and development manager, a production manager, and a marketing manager.
4. Organize a work force. Your company will need people to work in the various departments. Since your company will be small, each of you may work in several of them. Your company might include the following departments:
   - Research and development. This department helps determine what to make, develops plans and models for the product, and produces the working drawings.
   - Production tooling. The production tooling department designs and makes the jigs and fixtures needed to produce the product. Jigs and fixtures are devices that guide tools and hold workpieces. For example, if your product includes drilled holes, you may need to design a jig to guide the drill to the correct position on the workpieces.
   - Production planning and control. This department prepares the bill of materials and plan of procedure. It also prepares flow charts to show every step of the production process from beginning to end. It is the responsibility of the production planning and control department to set up the production

line and help keep it running smoothly.

- Quality assurance. This department inspects the product and recommends changes in materials or procedure to correct any defects.
- Manufacturing. This is the department that produces the product.
- Marketing. In the planning stages of your project, the marketing department conducts research to determine whether a proposed product might sell. It also develops an advertising strategy and helps create the packaging for the product. After production, the marketing department is responsible for distributing the finished products to the customers.
- Personnel management. This department is responsible for selecting and training the workers. One of its most important tasks is to make sure working conditions are safe and workers are following the recommended safety rules.

5. Choose a name for the company. The entire class may wish to submit ideas, or the marketing department may be asked to think of some names. The final choice should be made by a class vote.

Intermediate Activity #3: Research and Development

One of your company's first tasks will be to decide what to make. Your product should be fairly simple and inexpensive to produce, and it should relate to communication technology. The *Student Workbook* that accompanies this text describes one such product—a videocard. Here are some other ideas.

- Publish a newsletter about your school's technology education department. You could describe the courses, interview some of the people, and highlight the department's activities and accomplishments.
- Produce a video sports documentary about one of your school's athletic teams.
- Develop, test, and build a communication device. For example, you might make a device that signals people when there is mail in their mailbox.

Many communications companies do not produce products at all. Instead, they provide a service. For example, advertising is a service industry. Your company may choose to provide a service, such as advertising, rather than produce a product.

In determining what your company's product or service will be, follow the procedure outlined here.

Procedure

1. As a class, brainstorm ideas. Select the three or four best ideas.
2. The marketing department and the research and development department should research these ideas. Questions to answer include (1) Is there a market for the product or service? (2) How much will it cost to pro-

duce? (3) What kinds of materials and equipment will be needed? (4) What skills will be needed? (5) How much time will it take? The marketing department may conduct a market survey of potential customers. The research and development department may draw plans and test prototypes.
3. The results of the research should be brought before the board of directors (the class), and the board should then decide on one product or service.
4. If a product was selected, the research and development department should prepare final, working drawings and specifications.

Intermediate Activity #4: Production and Distribution

This activity lists the steps needed to produce and distribute a product. For a service company, the steps will need to be modified. Refer to "Intermediate Activity #2" for additional information about the work of each department.

Procedure

1. The production tooling department should prepare any jigs and fixtures needed to make the product. Test them to make sure they work properly. Make any necessary changes.
2. Production planning and control should consult with the research and development department to learn what materials and supplies are needed. Prepare the bill of materials and order the items that are needed. Prepare a plan of procedure and flow chart. Do a "dry run" of the production line to make sure things work according to plan. The company treasurer should keep records of all expenses.
3. The manufacturing department should produce the product.
4. (Some of these tasks can begin before production.) Marketing should prepare the packaging for the product. Plan and carry out the advertising campaign. Establish the distribution system (who will sell the item and where).
5. Sell the product, keeping careful records of sales.

Advanced Activities

Advanced Activity #1: Documentary

In this activity, you will produce a 10-15 minute video documentary. If you do not have access to a video camera and video cassette recorder, you can prepare a booklet instead.

Materials and Equipment

video camera
video cassette recorder
television set
blank cassette tape

Procedure

1. In Chapter 3 you learned that technology has an impact on many areas of our lives. Listed here are the categories of impact.
   - political
   - social
   - economic
   - environmental
   - cultural
   - ethical

Select a communication device, such as television, as the subject of your documentary. The purpose of the documentary will be to explore how the communication device has had an impact on the lives of people in your community.

2. Because you probably won't have access to editing equipment, your video will have to be filmed in sequence. Therefore you'll need to plan carefully. Your documentary should include narrative as well as interviews. You'll need to go on location for many of the segments.
3. Set up your interviews and plan what questions you will ask. You may want to let the people you're interviewing preview the questions so that they can think about their responses before the actual taping.
4. Make sure you know how to operate the video camera. Make some short scenes for practice and replay them. Are your camera angles good? Do you need to hold the camera more steady?
5. Tape your documentary. Remember, if you run into a problem, you can rewind and tape the scene over.
6. Show the documentary to the class.

Advanced Activity #2: What if...?

Here's a chance to exercise your imagination. In Chapters 1 through 3 you learned something about past developments and current trends in communication. Now consider what the future might hold. Imagine a new invention. What impact—good and bad—might it have on society? What other inventions might result from it?

You may want to do some brainstorming with the class, or think about it on your own and then share your ideas later. Here are some possible topics, but feel free to select one of your own.

- It is possible to build a telephone that will automatically display the number of the caller. What if all telephones had this feature?
- What if the design and manufacturing processes became totally automated? For example, a designer working at a computer might input some ideas for a car. Within hours, the actual automobile would roll off the automated assembly line.
- What if someone invented a tiny computer that could be implanted in one's brain?
- What if we perfected communication with animals, and they could tell us what they were thinking and feeling? Fig. I-6.
- What if holography were perfected so that a hologram image would look as realistic as the actual item? How might such technology be applied?

Fig. I-6.

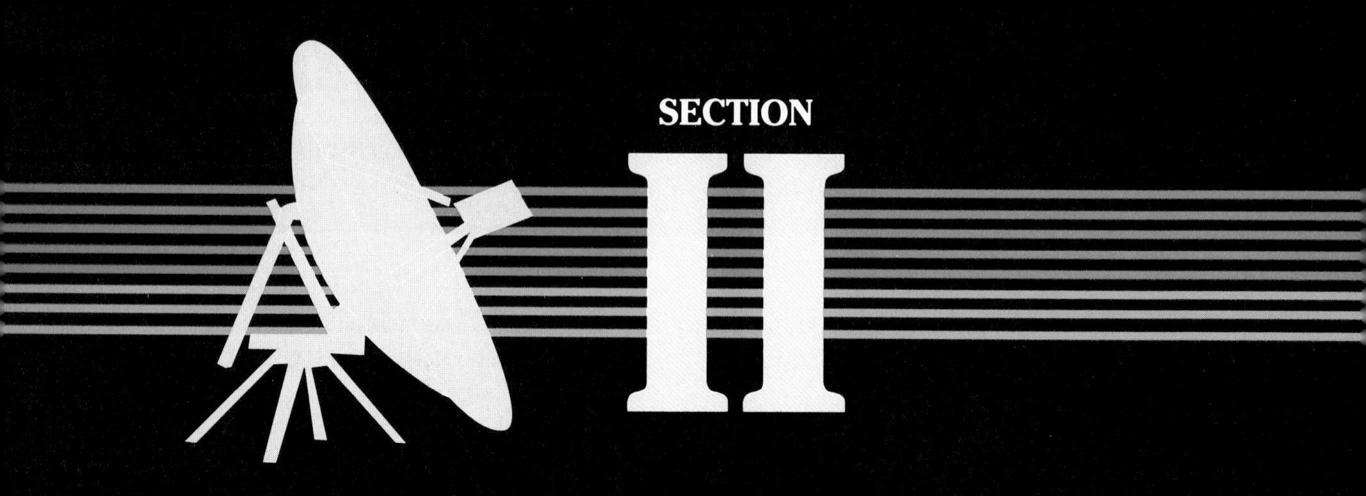

SECTION II

Data Communication Systems

Chapter 4: Introduction to Computers
Chapter 5: Computer Hardware
Chapter 6: Computer Applications

In 1944, Thomas Watson, then President of International Business Machines (IBM), predicted there would be a need for only about five computers in the entire world. How wrong he was! Now, not only do most businesses rely on computers, but a great many people have computers in their homes.

A computer is a machine that makes calculations and processes information very quickly. Computers are amazing tools because they can do so many different things. Consider that a typewriter can be used only for typing. A calculator can be used only for calculating. A T-square and drawing board are used for drafting. An artist creates illustrations with pen and ink on paper. An accountant balances financial records using a record keeping book. Friends communicate over long distances using a telephone. Video game arcades provide entertainment. A computer can be used to do *all* of these things, and it does them with far greater ease than ever before imagined.

Three words describe what computers do: input, process, output. You put some information into a computer (input). It works on that information (process). Then the computer displays the results (output).

Let's use a pocket calculator as an example. An electronic calculator is a small, simple version of a computer. Assume you want to multiply two numbers. What do you do? You punch in the first number, then a multiplication sign, and then the second number. That's the *input*. The calculator multiplies the two numbers. That's the *processing*. It shows you the answer on its display or prints it on a roll of paper. That's the *output*.

Most computers are bigger, cost a lot more, and have printers and other equipment hooked up to them. In principle, though, they are much the same as calculators.

In this section, you'll learn how computers work. You'll also have a look at how all of a computer's input and output devices work as well. Finally, you'll get an idea of the many different ways computers are used to communicate information. We are living in what many people call the "Information Age," and in this section, you'll begin to see why.

CHAPTER 4

Introduction to Computers

It's hard to believe that only a decade ago, a computer was a rare sight. Until recently, only large organizations could afford computers. They were very expensive and took up a great deal of space. Today they are almost as common as TV sets.

The microcomputers you have in your school are, in many ways, more powerful than the huge, costly, early computers. Your school computers can accept input from more sources, process information more quickly, and output to a wider range of devices than could those early monsters.

The computer is one of the most astounding inventions in history. Nobody really knows what will be possible with computers in the future. Yet, the principles upon which the computer is based are fairly simple. In this chapter, you will learn about these principles and how they apply to the computers in your school.

Terms to Learn

applications software
binary system
bits
bus
byte
central processing unit (CPU)
entrepreneurs
hardware
microchips
microprocessor
programming language
random-access memory (RAM)
read-only memory (ROM)
software
transistor

As you read and study this chapter, you will find answers to questions such as:

- Who invented the computer?
- How do computers work?
- How do microchips process information?
- What is a programming language?

THE TYPES OF COMPUTERS

Computers can be analog, digital, or hybrid. The difference is in the kind of data they use. Analog computers work with data that vary in a continuous way, such as temperature, pressure, or speed. They express one physical quantity in terms of another. You probably don't have an analog computer in your home, but chances are you do have an analog thermometer. Such a thermometer has a tube that contains a liquid. As the temperature increases or decreases, the liquid in the tube moves up or down. That's an analog device — one physical quantity (temperature) is expressed in terms of another (movement of liquid).

Digital computers require data to be in precise amounts. That is why all data used by digital computers are expressed as whole numbers (digits). The computer performs arithmetic operations using these numbers. Because many kinds of data, including words, can be changed to a number code, the computer can process many kinds of information.

Some computers combine digital and analog operations. These are called hybrid computers. Some flight simulators used to train pilots are controlled by hybrid computers.

Which type of computer to use depends on the kind of work to be done. Many scientific and engineering problems are best solved with analog computers. However, the most common type is the digital computer. It is used in businesses, factories, schools, and homes as well as in many science labs and engineering firms. Because digital computers are the most common, this book will focus on them. You will learn how they developed, how they work, and how people use them to perform a wide variety of tasks.

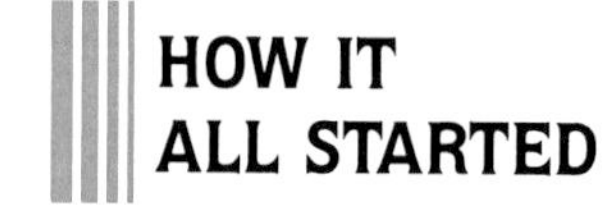

HOW IT ALL STARTED

The story of computers is the story of creative people solving one problem after another. Some of the inventors who developed the microcomputer were not much older than you are when they first started testing their ideas.

Early Calculators

We can trace the beginnings of the computer back about 5,000 years to the abacus. Fig. 4-1. An abacus is a calculator still used in some parts of the world, such as China. By sliding beads along its wires, the user can make calculations quickly.

In the 1600s, more complex calculating devices were developed in Europe. Blaise Pascal, a Frenchman, built fifty different experimental calculators. A wooden calculating machine he completed in 1652 finally worked, but it was too costly to be practical. In 1694, German mathematician Gottfried Leibniz designed the first calculator that could multiply and divide. Fig. 4-2. Another major breakthrough came in 1823. Englishman Charles Babbage built his "Difference Engine" that could solve algebra problems. Ten years later, he designed an "Analytical Engine" that was to use punched cards, but it never made it past the model stage. Fig. 4-3.

Important breakthroughs were made by Americans in the late 1800s. In 1885, Dorr Felt developed the first calculator to be operated by striking keys. Three years later, William Burroughs developed and sold a calculator that printed its results.

In 1880, the U.S. census took seven years to compile. Using a counting machine he built,

Fig. 4-1. The abacus, invented about 3,000 B.C., was perhaps the first calculating machine. It is still in use in some parts of the world.

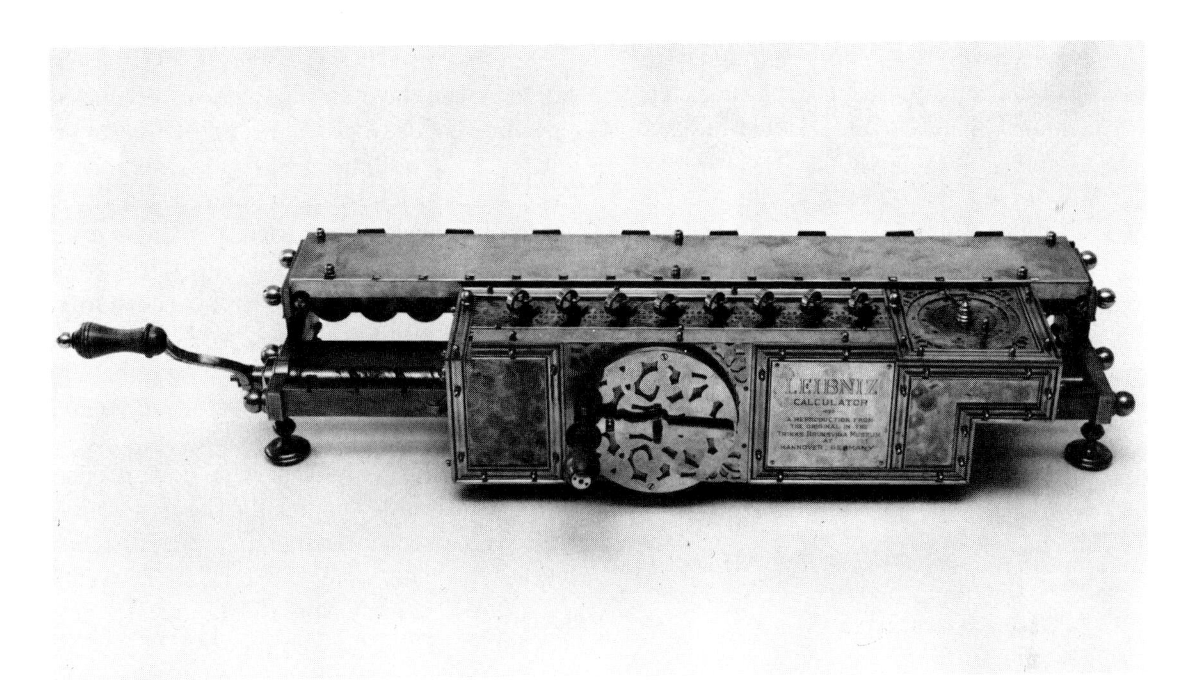

Fig. 4-2. Leibniz's calculating machine was called the "Stepped Reckoner." It could add, subtract, multiply, divide, and extract square roots.

Fig. 4-3. Charles Babbage's "Difference Engine" (left) and his "Analytical Engine" (right) were ancestors of the modern computer. However, these machines were never completed because of lack of money and because the technology of the 1800s could not produce metal parts to the required precision.

Herman Hollerith was able to announce the results of the 1890 census just six weeks after census day! Fig. 4-4. In 1924, Hollerith's company became International Business Machines (IBM). Today it is the largest computer manufacturer in the world.

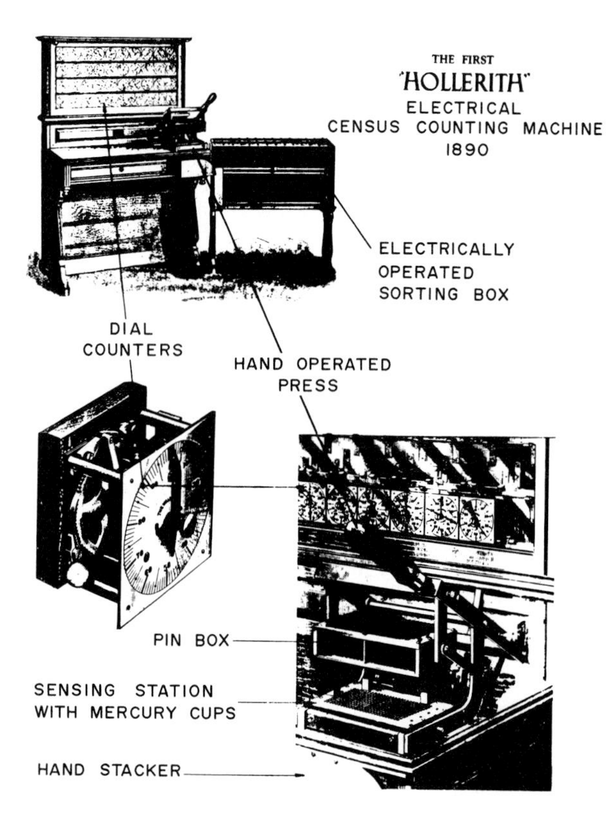

Fig. 4-4. Programmed with punched cards, Herman Hollerith's "Tabulator" was used to compile the 1890 United States census data in record time.

Computers Become Logical

All of these early machines were wonderful but they were limited in their ability to solve problems. They were calculators rather than true computers. A computer uses logic to solve problems. Logic is the ability to see the connection between two parts of a problem. The early machines were incapable of logic. The technology needed had yet to be invented.

Another piece of the puzzle fell into place in 1937. Claude Shannon, an American, developed electrical circuits that could handle binary arithmetic. Binary arithmetic uses only two numbers. This was crucial, because binary math is the basis of the logic system used by computers. In 1943, the British built one of the first true computers. It was designed to decipher coded messages used by the Germans in World War II. "Colossus," as it was called, used vacuum tube electronic parts. However, Colossus was limited because it could only solve codes.

In 1944, after five years of work, IBM unveiled the first full-scale computer, the Harvard Mark I. It used relay switches rather than vacuum tubes because switches were more reliable. Two years later, the Electronic Numerical Integrator and Computer (ENIAC) was introduced. Fig. 4-5. It could add five-digit numbers 5,000 times per second and could solve problems. However, it had very little "memory" in which to store different programs. (Programs are sets of operating instructions.) The British solved this problem with the Manchester Mark I. Introduced in 1948,

Fig. 4-5. ENIAC was the first all-electronic general-purpose computer. It was more than a thousand times faster than the electromechanical computers, such as the Harvard Mark I.

it was the first computer to run a program from its memory. UNIVAC I, brought out by the Sperry Rand Corporation in 1951, was ten times as fast as ENIAC and had one hundred times as much memory.

All of these early computers were huge and costly. The vacuum tube electronics they relied upon were bulky. In addition, they gave off a lot of heat. Some of the early computers were cooled with water cooling systems like those found on automobiles.

The Tiny Transistor

In 1948, the invention of the transistor brought about the second generation of computers. A **transistor** is a device used to amplify (increase) or control electric current. It works much like a vacuum tube but is smaller and generates less heat.

The transistor replaced the vacuum tube, and this increased speed. Fig. 4-6. The first computer to use transistors was the TX-O, built at the Massachusetts Institute of Technology in 1956. By 1960, transistors were commonly used in all new computers.

Electrical engineers figured out how to place large numbers of tiny transistors onto a piece of silicon only one-quarter-inch square. Silicon is the same material found in beach sand. These tiny squares became known as **microchips** or integrated circuits. Microchips were first used in calculators. When they began to replace regular transistors, the size and cost of computers were greatly reduced. A third generation of computers appeared in less than two decades!

SCIENCE FACTS

Silicon

Silicon is the second most common element in the earth's crust. (Oxygen is the first.) It is found in clay, sand, and many rocks. Silicon is an ingredient in many products, including glass, silicones, and microchips.

Silicon is used in microchips because it makes a good semiconductor. A semiconductor is a substance that conducts electricity better than an insulator (such as rubber) but not as well as a conductor (such as metal). The degree to which a semiconductor conducts electricity depends on the temperature and the materials from which it is made. By adding different ingredients to silicon, engineers can produce semiconductors for use in various electronic devices.

The Birth of the Apple® Computer

If the story ended here, you would probably not have computers in your school today. In the 60s, computers were still too large and costly for anyone but government and other large organizations. However, a couple of enthusiastic kids were about to turn the computer world on its ear.

Fig. 4-6. Transistors were much smaller than vacuum tubes, but computer chips are smaller still. Integrated circuits now contain more than a million electronic components on a single chip of silicon!

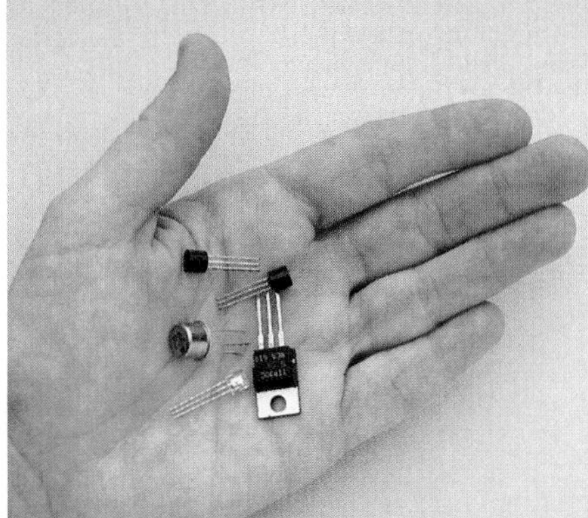

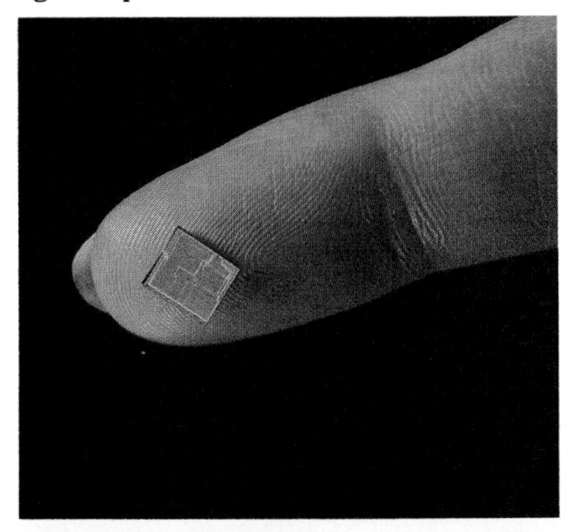

It all started in 1969, when the Intel Development Corporation produced the world's first programmable microchip. "Programmable" means that the microchip can carry a set of instructions. Known as the "4004"microprocessor, it was, in a sense, a self-contained computer. Since this microchip was programmable, it opened up brand new design possibilities.

The center of activity for the development of microprocessor chips was in the Santa Clara region of California, just south of San Francisco. The area soon became known as "Silicon Valley." Many electronics engineers were employed by companies in the area. Electronics hobbyists who lived there also experimented with the new microprocessors. On March 5, 1975, about thirty of them gathered in a garage in Menlo Park, California, for the first meeting of the "Homebrew Computer Club." By the third meeting, the club had several hundred members.

Steve Wozniak was a regular Homebrew Club member. Like many others in the club, he was a college kid excited by what the new electronic components (parts) could do. After MOS Technology introduced its "6502" microprocessor in 1976, Wozniak built a computer that used this chip. Within a matter of weeks, he had devised a code (programming language) to program it. He showed off his new computer at a Homebrew meeting and passed out copies of its design so others could build one as well.

Although the computer had no keyboard, power supply, or even a case, an old friend of Wozniak's saw its potential. Steve Jobs convinced Wozniak they should start a company that would produce and sell his new computer. To come up with the money they needed, Jobs sold his Volkswagen bus and Wozniak sold his two calculators. The calculators were worth hundreds of dollars at the time. The men called their new computer the "Apple," and convinced a local electronics dealer to buy 50 of them. The microcomputer revolution had begun! Fig. 4-7.

Jobs and Wozniak were **entrepreneurs** — people who organize and manage a new business. When they started the Apple computer company, these young men took a risk. Few people imagined there would be a market for computers. At that time, IBM was king of the computer market. It produced mainframe computers, which have a large capacity but are also bulky. The "brains" of a mainframe require a separate, climate-controlled room. A single mainframe computer can serve many users.

At first, IBM decided not to make microcomputers. However, it soon changed its mind. In 1981, when microcomputers were appearing everywhere, IBM introduced its IBM PC® line. These microcomputers met with great success, particularly among businesses.

Soon, other companies made IBM PC "clones," computers that could run the same programs as the IBM PC. In 1983, Apple introduced its highly successful Macintosh®. IBM came out with its Personal System/2® line of microcomputers in 1987.

Fig. 4-7. Steve Wozniak and Steve Jobs founded the Apple Computer Company in 1976.

HOW COMPUTERS WORK

What most people call a computer is actually a complete computing system. The various parts of the system are known as **hardware**. The main piece of hardware is the processing unit. Other hardware items, such as the printer, are called peripherals.

A computer system has ways to:

- input data
- process the data, once the data has been input
- store the data, both before and after they have been processed
- output the data in some usable format. Fig. 4-8.

In the next chapter, you will learn how computers input, output, and store data. Right now, let's take a closer look at how computers process the data they receive.

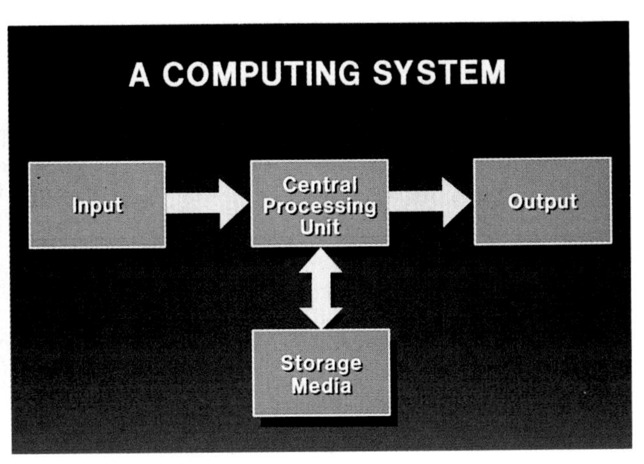

Fig. 4-8. A fully workable computing system includes input, processing, storage, and output components.

The Binary System

Digital computers process information using a counting system. The counting system we use in our daily lives is based upon our ten fingers. This system includes "place holders" for ones, tens, hundreds, and so forth. When you have to represent the number eleven, for example, you put a 1 in the "ten's" place and a 1 in the "one's" place. That means you have 1 ten and 1 one, which totals eleven. You've been using this base 10 system for a long time, so it makes perfect sense to you.

In a way, digital computers are rather simple minded. They can keep track of only two numbers—0 and 1. It's as if they had only two fingers. A computer's system of counting is known as base 2, or the **binary system**. In the binary system, instead of a one's place, ten's place, etc., there is a one's place, two's place, four's place, eight's place, etc. Reading from right to left, each new place represents double the number before it. Therefore, the binary number 111 is equal to 1 one, 1 two, and 1 four, or the number 7. The binary number 1101 represents 13. Fig. 4-9.

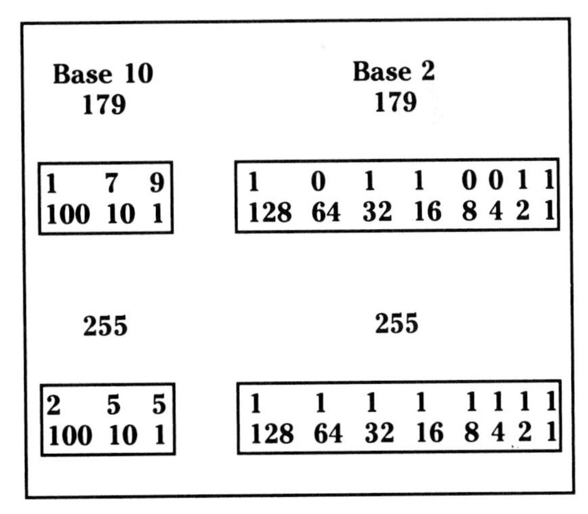

Fig. 4-9. In base 10, the counting system we generally use, the number 179 represents 9 ones, 7 tens, and 1 hundred. The same number is shown above in the base 2 system the computer uses. Another example is given using the number 255.

Bits and Bytes

The way a computer keeps track of these 1's and 0's is with small electrical pulses. A small amount of voltage represents a 1, and no voltage represents a 0. Thus, when we talk about computer data, we are really talking about the storage of 1's or 0's. These 1's or 0's are called data **bits**. The word *bit* comes from the term *b*inary dig*it*.

A bit is a tiny amount of data. Most computers can easily manage at least eight bits at a time. This chunk of eight bits is called a **byte**. When you put together eight bits of data to form a byte, you can represent numbers as small as 00000000 or as large as 11111111 (255).

All computer input is first translated into a number code. For example, consider how the computer handles letters that you type on the keyboard. Each letter has been assigned a binary number. This binary code for letters and characters is known as the American Standard Code for Information Interchange (ASCII) character set. Fig. 4-10A. The letter "A," for example, is the 65th character, so it is represented by the binary number for 65, which is 01000001. The letter "B" is 01000010, because it is the 66th character on the ASCII chart, and so forth.

How does the change from letter to number take place? To understand that, you need to know more about microchips.

Microchips—the Basic Building Blocks

Computers start with transistors used as electrically operated switches. They either allow current to flow through them, or they don't. If you combine them, you can make a logic circuit or "gate." An "OR" logic circuit, for example, operates like a doorbell on a house. There might be one doorbell switch at the front door and another at the back door. When either circuit is closed by pushing the button, the doorbell rings. Fig. 4-10B.

In computers, these logic circuits produce binary code output — either 0's or 1's. In one common type, 2.5 volts of current represents a 1 and a few tenths of a volt is a 0. In the OR gate just described, if either transistor (switch) passed 2.5 volts (a 1), a 1 would come out. That 1 could be the signal to your computer to ring the doorbell.

An "AND" logic gate requires two or more switches to be in the 1 state in order to output a 1 from the circuit. In your car, for example, each seatbelt could close a switch when buckled. Your key could close a switch when turned in the ignition. When all of the seatbelts were buckled, this logic circuit would output a 1, which would start your car. If one seatbelt were not buckled, a 0 would be output, which would not allow the starter motor to operate.

Fig. 4-10A. The ASCII code uses binary numbers to represent 255 different characters/functions. This table shows how the computer interprets certain keystrokes through ASCII code by using binary numbers.

When you type this character on the keyboard	it sends this binary number to the central central processing unit	which represents this ASCII code #	and results in this . . .
^G*	00000111	07	Sounds the "bell" or beeper in the computer
^M*	00001101	013	Moves the cursor on the monitor down to the next line like a carriage return on a typewriter
3	00110011	051	Displays a "3" on the monitor
B	01000010	066	Displays a "B" on the monitor
t	01110100	116	Displays a "t" on the monitor
Delete key	011111111	127	Deletes the character to the left of the cursor

*Note: The^ represents the "control" key on the computer keyboard.

AN "OR" LOGIC CIRCUIT

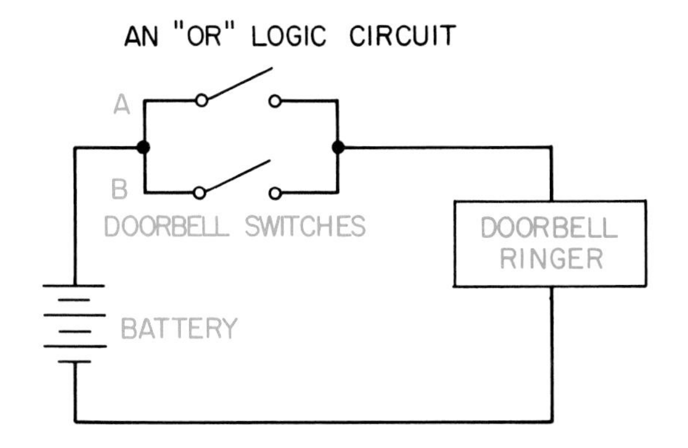

AN "AND" LOGIC CIRCUIT

BATTERY
STARTER MOTOR
SEAT BELT
A
IGNITION
B

LOGIC TABLE		
SWITCH A	SWITCH B	DOORBELL
0	0	SILENT
1	0	RINGS
0	1	RINGS
1	1	RINGS

LOGIC TABLE		
SWITCH A	SWITCH B	MOTOR
0	0	DOESN'T START
0	1	DOESN'T START
1	0	DOESN'T START
1	1	STARTS

Fig. 4-10B. Logic circuits are used in computers to make decisions. Here you see examples of an "OR" logic circuit and an "AND" logic circuit.

SCIENCE FACTS

Nanoseconds

To us, a second seems like a short time, but not to a computer. Computer scientists measure time in *nanoseconds*. One nanosecond is one billionth of a second. Some of today's supercomputers can make more than 20 calculations per nanosecond. That's over 20 billion calculations per second.

Over a million transistors and electronic components may be combined with one another on a single quarter-inch microchip made of silicon. When transistors are arranged as a series of logic gates on a microchip, they can store and process input, such as that coming from a keyboard, at amazing speeds. The speed of these logic calculations is measured in nanoseconds.

Microchips may be engineered to perform different functions. The three most common microchips are microprocessors, ROM chips, and RAM chips. A **microprocessor** processes all input and provides control for the computer. In microcomputers, this single chip serves as the **central processing unit (CPU)**. The CPU is the part of the computer which contains the logic circuits that control the computer's functions. Fig. 4-11.

When you type "B" on the keyboard, for example, small voltages are sent to a microprocessor to represent the number 1000010. The microprocessor interprets the number and calls up a picture of the letter "B" from the computer's memory. The letter is then sent to the display monitor where you can see it. That sounds like a lot to do in a short time. Fortunately, a typical microprocessor can handle hundreds of thousands of such operations each second!

Read-only memory (ROM) chips provide permanent storage of information. This information is "burned" into the ROM chip when it is manufactured. The information may be read *from* these chips, but, no additional information may be written *to* them. In other words, they cannot be added to or changed. However, there are ROM chips that may be programmed. These are known as PROM's (the "P" stands for programmable) or EPROM's (the "E" stands for erasable). They are used when a manufacturer believes there may be a need to change the information stored in ROM at some future date.

Random-access memory (RAM) chips store information temporarily. When the computer is turned off, the information stored in the RAM chips is erased. For example, when you load a program into a computer, you transfer the program from a disk to the computer's RAM chips. The program is automatically erased from RAM when the computer is shut down. It must be reloaded to be used again.

Putting It All Together

When you combine the three basic types of microchips (microprocessors, RAM, and ROM) with a power supply, an input device, and an output device, you have a microcomputer. Of course, you need a way to connect these parts.

Fig. 4-11. This photo shows the inside of a Macintosh microcomputer. Note how small the CPU is. The SCSI port connects the computer to equipment such as hard disk drives, scanners, and printers. The SIMMs (Single Inline Memory Modules) contain the random access memory (RAM) chips.

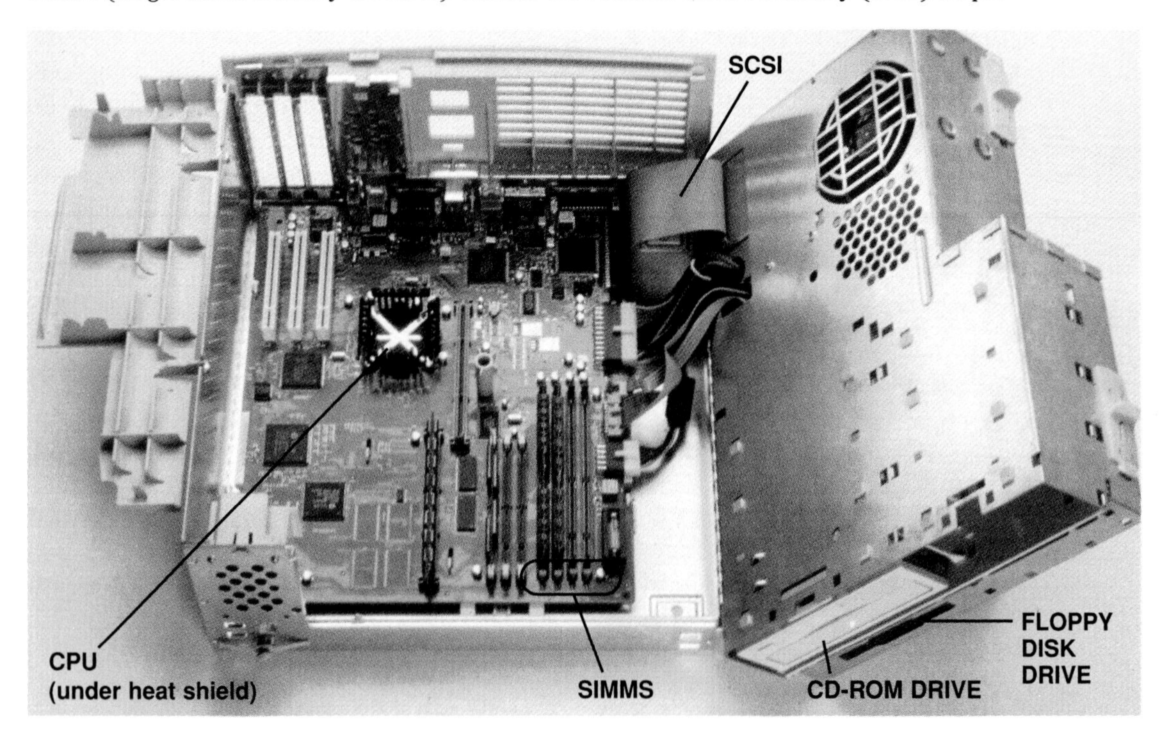

Buses

Bits of data constantly travel from one component to another inside the microcomputer. A system known as a **bus** handles this task. You might think of the bus as a series of wires that connect the components. Fig. 4-12.

Data bits travel along the bus as electrical pulses. In an "8-bit" microcomputer, eight data bits travel along the bus at a time, the way eight people might walk side by side through a wide hallway. The 16-bit and 32-bit microcomputers allow even more data to flow at one time through the bus. The more bits that can travel at one time, the faster the computer processes information.

Ports

Devices such as disk drives and printers are also connected to the bus system. This is done by connecting a cable from the device to a port, or connection, on the computer. It's like plugging an appliance into an electrical outlet. A port acts as a bridge between the computer and the device to which it is connected.

There are two different types of ports: serial and parallel. A serial port is like a one-lane bridge. Data can travel across a serial port only one bit at a time. This makes serial ports relatively slow. Parallel ports are quicker because they allow bits of data to travel side by side as in the bus system. Fig. 4-13. A parallel port is like an eight-lane bridge instead of a one-lane bridge. Since the internal bus system is already carrying data eight bits (or more) at a time, parallel ports are more efficient than serial ports. However, a parallel port is usually more expensive than a serial port.

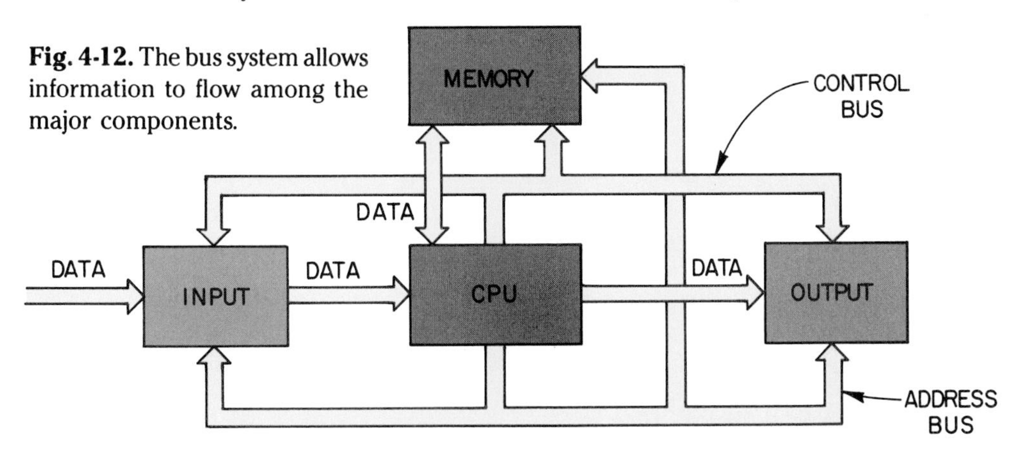

Fig. 4-12. The bus system allows information to flow among the major components.

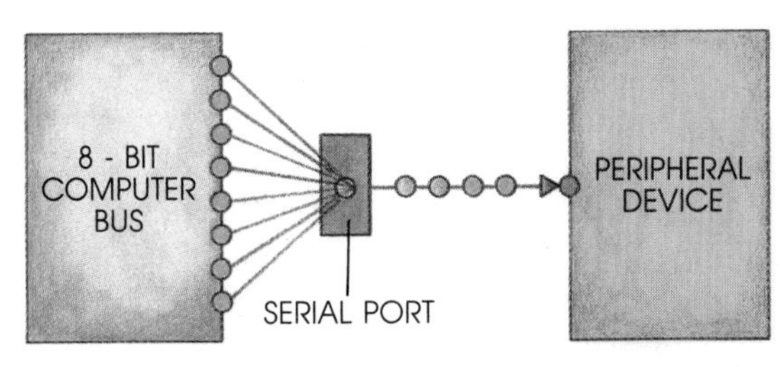

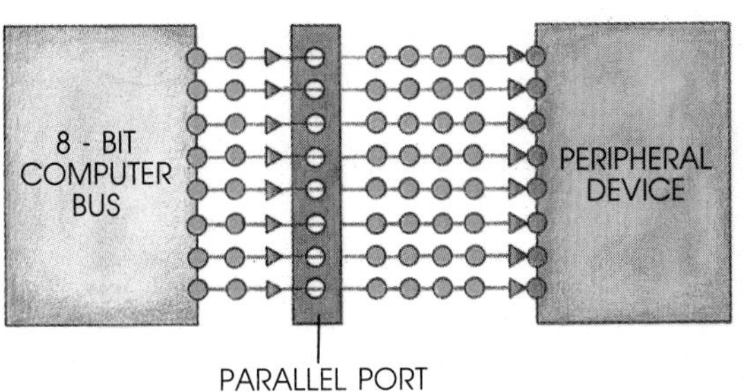

Fig. 4-13. In an 8-bit computer, data travel inside the CPU in 8-bit pathways. A serial port allows only one data bit at a time to flow to external devices. A parallel port allows all 8 data bits to travel to the external device at the same time. Therefore the parallel port is faster.

Peripherals

It takes more than a central processing unit to make a microcomputer. Almost all microcomputers include a keyboard, monitor, and one or more disk drives. Fig. 4-14. The keyboard is simply a way of inputting data, whether it be text or numbers. The monitor displays the input either in one color (monochrome) or in many colors.

Because RAM chips inside the computer store information on only a temporary basis, material you wish to keep is put on a magnetic disk. Magnetic disks may be either flexible (floppy) or hard. They come in various sizes and hold differing amounts of data. A disk drive is a device that reads data from and writes data onto a disk. Disks are discussed in more detail in Chapter 5.

OPERATING SYSTEMS

There are a number of tasks a computer must take care of on a regular basis. For example, it must coordinate all of the data input and output functions. These include displaying keyboard input as text on the monitor, communicating with the disk drives, outputting data to the printer, etc.

To handle all of these "housekeeping" chores, every computer has an operating system. The operating system is a program (set of instructions) that tells the computer what to do with the data it receives. Another word for program is **software**.

If the computer is to store the input in its memory, for example, the operating system tells it where to store it. It also keeps track of the location. If the input is to be moved to a storage disk, the operating system channels the input to the disk drive.

Part of the operating system is stored in the computer's memory chips. Another part of it is usually stored on a storage disk. When you start up, or "boot," a computer, it automatically loads

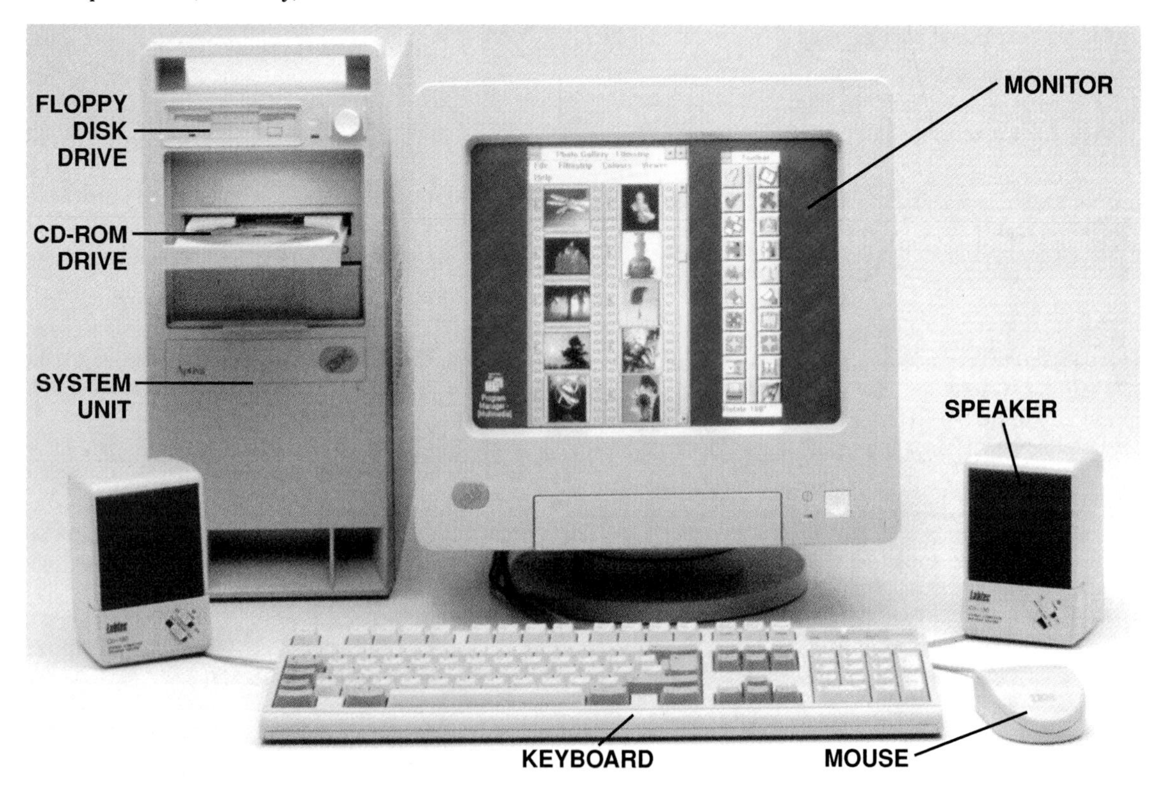

Fig. 4-14. A typical computer includes the components shown here. The system unit houses the microprocessor, memory, and built-in disk drives.

the operating system into its random-access memory (RAM). There, it can easily call upon the system to handle routine tasks.

While there are dozens of different computer manufacturers, there are only a few widely used operating systems Windows, Macintosh, and Unix, for example, are popular operating systems. By using common operating systems, computers made by different companies can often run the same software programs.

Programming Languages and Software

Some people think computers "have a mind of their own." Nothing could be further from the truth. A computer is really just a mass of tiny electronic parts that can process information.

There's a saying that dates back to the very first computers: "Garbage in, garbage out." This means that computers do only what we tell them to do. While they are extremely efficient, they are only as good as the input they are given. Computers tend to be quite reliable. When errors occur, more often than not they are the result of a human mistake.

In order to process information, the computer must be given a detailed set of instructions, or program, which tells it exactly *how* the processing should be done.

It would be nice if we could just type in a description of what we want the computer to do. Unfortunately, computers do not understand the English language. However, they do understand those known as **programming languages**. One of the simplest programming languages is called BASIC.

Following is an example of a BASIC program that would first erase any existing programs. Then it would print the saying "Garbage in, garbage out!" on the monitor:

```
10 NEW
20 PRINT "Garbage in, garbage out!"
30 END
```

The words "NEW," "PRINT," and "END" are all commands in BASIC that the computer understands. Each statement in a program written in BASIC is given a number. That's why 10, 20, and 30 appear before each statement. If you wanted to add a line to this program, you might write:

```
15 PRINT "Computers only do what you tell them to do . . ."
```

The computer would know to insert this statement between lines 10 and 20. Therefore, when the computer runs this simple program, it would print "Computers only do what you tell them to do . . . Garbage in, garbage out!" on the monitor.

BASIC is a general purpose programming language and a fairly simple one at that. Other commonly used general purpose programming languages include PASCAL, C, FORTRAN, and, ASSEMBLY. Some languages are designed for special uses. LISP, for example, is used to write artificially intelligent programs. Artificially intelligent programs have the ability to store user input, process it, and react differently in the future based upon this input.

By combining a large number of program statements like those shown above, computer programmers can build long programs that handle many different tasks. These programs are known as **applications software**, in contrast to the operating system software you read about earlier. Some common types of application software packages include:

- word processing programs that allow you to write more efficiently than you could on a typewriter
- spreadsheet programs that allow you to maintain financial and numerical records
- database programs that allow you to store and retrieve all kinds of text records, such as mailing addresses

In Chapter 6, you'll learn more about these programs.

TECHNOLINKS

HERE COME THE MICROCHIPS!

Microchips (integrated circuits) are small but mighty. The first integrated circuits were invented to replace the tangle of wires ordinary circuits required. Today most integrated circuits consist of a tiny chip of pure silicon that contains small amounts of certain impurities. It's the impurities that make a chip so versatile. Depending on where the impurities occur, a chip can perform many functions. For example, some impurities may act as amplifiers or switches. These amazing microchips have enabled scientists and inventors to come up with devices that could not have been built using conventional circuits.

You know that microchips are used in computers. Many other devices also contain microchips. The following list names some of them. See if you can think of others.

- greeting cards that light up or play a tune
- talking dolls and other toys
- clocks and wristwatches
- Christmas tree ornaments that play a tune
- thermostats that control heating and air conditioning
- smoke alarms
- pocket TVs and radios
- videocassette recorders
- microwave ovens
- electronic "beeper" signaling devices
- medical imaging machines (similar to X-ray machines)
- sensors for insulin pumps for diabetics
- aircraft and spacecraft guidance systems
- electronic typewriters
- arcade games
- "smart" credit cards linked to a computer system
- hotel room "key" cards that unlock doors
- film negative analyzers and processors
- clothing tags that sound an alarm if taken out of the store
- lottery systems
- home security systems
- supermarket checkout systems
- police vehicle monitors
- telescope guidance for astronomy
- sewing machines
- controls for automotive systems (fuel, emissions, cooling, braking, etc.)
- irons
- cameras
- telephones
- motorized wheelchairs
- blood pressure monitors
- scuba-diving gear

CHAPTER 4 REVIEW

Review Questions

1. Name some of the early inventions and discoveries that led to the development of the modern-day computer.
2. What were the general characteristics of first-, second-, and third-generation computers?
3. What are the differences (if any) among the following: transistor, microchip, integrated circuit, microprocessor, RAM chip, and ROM chip.
4. What are the four basic things that microcomputer systems can do with data?
5. Change the following binary numbers to the base ten system: 00000000, 11111111, 10101010, 10011100, and 00000111.
6. What is the basic difference between an 8-bit microcomputer and a 16-bit microcomputer?
7. What is the primary function of a computer bus?
8. What is the difference between a serial and a parallel port?
9. What does an operating system do?
10. How does applications software differ from operating systems software?

Activities

1. Research the abacus and learn how it works. Build one of your own using simple materials. Then, use it to perform some basic calculations.
2. Conduct a survey of the students in your class, grade, or school to find out what computers they have at home and what they are used for. Prepare a report that describes your findings.
3. Under the supervision of your teacher, remove the cover from a microcomputer. (Be sure it is disconnected from the power source first.) Try to identify as many different components inside it as you can.
4. With the help of a user's guide, format a floppy disk for a computer used in your school. Then, copy several files from a disk provided by your teacher to the newly formatted disk.
5. Programming a computer requires very small, logical steps. Practice "programming" a computer to do a simple task, such as brushing teeth. Break the task down into as many individual steps as you can. Make a list of the steps.

CHAPTER 5

Computer Hardware

There are many ways to put information into a computer and nearly as many ways to get it out again. In addition, the computer has to have some way of storing the information that it processes. Before this book was printed, for example, the text was stored in a computer as electronic data.

This chapter discusses how computers shuttle information in and out and how and where they store it.

Terms to Learn

cathode ray tube
cursor
digitizing tablet
joysticks
magnetic media
mouse
optical storage disks
pixels
plotter
potentiometer
resolution
scanner
synthesizer
touchpad
touch screens

As you read and study this chapter, you will find answers to questions such as:

- In what ways is information put into a computer?
- How does a monitor work?
- What are the different types of printers that a computer can use?
- What media do computers use to store information?

INPUT DEVICES

Suppose you were asked to put together a music tape that could be used at a school dance. A number of different music sources could be used to create this tape. You could record music from a record album or compact disk, a song played on the radio, or music already on another tape. If you had some friends who play music, you could set up a microphone and record them as well. All these different methods could provide input for the tape recording.

Just as there are different ways to put music on a recording tape, there are many different ways to feed information into a computer. The tools used to provide the information are called input devices.

Keyboards

A keyboard is simply a group of keys like those on a typewriter. When struck with your fingers the keys send data to the computer. Keyboards are, by far, the most common way of inputting data. In fact, when we think of a microcomputer, we assume that it includes a keyboard of some sort.

Most computer keyboards are based upon the standard typewriter "qwerty" keyboard. If you look at a typewriter keyboard, you will see the letters Q, W, E, R, T, and Y, all in a row near the top left corner. Fig. 5-1. That's where the name "qwerty" comes from.

The inventor of the typewriter, Christopher Sholes, didn't use a qwerty keyboard on his first typewriter in the late 1800s. He arranged the keys alphabetically at first, but that caused a problem. Most of the internal parts of the typewriter were wooden. If the keys were struck too quickly, the wooden mechanism didn't work properly. Therefore Sholes scrambled the letters around the keyboard to slow down the typist. As the typewriter became popular, people got used to the qwerty arrangement. Now nearly all computer keyboards have the letters arranged in the qwerty style. Some, though, allow the user to flip a switch that changes the keyboard to a style that is faster to use.

In addition, many computer keyboards have a separate numeric keypad that looks like the keypad of a simple calculator. This allows the user to input numbers more quickly than with the number keys on the top row of the qwerty keyboard.

Fig. 5-1. The qwerty keyboard is standard on typewriters.

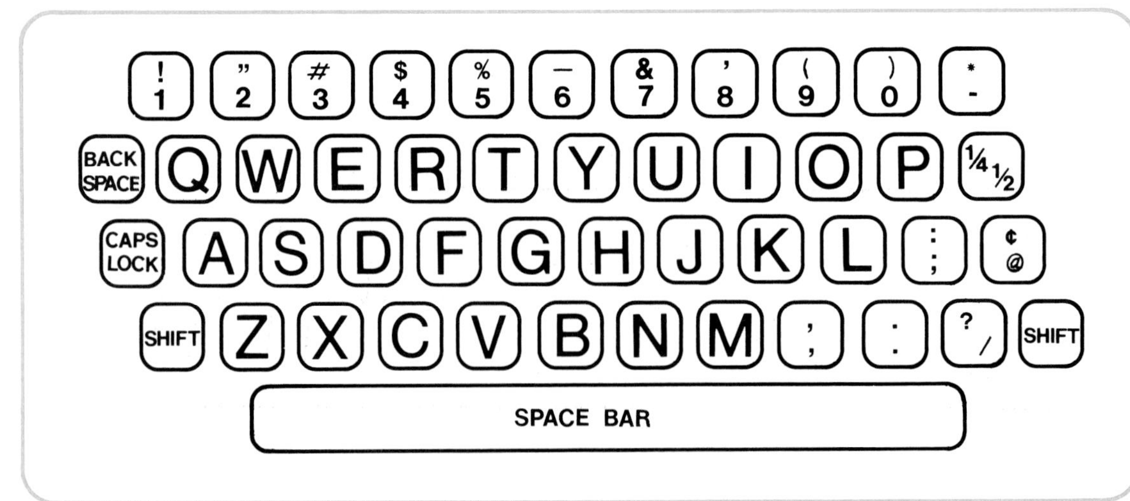

There are often several keys on a computer keyboard that are not found on typewriters. Fig. 5-2. For example, there is generally a "delete" key to remove characters that have been typed. "Arrow" keys allow you to move the cursor around on the computer screen. (The **cursor** is usually a short blank line that blinks and indicates position on the screen.) A "control" key lets you call upon special features of the computer and software. The control key is pressed at the same time as another key. For example, you might press the control key and the letter "C" to cancel an operation.

On a typewriter, what you see is what you get. When you strike a key, the typewriter produces that letter or number. On a computer, that's not always the case. When you strike a key on a computer, you are really turning on an electrical switch. This switch sends one or more bytes of data to the CPU, where it is interpreted by the software you're using. Programmers who write the software, therefore, can program certain keys to perform specific tasks.

For example, programmers can decide what the function keys will do. The function keys are often marked on the keyboard as "F1," "F2," "F3," etc. Each of these keys performs a special function when struck. For example, F6 might type characters in boldface, F8 may underline characters, and so forth. Just what these function keys actually do depends upon the software being used.

Joysticks

In the early 1980s when microcomputers were a novelty, a surprising number of people bought one for use at home. What were they doing with these powerful devices? They were playing games! Games were a great way for beginners to learn about computers. Computer video games were quite the rage, both at home and in video arcades.

For most of these games, a keyboard would have been more complex than necessary. All that was needed was something to steer a car around on the screen or shoot down "enemy" planes. A joystick became the tool of choice.

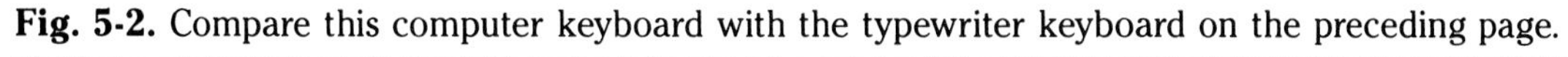

Fig. 5-2. Compare this computer keyboard with the typewriter keyboard on the preceding page.

The first **joysticks** were little more than a box with four electrical contacts inside. Fig. 5-3(a). When the stick was rotated, one of the contacts was touched. This moved an object around the screen in one of four directions: up, down, left, or right. If the stick was moved so as to contact two of the points at once, the object moved on the diagonal. A button on the joystick could be programmed for different functions, such as shooting "bullets" across the screen.

Later, joysticks used potentiometers to move objects in all directions around the screen. Fig. 5-3(b). A **potentiometer** is an electrical device that varies the amount of voltage passing through it. Volume control knobs and dimmer switches on lights are examples of potentiometers. By moving this type of joystick, varying amounts of voltage are sent to the computer. The computer translates these voltages into different locations on the screen.

Fig. 5-3. Joysticks were first used for computer games. Early joysticks (a) had four contact points. Later models (b) had a potentiometer for more accurate control.

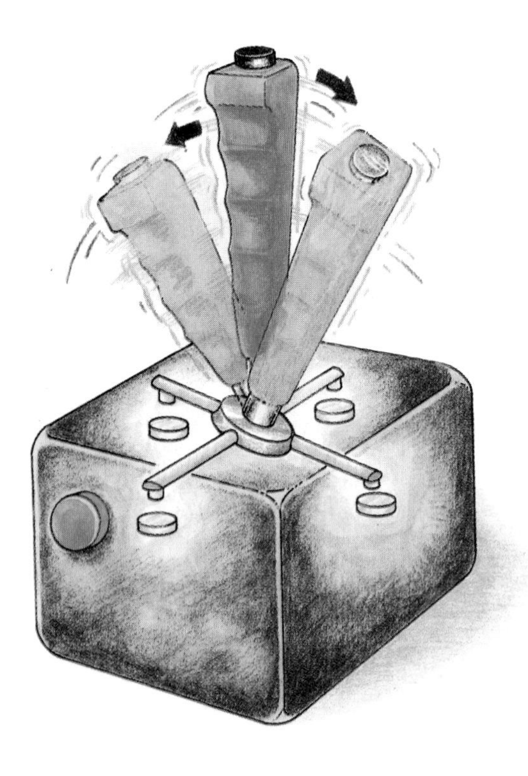

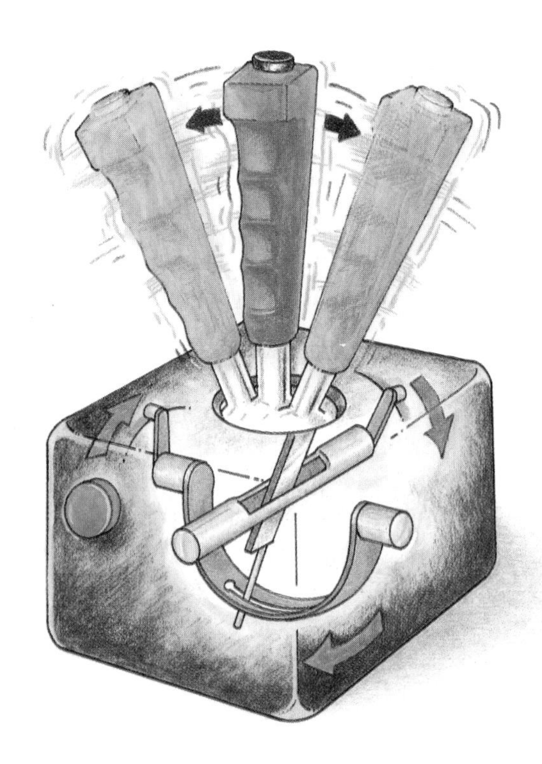

Digitizing Tablets and Touchpads

Graphic designers have found a wide range of uses for the computer. However, this did not happen right away. When microcomputers were first introduced, designers were used to working with pencil and paper, not keyboards and joysticks. This was true whether they were graphic illustrators or technical drafters. Most found the keyboard and joystick of little use to them in their work because it was awkward to draw with these devices.

Computer manufacturers, however, recognized that there was a lot of potential for computers in the design field. They developed the digitizing, or graphic, tablet. Fig. 5-4. A **digitizing tablet** is, in effect, electronic paper and pen. The tablet itself is an electronic drawing board. It may be a small hand-held device or as big as a table. An electronic pen, connected with a wire to the tablet, allows the user to create the image.

Today's computer-aided drafting (CAD) software takes advantage of what the digitizing tablet can do. CAD software is commonly used by drafters who create technical drawings. (See Chapter 9.) Graphic artists have made use of digitizing tablets as well.

The "pen" used to draw on the tablet can be one of three types: stylus, pen, or puck. Each is connected to the tablet with a wire and sends signals to the computer that move the cursor on the screen. The drawing created by the artist appears on the screen, not on the tablet. The puck slides around the tablet when pushed. It has a plastic lens with two crosshairs that can be aligned with lines on a drawing. The puck is often used by drafters when they need to input an existing drawing into the computer. (See Chapter 9.)

Not all digitizing tablets work the same way. The most common is the antenna-transmitter type. The tablet has a fine wire grid which acts as an antenna. As the stylus or pen moves across the grid it sends a signal to the grid/antenna. The location of the stylus is communicated to the computer as X,Y coordinates. The computer

Fig. 5-4. This digitizing tablet is equipped with both a stylus and a puck.

translates the location into cursor movement on the screen. Fig. 5-5.

An acoustic tablet uses a stylus that emits a high-frequency sound wave that cannot be heard by the human ear. Tiny microphones at the edge of the tablet pick up these signals and send them to the computer. From these signals, the computer can quickly calculate the location of the stylus. This is translated into cursor movement on the monitor. Fig. 5-6.

A third type of tablet uses a stylus mounted on a mechanical arm. As the stylus is moved, potentiometers are activated. The signals from the potentiometers are then translated into cursor movement.

A **touchpad** is similar to a digitizing tablet but is cheaper and less accurate. It, too, has X and Y coordinates. Any blunt instrument, even your fingertip, may be used to draw an image. Fig. 5-7. The drawing pad is really a two-layer sandwich of materials that conduct electricity. There is a small gap between these two layers. When you push down on the pad, the two layers touch each other, and a small amount of current flows through. The amount of current varies depending upon where the pressure is applied. This varying current translates to a changing cursor location on the monitor.

Fig. 5-5. The antenna-transmitter digitizing tablet is the most common type.

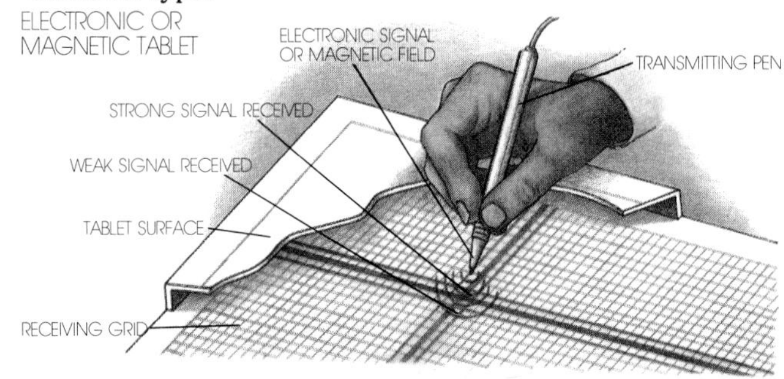

Fig. 5-6. Acoustic tablets work with sound waves.

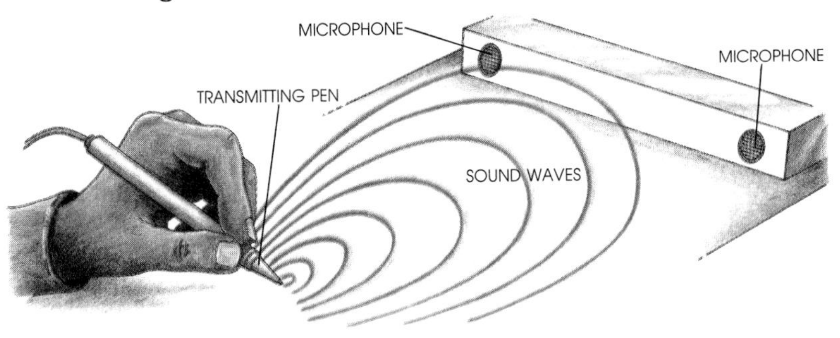

Fig. 5-7. A touchpad lets you draw with your fingertip.

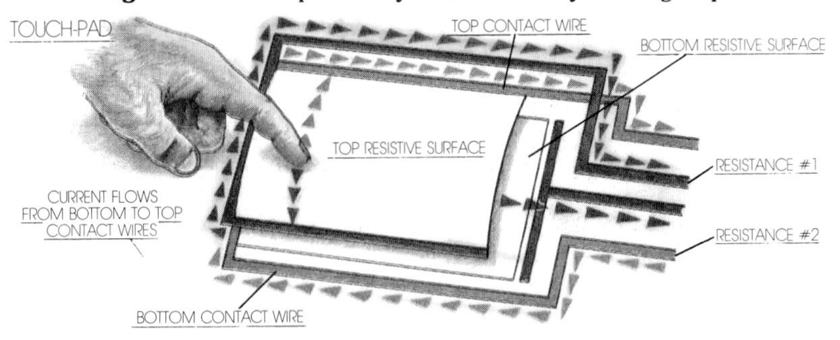

Mice and Trackballs

Soon after joysticks and digitizing tablets made a hit with computer users, mice crept into the picture. A **mouse** is a hand-held device that is moved around on a surface and in turn moves images on the computer monitor. A mouse generally has one or more pushbuttons that can be used for different functions. For example, the mouse first locates an object on the screen. When the button is pushed, the object can be moved as the mouse is moved. Next to the keyboard, the mouse is the most commonly used input device.

Mice are popular because they are easy to use. Apple Computer recognized their usefulness and included one with each Macintosh® computer they sold. The Macintosh® was a huge success, in part, because the mouse helped make it easy to use. All Macintosh® software can take advantage of the mouse's capability.

The mouse is similar to the joystick. There are three different kinds of mice: electro-mechanical, opto-mechanical, and optical. Fig. 5-8.

The electro-mechanical mouse has a round rubber ball in it that rolls as the mouse is pushed along. Inside the mouse are two rollers that turn as the ball rotates. These rollers are each connected to a disk which also rotates. As the disks rotate, contact points on their edges connect with stationary contact bars. Each time the two touch, a signal is sent to the computer. The computer translates the number of contacts made into distance travelled. Objects on the screen are moved accordingly.

The opto-mechanical mouse is similar. The disks have slots instead of contact points. Light from a light emitting diode (LED) on one side of the disk passes through the slots. It is picked up by a sensor on the other side of the disk. The more pulses of light sensed, the greater the distance the mouse has travelled.

An optical mouse is moved around on a special mirror-like pad that has fine grid lines in it. A light emitted from the mouse is reflected from this surface. It is picked up again by sensors

Fig. 5-8. Mice are popular input devices. The diagrams show the three kinds of mice.

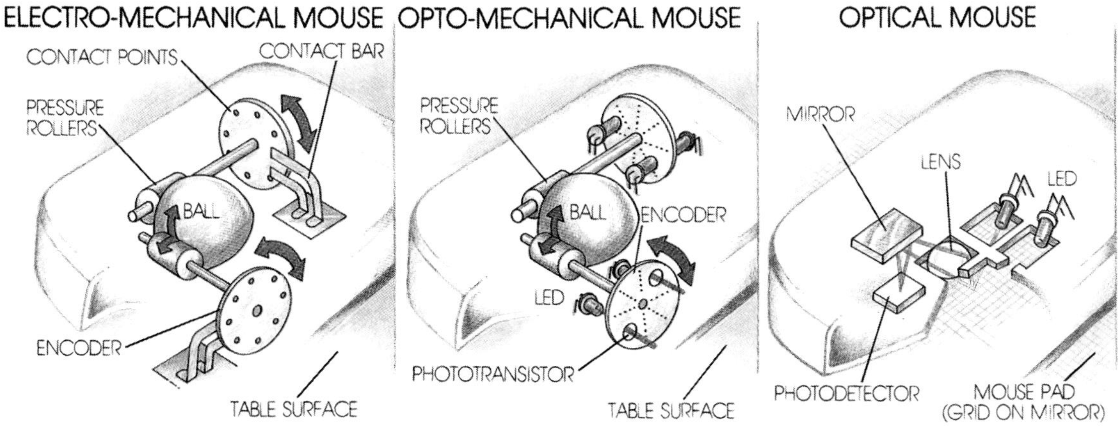

in the mouse, but the grid lines have absorbed some of the light, creating pulses in the reflection. These pulses can be translated by the computer as a distance moved. There are no moving parts in an optical mouse.

If you turned a mouse on its back, you'd have a trackball. A trackball works the same way, except that the ball is spun by hand rather than by moving it around on a surface. Its main advantage is that a trackball requires less space to use. Some keyboards have built-in trackballs. Another space-saving option is the trackpad, a type of touchpad. Fig. 5-9.

Fig. 5-9. Instead of a mouse, many portable computers have either a trackball or a trackpad (touchpad). The trackpad has a touch-sensitive surface. Gliding a finger over the surface moves the cursor on the screen.

Video Digitizers

A video digitizer is a device that converts standard (analog) video signals into digital information. Conventional video displays 30 frames (screens of still information) per second. This rate is fast enough to create the illusion of "full motion." Video digitizers convert all or some of these frames to computer graphics, which are then displayed quickly on the computer screen to recreate the illusion of full motion.

While full-motion, full-screen digital video requires large amounts of computer memory, smaller "windows" of digital video, for example, 1/8 of the total screen size, can be captured from the original video source and displayed on most desktop computers. If the computer doesn't have enough memory, these digital video segments may also be captured/displayed at fewer than 30 (such as 12 or 15) frames per second. In this case, they appear choppier than conventional television.

Apple Computer introduced the QuickTime® digital video format in the mid-1990s. QuickTime video digitizers immediately became popular in the Macintosh environment, since the QuickTime "movies" they captured could be displayed on any Macintosh computer. As a result, multimedia became an immediate sensation in the computing industry. *Multimedia* refers to the combining of several media, such as text, sound, animation, and video.

Touch Screens

When microcomputers first appeared, many people viewed them as complex monsters. While computers are the result of complex engineering, designers have tried to make them as easy to use as possible. The term "user friendly" describes a computer or program that is easy to use.

User friendliness is very important when the computer is used in public places. For example, computer information systems are now common in shopping malls. Shoppers can stroll over to a computer system and get different kinds of information from it.

Since keyboards are impractical for this purpose, touch sensitive screens were developed. **Touch screens** are similar to digitizing tablets. They sense where a finger has touched them and send this information to the computer. Fig. 5-10. Using a touch screen and the proper software, you can draw a picture on a monitor with your finger.

Computers used by the public are often equipped with touch screens. A computer in a mall might display the question: "What are you shopping for today?" Then a list of items, such as clothes, shoes, records, food, and so forth, appears. Each of the items on the list has a touch area next to it. A touch area is usually indicated by a box drawn on the screen. When the user touches one of these areas, the touch screen sends the information to the computer. An answer then appears. If you select "food," the computer might list different types of restaurants in the mall and show a map giving locations.

One type of touch screen has a row of light emitting diodes (LEDs) mounted just outside the top and one side of the screen. Fig. 5-10(a). Sensors on the bottom and opposite side detect the light beams. When you touch the screen, some of these light beams are blocked by your finger. By noting which light beams were blocked, the computer calculates the location of the touch.

The other type of screen has a special coating which conducts electric current. Fig. 5-10(b). When you touch this screen, a tiny amount of current is drained. Electrical sensors at the screen's four corners can detect how much current is drained. From this information, the computer can locate where the touch occurred.

While touch screens are easy to use, they are not without flaws. Fingerprints building up on the screen tend to make it look messy. Your arm can get tired from holding it outstretched for too long. Sometimes it is hard to touch in the center of a small touch area. Touching outside the programmed area sends the wrong message to the computer.

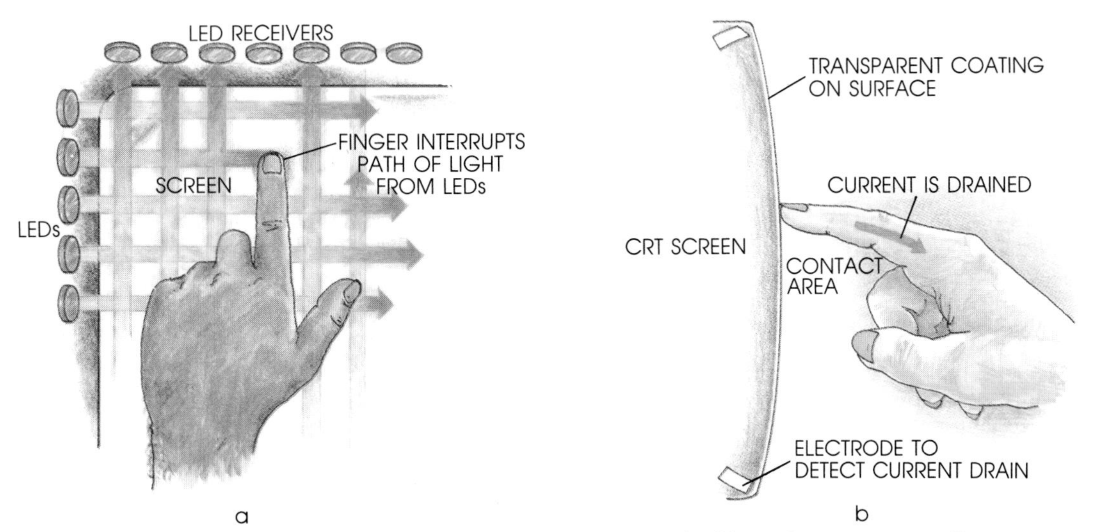

Fig. 5-10. Touch screens may use LEDs (a) or a special coating (b) to detect contact. Because touch screens are so easy to use, they are found on many computers to which the general public has access. For example, computers in stores and museums often have touch screens.

Scanners

You can type information into a computer with a keyboard. You can also create graphic images with tools such as a mouse or a graphic tablet. What if the words or illustrations already exist on paper somewhere? Is there any way of feeding this into the computer?

A **scanner** does just that. Fig. 5-11. It sends a "picture" of the text or artwork to the computer. Once stored as data, the text or artwork may be edited and/or output like any other information.

Scanners that are designed for text (both letters and numbers) are known as optical character recognition (OCR) devices. Generally, they can read only uniform characters, such as those that have been printed in a book or typed on a typewriter. A scanning head moves back and forth over the text and "reads" it. Each time it passes over a line, it sends this information to the computer. With the proper software, some text scanners can also input handwritten text. This is still a slow process, and it has not yet been perfected.

When color artwork is to be scanned, a drum scanner is generally used. The artwork is mounted on a cylinder that spins around at high speed. The artwork may be transparent, like a color slide, or opaque, like a photograph. As the artwork rotates, a scanning head slowly moves across the image area. The scanning head projects a small beam of light that goes through a slide or is reflected off a photograph. A sensor picks up this light beam and it is translated into computer data. The data that represent artwork may then be stored, edited, or output in different ways.

Voice Input

Descriptions of the "office of the future" generally include voice input devices. Imagine how easy it would be to write a letter if all you had to do was talk to your computer. You wouldn't be slowed down by the keyboard. In fact, there might no longer be a need for a keyboard!

Voice input devices already exist. However, changing the spoken word into data a computer can work with is difficult. As you know, everyone's speech is different. That makes it hard for the computer to understand different people. However, the commercial potential is there. A number of companies have spent large sums of money to perfect the technology. No doubt, it will some day be as common to talk to your computer as it is to speak into a telephone.

Fig. 5-11. A scanner sends an image to a computer. The scanner at left is used with a personal computer to input text and graphics. Such scanners are popular for desktop publishing. At right is a drum scanner used by graphic production firms to scan color artwork.

OUTPUT DEVICES

Computers are able to output information in many different ways. Various forms of output include visual displays, printed material, and the spoken word.

Monitors

The most common output device is the computer monitor. It generally displays something every time we provide input, no matter which input device is used.

A computer monitor is similar to a television set. The heart of the monitor is a **cathode ray tube (CRT)**. Fig. 5-12. The CRT is a large glass vacuum tube that is flat on one end. The inside of the flat end is coated with a phosphorescent salt. At the other end of the vacuum tube, an electron gun emits a narrow beam of electrons. Wherever the electron beam strikes the coating, it glows for a few thousandths of a second. This glowing creates the image we see on the monitor.

As the beam leaves the gun, deflector plates direct it both horizontally and vertically. The plates deflect the beam across the entire screen, one line at a time, about 25 times a second. As it scans across the screen, the beam is turned on and off. When the beam is on and strikes grains of phosphor, the grains glow. When it's off, the phosphor remains dark. In this way little "dots" of light are created. The glowing dots that make up the image we see are really flashing on and off. However, this happens so quickly, we can't really see it. This type of a scanning beam display is known as a raster scanned image.

SCIENCE FACTS

Electrons

Until about one hundred years ago, scientists believed that atoms were the smallest particles of matter. Then in 1897 a British physicist, J.J. Thomson, proved the existence of electrons—tiny, negatively charged particles that make up part of an atom. The equipment Thomson used for his experiments included a cathode ray tube with a fluorescent screen at one end. This device was the forerunner of the cathode ray tubes used in television sets and computer monitors.

Fig. 5-12. In a CRT an electron gun shoots a beam of electrons toward the screen.

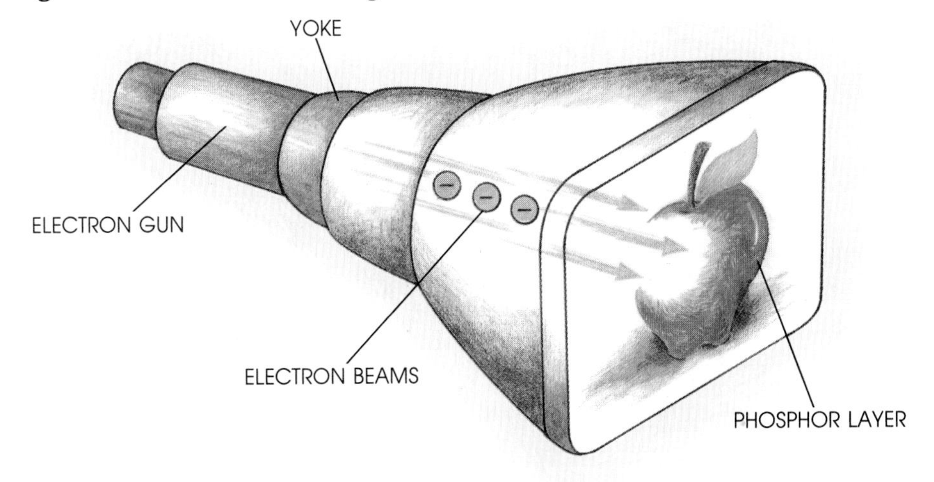

Resolution refers to the degree of sharpness of an image on a monitor. The glowing dots that make up the image are known as **pixels**, which is short for "picture elements." The more pixels a monitor has, the higher its resolution and the sharper its image. Fig. 5-13.

Monochrome monitors project the image in only one color—usually white, green, or amber. Color monitors are more complex. Each pixel is made up of three different colors: red, green, and blue. The electron gun can turn on any of these colors in the pixel, creating different colors on the screen. The intensity of the electron beam can also be varied. This changes the brightness of the pixel. By turning on different color pixels and varying their brightness, color monitors can create thousands of different colors.

Computer manufacturers have also developed flat screen monitors. These monitors use light emitting diodes (LEDs) or liquid crystal displays (LCDs) to represent the pixels on the screen. LEDs are tiny lights and LCDs are tiny cells of liquid. Either can be made to glow. The computer turns on the right ones to create the image. Flat screens are used in portable computers because they are much smaller and lighter than CRT displays.

Printers

Most printers used with microcomputers are single-color printers. Color, however, is becoming more and more important. We have come to expect color in our magazines, newspapers, and in computer output as well.

In the 1980s, microcomputers gave rise to "desktop publishing." (See Chapter 13.) Users could do work they previously had to hire a typesetter and commercial printer to handle. Likewise, in the 1990s microcomputer users will be able to do many things with color that, in the past, were handled only by commercial printers.

Dot Matrix Printers

The most common printers used with computers are dot matrix printers. They are inexpensive and can print both text and graphics. Dot matrix printers vary in quality. Their output ranges from "quick and dirty" to "letter quality." In general, the better the quality of output, the longer it takes to print the job.

The key part of all dot matrix printers is the print head. It is located right next to the paper, separated only by a ribbon like that found in

Fig. 5-13. The more pixels a monitor has, the sharper its image.

a typewriter. The print head is made up of a series of thin blunt-ended metal pins, called wires. Fig. 5-14. As the print head moves back and forth, the computer activates each of the wires individually. When activated, the wires drive the ribbon against the paper. This strike-on process transfers tiny dots of ink or black carbon onto the paper. The dots form the image.

The more wire pins in the print head, the better the image quality. Inexpensive dot matrix printers may have as few as nine wires. Better quality printers may have 24 or more. Color printing is possible on these printers by using multicolored ribbons.

Ink Jet Printers

The print head of an ink jet printer is made up of tiny nozzles that spray ink onto the paper. Fig. 5-15. Ink jet printers are quiet, fast, and versatile. For example, they can print on uneven surfaces, since the nozzle never touches the printed surface. High speed black and white ink jet printers are used to print bulk mail labels. Color ink jet printers, using cartridges of yellow, magenta, cyan, and black ink, offer relatively low-cost color output. (See pages 306-307 for a discussion of how dots of these four colors can create the illusion of many colors.)

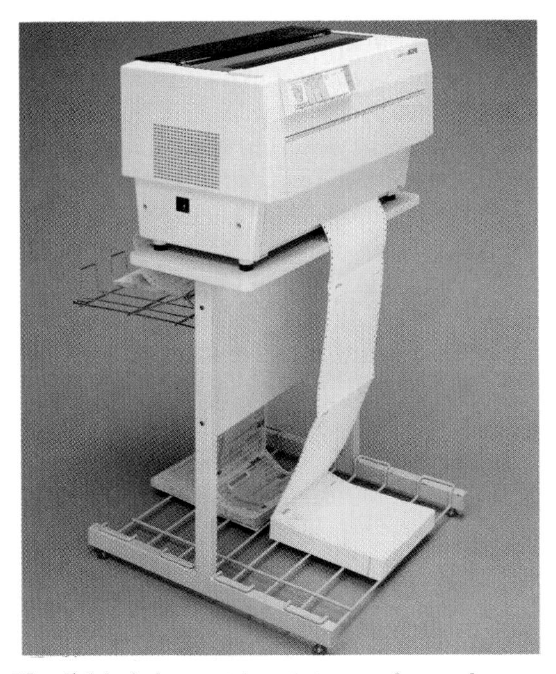

Fig. 5-14. A dot matrix printer produces characters by forming patterns of dots.

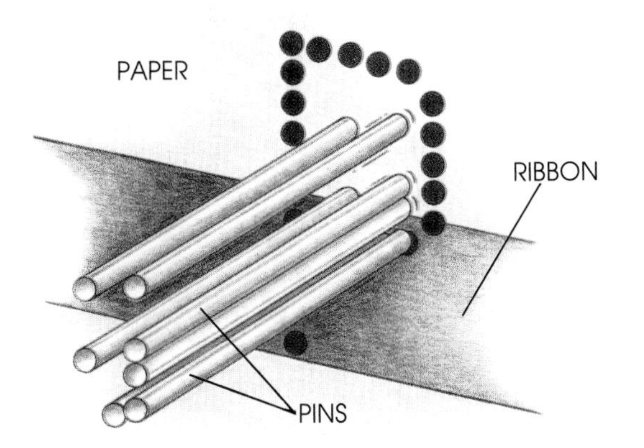

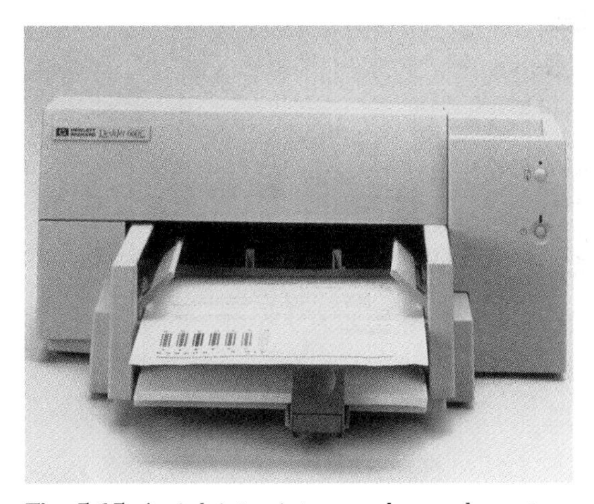

Fig. 5-15. An ink jet printer produces characters by spraying tiny droplets of ink onto paper.

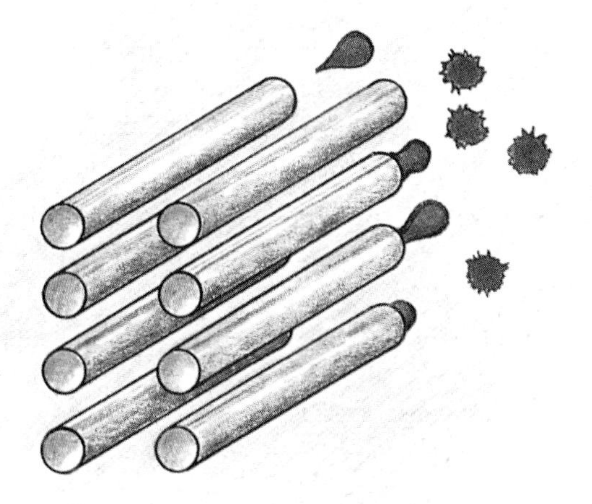

Laser Printers

Laser printers, first introduced in the mid-1980s, use the same technology found in photocopy machines. (See "Electrostatic Message Transfer" in Chapter 15.) A computer-driven laser beam leaves an electrostatically charged image on a drum. Fig. 5-16. The charged image area is transferred to the paper. The image attracts tiny particles of toner. The toner makes the image visible and is fused with heat.

Laser printers produce very high quality images. Since the images they create are more a function of software than hardware, they can print any graphic or typestyle a computer can create.

The resolution of a laser printer is measured in dots per inch (dpi). Resolutions of 300-600 dpi are common for general office work. Quality printing generally requires 2400 dpi output.

Color laser printers use yellow, magenta, and cyan toners (pigments) in addition to black to create full-color images. Color laser printing costs considerably more than black and white output, but some applications justify the added expense.

Wax Thermal Printers

Wax thermal printers are a cost-effective way to produce color output. A highly focused heat source is used to transfer yellow, magenta, and cyan wax from a ribbon to the page. As with most other output devices, the images are built up as dots on the page.

Dye Sublimation Printers

Dye sublimation printers use heat to vaporize solid colors. This colored vapor then diffuses into a special paper. This results in a very high quality "continuous tone" (rather than a dot pattern) appearance on the paper. Since you cannot really see the tiny dots on the page (as you can with ink jet, laser, or wax thermal printers), dye sublimation output looks like a color photograph.

Plotters

A **plotter** is different from a printer in that it actually draws, or "plots", the images on paper with a pen. Plotters are the standard output device for computer-aided design (CAD). Fig. 5-17. This is because they can produce continuous lines very efficiently.

Plotters are of two types: flatbed and drum. In a flatbed plotter, the paper is held in position while a pen is moved along horizontal (X) and vertical (Y) axes. By lifting or lowering the pen as it moves, the computer can draw images. Flatbed plotters often sit right on a desktop, and are used for fairly small drawings.

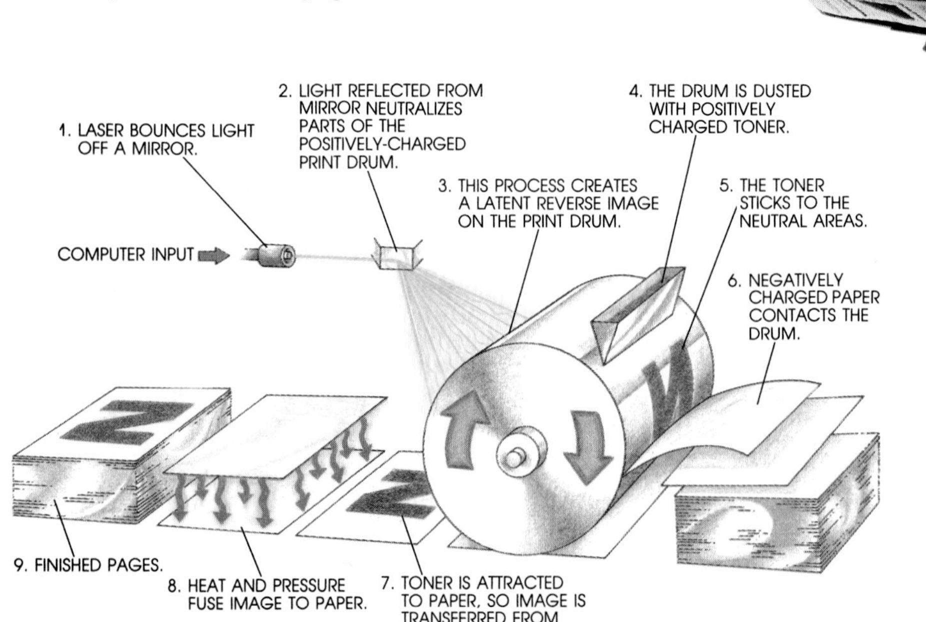

Fig. 5-16. Laser printers can print several pages per minute, and they print both text and graphics.

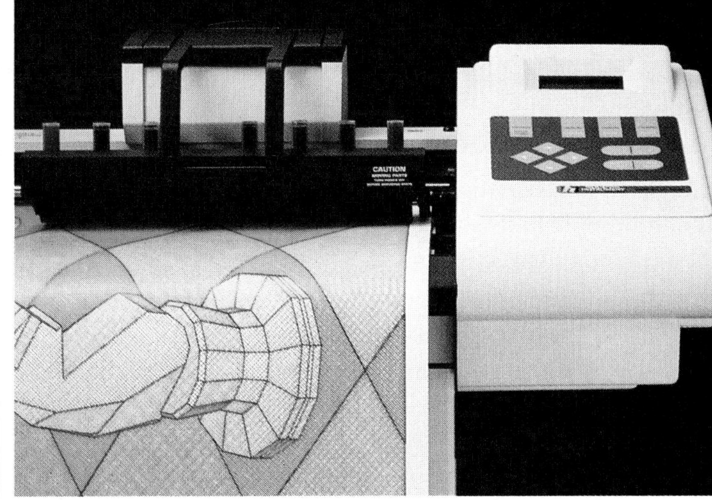

Fig. 5-17. Flatbed plotters (left) and drum plotters (right) are the standard output devices for computer-aided drafting.

Drum plotters move the pen along the X axis while the paper is moved along the Y axis (or vice versa) to create the drawing. They are more expensive but more convenient for producing large drawings. They generally stand on the floor. Because the paper hangs down in front and back, these plotters take up less space than would a flatbed plotter of equal drawing capacity.

Plotters often have a variety of pen colors to allow them to output drawings in multiple colors. Some plotters can hold only one pen at a time; others can hold several. The multipen plotters are more convenient because there's no need to manually change pens each time a different color is to be plotted.

Film Recorders and Imagesetters

Most of the output devices mentioned so far create images on paper. Film recorders are designed to output to light-sensitive films, like those used in photography.

One common type outputs to color slide film. It is used to generate color slides for presentations. Using data created with a computer program, the film recorder "draws" on the film with a laser unit. This produces a high quality color image. The slide is then processed like ordinary film. Film recorders can output to different size films.

An imagesetter works on the same principle as a film recorder. Data from the computer drive a laser, which exposes an image onto a photographic material. This material is then processed in photographic chemicals, resulting in a very high quality black and white image. Precise text, like that found in this book, can be created this way at resolutions of 2400 dpi or greater. Likewise, graphics of any kind and halftones (dot representations of photographs) may be created at very high resolutions. Imagesetters can output to photographic papers, films, or printing plates for use in the graphic production process.

Synthesizers

A **synthesizer** is a device that creates sound from computer data. There are music synthesizers and voice synthesizers.

A music synthesizer is much like an electric organ attached to a computer. An electric musical keyboard is connected to the computer with a device known as a "midi" interface. Signals sent from this keyboard are processed by special software and computer circuitry. Then they are output through a speaker. The result is electronic music.

We have already discussed the potential for voice input in offices. Synthesized voice output also has possibilities. For example, if you call telephone information and ask for a number, a synthetic voice output device gives it to you. Common telephone answering machines use a cassette audio tape to store the message played over the phone. More and more answering systems, however, use a computer to store and generate audio messages. There is no need for an operator or audio tape deck at the answering end. As voice output devices become more advanced, their use will increase.

STORAGE DEVICES

When you turn off a computer, all of the data temporarily stored in RAM are lost. In order to save these data, they must be stored somehow. The most common storage media are magnetic disks and tape. Optical disks are another means of storing large amounts of data.

A device must be connected to the computer that allows data to be *written to* and *read from* the storage medium. This device is usually a drive of some kind.

Magnetic Storage Media

Magnetic media all rely upon tiny particles of metal called oxides. These oxides are coated onto a plastic base. A magnetic head passes over these oxides in such a way as to magnetize them in a precise pattern. The pattern represents the digital (binary coded) data created by the computer. This is known as writing data to the magnetic medium.

Similarly, the magnetic head may read data from the magnetic medium and send it back into the computer. The magnetized oxides remain as they were unless the magnetic head writes new data over the old. In fact, an ordinary magnet, if brought in contact with the magnetic medium, will destroy any stored data!

SCIENCE FACTS

Magnetism

Magnetism is caused by the movement of electrons. In an atom, the electrons orbit the nucleus. They also spin (rotate). Both these movements contribute to magnetism. In most atoms, the magnetic force is very weak. However, some atoms, such as those in iron, have strong magnetic properties.

There are three kinds of magnetic media: floppy disks, hard disks, and tape. The following sections describe each of these.

Floppy Disks

Floppy disks are perhaps the simplest and least costly magnetic storage medium. One common type consists of a 5¼″ thin plastic disk coated with oxides. The disk is protected by an outer sleeve. Because the disk is so thin, it is flexible. This is how it got the name "floppy" disk.

A disk drive spins the disk while the magnetic read/write head moves back and forth slightly above the disk. Fig. 5-18. In this way, the read/write head can quickly locate any spot on the disk. A typical 5¼″ floppy disk can hold up to 360K (kilobytes) of data. A kilobyte contains about 1000 bytes.

A smaller 3½″ version has oxides coated on a thicker piece of plastic. This disk is therefore a little more rigid. Despite its smaller size, the 3½″ disk is capable of holding up to 1.4 megabytes (1.4 million bytes) of data, or nearly four times as much as the 360K 5¼″ disk. Fig. 5-19.

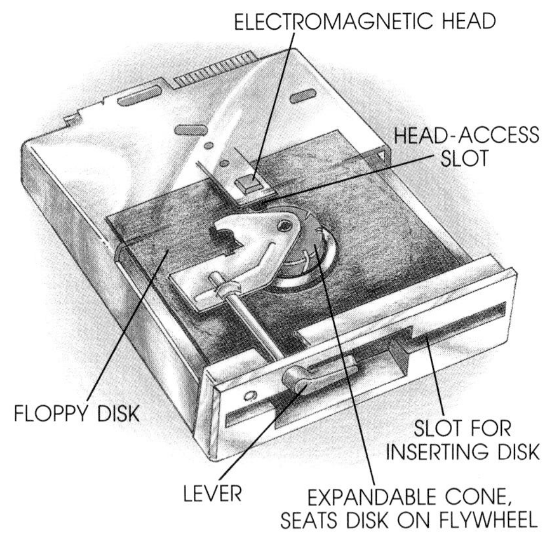

Fig. 5-18. The electromagnetic read/write head moves back and forth above the floppy disk.

Fig. 5-19. The 3½″ disks hold more data and are more durable than 5¼″ disks.

Hard Disks

Hard disk drives work on the same principles as those for floppy disks. However, the disks are rigid "platters," and they usually are permanently located inside the drive. Fig. 5-20. The hard drive spins one or more of these disks constantly (usually at 3600 rpm). This makes the hard disk much faster than the floppy disk, which spins only when read from or written to. Hard drive users can easily store tens or even hundreds of megabytes of data.

Some hard drives use removable platters, known as cartridges. The information is written to the cartridge, and the user may then remove it. In practice, this works much like a floppy drive. A typical cartridge, however, can store 20 megabytes of data.

Magnetic Tape

Large quantities of data may be stored on magnetic tape, which is wound on a reel. Tape reels are mounted on the drive or placed in storage when not in use. To access a particular part of the tape, the user "fast forwards" to the proper spot. This takes far longer than finding data on a disk. When tape is used, it is generally to store a back-up copy of data.

Optical Disks

Three different types of **optical storage disks** are commonly used to store computer data: CD-ROM's, WORM's, and erasable optical disks. They are called *optical* disks because a laser (an optical device) is used to write and read the data on the disks. In addition, there exist videodisks designed to store video data and CD-I disks that store both computer and video data.

Fig. 5-20. A hard disk drive is similar to a floppy disk drive. Hard disks can store much more data than floppies, and many of today's more powerful programs require hard disks.

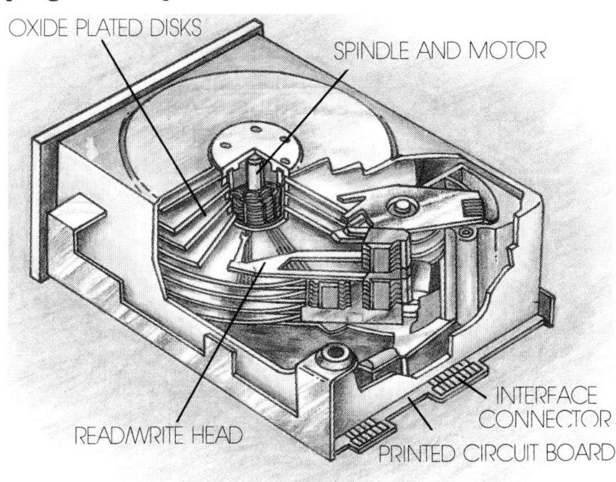

CD-ROM Disks

CD-ROM stands for "compact disk, read-only memory." These disks are made in much the same way a compact disk of music is made. In one method, a laser beam cuts millions of tiny holes known as micropits in a master disk. The micropits represent the data. From this master disk, plastic copies are pressed and then coated with a mirror-like aluminum coating. Finally, the disk is covered with a protective layer of clear plastic.

A CD-ROM disk drive uses a laser beam to scan the disk. Fig. 5-21. The reflection of the laser

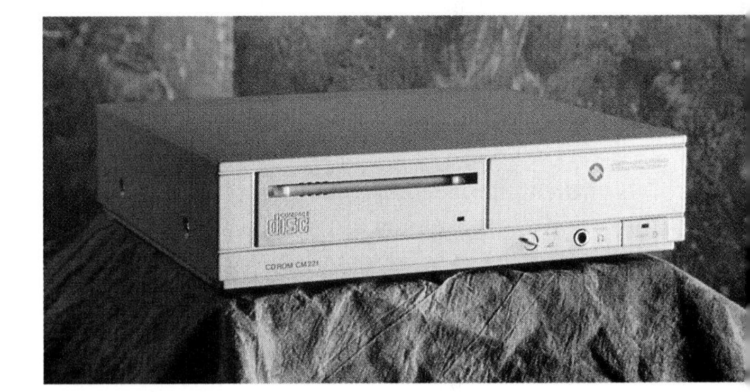

Fig. 5-21. With a CD-ROM disk and drive, it's possible to store vast amounts of data.

Courtesy of Mobay Corporation, a Bayer USA Inc. Company

from the micropitted surface sends the data to the computer. CD-ROM disks can store huge quantities of text. For example, the entire *Encyclopedia Britannica* has been stored on one 5″ CD-ROM disk. However, data cannot be erased from a CD-ROM disk, nor can the computer user change or add to the data.

WORM Disks

WORM stands for "write once, read many." The user of a WORM disk can store information on the disk, but only once. Once written to the disk, the data cannot be changed or erased.

WORM disks have a thin metal film on their surface. A laser in the WORM drive cuts tiny holes in this metal film to represent data. This surface can then be read by a reflective laser in much the same way a CD-ROM disk is read.

Erasable Optical Disks

Erasable optical disks use a different method of recording the data. This method is what makes them changeable. They rely upon a combination of magnetism and optics. A magnetic write head is used to create two different reflective states on the surface. One state represents an "on" bit and the other an "off" bit. Since these states are changed magnetically, the disk can be erased and rewritten over and over. Unfortunately, systems that use erasable optical disks are very costly. They are not yet commonly used.

Videodisks and CD-I Disks

Videodisks are optical disks that closely resemble CD-ROM disks. However, the laser that etches the micropits into the master is driven by a videotape. The data stored on the disk are in video format. Fig. 5-22.

One video format, called CLV, allows up to two hours of video on each side of the disk. This format is used primarily to record movies. The other format, CAV, can hold up to one-half hour of motion video on each side. Since video motion is shown at 30 frames per second, CAV videodisks can hold up to 54,000 frames of video on each side. Using a computer or a handheld remote control, any of these 54,000 frames can be shown at random.

Once the master has been created, videodisk copies may be produced rather cheaply. Videodisks are available in both 8″ and 12″ diameters.

The optical disk used for CD-I (compact disk-interactive) combines CD-ROM and videodisk technology. CD-I disks can store both computer data and video images. For example, the CD-ROM version of the *Grolier Encyclopedia* has no graphics. A CD-I version could have not only full color graphics but video motion as well!

Fig. 5-22. Videodisks can store video, audio, and computer data. This capability allows them to be used with computers in a variety of ways.

User's Guide to 

BUYING A COMPUTER

You know the old saying, "Good things come in small packages." When it comes to computers, that's often the case. However, there are a lot of things to consider when buying a computer.

First, decide what you want the computer to do. Some computers are designed to handle one function, such as graphics, better than another. You will want a machine that has the capabilities you need.

Next, find out what software is available for doing what you want to do. Your local computer dealer and computing magazines are both good resources. You may discover that some software runs only on a particular computer.

Cost is usually a consideration for most people. Simply paying more money will not guarantee better quality, however. For instance, a "name brand" computer will cost more than a "compatible." Compatibles run most of the software made for a name brand computer but generally cost less. Used computers are another option.

Compare manufacturers' warranties as well as the reputations of the companies who distribute the computers. Some computers come with only 90-day warranties. Others have warranties that last a year or more. Some companies will promptly replace defective computers; others will not.

How long will it take for delivery of the computer? Is it in stock, or will it take a month or longer to get it?

Find out if local service is available for the computer you are considering. As with any complex device, things can go wrong with computers. Local service is important.

After you have decided on the brand of computer you want, you must decide what components to purchase. Choices include:

- *Type of monitor.* Monochrome monitors are fine for word processing and "number crunching" software. When it comes to graphics, color is better.
- *Type of disk drives.* One floppy drive is the bare minimum. Two floppy drives will save you a lot of time in the long run. A hard disk drive is needed for more complex programs. It is also convenient if you are using several different pieces of software. Optical drives (CD-ROM drives) offer huge amounts of storage but are slower in retrieving data.
- *Amount of RAM.* Some programs require a megabyte or more of RAM. Be sure you have enough RAM to run the software you will be using. Also, try to estimate the size of files you'll create. Large files require more RAM.
- *Type of printer.* Dot matrix printers are good general-purpose printers. Check the speed, quality of output, warranty, and replacement ribbon costs. If you need very high quality output, a laser printer is a good (though expensive) choice. When purchasing a laser printer, see if it supports a page description language, such as PostScript™.
- *Modem.* Will you need to send or receive data via phone lines? If so, you will want to purchase a modem. Here again, speed is a primary concern. Modem speeds are measured in bits per second (baud rate). The higher the baud rate, the faster the modem.
- *Portability.* Some computers are portable, others are not. If you will need the computer in more than one location, consider a portable.
- *Speed.* The faster the computer, the less time you'll have to wait for it to process your work.

HEALTH & SAFETY

DON'T LET HARDWARE GIVE YOU HARD WEAR

Here are tips for avoiding muscle aches, eyestrain, and fatigue when using a computer.

- Allow plenty of room for the computer equipment plus reference manuals. You'll be more comfortable, and the equipment will be less likely to overheat or get bumped.
- When you're using the keyboard, your lower arms should form a 70- to 90-degree angle with your upper body. Your wrists should not be bent.
- Use a chair with adjustable back support. Your feet should touch the floor, and your legs and thighs should be parallel with the floor.
- The computer monitor should be 22 to 26 inches away from you. The angle between your eyes and the screen should be 0 to 45 degrees. At 0 degrees, the screen would be at eye level. At 45 degrees, you would be looking down toward the screen. Some monitors have bases that tilt and swivel so that you can adjust them for the best viewing angle.
- The right lighting can help prevent eyestrain. Small lights near the work area are better than bright overall lighting. The lights should not be directly behind or directly in front of you. Consider using an anti-glare filter over the monitor screen. If you wear glasses, ask your eye doctor about tinted lenses developed for computer users.
- Long sessions at the computer can cause muscle aches in your neck, shoulders, back, and wrists. Take breaks; get up, stretch, and move about. The National Institute of Occupational Safety and Health recommends a 15-minute break after every 1 to 2 hours at the computer.

CHAPTER 5 REVIEW

Review Questions

1. Describe a qwerty keyboard.
2. How does a joystick work?
3. How are touch screens used as input devices in shopping malls?
4. What device enables a computer to read printed or handwritten text or graphics?
5. How are colors produced on a color monitor?
6. Name five different types of computer printers. Give an advantage and/or disadvantage of each type.
7. What is a synthesizer?
8. Describe three different types of magnetic storage media used with computers.
9. What is the difference between floppy and hard disks?
10. Describe four different types of optical disks.

Activities

1. Following the instructions in the owner's manual, perform routine maintenance, such as cleaning, on a mechanical mouse input device.
2. Collect output samples from different types of printers. Compare the prices and features of these printers. Make a written recommendation to your instructor or school principal as to the best kind of printer for your class.
3. Ask if your school's music teacher has a music synthesizer. If so, arrange to try it out.
4. A keyboard character requires 8 bits (1 byte) of information. The average word is six characters long. A typewritten page contains about 300 words. Figure about how many pages of typewritten text can be stored on an 800K floppy disk. About how many pages of text could you store on a 100 megabyte hard disk?
5. Many companies are concerned about keeping computer data confidential. Find out how access to computer disks and tapes can be controlled and write a report about this.

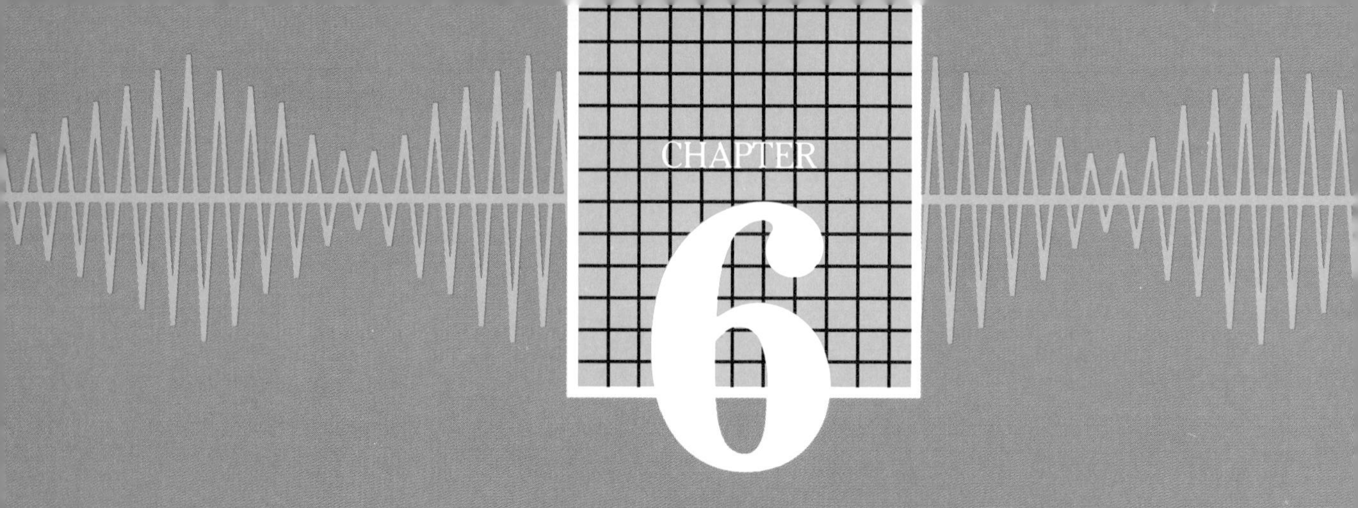

CHAPTER 6

Computer Applications

It used to be said that about one American in seven worked a job having to do with making automobiles. Now, about one American in seven uses a computer in his or her work. That number grows each day. Computers are the workhorses of the Information Age.

From tiny microprocessors to large "supercomputers," computers are used to manage vast quantities of information. In this chapter, you will take a look at some computer applications.

Terms to Learn

analog signals
computer-aided design (CAD)
computer-aided manufacturing (CAM)
computer-integrated manufacturing (CIM)
database
data transmission channel
facsimile transmission
interactive video
interfaced
local area networks (LANs)
modem
networking
universal product code (UPC)

As you read and study this chapter, you will find answers to questions such as:

- How do computers handle analog signals?
- How are people making use of computers in offices?
- What is a computer network, and what can it be used for?
- How are computers being used in manufacturing?
- What do arcade games and interactive video have in common?

COMMUNICATING WITH COMPUTERS

By themselves, computers are extremely useful. When used to communicate with other devices or with one another, their usefulness increases. Often, the data generated by other devices are in a different form than a computer can handle. Then the signals must first be converted to digital data.

Changing Analog Signals to Digital Data

Earlier, you learned that digital computers use the binary number system. Every bit of data is put into number form—either a 0 or a 1. The numbers are represented by very low voltages of electricity. The presence of this low voltage indicates a 1 and its absence indicates a 0.

Some information, like keyboard characters, can easily be assigned a number with which the computer can work. What about other kinds of information? Take sound waves, for instance. In the natural world, energy does not occur in an on/off pattern. Sound waves are continuous, even though their intensity may vary. For example, when you hum a tune you don't usually jump from one note to the next in a start and stop rhythm. You keep going. You glide from note to note. If computers are to read information such as sound waves, sound must be changed to data they can handle.

Devices known as sensors can capture these continuous data as **analog signals**. An analog signal may be thought of as a constantly changing voltage of electricity. Digital computers can work with analog signals, but first they must convert these signals to digital (on/off) data. This is done by an analog-to-digital (A to D) converter. Fig. 6-1. The converter takes readings at intervals and assigns each reading a number. The song you hummed, for example, would be converted to individual notes. Each of these notes could then be represented by a digital number. Many computer applications rely upon this kind of conversion.

Fig. 6-1. An analog-to-digital converter takes readings of the analog signal and assigns each reading a number. The stronger the voltage, the higher the number.

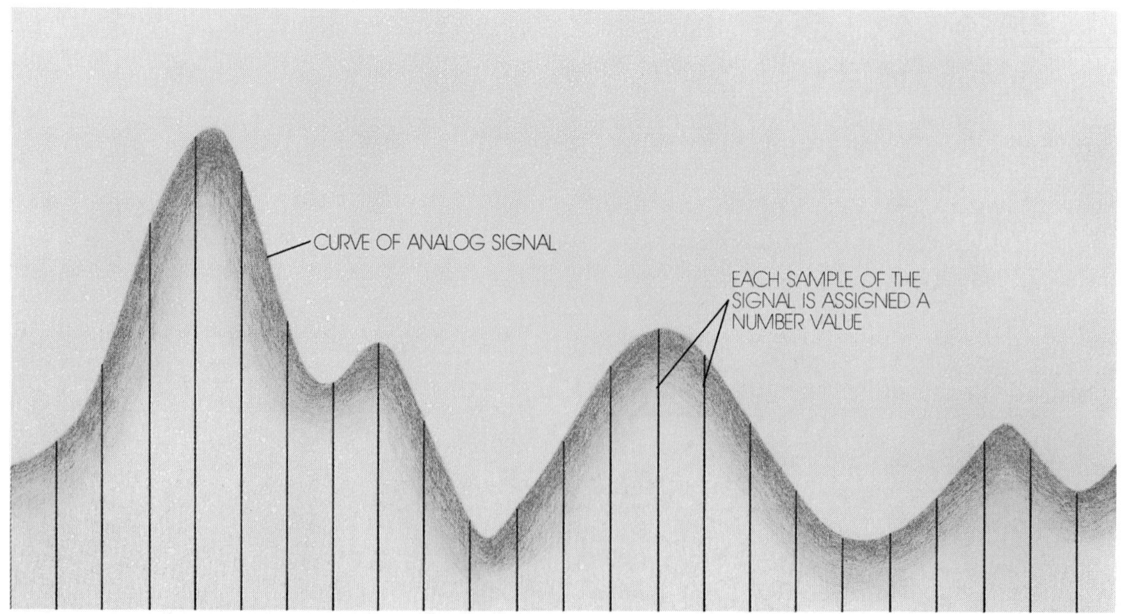

Data Transmission Channels

One common use for computers is to transfer information from one computer to another. However, in order for two or more computers to communicate with each other, there has to be some means of carrying the message between them. We refer to this pathway as a **data transmission channel**.

There are two types of data transmission channels: physical and atmospheric. Physical channels connect computers to one another with a cable. The cable is typically copper wire or optical fiber. (Optical fibers are made of glass.) The data are carried through the cable by a carrier signal of some kind. When optical fiber is used, for example, the carrier signal is a beam of laser light. With the atmospheric channel, the data signal is usually a microwave. The signal uses no cable but travels through the atmosphere. These microwaves may be sent over short distances or relayed off satellites to almost anywhere in the world. Transmission channels are discussed in greater detail in Chapter 17.

Computer Networks

Linking computers to one another is known as **networking**. Computer networks create all sorts of new ways for companies or individuals to communicate.

In most cases, in order to link one computer to others, a **modem** is required. Fig. 6-2. A modem changes the digital data output from the first computer into a series of analog tones. The tones may then be transmitted over telephone lines. The process is called modulation. At the receiving end, the analog tones must be changed back to digital data that the second computer understands. This demodulation, as it is known, is once again handled by a modem. "Modem" is just a combination of the terms *mod*ulation and *dem*odulation. It is one example of an analog to digital converter.

Modems are actually computer boards. They contain microprocessor, RAM, and ROM chips. In addition, they have a speaker with volume control and a jack that connects them to a telephone. Internal modems are installed in expansion slots inside the computer. An expansion

SCIENCE FACTS

Microwaves

Microwaves are waves of electromagnetic energy from 1 millimeter to 30 centimeters long. (See Chapter 16 for a discussion of electromagnetic waves.) Microwaves are longer than visible light waves; our eyes cannot see them. They are widely used in communications and in scientific research, but perhaps the most familiar application of microwaves is found in the kitchen. A microwave oven generates these waves to heat food. The same kind of energy that heats pizza and pops corn can be used to send data from one computer to another or relay a newscast from Asia to America.

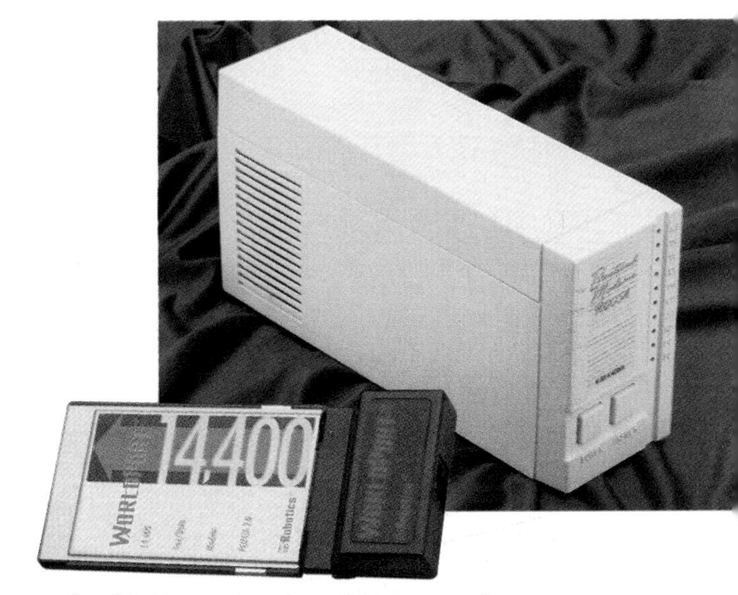

Fig. 6-2. External modems (shown at top) are connected to a computer by a cable. Modem cards (bottom) fit inside a computer.

slot is a place inside a computer in which specialized computer boards may be mounted. External modems may be plugged into a port on the computer case.

Modems are classed as to the speed at which they transmit data. Modems sold in the early 1980s transmitted at a 300 baud rate. "Baud" means bits per second. Modern modems are capable of 9600 or more baud, which is 32 times faster!

Computers may be connected through either a bus network or a switching network. Fig. 6-3. In a bus network, data are sent along a communication line to which a number of computers are connected, or **interfaced**. Each interface sifts the constant stream of data to see what is intended for its computer. When the interface recognizes signals that are meant for it, the data are taken in.

In a switching, or star, network each computer is linked to a central switching unit. All messages are routed through this central point. This network works the same way as a telephone exchange. The first three digits in your telephone number represent the exchange number. The last four digits allow up to 10,000 different phones to be connected to that particular exchange. In the same way, one computer in a star network may send information to other computers in the network by using identification numbers or names. Switching networks may be set up locally, or they may rely upon the telephone system to switch computer information just as it does telephone calls.

Local Area Networks

Computer networks that send information over short distances are called **local area networks (LANs)**. A LAN joins microcomputers to each other and to devices such as a hard disk drive or a printer. The computers share the drive or printer, and costs are kept down. Often, one microcomputer is used only to store and deliver files to the others in the network. This computer is referred to as a "file server."

Businesses and universities sometimes connect microcomputers to a central mainframe

Fig. 6-3. In a bus network, each interface "listens" for and forwards signals intended for its computer. The switching, or star, network routes all data through a central unit that channels the data to the intended receiving computer.

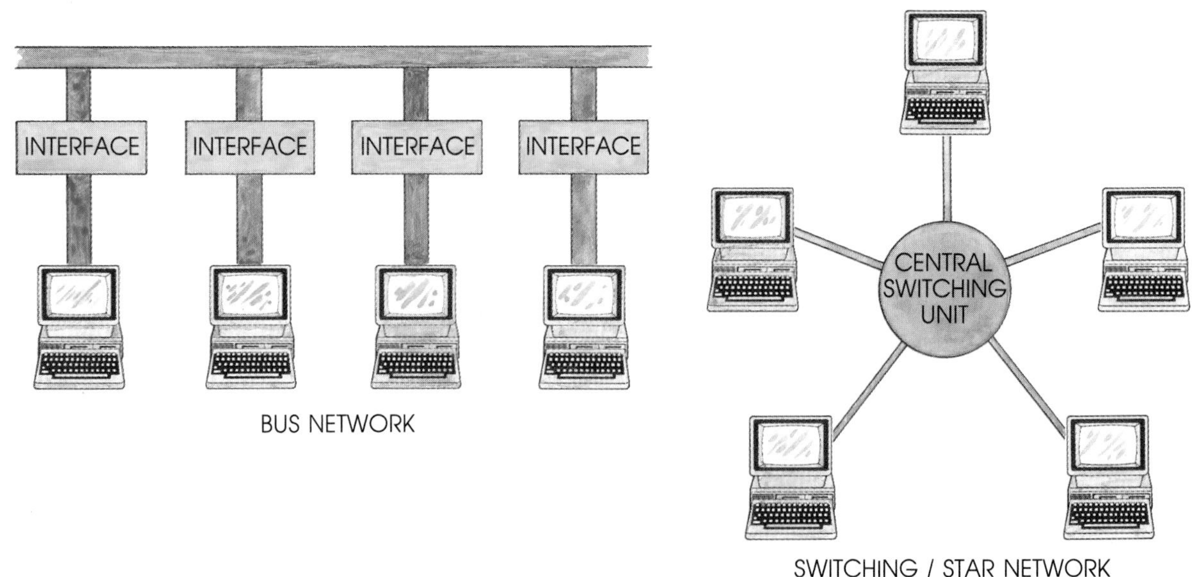

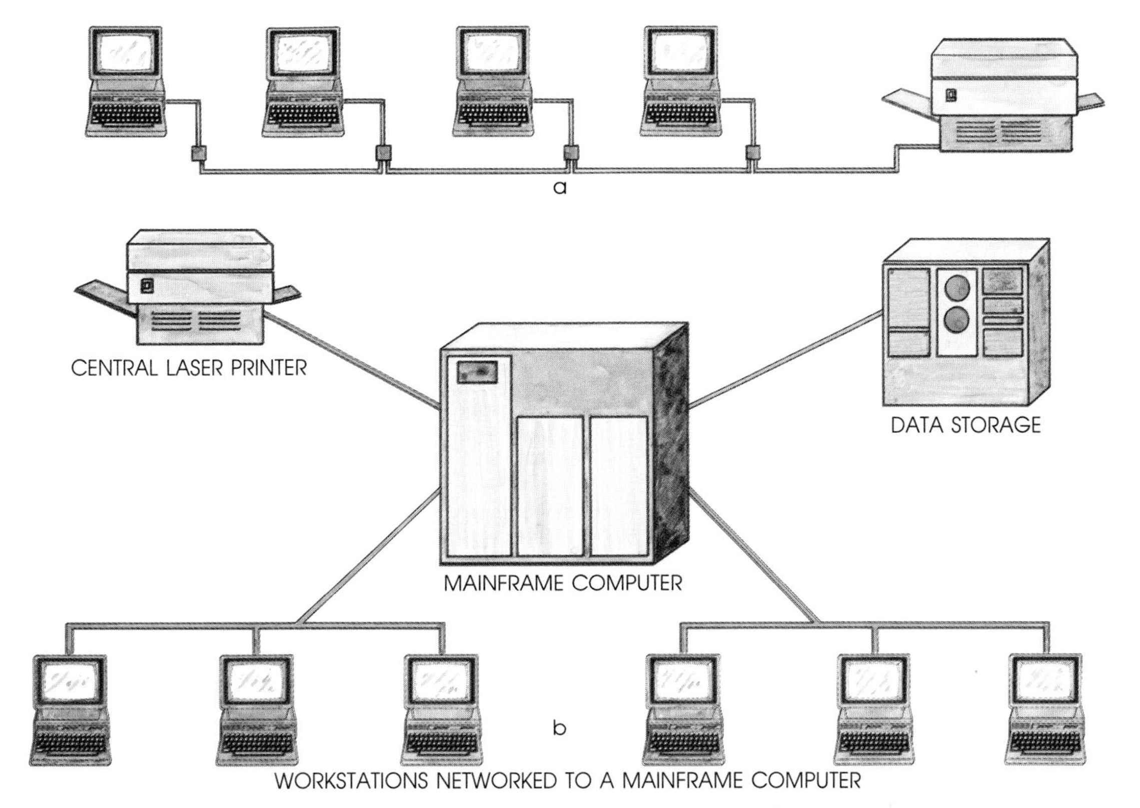

Fig. 6-4. The simple LAN in part a of this drawing connects four workstations to each other and to a small laser printer. The network in part b is typical of a large university campus. Individual workstations are connected to a mainframe computer that has a large memory and a high-speed laser printer that serves the entire university.

computer. Fig. 6-4. This network allows them to use the advanced software developed for the mainframe as well as its costly output devices. LANs such as these are changing companies and universities across the country.

Long Distance Networks

Long distance networks work much the way LANs do. Computers are used at each end of the network. Modems allow data to be sent over telephone lines. Satellites are used to relay data over great distances.

Sending data over long distances has had an impact on many industries. In the newspaper industry, long distance networks mean more news in less time. The publishers of USA TODAY, for example, practice a new kind of journalism. USA TODAY is advertised as "America's Newspaper." The same newspaper appears throughout the country. People in Los Angeles read the same news as do those in Washington, DC. All news for the paper is telecommunicated to the home office in northern Virginia. There, the entire newspaper is assembled electronically. That is, the stories, artwork, headlines, and ads are all arranged using computers and monitors. No paper images are "pasted-up" as is the case with other newspapers.

These data are then relayed to a satellite and on to different printing plants scattered throughout the United States. The paper is printed and immediately distributed throughout each region. Fig. 6-5.

There are a number of advantages to creating a newspaper this way. By assembling the paper electronically in one location, a lot of time and expense are saved.

Normally, color printing is also time consuming and expensive. However, the USA TODAY system makes it fairly easy to include lots of color. This is because the color images are produced centrally in one location. They then appear in

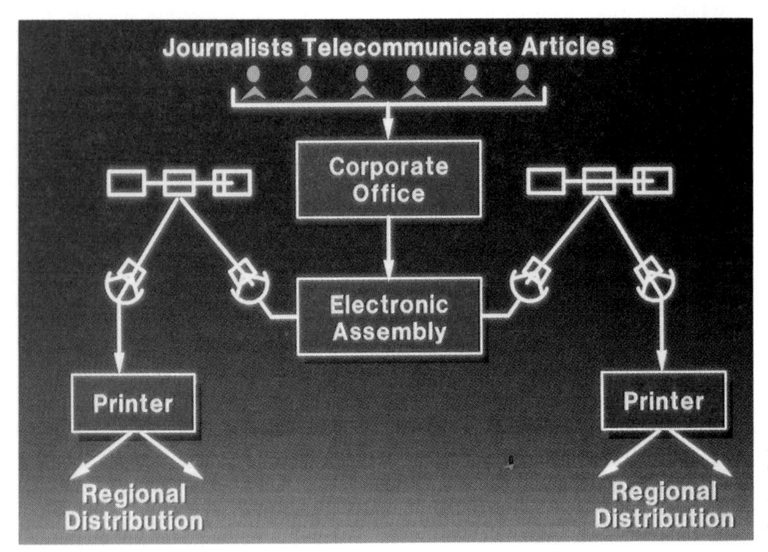

Fig. 6-5. This schematic shows how publications such as USA TODAY are produced.

all the USA TODAY papers around the country. This is much less expensive than if each newspaper publisher makes color images locally. USA TODAY therefore has much more color than any other newspaper.

Internet and the World Wide Web

The Internet is a network of computers spanning the entire globe. Each computer connected to this network has a unique IP (Internet Protocol) address. Moreover, each uses software that allows the computers to send and receive data using the same "language." The language is called TCP/IP. Thus, it is relatively easy for any one of these computers to communicate with any other that is connected to the Internet.

With the proper software, it is also fairly easy to turn any one of these Internet-connected computers into a server that stores files for distribution over the network. Typically, people on the network log into the server with "client" software and then download the files they want. In this way, digital text, graphics, audio, and video files can be distributed across the Internet.

The World Wide Web (WWW) is a network of servers like those described above. Each of these servers is set up with WWW server software and is registered on the WWW. Anyone with access to this network and the appropriate client software (known as a WWW browser) may look at (browse) any of the millions of files stored on these WWW servers. The browser software also allows people to download these files to their own desktop computers with the click of a button.

The WWW is a hypertext network. This means that a link can be established in any file that connects it to any other file stored on the WWW server. These links can be automatic, or they may be activated with a click of a mouse on a word or "button" appearing on the computer screen. In this way, WWW users navigate the globe looking for and acquiring information (text, graphics, audio, and video) without leaving their seats. Best of all, "keyword searching" allows users to find the information they want quickly and easily.

The Internet was developed with government support rather than with private money. So, until the mid-1990s, all of the information on the Internet was essentially free. At the same time, commercial on-line services were provided by companies such as CompuServe® and America On-line®. In 1995, the commercial services began to offer their subscribers access to the Internet and the WWW, and the trend toward commercial use of these networks began growing very rapidly.

USES FOR COMPUTERS

The ways in which computers can be used continue to grow in number. The following pages discuss some common ones. The uses are broken down into major areas, but the areas overlap. For example, businesses use word processing, but so do colleges and other schools. As you read about these applications, keep in mind that the computer is really being used as a *communication* device in each case.

Business

Computers have really changed the way many tasks are now handled in business offices. Secretarial and accounting tasks have been affected most. Fig. 6-6.

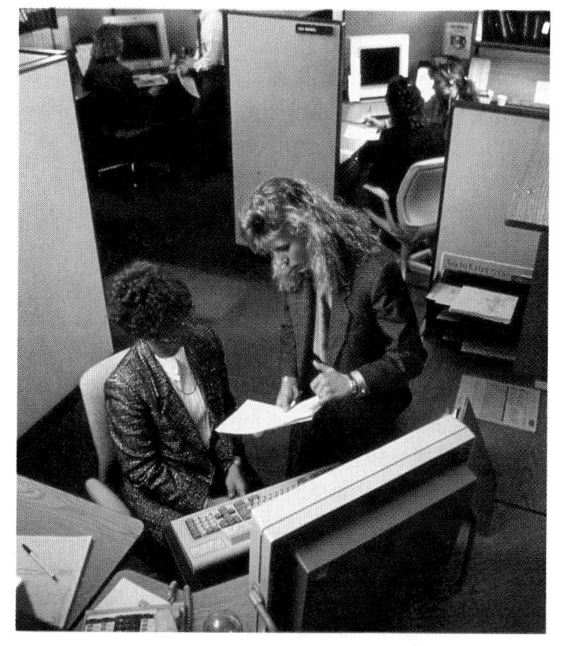

Fig. 6-6. Today many companies consider computers to be essential office equipment.

MAINTAINING YOUR COMPUTER SYSTEM

Perhaps yours is among the millions of homes that have a personal computer. If so, here are some tips on caring for your system.

- Place all components on a sturdy, flat surface.
- Make sure the CPU has at least 4 inches of space around the sides and back for heat to escape. Monitors need good ventilation, too.
- Do not place the computer in direct sunlight or next to heating devices.
- Ideally, the computer should be in a room that is relatively dust-free and can be air conditioned during hot weather.
- Use a surge protector. These can be bought as part of an electrical outlet strip. There also are whole-house protectors that an electrician can install. The surest way to protect the computer is to unplug it when not in use.
- You may clean the monitor screen with household glass cleaner, but don't spray it on the monitor. Put it on the cloth first; then wipe the glass clean. The monitor case and the outside of other parts, such as the CPU, can be cleaned with a damp (not wet) cloth. Don't get moisture into any of the openings. *Turn off and unplug the computer before cleaning it.*
- Do not let any liquid get inside the computer components. It can cause serious damage.
- Store floppy disks and CDs in a clean dry place. Avoid high humidity and extreme heat or cold.
- Keep floppy disks away from magnets. Monitors and TV sets, telephones, loudspeakers, and even some light fixtures have magnets in them.
- Read the owner's manual that came with the computer. The maintenance advice it provides could save you much time and money.

Word Processing

With a typewriter, keystrokes are recorded immediately on paper and cannot be easily changed. With a word processor, keystrokes are stored in computer memory and on a storage disk. At any time, the text may be edited (changed) and a clean copy printed. Word processing programs can make computers as easy to work with as a typewriter. However, they are far more efficient. As a result, word processing is rapidly replacing the typewriter in the modern office.

Advanced software and fairly high quality output from laser printers have also made desktop publishing possible. Organizations can produce newsletters, brochures, and other simple publications with their microcomputers. These systems are also being used to create "desktop" presentations. These are high quality slides and overhead transparencies required in many presentations. For example, slides of sales figures might be shown at a board meeting. With computers, people can prepare colorful charts that present information in interesting and effective ways.

Accounting

The microcomputer that uses word processing software may also be used to manage accounting chores. This is done with spreadsheets. A spreadsheet is software that keeps track of budgets and other financial matters. After setting up the spreadsheet, the user simply enters amounts of money spent and received, and the computer calculates the money left in the account. This is only one example. Spreadsheets are quite flexible and can handle a range of recordkeeping tasks.

Database Management

Another common use for computers is database management. A database is a long list of short records. Perhaps the most common example is a list of addresses. Database software can easily maintain files that contain thousands of different mailing addresses. Telephone numbers and other specific information can also be kept for each entry. The computer automatically sorts and prints out whatever information the user needs. For example, a set of address labels for all teenage customers living in Arizona could be printed from a database that contained addresses, telephone numbers, and ages of *all* the company's customers.

Facsimile Transmission

Earlier you learned that data can be sent from computer to computer using modems at each end. **Facsimile transmission**, or fax, is yet another way of sending data, such as text or photographs, from place to place. Fax machines can operate without a computer. However, by combining fax devices with computers, users can transmit the data with more ease and flexibility. Fig. 6-7.

The idea behind facsimile transmission is not all that new. The newspaper wire services have been sending photographs "over the wire" for decades. A page of text or black and white illustrations is first digitized, or turned into number

Fig. 6-7. Fax machines may be stand-alone devices, such as this one, or they may be a card installed in a microcomputer.

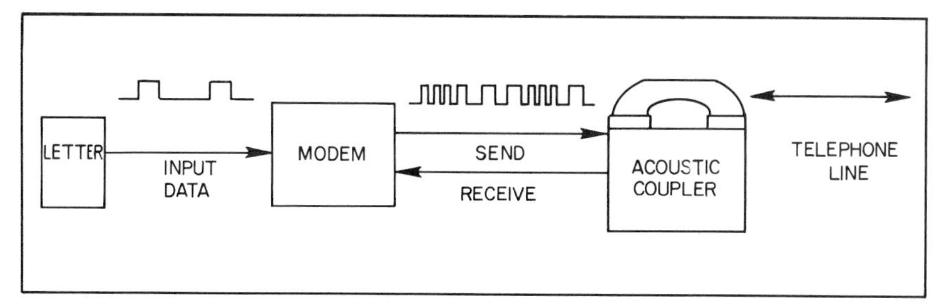

Fig. 6-8. Fax can be used to send words or pictures over telephone lines. Note that a modem is used to change digital data to analog data for transmission over the telephone lines. At the destination, the data are changed back to digital form by another modem.

code. This is done by a scanner built into the fax device. The data are then sent via telephone to one or more receiving fax machines. At the receiving end, the fax machine outputs the document by means of a thermal or laser printer. Fig. 6-8.

There is a standard for facsimile transmission developed by a committee established by the United Nations. This is the ICCTT: International Cooperative Committee for Telephone and Telegraph. The standard made fax designs compatible with one another, which increased their use. "Group 1" facsimile devices were developed in the mid-1960s. They could send one document page every six minutes. Now the most common "Group 3" fax machines can transmit one page every 30 seconds or less. They are capable of 200 dots per inch resolution. A "Group 4" type can manage 400 dots per inch resolution.

Fax machines may be connected to a computer as an external device. They may also be simply a card in an expansion slot inside the computer. While fax machines do not have to be linked to computers, there are a number of advantages if they are. ASCII files and some graphic files stored in computers can be converted and "faxed" to different locations. This removes the need for the scanner input part of the fax machine. The computer can also automatically retrieve and send documents to a large number of fax machines. In addition, the computer may be used to store the document on a disk or preview it without printing it.

Since fax is electronic, the document arrives at its destination almost immediately. This is important to many businesses that rely upon rapid communication. Fax is faster and, if done when phone rates are low, less costly than overnight mail service.

Industry

Computer-Aided Design (CAD)

More and more technical design is computer-aided. A typical **computer-aided design (CAD)** system includes a microcomputer, keyboard, monitor, hard and floppy disk drive, mouse or graphic tablet with light pen, a plotter, and the right software. Fig. 6-9. The user selects various functions from menus, generally by using a mouse or digitizing tablet.

Fig. 6-9. Computers have created a revolution in technical design and drafting.

For example, to draw a line, the user first chooses the type of line (object line, dimension line, etc.). The line is then drawn in the desired location by using the pointing device (mouse, stylus, etc.) to indicate the beginning and end of the line. The drawing appears on the monitor as the operator works. All work is stored in computer memory and on disk. The operator can output the drawing to the plotter at any time.

CAD systems have many advantages over drawing by hand. CAD designs may be changed at any time without destroying the original drawing. Dimensioning is automatic, saving the designer a lot of time. CAD plotters output drawings faster, and often with more precision and quality, than can be done by hand.

Computer-Aided Manufacturing (CAM)

Once a design has been approved, the product is ready for production. Before production can begin, however, a manufacturer must "tool up." This means special equipment, molds, jigs, and fixtures to handle raw materials must be made. Actual production involves the use of various machines, such as lathes and mills, to cut and shape the product.

Computers are changing the workplace by handling many of these steps. The use of computers to control manufacturing processes is called **computer-aided manufacturing (CAM)**.

Designs created with CAD can contain *all* of the information needed to produce a final product. Using this information, additional computer programs can be written to control the equipment that makes the product. For example, a door knocker can be drawn on a computer and then directly machined from steel. Fig. 6-10.

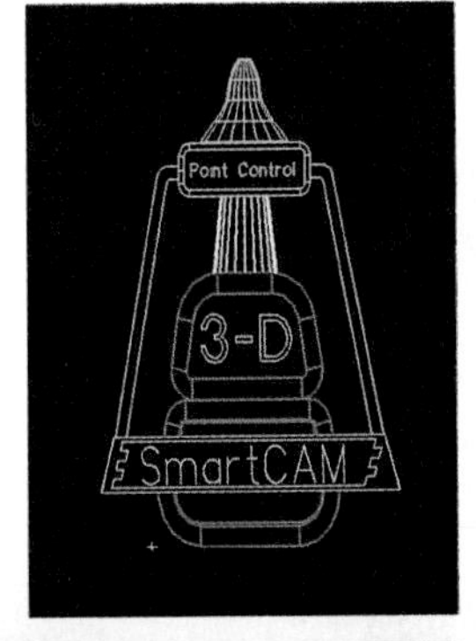

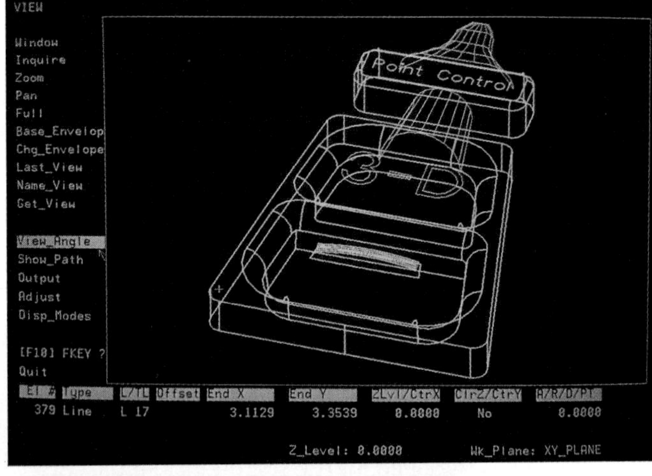

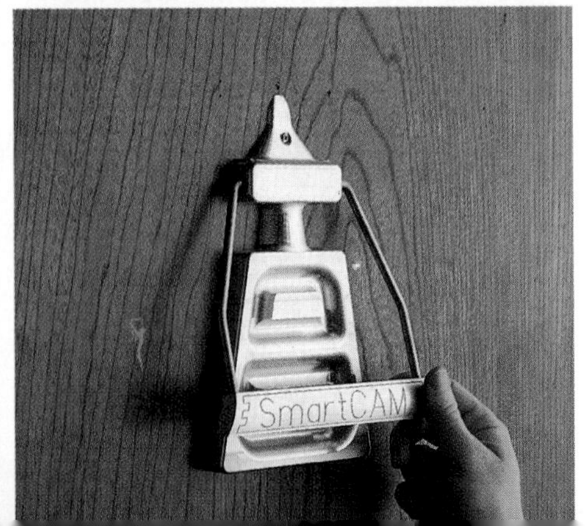

Fig. 6-10. Computer-aided manufacturing links many operations, such as design, drafting, and production, together.

Computer-Integrated Manufacturing (CIM)

The *complete* manufacturing process can be overseen by computers. Fig. 6-11. This is known as **computer-integrated manufacturing (CIM)**. In addition to CAM operations, computers are used to handle materials ordering, inventory, management operations, production control, shipping, and so forth. CAM and CIM systems are replacing certain workers in the production process and changing the nature of the modern factory.

What you should remember about CAM and CIM is that they are really information handling systems. They interpret CAD designs and *communicate* with other equipment, such as robots. All of this is the result of advanced programming, which human workers must create.

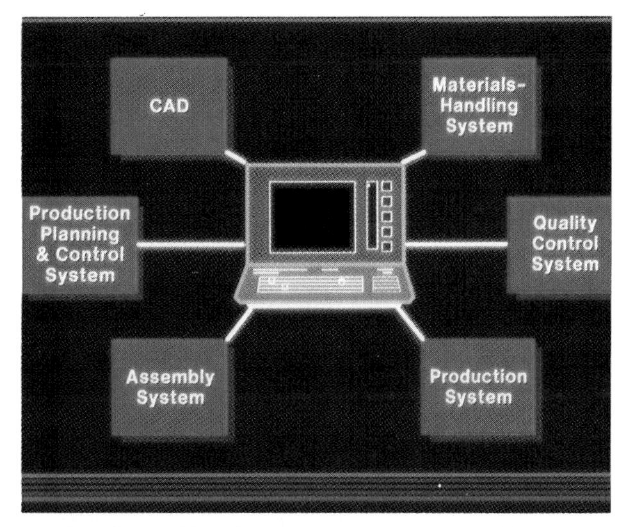

Fig. 6-11. With CIM, all operations in the factory are guided by computers.

Sensors and Feedback

A computer sensing device picks up input such as heat, light, or pressure and communicates it to a computer or microprocessor. The computer processes this information and provides feedback.

Consider, for example, an industrial robot. The most common industrial robot is the "pick-and-place" type. These are designed to pick up a certain part or product and move it to a nearby spot. You do the same thing with your hands all the time, but your brain directs your actions. When you pick up and move an egg, you do it more gently than you move a bicycle. Nerves in your fingertips communicate with your brain, telling it how firm your grip is. Your brain sends a signal back to your hand telling it to be gentle with the egg.

Sensors in the grippers of a pick-and-place robot determine the amount of pressure the gripper applies. The computer controlling the robot can adjust the pressure so that it is right for the part being handled. The sensors act very much like the nerves in our nervous system.

A simpler sensing device may be used to regulate temperatures or lighting throughout buildings. The important thing to remember is that there are always communication and feedback between the sensor and a computer.

Inventory Control

Computers are used in the retail sales industry to keep track of merchandise. Old-style cash registers used to be used just to ring up the sale and store the cash. Now, most cash registers are also computers that keep track of the store's inventory.

You've probably noticed the black stripes or bars printed on most packages in the grocery store. These bars are known as the **universal product code (UPC)**. Fig. 6-12. The bars represent numbers. By varying the thickness of the

Fig. 6-12. Bar codes appear on many items sold in retail stores.

TECHNOLINKS

BAR CODES: BLACK & WHITE & READ ALL OVER

You're familiar with bar codes. Those black and white stripes appear on nearly every product you buy. You also know that the bar codes are scanned at the checkout counter, causing the item's price to appear on the cash register screen. Did you know, though, that bar codes can do a lot more than speed your way through the checkout line?

- The bar codes on products we buy enable stores to keep track of how well those products are selling. They also can help a store measure the success of its promotions. For example, did Superfood's Sunday TV commercial for its peanut butter result in higher sales the following week? If people were buying more peanut butter, were they also buying more jelly?
- Manufacturers use bar code scanners to enter items into their inventory. Suppose a company has just received a large shipment of steel rods. To enter these in the inventory, the bar code is scanned, using a hand-held scanner. The amount is also entered in the scanner, and then the information is transferred to the company's computer. This method is much faster and more accurate than hand-written notes or typed records.
- The overnight package-delivery service Federal Express uses bar codes to track a package's progress from the sender to the receiver.

- One scientist is even gluing bar codes to the backs of bees to identify them. Laser scanners placed at the hive entrance read the bar codes. By keeping track of the comings and goings of individual bees, the scientist hopes to learn more about honey production. Knowing how far bees go to gather pollen and how many trips they make could eventually help beekeepers predict honey yields.

These are just a few of the ways bar codes are being used today. In the future, we will see even more applications. Can you think of some ways to use bar codes?

bars and the blank spaces between them, the numbers one through nine can be represented. You can't easily read the number by looking at the bars, but a laser scanning device, or bar code reader, can. Fig. 6-13.

When the sales clerk moves the bar code across the laser scanner built into the countertop, several things happen. First, the computer checks its files to see how much that product costs. (The cost of each item has already been programmed into the computer.) The computer then displays this cost on the cash register. Next, the computer writes that product code number to a file that keeps track of how many of that product have been sold. In this way, the store manager knows exactly how many units of every item in the store have been sold altogether. It is then a simple task to order more goods to restock the shelves. In fact, the computer can also do that automatically. The manager just keeps an eye on the system to make sure it's working properly.

Today, inventory control systems are common in all retail sales areas. When you order a hamburger from a fast-food restaurant, the computer records the sale. In fact, the cash register doesn't have number keys on it. It just has a key for a hamburger, a key for a cheeseburger, a key for a milkshake, and so on. When the clerk presses the milkshake key, the computer displays the cost and records another milkshake sale in the inventory file.

If there are no bar codes or keys for specific items, a clerk can simply key in the product number on the cash register. This method is slower but works just as well. Smaller stores, such as a local bicycle shop, are more likely to have this sort of system.

Consumer Services

Computers have also helped provide new services to consumers. These include bank cards and on-line services.

"Credit Card" Systems

Computers are also being used along with plastic cards like credit cards to aid consumers. Electronic banking, for example, involves plastic cards and an automatic teller machine (ATM). Fig. 6-14. The bank issues the plastic card bearing an identification number to each person who has an account. The card can be inserted into the ATM and money withdrawn on the spot. The computer checks the number on the card and subtracts the money from the user's account. Bank cards may also be used to check how much money is in the account and to transfer money to a different account. Some bank cards may be used to pay utility bills.

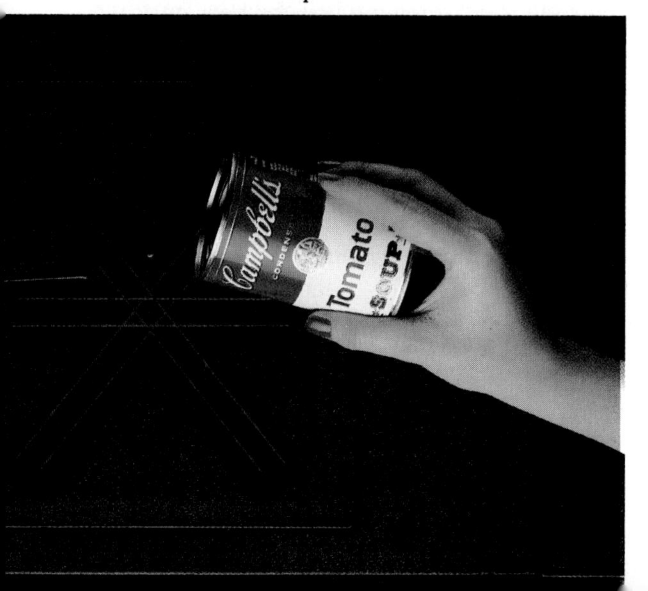

Fig. 6-13. The scanner reads the bar code and sends it to the computer.

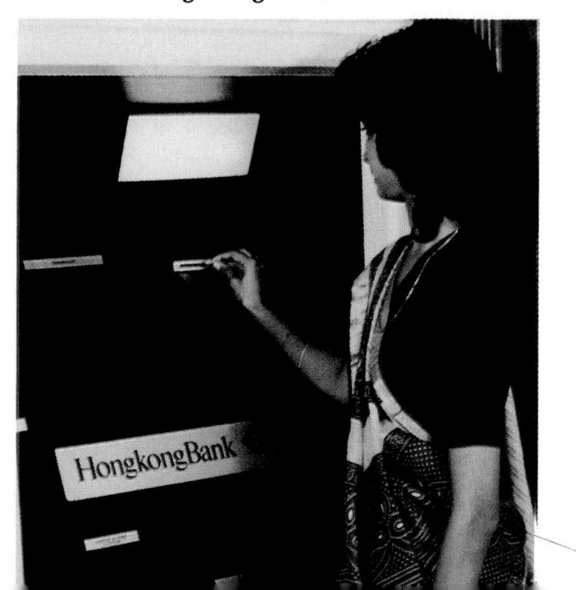

Fig. 6-14. Automatic teller machines are used worldwide. What do you think their advantages and disadvantages might be?

Banks aren't the only businesses to adopt plastic cards. The concept can work for any company. Some gas stations, for instance, have machines that allow customers to insert a credit card and pump their gas. The cost is automatically charged to the user's account.

On-line Consumer Services

Commercial on-line consumer services offer vast amounts of information and resources. These are easily accessed via computer/modem over standard telephone lines. These commercial services charge a monthly fee for basic service, with additional costs for extra services.

The options are seemingly endless. Here are a few examples of what you might find on-line:

- News: Highlights of the day's news
- On-line magazines
- Clubs: Large selection from which to choose
- Education: Resources for teachers
- Finance: Stocks and related information
- Travel: Reservations or planning
- Sports: Video clips, statistics, sports talk
- On-line shopping: Wide variety of products
- Children's services: Cartoons, games, kid talk
- Entertainment: Movie & music reviews, artist profiles, tour dates, radio/TV information
- Electronic mail
- Conversation: People can converse with others who are on-line at the same time.
- Internet access: Gateway to the Internet

On-line services such as those described above are market driven. As in any other commercial venture, if people are willing to pay for a service, companies will provide it. These services have expanded very rapidly in recent years, and the future for commercial network services looks very promising.

Entertainment and Education

Arcade Games

In the early 1980s, microcomputers were just catching on in the U.S. Businesses were learning they could save time and money by using them for word processing and financial recordkeeping. At the same time, the American public discovered what some people had known for years — computers could be fun. Software companies created computer games that could be run on inexpensive home computers. However, the people who really saw the entertainment value were those who owned game arcades.

For decades, children and adults alike had enjoyed arcade games. The most popular, the pinball machine, lit up, made noise, and gave points whenever the steel ball hit a series of targets.

Then came PACMAN®. Fig. 6-15. PACMAN was one of the first fully computerized games to be successful. By today's standards, PACMAN's technology was crude. The object of the game was to gobble up dots with the PACMAN, which you controlled with a joystick. Some of the dots flashed. If the PACMAN ate one of those, it was "energized" for a few seconds. "Ghosts" came along that could absorb the PACMAN. If three ghosts got the PACMAN, the game ended. However, an energized PACMAN could eat a ghost. The more ghosts PACMAN ate, the more points the player scored.

As simple as the game was, kids all over the country loved it. The game cost 25 cents per play—about two or three times the cost of a pinball machine game. Nevertheless, PACMAN was a huge success.

Later, games such as "Dragon's Lair"® and "Space Ace"® replaced PACMAN as the most popular arcade games. They cost 50 cents per play, but they were far more realistic. The characters and scenery were high quality video images stored on a videodisk. A video scene flashed in front of the player, and input from a joystick could change the entire scene instantly. The realism of these games made them a great success.

Arcade games continue to be big business. Software companies spend huge sums of money to develop exciting games for the computer arcades. They know that a good game may cost millions to make, but it can turn a profit in a matter of weeks, if kids like it.

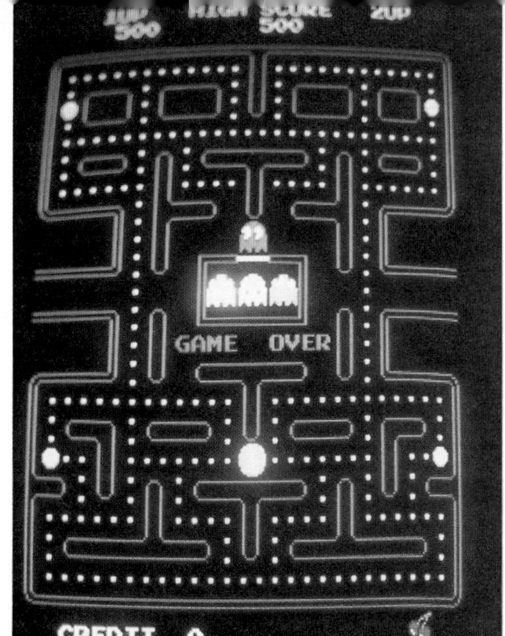

Fig. 6-15. PACMAN® (shown at right) was the first successful computerized arcade game. Today's games (above) have much more sophisticated graphics.

Interactive Video

Videodisks and computers have also been combined for educational purposes. **Interactive video** has been used to teach people how to fly airplanes, fire shells from a tank, and perform science experiments without a laboratory.

For example, interactive video is being used to train people in lifesaving techniques without a teacher. In the past, one technique, CPR, was taught to small groups of people using a dummy. The dummy is now connected to the computer, as is the video instruction program. Fig. 6-16. At certain points during the lesson, the students practice on the dummy. The computer program senses whether or not they are doing things correctly and lets them know. The computer tests their understanding and keeps track of their progress.

Interactive video systems are also being used as public information systems. The EPCOT Center in Florida is an exposition of modern technology. When you arrive, you can find out what may be seen there by using an interactive video system. This is done by pointing at things displayed on its touch sensitive screen. Different scenes are called up from the videodisk. It's just one more of the many ways computers are used to communicate information.

Fig. 6-16. Interactive video is used to teach CPR (cardiopulmonary resuscitation), a lifesaving technique used on people whose hearts have stopped beating.

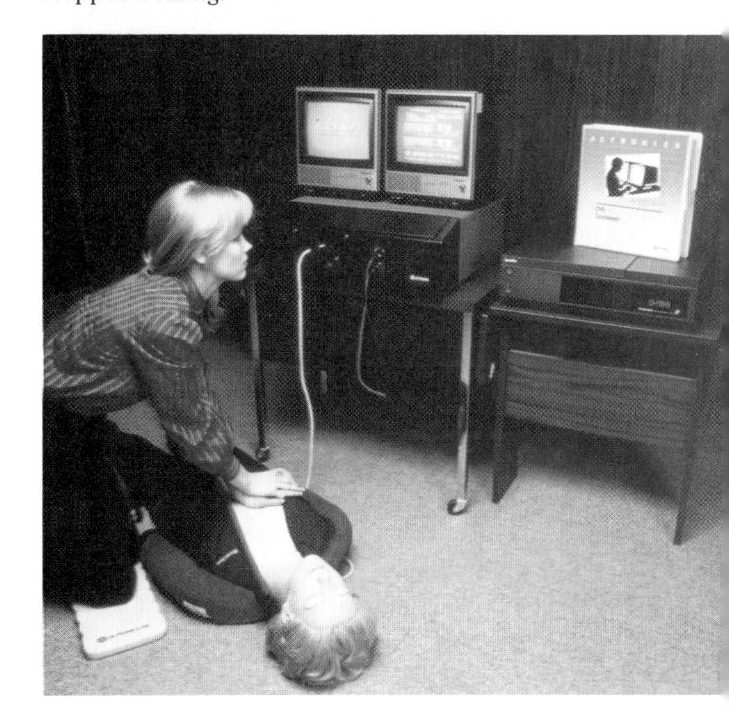

COMPUTERS ON CAMPUS

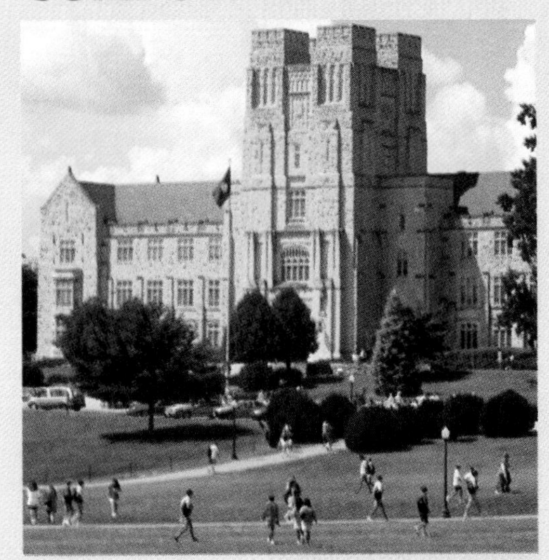

Virginia Tech/Bob Veltri

Computers have long been an essential tool for students at Virginia Polytechnic Institute and State University. A state-of-the-art LAN (local area network) connects thousands of computers all over the campus. A fiber optic cable "backbone" is at the heart of the network. It even extends to a number of apartment complexes off campus. Since the network and the phone system are both digital, modems are not required.

Computers can be connected to this high-speed network from any classroom, office, or dormitory room on campus with a simple cable. All students are issued a PID (personal ID). This allows them each to send and receive electronic mail from anywhere on the planet.

Students and faculty routinely use the network to communicate and to move information from place to place. Professors put course notes, syllabi (course outlines), and readings on-line, so that students may access them wherever and whenever convenient. Students submit work electronically, and professors evaluate and return it, sometimes without ever working with hard copy. Using the network, students can write papers as a group, with each individual seeing it on screen at the same time from different locations. The professor may or may not be part of the group. Similarly, group discussions take place on-line as students learn to work together in the "virtual classroom."

The Virginia Tech Library System (VTLS) is a very important part of the network. The electronic catalog for the library is on-line and accessible throughout the university, as well as from off campus. Similarly, CD-ROM databases can be accessed though the network. Many academic journals are accessible electronically through the library's information server. In some cases, no hard copy exists. The library simply provides access to the information, which may be stored on any server in the world.

CHAPTER 6 REVIEW

Review Questions

1. Explain the difference between analog signals and digital data.
2. What does a modem do?
3. Describe the two kinds of computer networks.
4. Describe how fax machines work and how they are used.
5. Why is color printing less expensive when computers are used?
6. How are computers being used in offices?
7. What do CAD, CAM, and CIM mean? Briefly describe them.
8. What is the World Wide Web?
9. How did PACMAN change the world of game arcades?
10. Describe how interactive video teaches lifesaving techniques.

Activities

1. Select three commercial on-line services. Compare their costs and features.
2. Find out what types of software are being used in three or four businesses in your community. What is the most popular software and the most common use?
3. Visit a local shopping mall. Find out what percentage of the stores are using some form of computer inventory control system. If none is visible, ask the managers.
4. See if you can tell which games in a local video arcade use computer graphics and which use videodisk images. Report on your findings to the class.
5. Identify a problem in your school that you believe could be solved using a computer communication system. Describe or design the computer system that you think would solve the problem.

Careers

Systems Analyst

The widespread use of computers has opened up many new career opportunities. One such career is that of a systems analyst.

Computer systems analysts are the people who figure out how a business can make the most of computers. They develop new ways to computerize various tasks within a company. They also figure out ways to improve tasks that are already handled by computers.

In order to understand a problem, analysts first discuss it with managers. Together, they break the problem down into small parts. Consider, for example, the problem of creating an inventory control system. The analyst must decide what information must be collected, how it must be processed, and what reports are needed.

After the analyst has designed the computerized system, he or she must describe it to the managers. The managers then decide whether or not to develop and use the system. If they choose to go ahead, the analyst provides instructions for the programmers. The programmers then write the computer code to make the system work. The analyst also specifies the hardware needed for the system.

Systems analysts usually work in cities where many large companies need their services. They often specialize in business, scientific, or engineering needs.

Systems analysts are well paid. Average starting salaries are more than $30,000. The job outlook for systems analysts is very good. The demand should be greater than the supply throughout the 1990s.

Education

While requirements vary, a bachelor's degree in computer science or a related field is usually necessary. (Related fields include information science, computer information systems, and data processing.) Analysts are expected to know a number of different programming languages. For some higher-level jobs, a graduate degree may be required.

Since systems analysts often specialize, coursework in business, science, or engineering is also very useful.

For More Information

To learn more about a career as a systems analyst, contact:

Association for Systems Management
24587 Bagley Rd.
Cleveland, OH 44138

Data Processing Management Association
505 Busse Highway
Park Ridge, IL 60068

Correlations

 *Language Arts*

1. In 1949, George Orwell wrote a science fiction book titled *1984*. In this novel, Orwell described a society watched and controlled by machines. Check out a copy of *1984* from your school library. Read it, and see if you think Orwell was on target. Do you think technology is being used to reduce our freedom? What do you think the role of computers will be in our society in the year 2084?
2. Do a survey of your school to find out how many different types of microcomputers are in use there. In what ways are all of the different computers being used? Write a report on your findings.

 *Science*

The microchips that are the basic components in computers all begin with a small wafer-thin chip of silicon. You learned in Chapter 4 that silicon is a semiconductor. Find out more about semiconductors. Why are they used in computers?

 *Math*

Computers are classed by the number of bits they can handle at one time. For example, there are 8-bit, 16-bit, and 32-bit microcomputers. The largest binary number that can be represented with 8 bits is 11111111, or 255. What is the largest number that can be represented with 16 bits? With 32 bits? Give your answers in base 10 as well as the base 2 system.

 *Social Studies*

Computers have had an impact on all of us. List ten common computer applications. Write a short description of one or more changes each has caused in the way we go about our lives. For example, you might list the automatic teller machine. Some of its effects on our lives include faster cash withdrawals and 24-hour access to accounts.

Basic Activities

Basic Activity #1: Calculator Exercise

Even basic calculators have some memory. That is, they can store the results of calculations until they are turned off. As long as they remain on, you can add or subtract from the number stored in memory without having to re-enter it.

Materials and Equipment

A simple pocket calculator and its instruction booklet.

Procedure

1. Follow the directions in the instruction booklet to learn how to add and subtract numbers that have been stored in memory.
2. Multiply 25 times 4 and press "Memory +". (The exact method may vary for your calculator.)
3. Divide 84 by 4 and press "Memory −".
4. Add 55 plus 19 and press "Memory +".
5. Press "Recall Memory" to display the answer.

Your answer should be 153.

Basic Activity #2: Introduction to the Microcomputer

Most microcomputers come with introductory software. This software helps you become familiar with the keyboard. It also gives you an idea of the kinds of things the computer can do.

Materials and Equipment

Microcomputer and introductory software package.

Procedure

1. Boot (start up) the microcomputer.
2. Insert and run the introductory software package.
3. Write your opinion of the program. Is it easy to use? What did you like most? Least? How would you improve it?

Basic Activity #3: Typing Tutor Program

Word processing is one of the most common uses for microcomputers. Can you type? Anyone can sit at a keyboard and "hunt and peck." Learning "touch" typing means you can type without looking at the keyboard. A number of software programs will teach you how to touch type.

Materials and Equipment
Microcomputer and a typing tutorial program.

Procedure

1. Using your current method, type for exactly five minutes. Count the number of words you typed all together. Circle the mistakes. Save the results.
2. Boot (start up) the computer and run the typing tutorial program.
3. Practice regularly with the program until you learn to type without looking at the keys. The drawing below shows which finger to use for each key. The black bars below the keys "ASDF" and "JKL:" indicate the "home" keys. Your fingers should always return to these keys after striking a character.
4. Using touch typing, type for five minutes. Count the words and mistakes. Compare the results to the first "test."

Basic Activity #4: Computer Game

Computers are often used for fun. Even highly paid executives in business and industry have been known to take a short break to play a computer game.

Materials and Equipment
Microcomputer and a software game.

Procedure

1. Boot (start up) the computer.
2. Run the software game.
3. Write your opinion of the game. Was it easy to learn? Why? Was it as much fun as you expected? How could it be improved?
4. Briefly describe a computer game you would design.

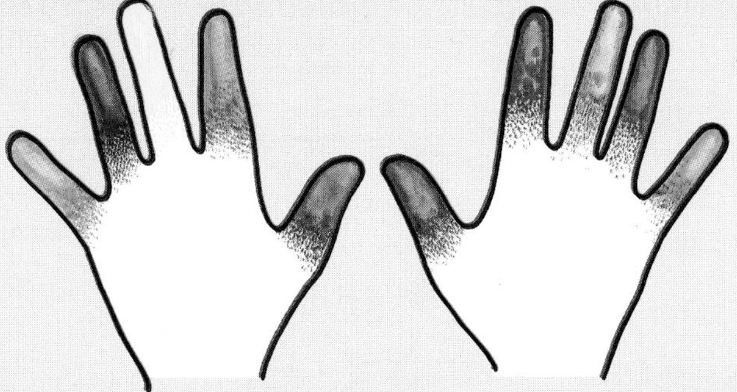

Intermediate Activities

Intermediate Activity #1: DOS Commands

Every computer has an operating system that handles commonly used functions. For example, the IBM PC® microcomputer uses the MS DOS® operating system. In this activity, you will have a chance to work with some common operating system commands.

Materials and Equipment

MS DOS® computer and MS DOS® disk or Apple II® computer and Apple DOS® disk
blank floppy disk
appropriate DOS manual
a data disk or software that may be legally copied. Make sure you have a back-up disk of anything important in case you accidentally erase a disk you are working with.

Procedure

1. Boot the DOS disk.
2. Read the instructions for each of the commands listed below before you attempt to use them.
3. Look to see what files are on the DOS disk by using the DIRECTORY (or CATALOG) command.
4. Format the blank disk using the FORMAT command.
5. Copy several files from one disk to another using the COPY command.
6. Erase one of these files using the ERASE (or DELETE) command.
7. Copy the entire contents of one disk onto another disk using the COPY command.

EXAMPLES OF TYPICAL DOS COMMANDS

Here are some examples of DOS commands and how they are used. You will find more detailed information in your system's DOS manual.

1. To see a list of what's on the disk, enter

DIR or **DIR/P**

Entering just DIR will cause the directory listing to scroll by, and you may not be able to read it. The /P will cause the listing to pause when the screen is full.

2. To format a floppy disk, make sure the disk you are formatting is either blank or contains information you don't want to keep. (Use the DIR command.) After you're sure you have the right disk, enter

FORMAT A:

The floppy disk drive is usually designated as the A drive. In systems with two floppy drives, one is A and the other is B.

3. To copy a file named Report from a disk in drive A to a disk in drive B, enter

COPY A: REPORT B:

4. To erase the file Report from the disk in drive B, enter

DEL B:REPORT

5. To copy the contents of a disk in drive A to a disk in drive B, enter

DISKCOPY A: B:

This will copy the entire contents of the disk in drive A to the disk in drive B. Anything previously recorded on the disk in drive B will be erased, so make sure the disk you put in drive B is either blank or has nothing that you want to keep.

Intermediate Activity #2: Analog to Digital Conversion

Computers often require that analog signals be converted to digital data. This activity demonstrates a basic analog-to-digital conversion.

Materials and Equipment

1.5-volt battery
1.5-volt light bulb
1K, 2-watt potentiometer
digital volt-ohmmeter
copper wire

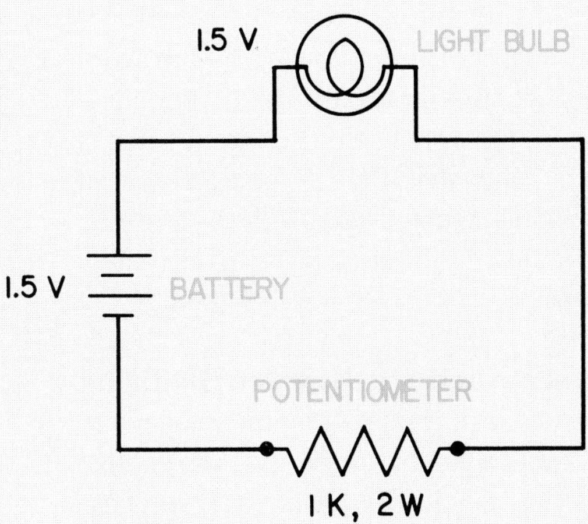

Fig. II-1.

Procedure

1. Wire the circuit shown in the schematic diagram, Fig. II-1.
2. Note how the lightbulb gets dimmer and brighter as you adjust the potentiometer in the circuit. This is because the voltage varies *continuously.* By adjusting the potentiometer, you vary the brightness of the light. An infinite number of brightnesses are represented by the bulb.
3. Hook up the digital volt-ohmmeter to measure the voltage across the lightbulb. The analog-to-digital converter (inside the volt-ohmmeter) will convert the range of voltages into a number of specific voltages.
4. How many different readings can be measured in this circuit by the digital meter? How many different voltages are there really?

Intermediate Activity #3: Word Processing

Word processing is one of the most common uses for microcomputers. With word processing programs you can type, save, edit, and print text.

Materials and Equipment

Apple Macintosh®, IBM PC® or compatible, Apple IIe®, or Apple IIGS® microcomputer
basic word processing software (such as MacWrite® for the Macintosh®, Writing Assistant® for the IBM PC®, or Bank Street Writer® for the Apple II®)
a formatted floppy disk
word processing documentation (instructions)

Procedure
1. Boot the computer and load the word processing software.
2. Type a paragraph of text.
3. Do as many of the following as the word processor will easily allow you to do: underline words, make words boldface, change size of letters, change style of type, type single spaced and double spaced, save, edit, and print.
4. Use the word processor for your next class report.

Intermediate Activity #4: Video Digitizing

An easy way to create a graphic image is to capture it with a video camera and then digitize it. Digitizing, as you will recall, is the process of converting the image to digital data. Once digitized, the image may be edited, combined with text, or printed and used in many ways.

Materials and Equipment

Apple Macintosh® and MacVision® software/hardware and MacPaint® software (or Apple II® and Computer Eyes® software/hardware and Blazing Paddles® software)
instructions
video camera and tripod or video cassette recorder
printer

Procedure
1. Read the instructions for the system you are using.
2. Ask a fellow student to pose. Be sure to light your subject well.
3. Using a video camera, scan in a picture of your subject. Adjust the brightness control. An alternative method is to use an existing photo and a scanner, as shown in the picture below.
4. When the image looks the way you would like it to look on the monitor, capture it electronically.
5. Save it to a floppy disk.
6. Enhance the digitized image using MacPaint® (Macintosh®) or Blazing Paddles® (Apple®) software.
7. Print out the image.

Advanced Activities

Advanced Activity #1: Macintosh HyperCard® Stack

HyperCard® is software for the Macintosh® that lets you make "stacks." A HyperCard® stack is like a stack of 3″ × 5″ note cards. You may put notes or illustrations on each "card" and can easily jump from one card to another. However, it is created on the Apple Macintosh® and may easily be stored on a floppy disk.

HyperCard® stacks are a good way to keep track of information. For this activity, you may need to do a little research. Your assignment is to create a HyperCard® stack on a topic approved by your instructor. One example may be to create a HyperCard® timeline on the history of computers. Each card could contain information and an illustration of an important invention.

Materials and Equipment

Apple Macintosh computer®
HyperCard® and *HyperCard User's Guide*
video digitizer (such as MacVision®)
video camera

Procedure

1. Work through the tutorials in Chapter 1 of the *HyperCard User's Guide* to learn how HyperCard® works.
2. Design a layout for a HyperCard® that can be used for a stack idea approved by your instructor.
3. Working in the "Author" mode in HyperCard®, experiment with the "Tools" to create the layout developed in Step 2 above.
4. Use the video digitizer to capture at least one graphic to include in your stack.
5. Include notes and an illustration on each card, as well as "home," "help," "forward," and "backward" buttons. You may build in any other options you wish.

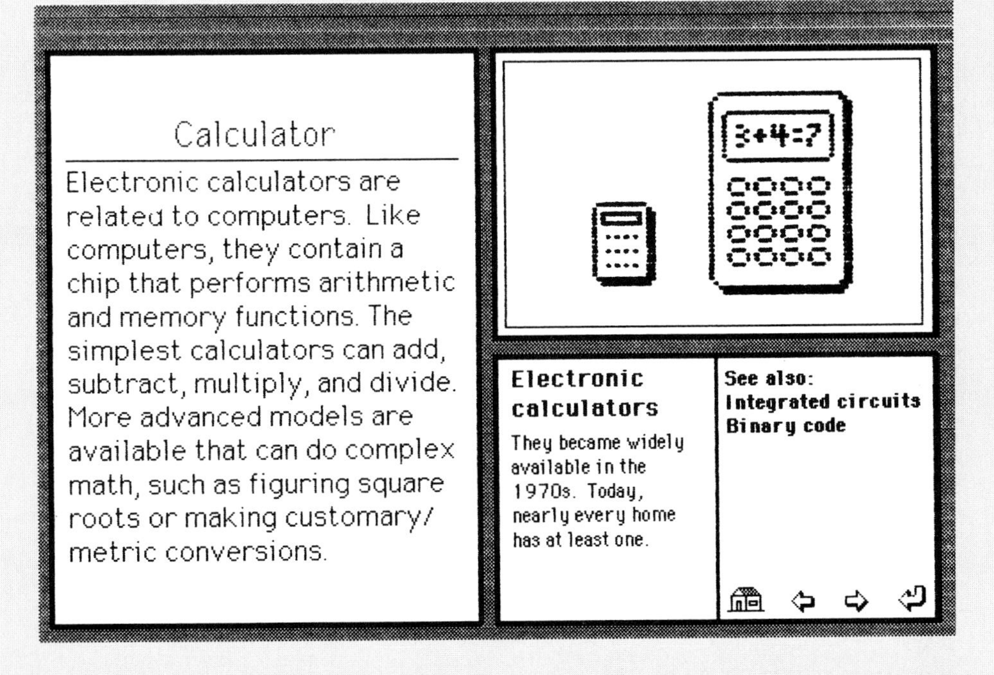

Advanced Activity #2: Technical Support from the World Wide Web and/or Internet

The Internet contains vast amounts of information on nearly every imaginable topic. The lab you are working in has equipment that often requires "technical support." Technical support might mean locating answers to "frequently asked questions" (known as FAQs on the Internet) or the latest software upgrade for a specific piece of equipment.

Materials and Equipment

computer with Internet access
a World Wide Web (WWW) browser, such as Netscape Navigator
access to the Internet/WWW

Procedure

1. Make a list of equipment manufacturers for some or all of the equipment in the lab. Include the names and model numbers.
2. Locate and open the search tool of a World Wide Web browser that allows "keyword searching."
3. Key in the name of one of the larger manufacturing company names (such as Apple) and search for a "Home Page" for this company. If you can't find one, try a different manufacturer.
4. Once you locate one of these home pages, see if you can locate any technical information that pertains to the specific equipment in your lab. Software updates, such as a new printer driver, are always useful. (A printer driver is software that allows a computer and printer to communicate with each other.)
5. When you find useful information, download it to a floppy disk and/or print it to a local printer.
6. Next, see if you can locate some useful new software or data (for example, clip art files) for the lab from an FTP server. FTP stands for File Transport Protocol. These servers store information for public access. You can locate a specific FTP site by using the "Open Location" option in your WWW browser. Enter one of the URLs (Uniform Resource Locators) shown below, or enter one that you have identified from another source.

URLs for popular FTP sites:
ftp://ftp.sumex-aim.stanford.edu
ftp://ftp.austin.apple.com
ftp://ftp.mcom.com
ftp://ftp.rascal.ics.utexas.edu
ftp://ftp.ncsa.uiuc.edu

7. Once you arrive at one of these FTP sites, your WWW browser will allow you to download the files that you want.

Advanced Activity #3: Electronic Portfolios

If you have access to the Internet, it is possible to create an "electronic portfolio" of your work that may be viewed from anywhere on the planet. For this activity, plan to display work from this class, particularly if it's already in digital format.

Materials and Equipment

computer with Internet access

access to a World Wide Web server (on which you may store your completed electronic portfolio)

WWW page development tools, such as:

- an HTML editor, such as World Wide Web Weaver
- an HTML file conversion utility (to convert a word processing file to an HTML file), such as RTF to HTML
- a graphic conversion utility (to convert graphic files to GIF files), such as Graphic Converter
- (optional) a color scanner with a digital editing application, such as Adobe Photoshop™

Procedure

1. Use a WWW browser to locate information on how to create World Wide Web pages. For example, the Help menu item in Netscape Navigator has an option called "How to Create Web Services." This option provides immediate access to lots of information about creating WWW pages. (Note: You may be able to find a book in the library with similar information.) Review this information to get an idea of how to proceed.
2. Using a word processor, create a simple resumé that includes your name and a brief list of the things you wish to display in your portfolio.
3. Save this resumé file in the HTML file format. If your word processor doesn't provide that option, use a file conversion utility (which you could find with a WWW browser as in the previous activity) to convert the file into an HTML file. Or, use an HTML editor to "mark up" this file with HTML "tags."
4. Convert graphic files you may have created to the GIF file format with a graphic conversion tool such as Graphic Converter. Note: Some graphic editors, such as Adobe Photoshop™, can save files in the GIF format.
5. (Optional) Scan pictures of your work and save them as GIF files (or convert them to GIF files).
6. If facilities exist to create digital video/audio, consider creating digital video/audio clips as well. Save digital video in the .mov or .mpeg file format. Save audio files in the .aiff or .wav file format for display on the WWW.
7. Use the HTML editor to create links from your resumé file to other text, graphic, or video/audio files.
8. Telecommunicate these files (or put them all together on a floppy disk and provide them) to whoever manages the nearest WWW server in your school, district, or region.
9. Be sure to tell your friends and relatives (and possible employers, of course) to check out your new electronic portfolio on the World Wide Web.

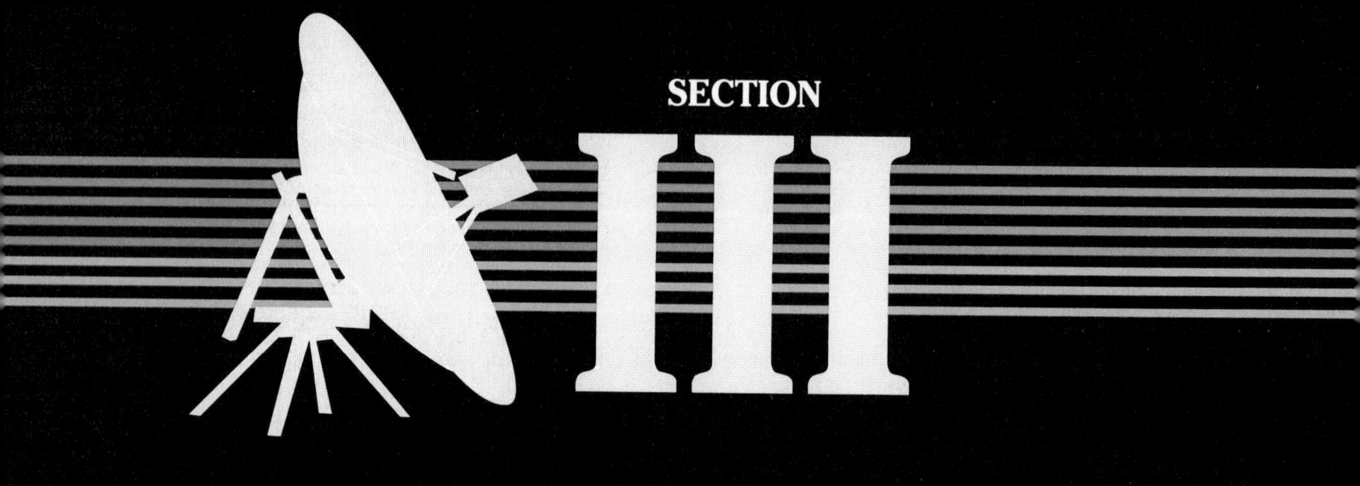

Technical Design Systems

Chapter 7: Principles of Technical Design
Chapter 8: Technical Design Processes
Chapter 9: Computer-Aided Design

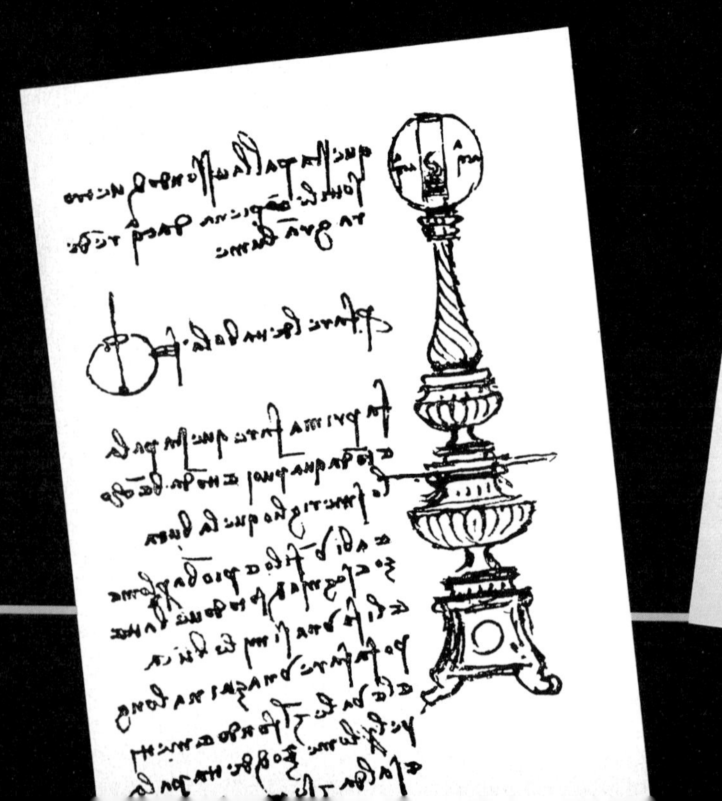

Throughout history, drawings have been used again and again to communicate ideas. In fact, the advancement of civilization, industry, and technology can often be traced through the evolution of graphics.

Stone-age people used cave drawings to record their achievements, rituals, and day-to-day activities. In ancient Mesopotamia, Egypt, and China, drawings were used for calendars and to aid in building structures. Some of the earliest recorded plans were for a fortress; they were drawn in clay.

Throughout the ages, drawings were used to show design ideas. For example, the drawing on the left-hand page shows Leonardo da Vinci's design for a lamp. Da Vinci and others also used drawings to test new scientific and mathematical theories. In fact, all geometry and trigonometry concepts were first tested using drawings.

During the Industrial Revolution, the science of drafting, then called descriptive geometry, was invented in France. As a result, France leaped into the forefront of industrial science and technology.

More recently, a second revolution occurred with the introduction of the computer. Design and drawing can now be done in three dimensions. Other changes are taking place so fast that it is hard to keep up with them.

Though the ways in which designs and drawings are prepared have changed, the system of technical design has not. In the following chapters, you will learn about the principles and applications of technical design.

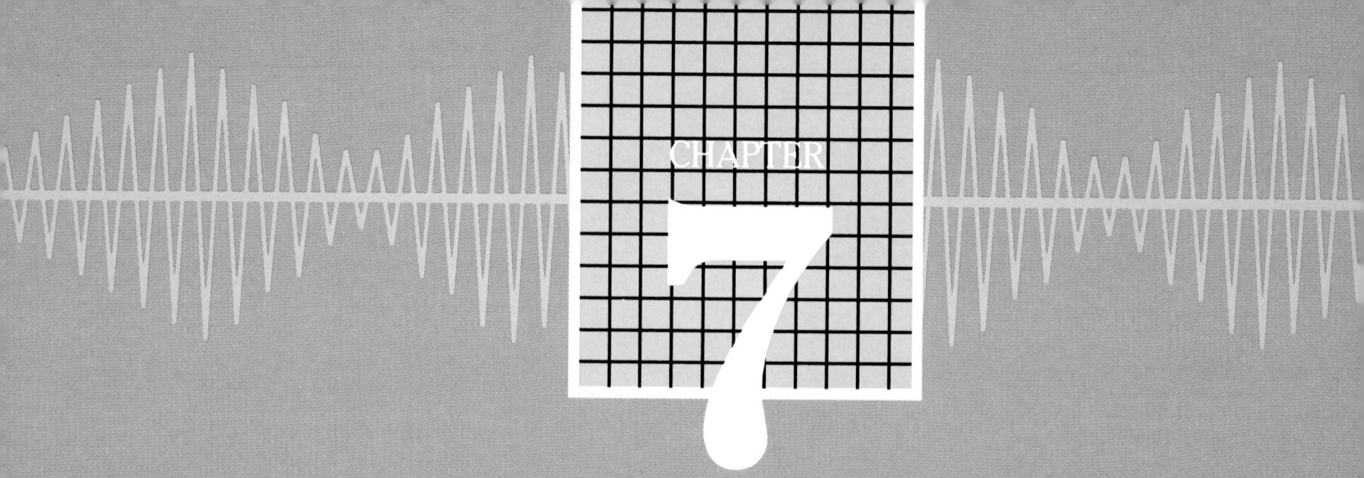

CHAPTER

7

Principles of Technical Design

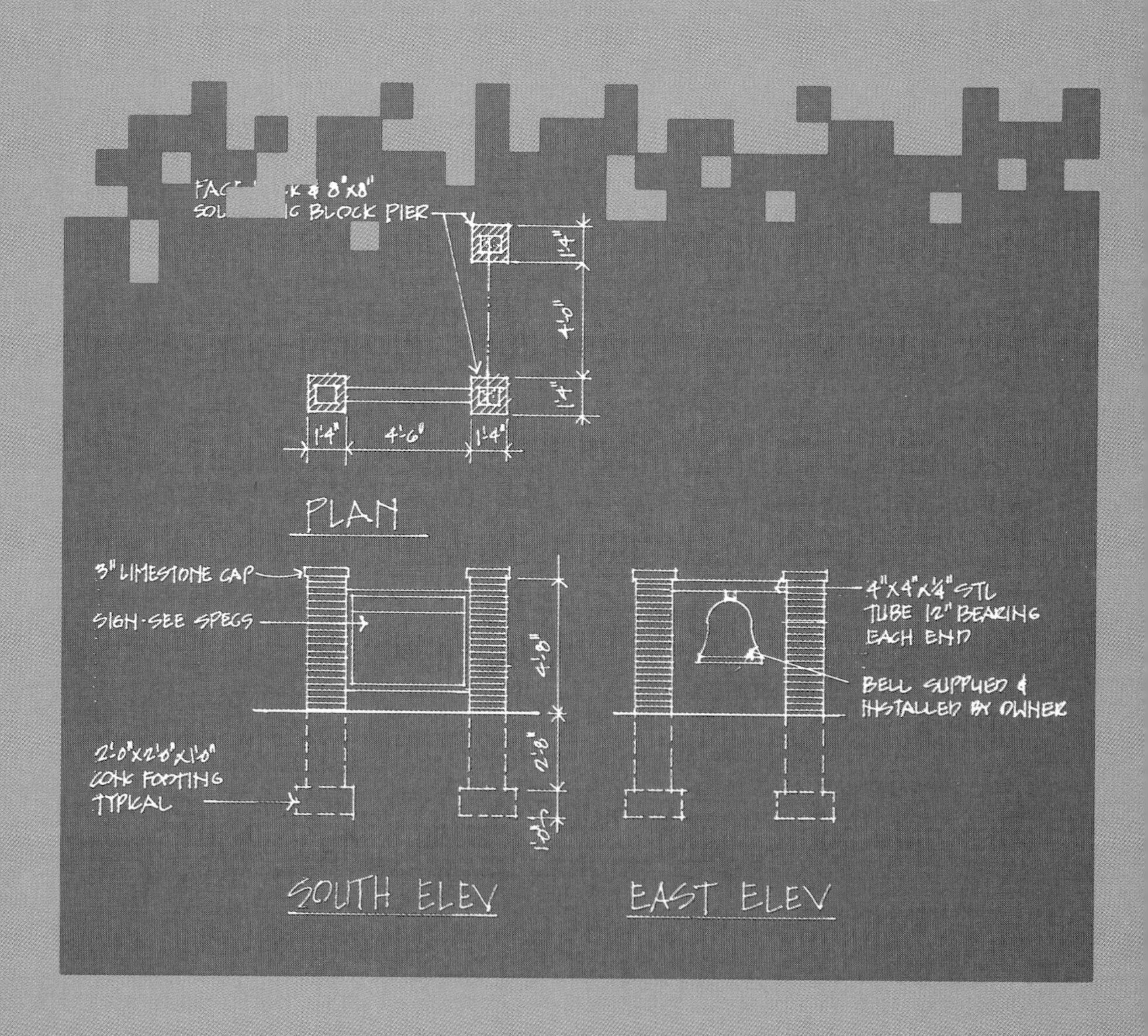

The drawing on page 140 describes the construction of a bell tower. Do you understand exactly what it's showing? To be able to read this drawing and others like it, you have to learn the language in which they are written—the graphic language.

The drawing on page 140 is made with a system of lines, shapes, letters, and numbers. It can be understood by any trained person in any country in the world. Though the drawing was prepared in the United States, it can be read easily by someone who speaks only French, Swahili, Russian, or Japanese.

Terms to Learn

aligned dimensioning
dimension lines
dimensions
drafting machines
drawing to scale
extension lines
leader lines
parallel straightedge
scales
standards
stick lead holder
technical drawing pen
tolerance
unidirectional dimensioning
vellum

As you read and study this chapter, you will find answers to questions such as:

- What do the different lines and symbols in a technical drawing mean?
- What is dimensioning and what information does it give?
- What special equipment is used by people who prepare technical drawings?
- How are drawings copied and stored?

ANATOMY OF A TECHNICAL DRAWING

The best way to learn to make or read a technical drawing is to study its parts and the rules used to put the parts together.

Every drawing must do the following:

- Give sizes for all standard parts and features.
- Give correct distances.
- Have measurements, symbols, and other notes placed in an orderly way so they are easy to follow.
- Provide any needed information about production methods.
- Use correct line weights and spacing.

Standards

A technical drawing can be read by any trained person because international standards have been established. **Standards** are rules that say what symbols, and so on, should be used for different things. In the United States, the American National Standards Institute (ANSI) sets all drawing standards. International standards are set by the International Organization for Standardization (ISO). Because many of the drawings made in this country are also used in other countries, ANSI and ISO standards are similar. Both organizations make these standards available to those who need them.

Lines

Different types of lines are used in technical drawings. They vary in thickness, regularity, and direction. Both ANSI and ISO have standardized the different lines to mean certain things.

Lines used in technical drawings are shown in Fig. 7-1. This chart is called the "alphabet" of lines.

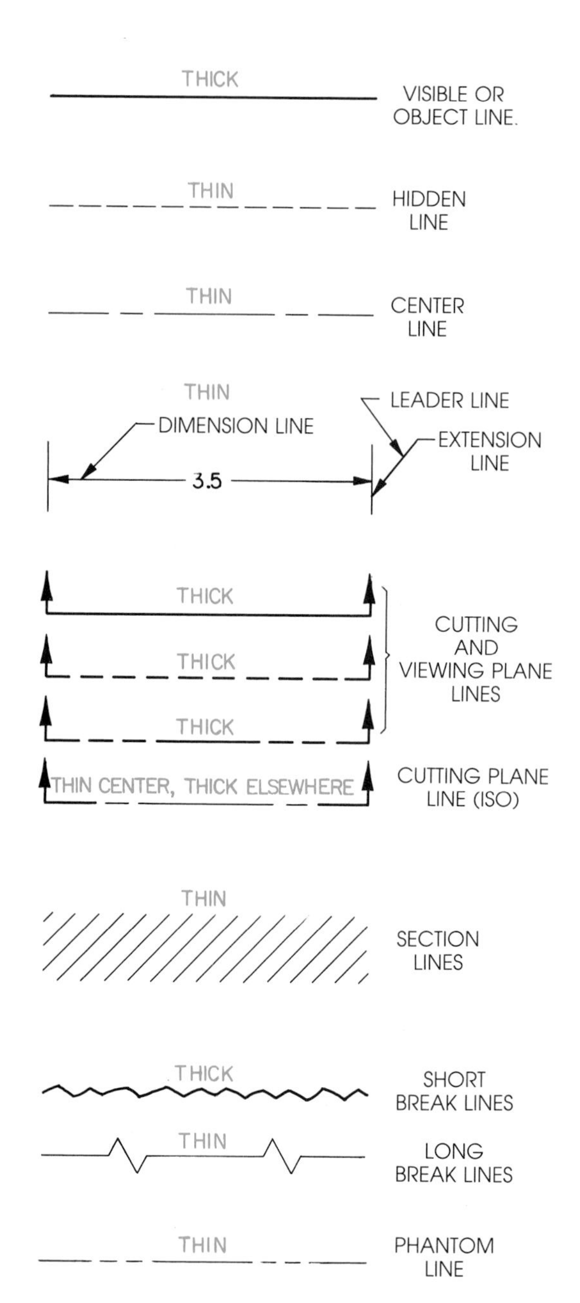

Fig. 7-1. In technical drawing, different kinds of lines are used for different purposes. For example, to show edges that are not visible to the eye, hidden lines are used.

Lettering

Words and numbers are most commonly used for indicating sizes and making notes on a drawing. However, they should only be used to make a drawing easier to understand. A drafter's work should be clear enough that few words are necessary. A title block, which gives the name of the drawing and other important information, is usually placed at the top or bottom.

Any lettering must be neat and easy to read. Poor lettering will make a drawing difficult to read and understand. For this reason, good drafters work hard to perfect their lettering skills.

Good lettering practice requires that all letters, figures, and symbols be made dark, clear, and upright. They should be of a readable size, even though the drawing itself is small.

Most lettering used for sizes, notes, and specifications is ⅛ inch (3 mm) in height. For lettering headings, a size between 3⁄16 and ¼ inch (5 to 6 mm) is recommended. Lettering up to ½ inch (13 mm) in height may be used for title blocks.

The most common lettering style used in drawings is the single-stroke vertical style. Fig. 7-2. In some cases, inclined or slanted lettering is used. However, it is not good practice to use both vertical and inclined lettering together.

All lettering is made with three basic movements of the pencil:

- Vertical strokes are made by the fingers only.
- Horizontal strokes are made by moving the hand and wrist, while the fingers move only slightly.
- Circular strokes are made mostly by finger movements and slight wrist action.

All letters and numbers are made by using one or more of the three strokes described.

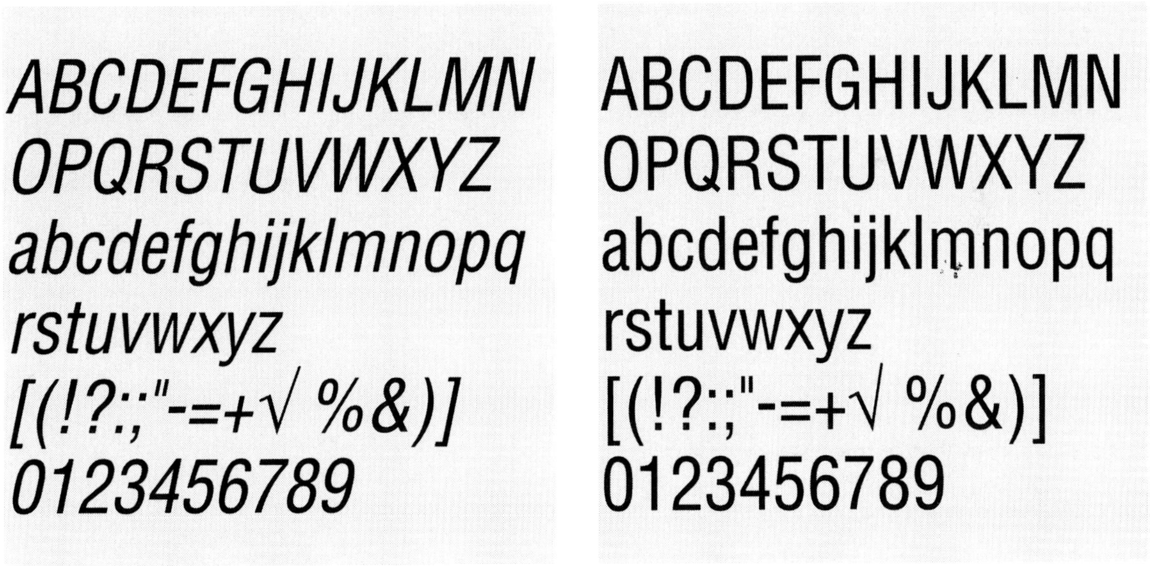

ABCDEFGHIJKLMN
OPQRSTUVWXYZ
abcdefghijklmnopq
rstuvwxyz
[(!?:;"-=+√ %&)]
0123456789

ABCDEFGHIJKLMN
OPQRSTUVWXYZ
abcdefghijklmnopq
rstuvwxyz
[(!?:;"-=+√ %&)]
0123456789

Fig. 7-2. Single-stroke vertical lettering is the most common style used on drawings.

User's Guide to Technology

SOME ASSEMBLY REQUIRED

Who needs to learn about drafting? People who plan to become engineers, architects, or designers need to study drafting, of course. Machinists, carpenters, and others whose work requires them to understand drawings should also study drafting. Even if your career plans have nothing to do with designing or with making products or building structures, you can benefit from a knowledge of drafting.

The words "Some assembly required" appear on the packages of many products, from furniture to toys. These products come with drawings that show how to put the product together. There are written directions also, but a drawing or two can be much more helpful than several paragraphs of directions such as these: "Place the module connectors over the module back panels and fasten with the screws from which you have removed the washer nuts." The drawings are helpful *only if* you know how to read them. Broken out sections, hidden lines, phantom lines, dimensions—all these are commonly found in such drawings. You need to know what they mean in order to understand the drawings.

Someday you may be a homeowner. If you decide to have the kitchen remodeled or a room built on, will you be able to read the contractor's floor plans? If you can, you'll be better able to communicate with the contractor and you'll be less likely to get unpleasant surprises, such as a window where you wanted a closet.

Did you know mapmaking is a type of drafting? In Chapter 8, you'll learn about maps and how they are drawn. Knowing how to read maps may not lead you to buried treasure, but you'll at least be able to find your way to your next vacation spot—and home again.

Almost daily, you encounter drawings of one sort or another. They may be assembly drawings, floor plans, maps, or some other type of technical drawing. Understanding these drawings will help you to understand your world a little better.

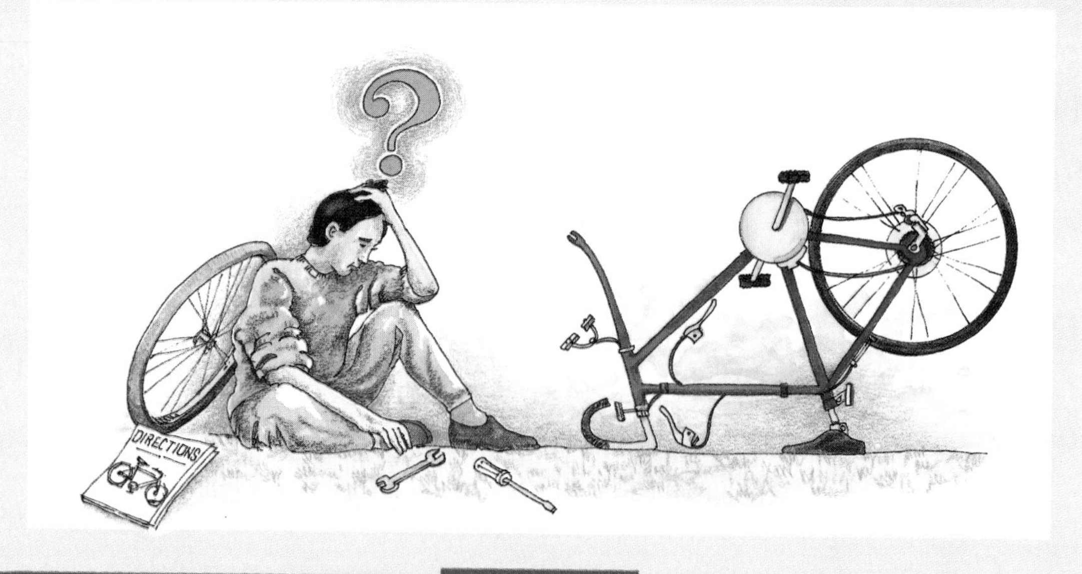

Shown in Fig. 7-3 is how each letter and number should be made using the proper sequence of strokes.

Dimensions

A drawing or plan that is made without any information about size and location of parts and features would be difficult to understand and of little use. Thus, measurements, or **dimensions**, are given either alongside the item or in a block of type in a corner of the drawing. Most dimensions are alongside and appear in number form.

Size dimensions are used to describe the shape and characteristics of a feature or part. Fig. 7-4. Examples include sizes of holes, slots, squares, circles, arcs, rounded areas, corners, and surfaces. Location dimensions are used to tell the position of features. One of the most common establishes the center of an item, such as a circle. Other dimensions give the distance between two items. The items can be points, lines, surfaces, or solids.

Fig. 7-3. This shows the proper stroke order for lettering.

Fig. 7-4. Size dimensions show how large an item is. Location dimensions show the position of certain features, such as holes.

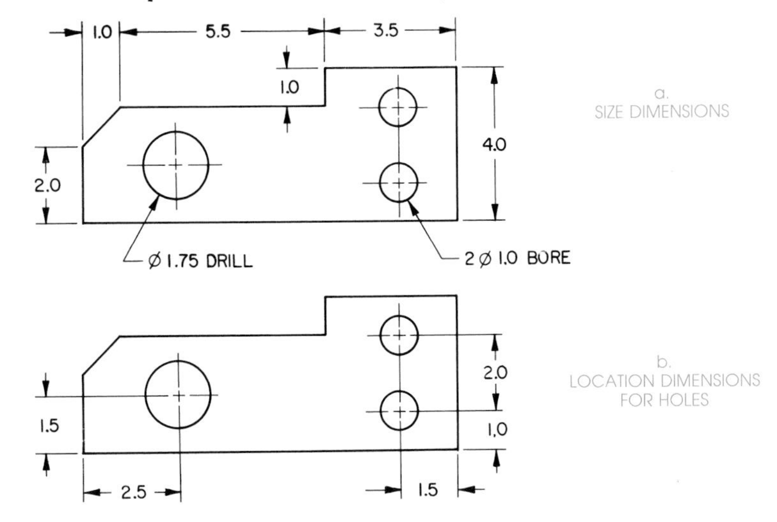

Though many drawings are still dimensioned using U.S. customary measures (inch-foot and ounce-pound), use of metric measures is gradually increasing. Drawings done for the international market must almost always use the metric system. In most drawings, dimensions will be given in one or the other. There are times, however, when both are used. In such cases, one of the units is noted in parentheses. Fig. 7-5.

Most metric drawings are figured in millimeters (mm). Some, such as architectural and civil engineering drawings, use the meter (m). Long distances found in cartography and topographic drawings are often given in kilometers (km).

To avoid confusion, it is recommended that all drawings using the metric system contain the following note: ALL DIMENSIONS ARE IN MILLIMETERS.

Placement

Either aligned or unidirectional dimensioning is used for drawings. Fig. 7-6. **Aligned dimensioning** places the dimensions so they are read parallel to either the bottom or right side of the drawing. It is the older of the two systems. **Unidirectional dimensioning** arranges the dimensions so that they are all read in line with the bottom of the drawing. Drawings using this system are easier to read and understand. Most drawings prepared today use unidirectional dimensioning.

Fig. 7-5. On a dual-dimensioned drawing, one unit (either the customary or metric) is shown in parentheses.

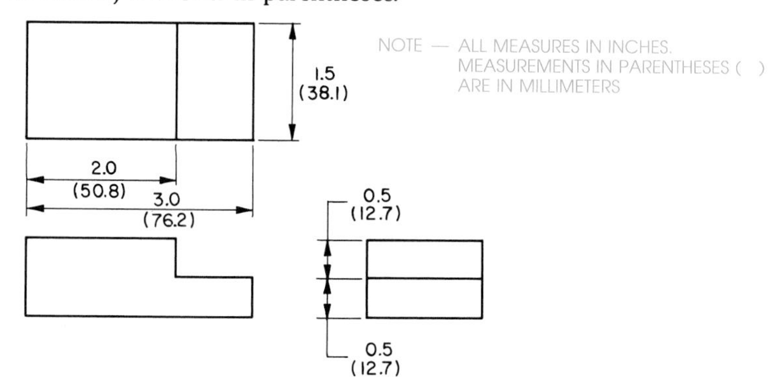

Fig. 7-6. Dimensions may be aligned. However, unidirectional dimensioning is preferred.

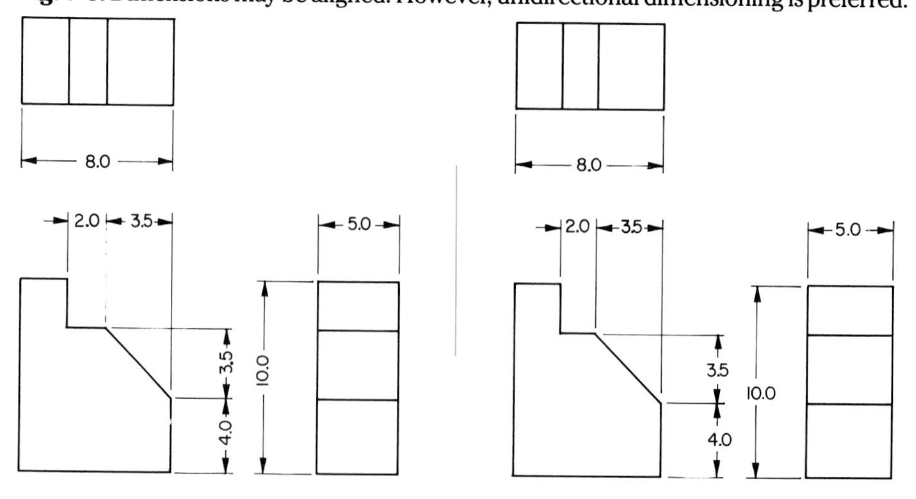

Lines for Dimensioning

The types of lines that are used especially for dimensioning are dimension, extension, and leader lines. Fig. 7-7. All are drawn dark, thin, and solid. They should be thinner than the lines forming the object itself.

At the end of the dimension and leader lines are arrows. Two types are used: open and solid. Fig. 7-8. The solid is perhaps the most common. Generally, the width of the arrowhead should not exceed one-third of its length. For example, if an arrowhead is drawn 3/16 inch (5 mm) long, its width should be no more than 1/16 inch (1.5 mm).

Extension lines are used to show the beginning and end of a dimension. They should not come in direct contact with the part or section that they refer to. It is good practice to start the extension line about 1/16 inch (1 mm) from the object. It should end about 1/8 inch (3 mm) beyond the last dimension. Once started, extension lines should not be broken, unless the space around them is very small. Fig. 7-9.

Dimension lines, drawn between extension lines, run parallel to the surface they measure. The dimension is written in a break in the line.

Leader lines are straight. They are used to help dimension or describe geometric shapes, such as a circle. It is best to draw the leader line at an angle of 45°. However, angles of 30° and

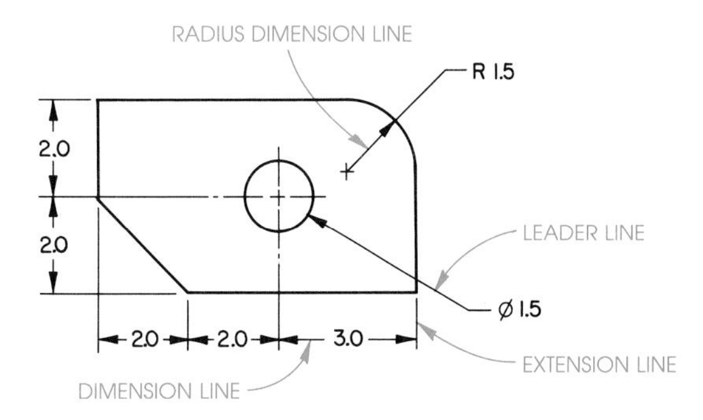

Fig. 7-7. This drawing shows examples of extension, dimension, and leader lines.

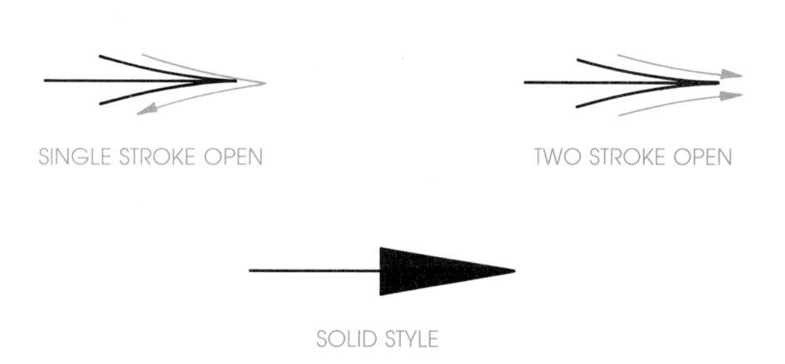

Fig. 7-8. Open arrowheads may be made with one stroke or two. Solid arrowheads are more common in today's drawings.

Fig. 7-9. Extension lines should be unbroken if space permits.

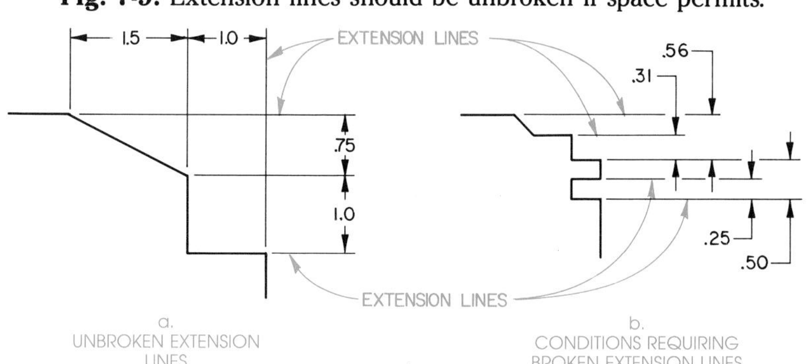

60° are also acceptable. In Fig. 7-10 leader lines are used to indicate the sizes of a hole, chamfer, and slot. Though the leader line itself ends at the edge of a circle or hole, it should be drawn in such a way that, if it continued, it would pass through the hole's center.

For dimensioning shapes, arrowheads are used to identify the outside boundary. When describing a surface, the leader is drawn to the middle of that surface. Then a dot is used in place of the arrowhead.

Numbers

Dimensions are usually written as numbers. Fig. 7-11. Most drafters usually leave a space in the middle of the dimension line for the number. In some cases, especially in European countries, the dimension is placed above the dimension line.

Numbers as dimensions are usually made between ⅛ and 3/16 inch (2.5 and 5 mm) in height. The exact height depends upon the size of the drawing and where the dimension is placed.

ANSI standards recommend that all dimensions up to and including 72 inches (6 feet) be given in inches. Any larger dimensions should be given in feet and inches. When both feet and inches are used, it is necessary to include the foot (′) and inch (″) symbols. When all dimensions are given in inches, the inch symbol is not used but a note is placed on the drawing.

Fig. 7-10. Leader lines are used to indicate a geometric feature, such as a hole.

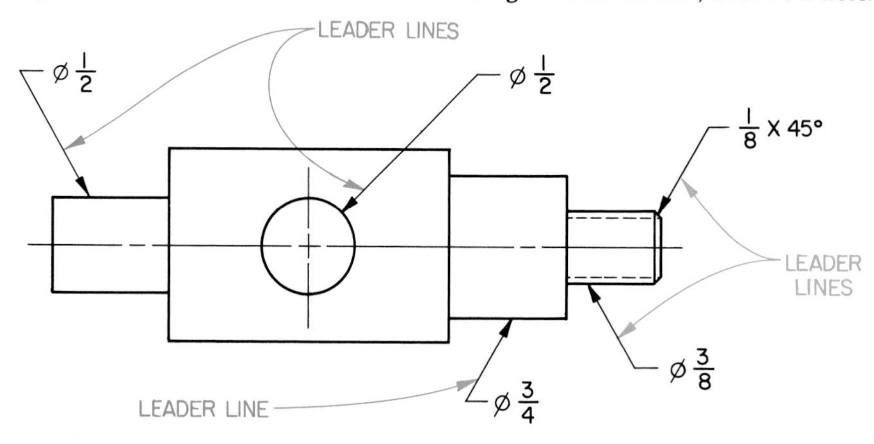

Fig. 7-11. ANSI and ISO standards differ in placement of dimensions.

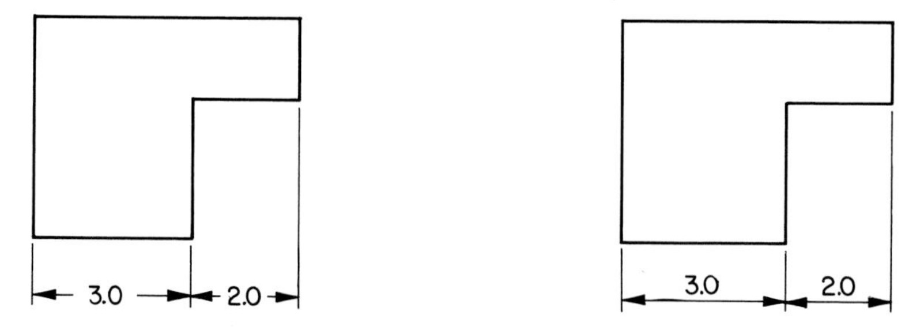

Numbers may be expressed as either fractions or decimals, such as ⅛ and ¼ or .185 and .25. Fractions or decimals are used in three ways. Common fraction dimensioning is used in many industries, such as manufacturing, architecture, and construction. All dimensional values are written as common fractions—4¾, ½, 3⁵⁄₆₄, and ³⁄₁₆. The fractions are normally no smaller than in sixty-fourths of an inch (¹⁄₆₄, ³⁄₆₄, 3⁵⁄₆₄, etc.).

Decimal fraction dimensioning uses only decimal values. Under ANSI standards, only two-place decimals (.18 for ⅛, .25 for ¼, and .50 for ½) are required for ordinary items. When greater accuracy is needed, any number of decimal places (.041, .0250, and .550) may be used. Machine tool processing and engineering are examples of fields that require greater accuracy in drawings.

Combined common fraction and decimal fraction dimensioning use both fractions and decimals in the same drawings. This method is often found in machine drawings.

Drawings made for the construction, electronics, and structural industries differ somewhat in use of fractions and decimals. However, the method is usually standardized within a given industry. Drafters and engineers must use the accepted method so that workers in the industry can use the drawing without difficulty. The choice of method depends upon factors such as manufacturing processes, the type of product, consumer requirements, and type of servicing and maintenance needed.

Letters and Symbols

For dimensioning, letters and symbols are used to identify special shapes and conditions. Fig. 7-12. In the past, these conditions were described in notes. Today, letters and symbols are used because they make drawings neater and easier to read. For example:

- For diameters the symbol Ø precedes the size, such as Ø2.25.
- For a radius, the letter R precedes the size, such as R1.185.
- For a square, the symbol □ is written before the size, such as □2. This means *each* side measures 2.
- When dimensioning a sphere, the word "sphere" or the letter S is placed before the diameter symbol and the size, such as SØ4.0. If SR is written before the dimension, it refers to the sphere's radius.
- A current technique is to use X as a multiplication symbol. For example, 4XØ.50 refers to 4 holes or circles, each with a diameter of .50.

Notes

Prior to 1982, all machining and dimension notes were in word form. For example, if a ½-inch hole was to be drilled, the following notation was used: ½ DRILL. Today, this type of note has given way to a series of symbols. Several

Fig. 7-12. These are some of the symbols used to indicate shapes and conditions.

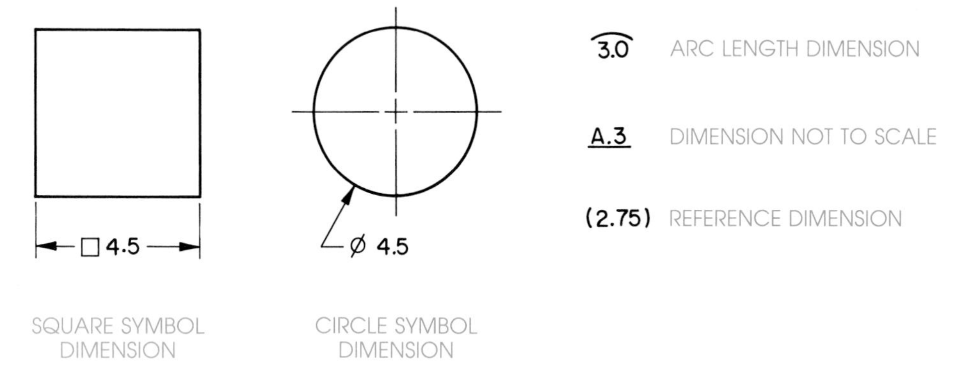

common machining and dimension symbols are given in Fig. 7-13.

A need for general work notes using words still exists, however. When used, they simplify or improve the dimensioning process. There are two basic types of word notes. General notes are used to describe something about the entire drawing. They are placed where they are easy to see, usually in the central part of the drawing. Examples include:

FINISH ALL OVER
ALL RADII 1/8 UNLESS OTHERWISE NOTED
ALL BEND RADII .075

Local notes are used to describe local features. They are usually connected to the feature by a leader line. Examples are:

WELD C.R.S. STRIP TO BOLT
10-28 TAP THRU
1/4 X 60° CHAMFER

Precision

All products require some degree of precision. The precision used in making a product determines its quality. But what is meant by precision varies with the product. For instance, a building needs to be precisely measured only within one-fourth inch. Some laboratory instruments, however, must be precise within 0.000001 (1/1,000,000 or one-millionth) of an inch.

Tolerances. **Tolerance** is the amount a measure can vary without harming the product. That is, tolerances show how much larger or smaller a part can be from a given dimension.

On drawings, tolerance is figured in one of two ways. In unilateral tolerancing, variation is allowed in only one direction. That is, the finished product can be made either larger *or* smaller than the dimension. Fig. 7-14A.

In bilateral tolerancing, variation is allowed in two directions. A part can be *both* larger or smaller than the dimension. Fig. 7-14B.

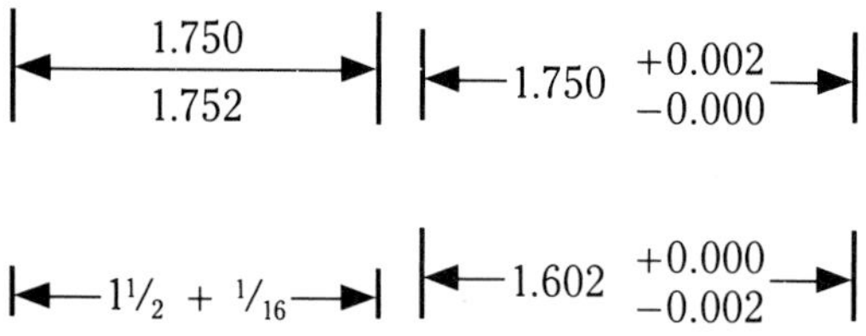

$\frac{1.750}{1.752}$ $\quad$ $1.750 \begin{smallmatrix}+0.002 \\ -0.000\end{smallmatrix}$

$1\tfrac{1}{2} + \tfrac{1}{16}$ $\quad$ $1.602 \begin{smallmatrix}+0.000 \\ -0.002\end{smallmatrix}$

Fig. 7-14A. Unilateral tolerancing allows a variation in only one direction. Although the tolerance in the example on the right gives both plus and minus figures, one is a zero.

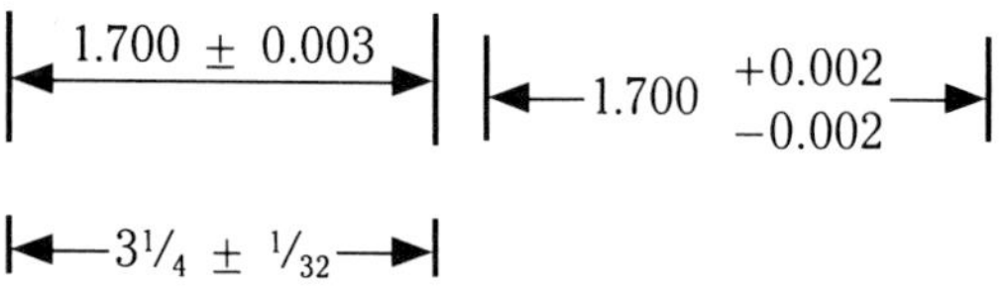

1.700 ± 0.003 $\quad$ $1.700 \begin{smallmatrix}+0.002 \\ -0.002\end{smallmatrix}$

$3\tfrac{1}{4} \pm \tfrac{1}{32}$

Fig. 7-14B. Bilateral tolerancing means that a measure may vary in two directions.

Fig. 7-13. Here are some of the symbols used for machining processes.

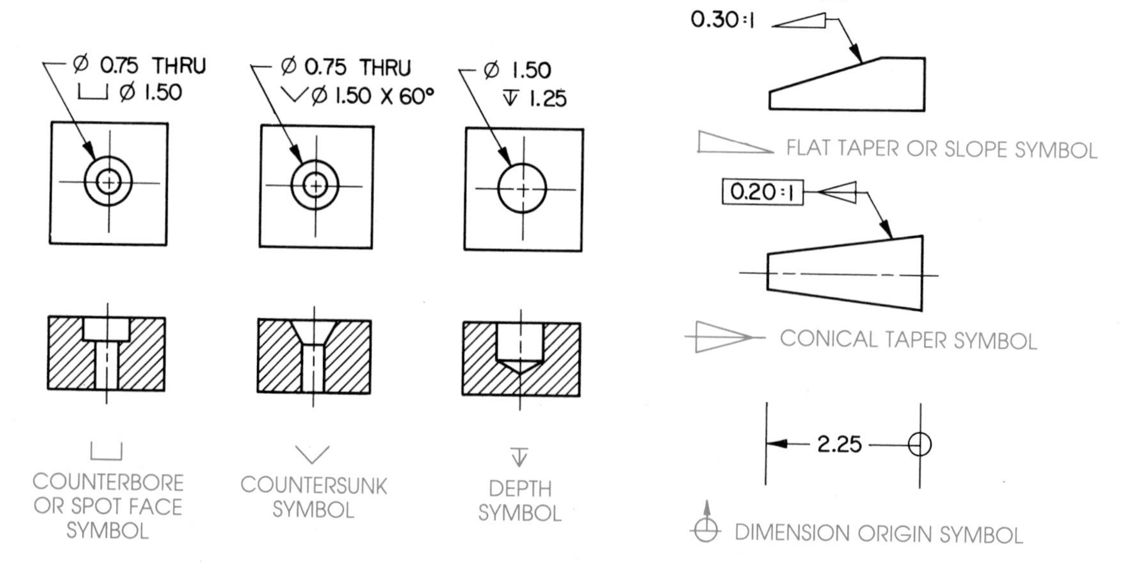

Tolerancing may also be shown in a note. Such notes make reference to specific dimensions. For example:

ALLOWABLE VARIATION ON ALL DECIMAL DIMENSIONS IS ±0.005 UNLESS OTHERWISE SPECIFIED

Limits and Fits. Similar to tolerances, limits and fits tell how parts will fit together for proper assembly. Clearance fits are needed where freedom of motion is required, such as for a shaft rotating in a hole. Interference fits are needed for parts that fit together without use of fasteners, welding, or adhesive. A heat-shrink fit between a rod and hole is an interference fit.

When the accuracy of dimensions is not critical, chain dimensioning is used. Fig. 7-15. Here, every dimension begins and ends with another dimension. There is no reference point or line.

When dimensioning must be more accurate, parallel, or rectangular coördinate, dimensioning is recommended. Fig. 7-16. All dimensions are taken from a common reference point or feature called a datum feature.

Chain and parallel dimensioning both have strong and weak points. It is common practice to combine the two. Fig. 7-17. This allows the drafter more flexibility.

Dimensioning Rules

A number of rules are followed when making a drawing. These rules tell where and how dimensions should be placed. The rules listed here are used for most industrial drawings.

- The smallest dimensions are located closest to the object.
- All other dimensions then follow for each successively larger feature. The last dimension shown represents the largest measurement. In this case, the dimension giving the overall size is the farthest from the object.
- An object or feature is dimensioned in a view that shows it most clearly.
- In most cases, at least two dimensions are needed to describe a feature accurately.
- Dimensions are not placed on the object unless no other place is available.

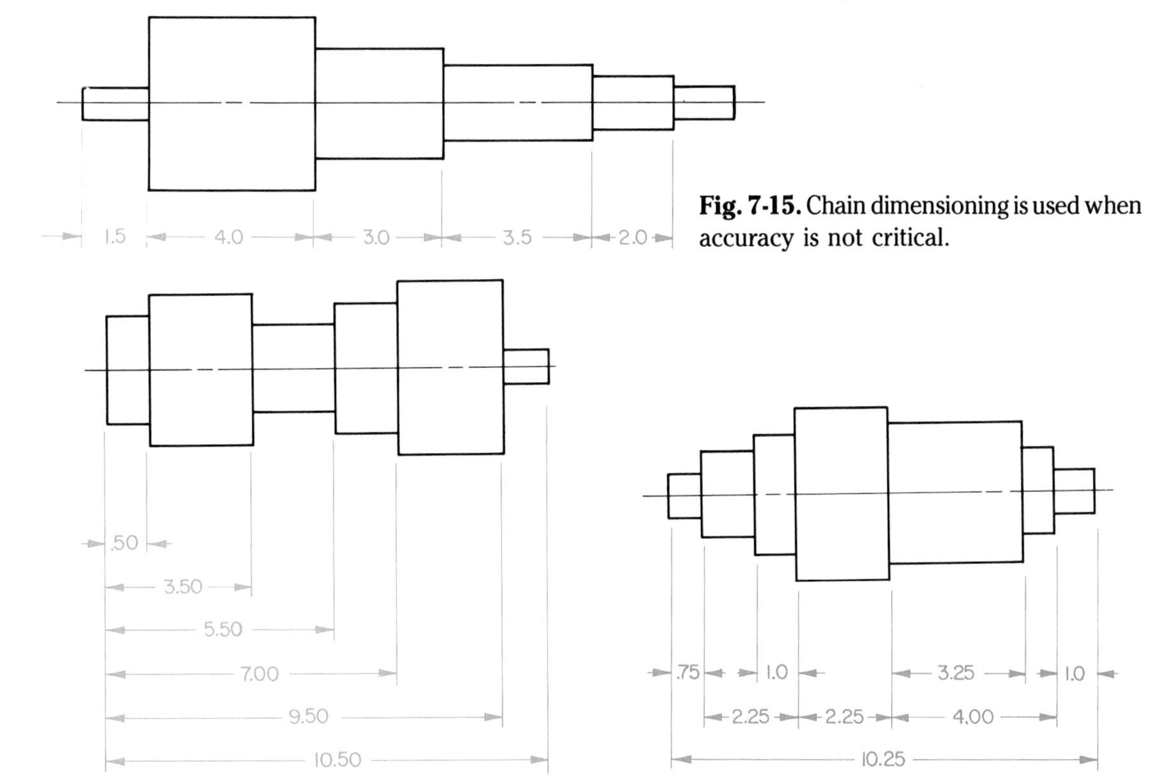

Fig. 7-15. Chain dimensioning is used when accuracy is not critical.

Fig. 7-16. Parallel, or rectangular coordinate, dimensioning is used when accuracy is critical.

Fig. 7-17. Both chain and parallel dimensioning may be used on a drawing.

- Most dimensions are placed between views.
- The dimension used for a hole or circle is the diameter.
- For other circular features, such as rounds, arcs, and fillets, the radius is dimensioned.
- The location of all round features or parts should be made with reference to the point where the two center lines meet.
- When two or more parts are shown on a drawing, dimensions should be grouped so that there is no confusion about which dimensions refer to which part.

Drawing to Scale

Before drafters start a drawing, they decide whether the drawing should be made full size, larger, or smaller. The choice will depend upon the size of the object, the amount of detail that must be shown, and the space available on the drawing.

Small parts are usually drawn full size to show clearly any details. Very small objects may be shown larger than full size. Larger members or assemblies often require drawings smaller than full size.

When the drawing is made, the drafter must be sure the object or features are shown in the correct proportions. This is called **drawing to scale**. For example one inch on the drawing may equal one foot on the real object. Metric scales are given in the form of a ratio, such as 1:10.

The choice of scale is noted on all drawings and is usually located in the title block. When an individual detail is shown at a different scale, the scale is noted near it.

Fig. 7-18. Most drawings today are made on either paper or film.

EQUIPMENT AND SUPPLIES

How good a drawing looks depends not only upon the skills of the drafter, but also on the quality and type of equipment and supplies used. Accurate, clear drawings cannot be made with broken or poor quality instruments. Good drawings require precision instruments and measuring scales, along with high quality drawing materials.

Drawing Media

Surfaces on which drawings are made include paper, film, and cloth. Fig. 7-18. Leads, lead substitutes, and inks are used to draw with. There is a variety of media available, and the selection made greatly affects the way a drawing looks.

Surfaces

Drafting paper, film, and cloth come in standard-sized sheets and roll widths. ANSI standardized sheet sizes are used throughout most industries:

- Size A — 8.5 × 11 inches
- Size B — 11 × 17 inches
- Size C — 17 × 22 inches
- Size D — 22 × 34 inches
- Size E — 34 × 44 inches

Some are preprinted with drawing formats, company names, and logos.

Most rolls come in standard lengths of 20 or 50 yards. Each is available in standard widths of 24, 30, 34, 36, 42, and 54 inches. Rolled material is not preprinted.

Perhaps the most familiar surface used in most industries is drawing paper. Drafting and engineering departments usually have two basic types. The first is a thick, opaque paper used for maps, charts, and drawings that will be photographed. In some cases, a cream or buff version is used for layouts and detailed drawings.

The second type of paper is by far the most popular. It is recommended for general technical drawings. This paper is known as **vellum**. In many cases drafters refer to vellum and tracing paper as the same thing, although there is a slight difference. Tracing paper is sometimes more translucent than vellum.

The newest drawing material used by drafters is drafting film. It is an excellent drawing material because it doesn't shrink or warp. It is also resistant to age and heat, and it is waterproof. The drawback to film is its higher cost. As a result, it is used for drawings where high quality and stability are important.

Ancient drawings and writings were produced on a cloth-base paper. Due to the excellent qualities of today's papers and films, there is little need for cloth. However, cloth is sometimes required for special jobs.

SCIENCE FACTS

Graphite

What does a lead pencil have in common with a diamond ring? Both contain a form of carbon. The "lead" in a pencil is actually a mixture of clay and graphite. Both diamonds and graphite are carbon, but their atoms are arranged differently. Therefore they have different properties.

In graphite, the carbon atoms are arranged in thin, flat crystals. The crystals slide easily over one another. That's why graphite is soft and feels greasy. Diamonds are made of crystals, too, but the crystals do not slide. In fact, a diamond is the hardest substance found in nature.

Leads and Inks

Most manual sketches and technical drawings are made with lead. For a long time drafters relied on pencils. Then came the mechanical lead holder and most recently the **stick lead holder**. Fig. 7-19. Today, the stick lead holder is used by most drafters because it needs no sharpening and keeps a constant line width.

The hardness of lead used in drawings is important. All leads are made of a combination of clay and graphite and sold in 14 different hardness ratings.

- 6B and 5B are very soft leads. They are too soft for use in technical drawing. They are used primarily for artwork.
- 4B, 3B, 2B, B, and HB are moderately soft. They can be used for drawing certain types of illustrations, renderings, and sketches.
- F, H, 2H, 3H, and 4H are hard leads used for industrial drafting work. F and H are often used for sketching. 2H through 4H are used for technical drawings.
- 5H through 9H are very hard leads popular for layout and light construction work. They are also used when the lines made should not reproduce.

Fig. 7-19. Shown here, from top to bottom, are a mechanical lead holder, a drawing pencil, and a stick lead holder.

Lead substitutes are made of a plastic that comes in one grade of hardness. They are designed for surfaces where lead wear and smearing is a problem, such as polyester drawing films.

Until the late 1970s, only such things as patterns and illustrations were drawn with ink. But today's synthetic inks are a big improvement over the old slow-drying India inks. They are of high quality, fast drying, and in some cases, easily erased. For these reasons, many companies are now requiring that all drawings be made in ink.

Ink drawings used to be done with ruling pens. In modern drafting rooms, ruling pens have been all but replaced by the **technical drawing pen**. Fig. 7-20. These pens come in a variety of standard inch and metric widths. Compasses and lettering sets are also available with both inking nibs and technical drawing pen attachments.

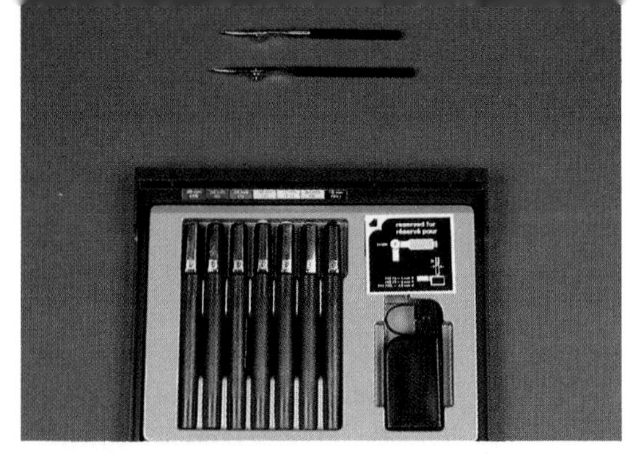

Fig. 7-20. Ruling pens (top two) and technical drawing pens (bottom) come in a variety of widths.

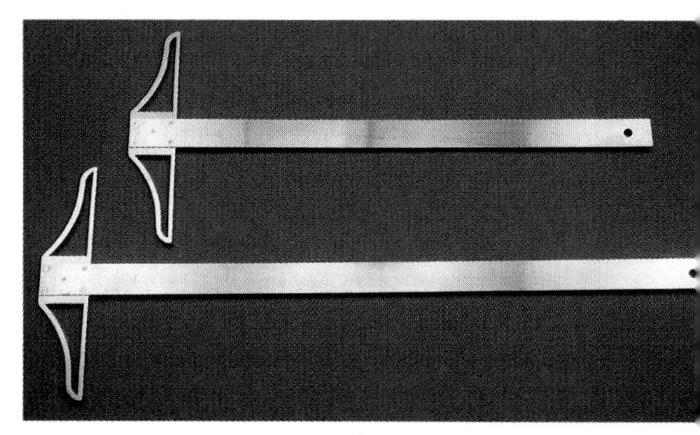

Fig. 7-21. The traditional T-square is seldom used in industry today.

Drafting Tools

To produce good technical drawings, the correct tools are needed. Today, a variety of these products can be purchased, including those for special architectural, mapping, topographic, electrical, vacuum, and piping drawings.

The T-square has always been associated with drafting. Fig. 7-21. Though it is still used for instruction, it is seldom found in industry. Today, most drafters use parallel straightedges and/or drafting machines.

Parallel straightedges are excellent for drawing parallel horizontal lines. Fig. 7-22. They come in sizes ranging from 30 to 96 inches. Parallel straightedges also serve as a reference surface upon which triangles sit for drawing vertical lines and angles. Parallel straightedges, sometimes referred to simply as "parallels," are found in many architectural, construction, and engineering firms where large layouts are prepared.

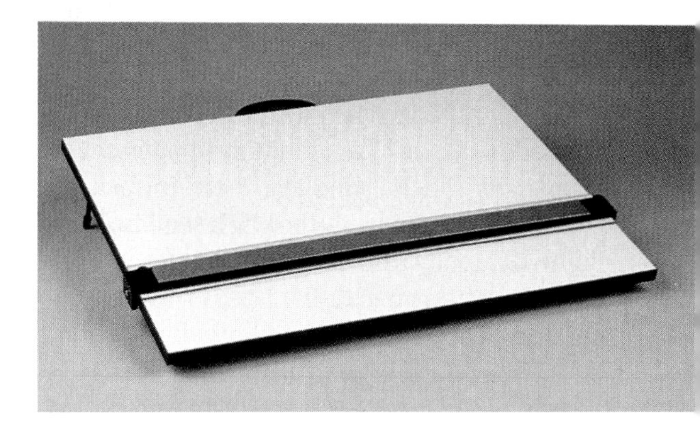

Fig. 7-22. Parallel straightedges are used to draw horizontal lines. When combined with triangles (Fig. 7-24), they can also be used to draw vertical lines.

Another important drafting tool is the **drafting machine**, which is used to draw horizontal, vertical, and angled lines. Fig. 7-23. The drafting machine comes with two perpendicular blades attached to a rotating head. The head is equipped with a precision protractor so the blades can be set at any angle. Because of this unique feature, a drafter can work without triangles.

HEALTH & SAFETY

KEEPING YOUR WORK ENVIRONMENT SAFE

You may think there are no health or safety hazards in a drafting room, but there are a few things of which you need to be aware.

- There should be enough space between drafting stations to allow people to move freely. Don't let chairs, boxes, or equipment block the aisles. Don't let drafting machines or T-squares extend into the aisles.
- Be careful when handling dividers, compasses, and other objects with sharp points.
- Store instruments in a drawer when not in use. If you put them on the slanted surface of the drafting table, they may slide off.
- Chemicals, such as those used in some print reproduction processes, should be kept away from the eyes and mouth. Make sure there is adequate ventilation in any area where such chemicals are used.
- When using a paper cutter, keep your fingers away from the path of the blade. When you've finished, lower the blade and lock it in place.
- If you're using a CAD system, see the health and safety tips regarding computer use in Chapter 5.

Fig. 7-23. Drafting machines such as these help drafters draw horizontal, vertical, and inclined lines.

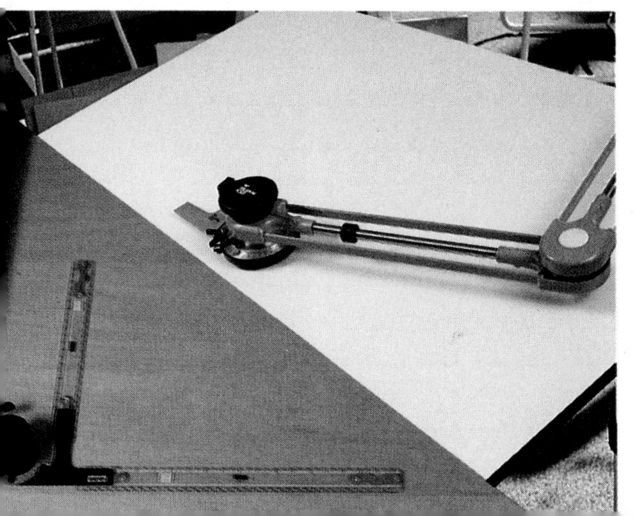

There are two types of drafting machines, standard and track-type. Standard machines are used for drawings prepared on E-size or smaller paper. The track-type drafting machine can be used for drawings larger than E-size. Both machines are widely used in the machine tool and manufacturing industries.

Another group of tools includes triangles, templates, and curves. As already mentioned, triangles are used to draw vertical and angled lines. There are three basic types: 45°, 30-60°, and adjustable. Fig. 7-24. The adjustable triangle comes with a protractor and a movable side. The drafter can adjust it to draw any angle to the nearest degree.

Templates are used to draw standard shapes and figures. Fig. 7-25. Examples of these are circle, ellipse, and square templates. Specialized

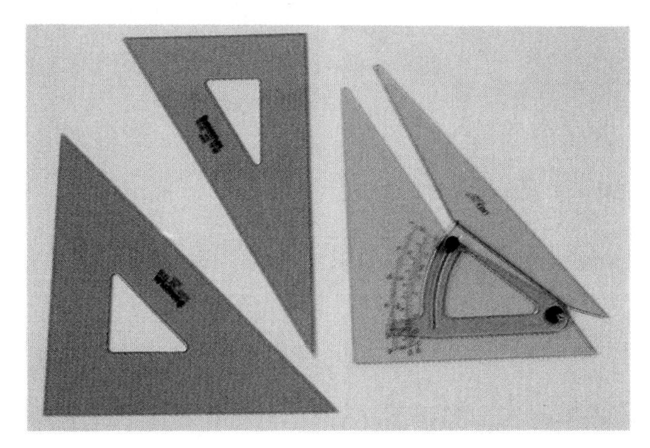

Fig. 7-24. Shown left to right are a 45°, a 30-60°, and an adjustable triangle.

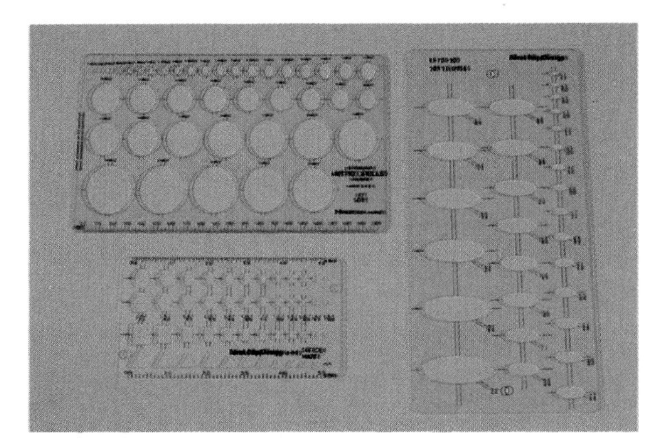

Fig. 7-25. Templates save time when drawing common shapes.

templates are made for architectural, electrical, plumbing, machine, and landscape drawings.

Curves are used by drafters to draw curved lines. Fig. 7-26. Curves come in three broad categories: French, radius, and flexible. French curves are also known as irregular curves and are used for general purpose work. Radius curves are used to draw true curves where there are parallel curved surfaces. Flexible or adjustable curves can be shaped to create any curved pattern.

One of the most important tools used in technical drawings is the scale. Fig. 7-27. **Scales** provide the means by which proportional measurements are made. There are four basic types.

Mechanical engineer's scales are used for drawings ranging from full size down to one-eighth size. Inches and fractions of an inch are divided to stand for inches.

Architect's scales are designed for work in foot measurements ranging from 3/32″ = 1′-0″ (1/128 size) to 1″ = 1″ (full size).

Civil engineer's scales divide the inch into decimals. One unit may then stand for a number of things, such as feet, rods, miles, time, and so on.

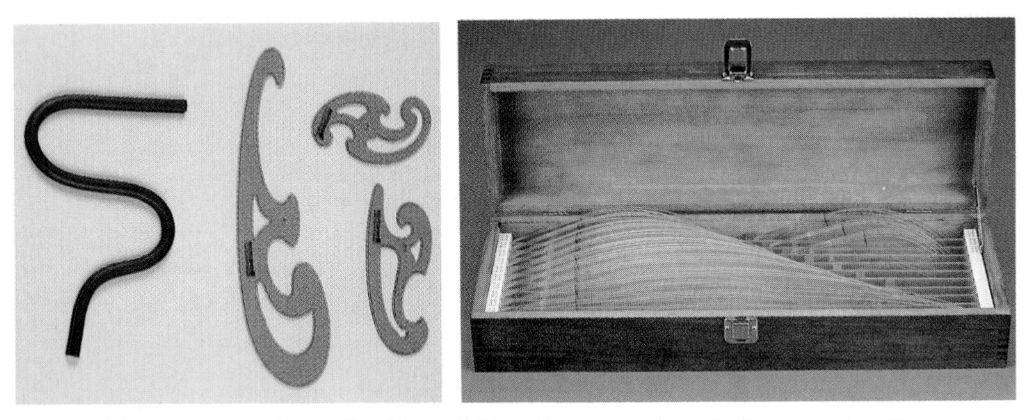

Fig. 7-26. At left are shown flexible and French curves. At right is a set of radius curves.

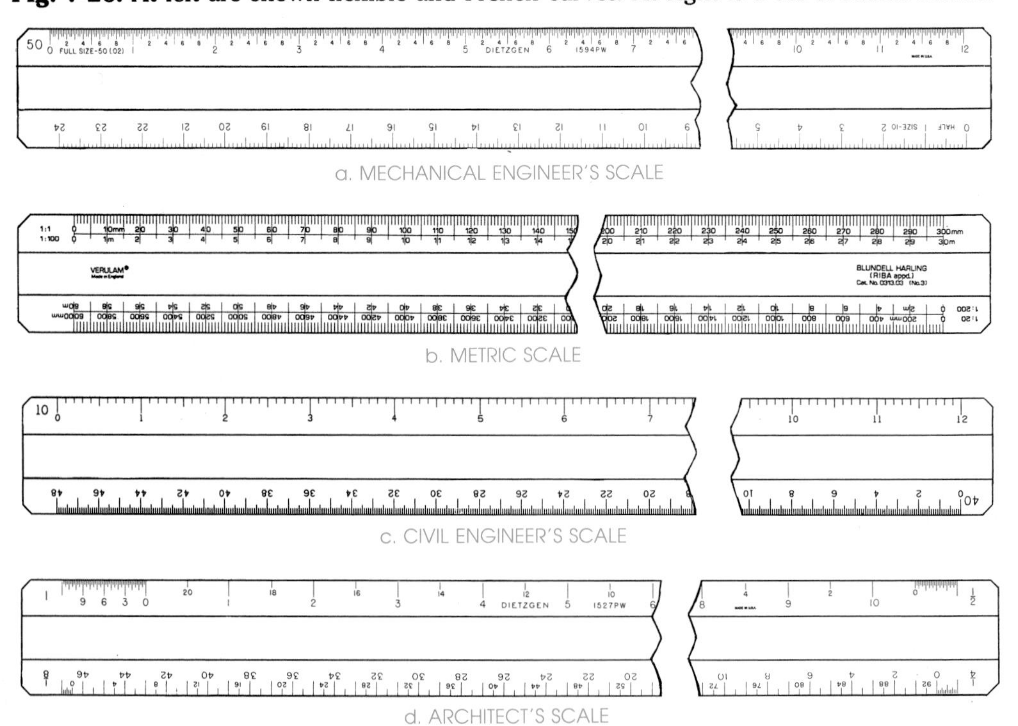

Fig. 7-27. Scales help drafters draw objects in the correct proportions.

SCIENCE FACTS

Parallax

Try this experiment. Hold a pencil out in front of you. Look at it first with one eye and then with the other. Does the pencil seem to move in relation to objects behind it? This effect is called parallax, and it's caused by the fact that one eye sees things from a slightly different angle than the other eye.

At close range, the parallax effect is not as obvious. However, when you are measuring with a scale, try to position yourself directly above the endpoint of the line. If your eyes are too far to the left or right, you may not measure accurately.

Metric scales are used for drawings made with metric measures. With these scales, drawings can be made from a ratio of 1:1 (full size) to 1:150.

Two instruments are basic to drafting—dividers and compasses. Fig. 7-28. Dividers are used to transfer measures from one location to another, such as from a layout drawing to a detail drawing. They are also used to check measurements made on the drawing against the final product. Compasses are used to draw circles and arcs. Most drafters have dividers and compasses of different sizes.

Fig. 7-28. The dividers (left) and compass (right) are basic tools for drafting. A compass holds lead for drawing; dividers do not.

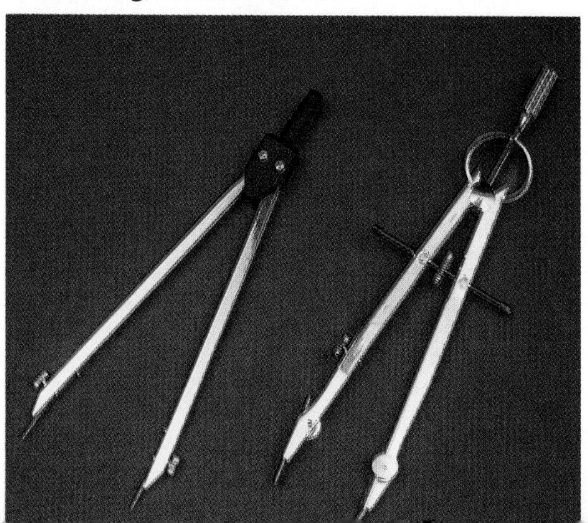

DRAWING REPRODUCTION

When most people think about how drawings are reproduced, the first thing that comes to mind is the blueprint. The blueprint with its white lines and deep blue background is easy to identify. However, in today's industries, the blueprint is rarely used.

Over the last 30 years, there have been many improvements in reproduction, and many techniques exist. Several things determine which technique is chosen. They include size and color of original, how many copies are needed, size and color of the reproduction, cost, and whether the reproduction must be workable or not. "Workable" reproductions are those that can be drawn on. Three common techniques are diazo printing, xerography, and photography.

Diazo prints have dark lines on a white background. They are sometimes called whiteprints. In the diazo process, a translucent master drawing is placed over a treated print paper and exposed to a strong ultraviolet light. The light does not pass through the lines of the master drawing. However, where there are no lines, the light passes through, and the coating on the print paper is burned off. The print paper is then developed, either by ammonia vapors (dry diazo) or liquid chemicals (moist diazo).

Xerography is a form of photocopying using an electrostatic charge and heat processing. (See also Chapter 15.) A recent use of xerography is to merge it with a computer system. A laser printer makes a high quality copy similar to an inked line drawing.

Photographic systems are important when precise duplication is necessary. The original is photographed and prints are made on matte paper. The prints may be drawn on with pen or pencil.

STORAGE

A company's policies and legal requirements often make it necessary to keep drawings in storage for periods of time. In many businesses, it is not unusual to find drawings made 80 to 100 years ago. The exact type of storage system depends upon company needs.

Short term storage systems are used when drawings have to be changed or referred to frequently. Drawer units, rack storage, and roll files are all forms of short term storage. Fig. 7-29.

Long term storage systems are used for drawings that are not often needed. One of the most common is the storage vault or room, where drawings are stored in tubes or on shelves. A more reliable and less costly form of long term storage is to photograph the drawings at a greatly reduced scale and store only the film. This process, called microfilming, makes it possible to store many drawings in a small space. A microfiche is a sheet of film with several rows of small drawings on it. Fig. 7-30.

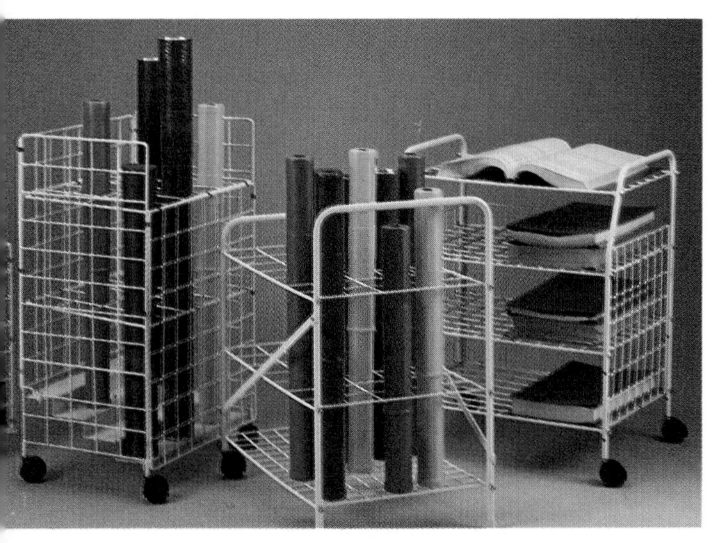

Fig. 7-29. Roll files (above) and racks (below) are used for short-term storage of drawings. Why do you think these would not be good for long-term storage?

Fig. 7-30. Shown above is a microfilm image recorder. Below is a reader/printer that can be used with either microfiche or microfilm.

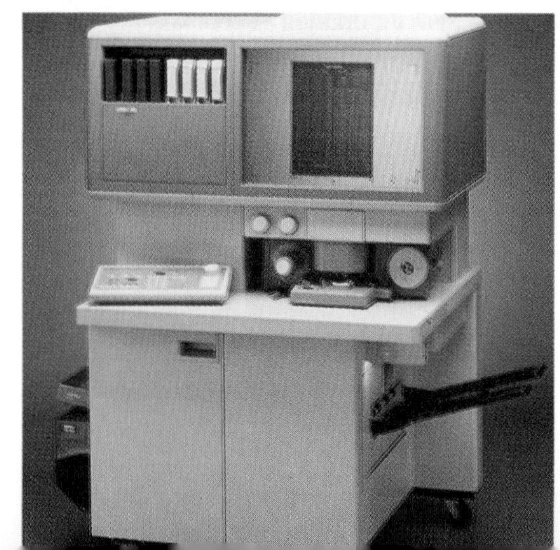

CHAPTER 7

REVIEW

Review Questions

1. Who sets drafting standards for the U.S.?
2. Briefly describe the following lines: visible, hidden, center, dimension, and cutting plane.
3. Give three examples of special symbols used in dimensioning and tell what they mean.
4. Explain the difference between aligned and unidirectional dimensioning.
5. What is the purpose of notes on a drawing?
6. What are dividers and a compass used for?
7. What is the type of paper most often used for making technical drawings?
8. What is the difference between a mechanical engineer's scale and an architect's scale?
9. Briefly describe how diazo prints are made.
10. What kinds of long term storage are used for drawings?

Activities

1. Obtain three different grades of leads and compare them by using them to make a simple drawing.
2. Research the manufacturing of drawing papers and film. Write a report on your findings.
3. Identify the different types of lines labeled A-H in Fig. 7-31.
4. Letter the following using ⅛ and ¼ inch vertical letters:
   ⅝ DRILL ⅜-16UNC
   1930 N/mm^2
   20% Ni-Ti-A1
   Weld 75 Ni-15 Cr-9 Fe+Mo+Ag
   COUNTERSINK ALL TAPPED HOLES
5. Find out the cost of having four drawings done on C-size paper reproduced by blueprint, ammonia-sensitized, and photographic techniques.

Fig. 7-31.

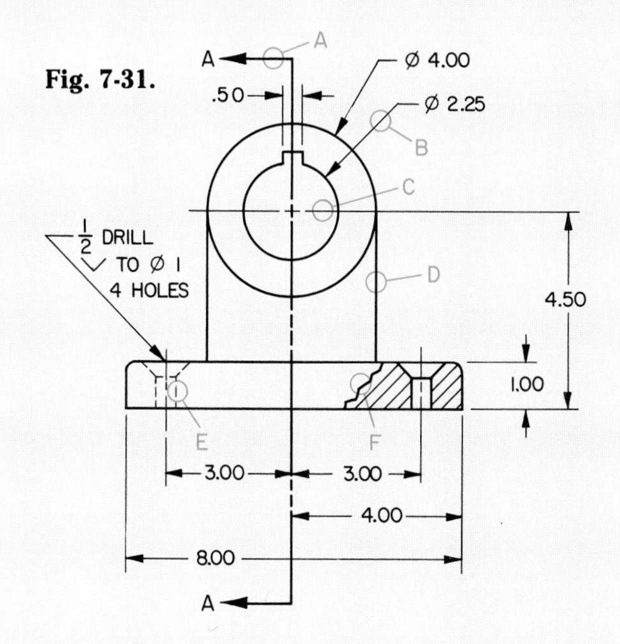

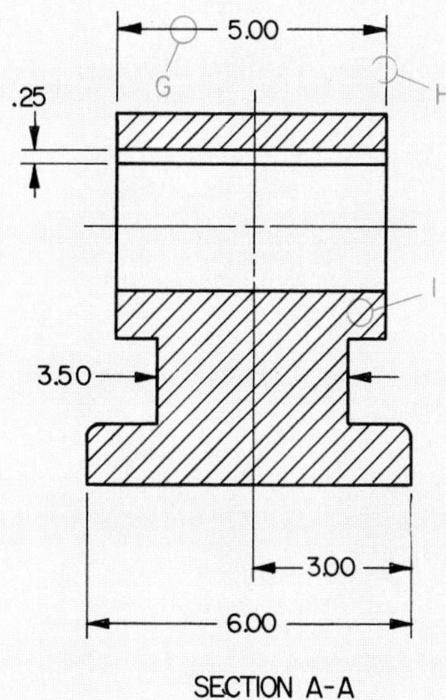

Technical Design Processes

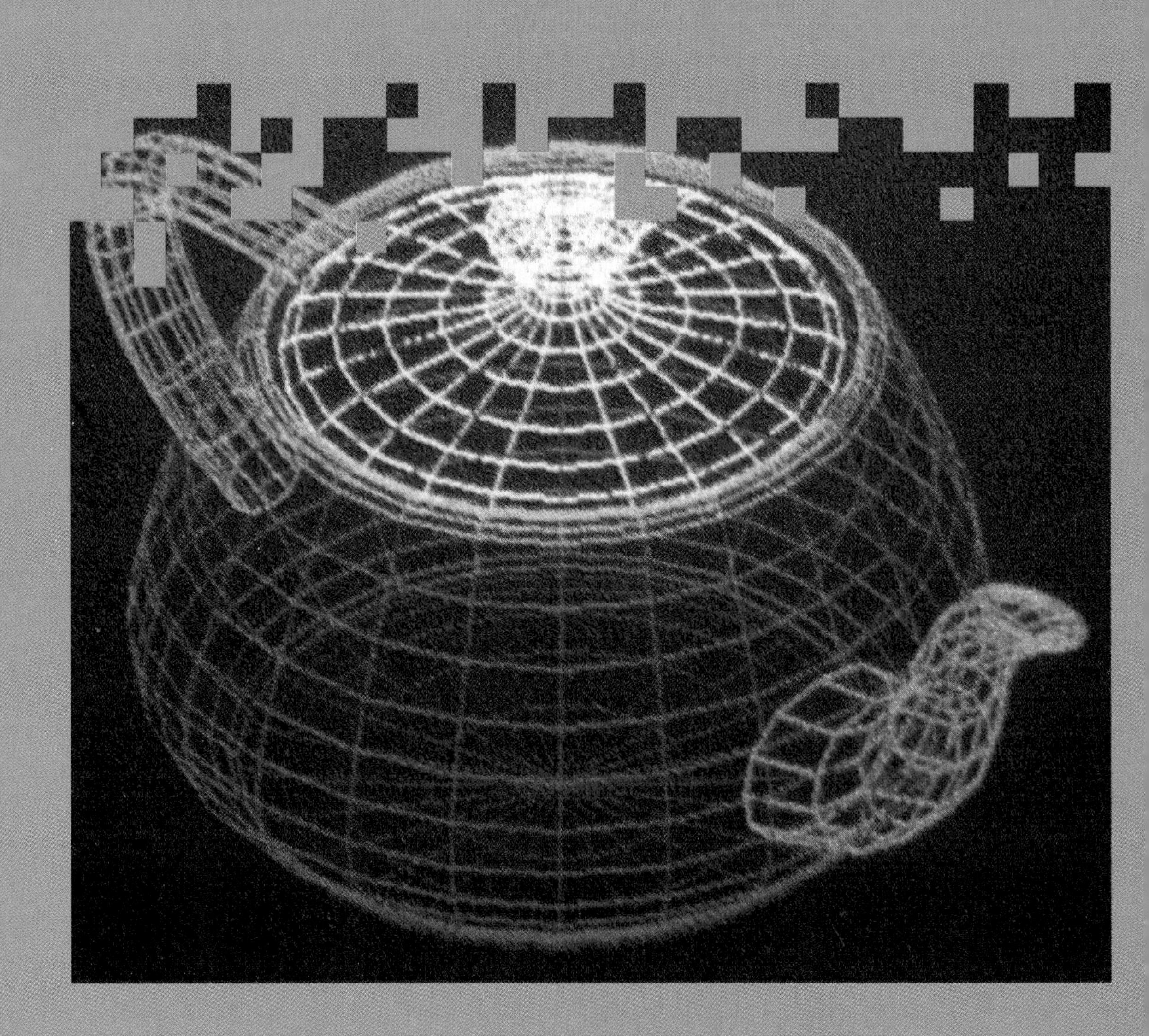

The field of technical design began as a form of mathematics. Problems in arithmetic, algebra, geometry, trigonometry, and calculus were solved by early mathematicians using drawings. In fact, the Greeks developed their theories of geometry by means of drawings.

Over the years, the same drawing methods were applied to other fields. Architects, scientists, mechanics, and artisans (skilled workers) all understood the advantages of drawings. Leonärdo da Vinci, the great fifteenth-century artist, architect, and engineer, used technical drawings to develop many of the projects that made him famous.

In this chapter, you will discover the basic principles used by technical designers and how drafting methods are used in different fields.

Terms to Learn

auxiliary views
conventions
detail drawings
diagrams
floor plans
isometric drawings
multiview (working) drawings
oblique drawings
orthographic projection
perspective drawings
pictorial drawings
planes of projection
projection lines
sectional view
topographic mapping

As you read and study this chapter, you will find answers to questions such as:

- What are the different kinds of drawings and how are they used?
- What is orthographic projection and why is it important?
- How can the inside of an object be shown in a drawing?
- How are drafting skills important to different fields such as architecture and electronics?

DRAWING PRINCIPLES AND TECHNIQUES

A picture can often explain an idea better than a lot of words. In fact, some written languages got their start as pictures. Chinese is an example. People in fields such as architecture, electronics, mapping and topography, and structural engineering often make use of sketches and drawings to communicate their ideas. Because these ideas are often complex, several types of drawings may be used. Each type conveys certain ideas better than others. The four basic types of drawings include: sketches, illustrations, multiview drawings, and pictorial drawings.

Sketches

Sketches are quick, freehand drawings that remain fairly rough and unfinished. Fig. 8-1. Most sketches start out without dimensions.

Sketching is useful not only to drafters and engineers but to anyone wishing to express an idea graphically. Many technical drawings have their start as sketches. Many times, mechanics and other workers make sketches to help clarify some part or procedure.

Sketches are usually done with plain or graph paper, a medium-soft pencil, an eraser, and a small scale. A skillful drafter should be able to sketch four types of lines and shapes. These are horizontal and vertical lines, oblique (angled) lines, circles and arcs, and ellipses.

Sketching Straight Lines

Though they need not look like lines drawn with a ruler, sketched lines should be fairly smooth and straight. For horizontal lines, the elbow is kept near the drafter's side. Fig. 8-2. Movement should be made with the wrist. The line is sketched with one steady stroke of the pencil.

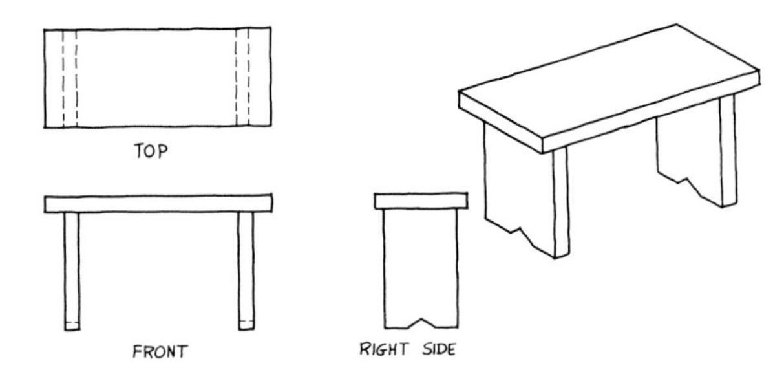

Fig. 8-1. Sketches are made quickly and remain rough.

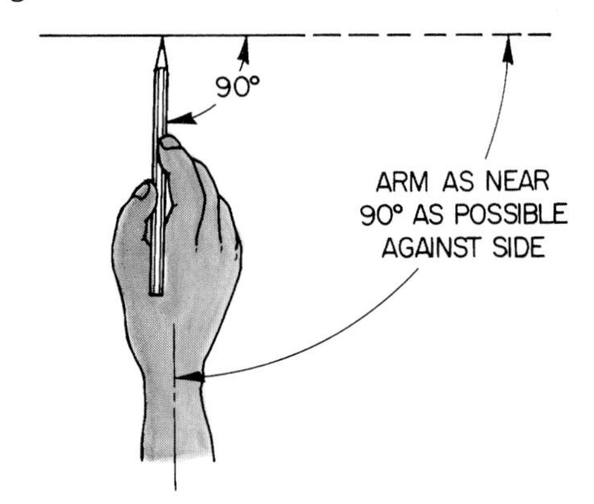

Fig. 8-2. This drawing shows the proper hand and pencil positions for sketching horizontal lines.

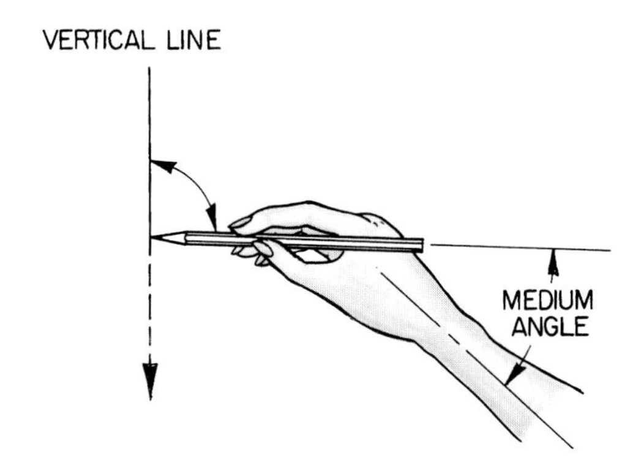

Fig. 8-3. This shows the proper hand and pencil positions for sketching vertical lines.

Vertical lines need to be sketched with more care. The hand holding the pencil is rested on the paper, and the elbow is held away from the side. Actual drawing is done with the fingers and thumb. Fig. 8-3.

For long lines, short, light connecting segments are first drawn. When the line is complete, it is darkened to the correct weight.

An oblique line is any line drawn at an angle. Fig. 8-4. For an oblique line, the elbow is held slightly away from the drafter's side. The line is drawn by moving the fingers and thumb. This technique can be used for lines angled to the left or right.

Sketching Curved Lines

The general position of the hand is the same for sketching curves as for straight lines. There are, however, some additional techniques used.

For sketching a curved shape, such as a circle, the axis method is recommended. Fig. 8-5. Equally-spaced axes are used for circles. On each axis the spot where the circular line should cross it is marked. Then the circle shape is sketched in by hand. Rectangular axes help create ellipses.

Another technique for sketching a circle is by inscribing it inside a square. Fig. 8-6. Once the square has been drawn, the circle is sketched inside so that the edges just touch the sides of the square.

A third method for sketching circular shapes is with preliminary movements. Fig. 8-7. The pencil is rotated several times above the drawing surface, then lowered for the sketch. Large shapes are drawn by rotating the entire arm. Smaller shapes are drawn with finger movement. Of all the techniques, this one requires the most drawing skill but is the quickest.

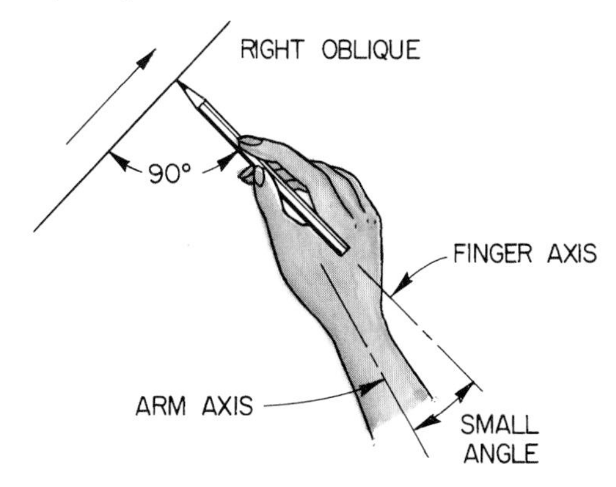

Fig. 8-4. This is the technique for drawing oblique lines. Some people find it easier to turn the paper and draw the lines horizontally.

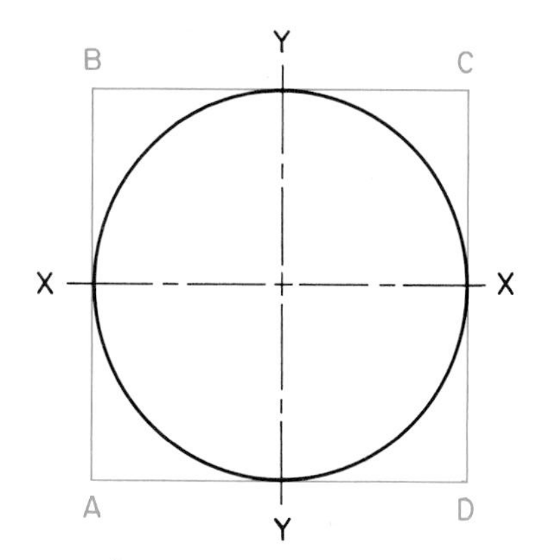

Fig. 8-6. A circle can be sketched inside a square. The length of each side of the square is equal to the diameter of the circle.

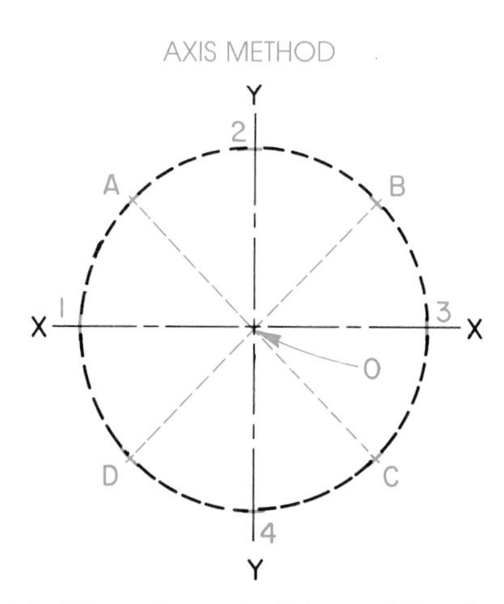

Fig. 8-5. The axis method is used for sketching circles and other curved shapes.

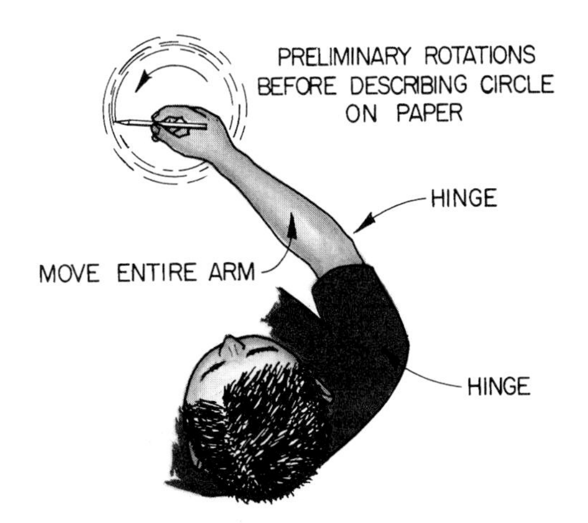

Fig. 8-7. This is another method for sketching circles.

Technical Illustrations

Technical illustrations resemble pictures more than they do typical technical drawings. Fig. 8-8. They are often used when companies are trying to sell an idea or product. They are designed for attractiveness rather than production needs.

The type of art media used depends upon the purpose of the drawing. The most common are pencil, pen and ink, colored pencil, watercolor paints, tempera paints, pastels, and charcoal. Sometimes photographic techniques are used. The skill of the drafter also affects the choice of medium. After several tries, drafters often find one medium has advantages over another.

Pencil and pen and ink are used for drawing sharp, fine lines to show small details and various textures. However, they are time consuming when large areas must be covered. But pencil or ink can be combined with a transparent watercolor wash. Thus the pencil and ink can be used for details while the watercolor fills in the large areas. Color is used to create emphasis and excitement. However, color requires skill in handling, especially on fine details.

Multiview Drawings

Multiview, or **working, drawings** are the type most often used in industry. They show an object from several different views, or angles. As a result, they describe the object completely. The more complex the object is, the more views are needed. Multiview drawings are used in manufacturing an object.

SCIENCE FACTS

Drafting

In the real world, objects have three dimensions: length, width, and height. Since drafting is done on a flat surface, a drawing can have only two dimensions: length and width. Some kinds of drawings create the illusion of three dimensions by changing an object's proportions. These are called pictorial drawings.

Drafting is both an art and a science. It requires a talent for drawing as well as an understanding of the principles of orthographic projection and perspective.

Orthographic Projection

The term "orthographic" comes from the Latin *ortho*, which means "at right angles to," and *graphic*, which means "drawn." The orthographic views of a multiview drawing are drawn at right angles, or perpendicular, to one another.

The method used to make a multiview drawing is called **orthographic projection**. You will understand orthographic projection if you imagine that the object to be drawn is inside

Fig. 8-8. Technical illustrations can show how the finished product will look and function.

Fig. 8-9. Imagine the object inside a clear box. Each side would give you a different view. If each view were drawn on the box's sides and the box were unfolded, the result would be a multiview drawing.

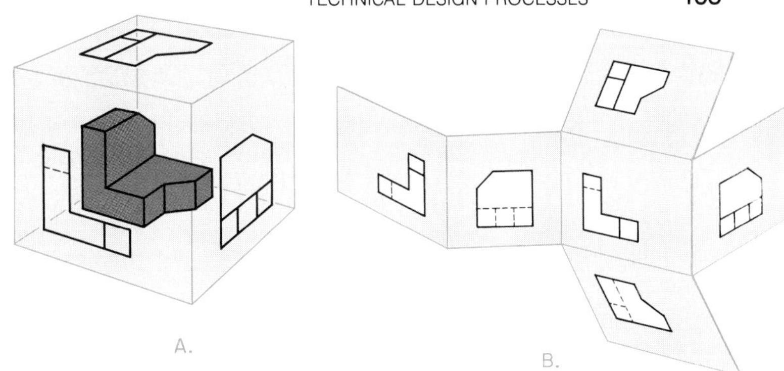

a transparent box. Fig. 8-9(A). If you looked through each side of the box, one side at a time, you would see six different views of the object. If you could "project" these views onto the box itself and then unfold it, you would have a multiview drawing. Fig. 8-9(B and C).

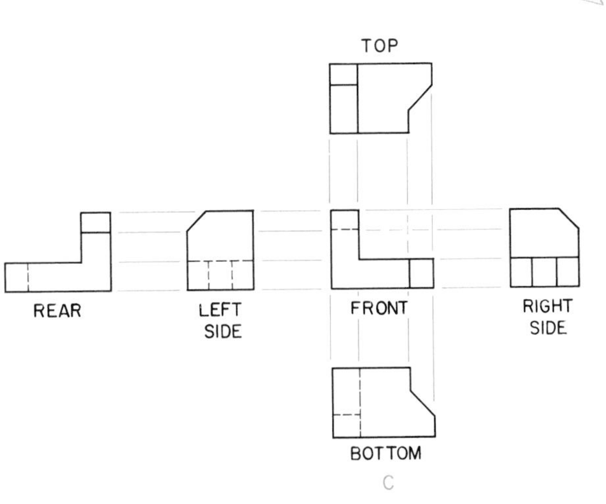

Planes of Projection

In all, there are six principal **planes of projection**. These correspond to the six sides of the box. Fig. 8-10. The views are: front, right side, rear, left side, top, and bottom. Most objects do not require all six views, however.

In a typical working drawing, Fig. 8-11, an object is presented in two or three views. The three major views, or *primary* planes of projec-

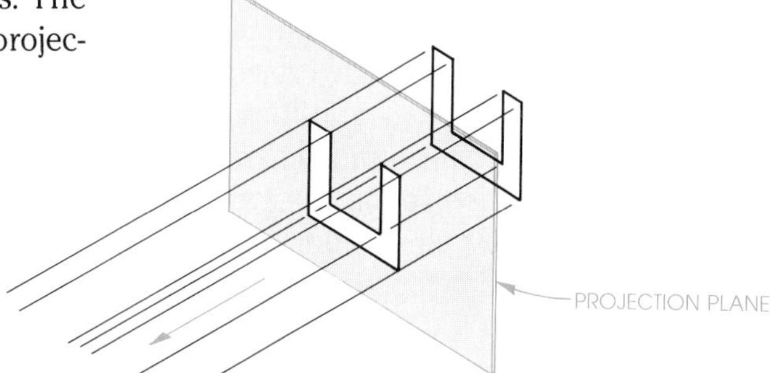

Fig. 8-10. Imagine the surface of an object projected outward at right angles to a clear, flat surface, such as the side of a glass box. This process is called orthographic projection, and the flat surface is the plane of projection.

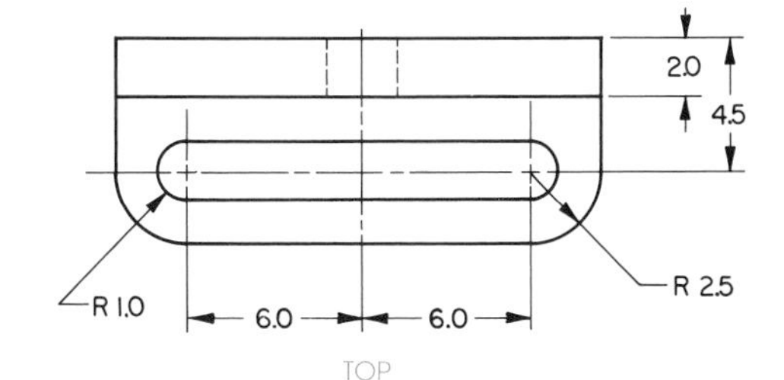

Fig. 8-11. This working drawing shows three views of the object. Some items — a can, for example — would need only two views.

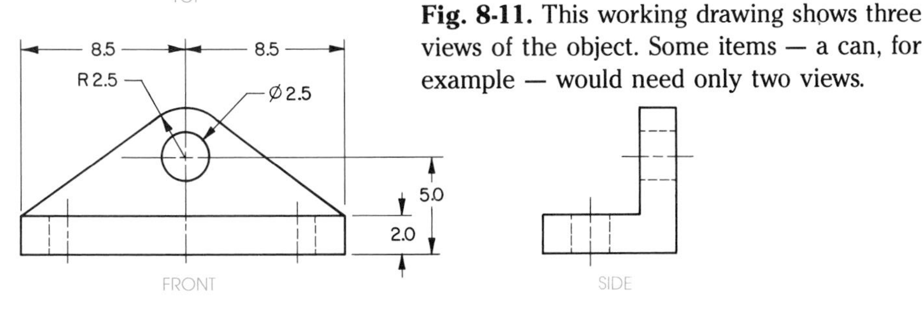

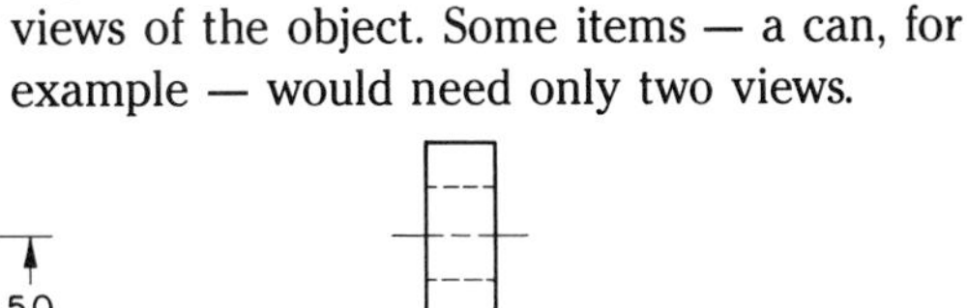

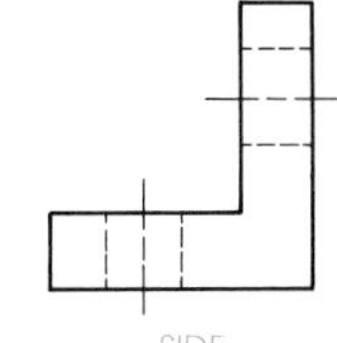

tion, found in most drawings are shown in Fig. 8-12.

- The front view shows an object from the front. This view gives length and height measurements.
- The top view is perpendicular to the front view and above it. It shows both length and width (depth) measurements.
- The right-side view is perpendicular to both the top and front views. It gives height and width dimensions.

These views usually provide all the information needed, including true size and shape. Sometimes only two views are needed, such as when two of the three views would appear the same, as for a cylinder. Rarely, rear or bottom views may be needed. The different views are always located in the same place on a drawing.

First and Third Angle Projections

Multiview drawings are called first or third angle projections, depending on how the views are placed on the drawing paper. In first angle projection, Fig. 8-13(a), the front view is on top, the left side view directly to the right, and the top view on the bottom. First angle projection is still used in some European countries.

Third angle projection, the arrangement we have been discussing, is used in the United States, Canada, and most other industrialized countries. Fig. 8-13(b). The top view is on top. The front view is below it, and the right side view is on the right.

Making a Multiview Drawing

Before drafters begin a multiview drawing, they select the views needed. Then the front view is chosen. The front view need not be the front of the object. It should be the side that gives the best view of the object and any details on it.

Next, size and scale must be determined. How many views must fit on the paper, and how much detail must be shown? The height, width, and depth dimensions must be figured to allow enough room. After all these decisions are made, the actual drawing is begun.

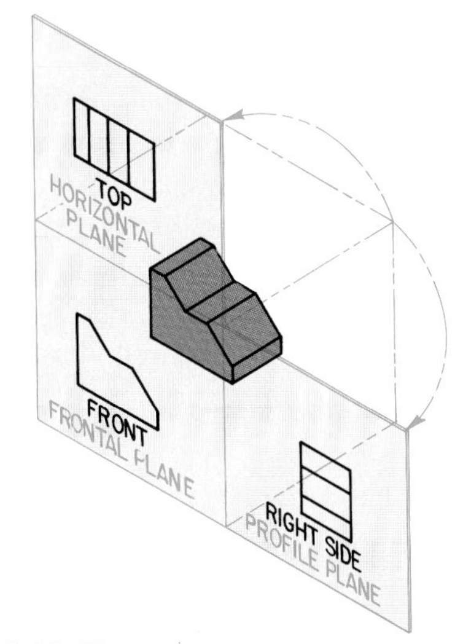

Fig. 8-12. The three primary planes of projection are horizontal, frontal, and profile.

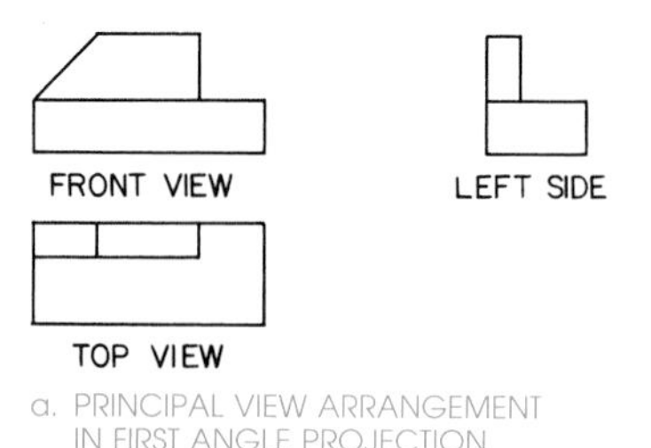

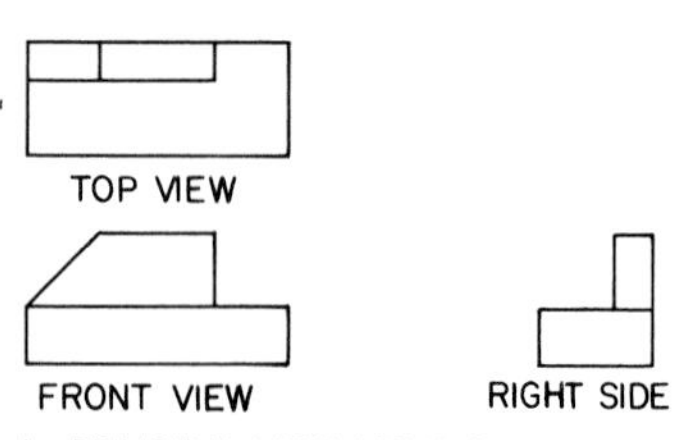

Fig. 8-13. Compare first angle and third angle projection. Which seems easier to you?

First, the object is sketched in with light construction lines. Then, using a parallel straightedge and triangles, projection lines are added. Fig. 8-14. **Projection lines** extend from all corners and edges on the front view to the other views. These lines save some remeasuring and help in aligning views.

By drawing a 45° line from the top corner of the front view, depth dimension for a side view can be obtained. Horizontal projection lines are drawn from the top view to the angled line. Vertical projection lines are drawn downward from where the lines intersect.

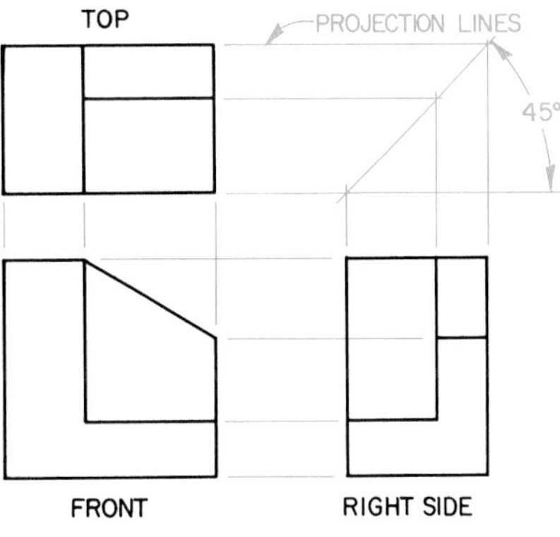

Fig. 8-14. The front view is usually drawn first. Next, projection lines are drawn outward from the front view. Finally, the other views are drawn.

Auxiliary Views

In most cases, one of the principal planes of projection will show any important edges or surfaces in their true size and shape. There are times, however, when this is not possible. When a feature does not appear in its true size and shape in one of the six principal planes, it is called an oblique feature.

To describe an oblique feature, drafters make use of **auxiliary views**. These are created on projection planes that are drawn at an angle, so that the oblique feature can be seen in its true size and shape. Fig. 8-15 shows the three basic types of auxiliary views.

Sectional Views

Most working drawings show interior details with hidden lines. However, sometimes the inner details are too complex for this method. Then, sectional views are made. A **sectional view** looks as though the object has been cut open. Fig. 8-16.

An imaginary cutting plane is passed through the object. The interior along this plane is then drawn. A cutting plane line on the drawing indicates where the cut has been made. Other, slanted lines highlight the cut surfaces.

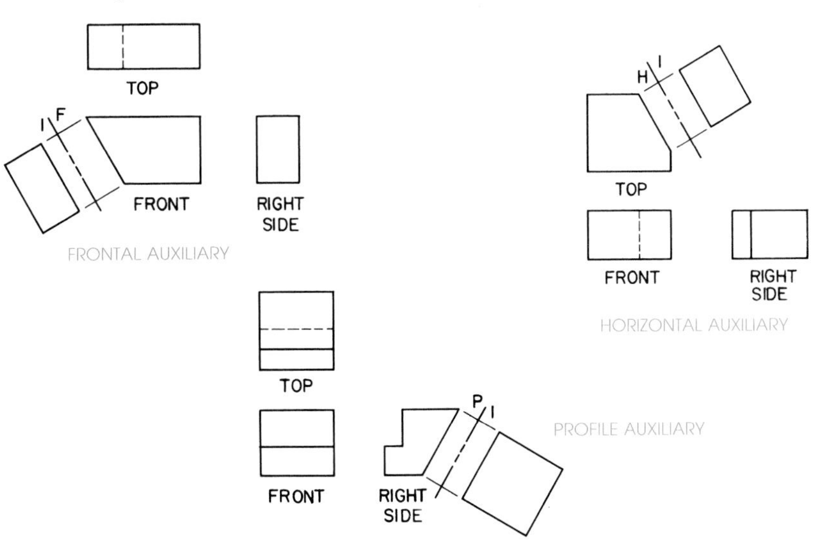

Fig. 8-15. Auxiliary views are used to show the true size and shape of features that would be distorted on the regular views. Note: On this drawing, the letter I stands for "inclined plane of projection." F is the frontal plane, H is the horizontal, and P is the profile plane.

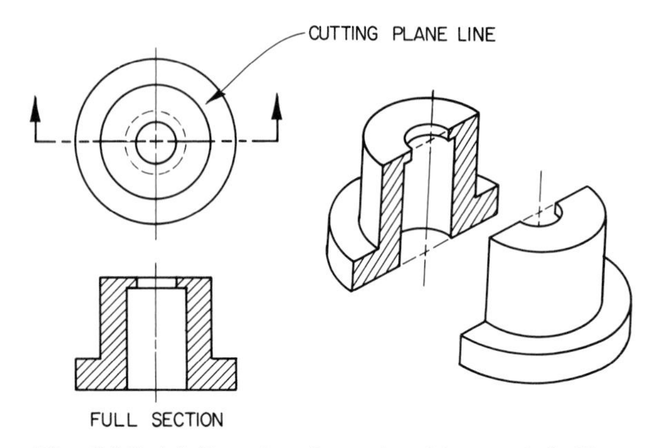

Fig. 8-16. A full section shows the object cut in half.

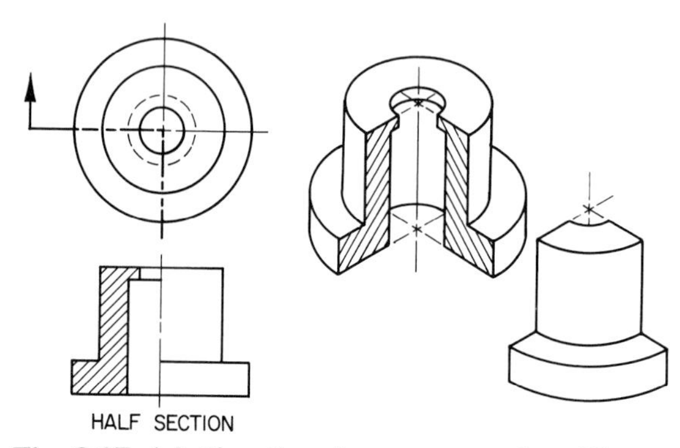

Fig. 8-17. A half section shows one-quarter of the object cut away.

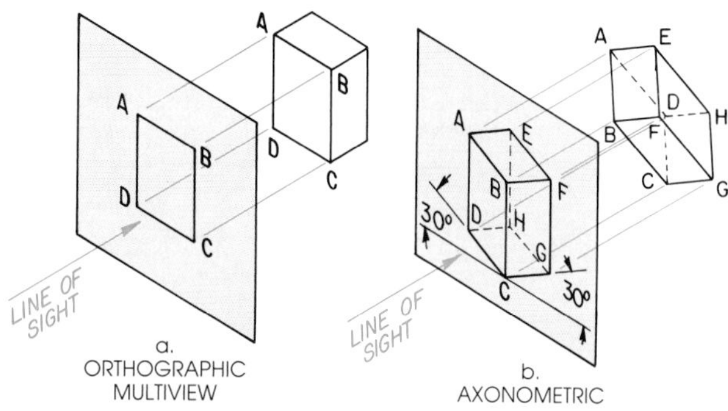

Fig. 8-18. Axonometric drawings show three sides of the object in a single view.

Two types of sectional views are common: full sections (Fig. 8-16) and half-sections (Fig. 8-17). In a full section the object is cut in half. In a half-section, only a quarter of the object is removed.

Pictorial Drawings

The oldest method of graphic communications is the pictorial drawing. **Pictorial drawings** portray an object in depth. Three sides of the object are visible in one view. They show how something "really" looks.

Today, pictorial drawings are specialized and technically complex. They are used in fine arts, commercial art, technical illustration, and industrial design. Of importance to drafters, however, are pictorial drawings that help them present an idea.

When pictorial drawings must be used, drafters select one of three types: axonometric, oblique, or perspective.

Axonometric Drawings

Axonometric projection is similar to orthographic projection except that the object being viewed is tilted or rotated. Fig. 8-18. By tilting the object, three sides of it are seen in the frontal plane or view.

Isometric drawings are considered the most important of the axonometric type. Isometric means "of equal measure." The object is tilted so its edges become axes that form equal angles. In Fig. 8-19 the edges of the cube form three equal angles of 120°.

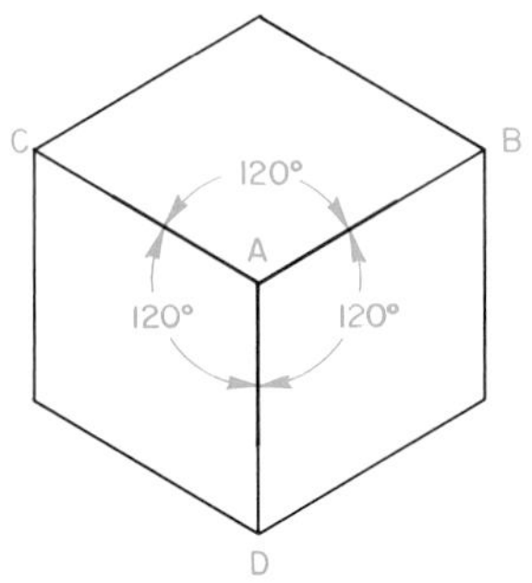

Fig. 8-19. In an isometric drawing, the edges of the object form three equal angles.

The basic procedure for making an isometric drawing is illustrated in Fig. 8-20. First, the three main axes are drawn using a 30-60° triangle. Using these axes as a starting point, a box equal in size to the *overall* dimensions of the object is blocked in. Then distances along each axis are measured and marked off. Lines are drawn where appropriate. At this point the drawing is compared with a multiview drawing for accuracy. The object lines are then darkened, and all construction lines are erased.

Two other kinds of axonometric drawings used in industry are dimetric and trimetric. Dimetric drawings show an object using two equal axes. Trimetric drawings use three axes. The procedures are the same as for isometrics, except that the angles are different.

Oblique Drawings

Sometimes curved or irregularly-shaped surfaces are difficult to draw in isometric. In this case drafters may use **oblique drawings**. The irregular side of the object (front, top, or side) is seen face on. Fig. 8-21. The other two sides are drawn along two axes and a look of depth is created.

Making oblique drawings is similar to making isometric drawings. There are, however, some differences. First, the side of the object to be drawn full size and shape is selected. Next, the scale to be used for all receding lines is determined. An oblique box, the dimensions of which are equal to the overall measurements of the object, can then be blocked in.

Oblique axes are usually drawn at standard angles of 30°, 45°, and 60°. When all surfaces and lines are drawn full-size, the drawing is called cavalier oblique. If the *angled* surfaces and lines are drawn half-size, the drawing is called cabinet oblique. Fig. 8-22.

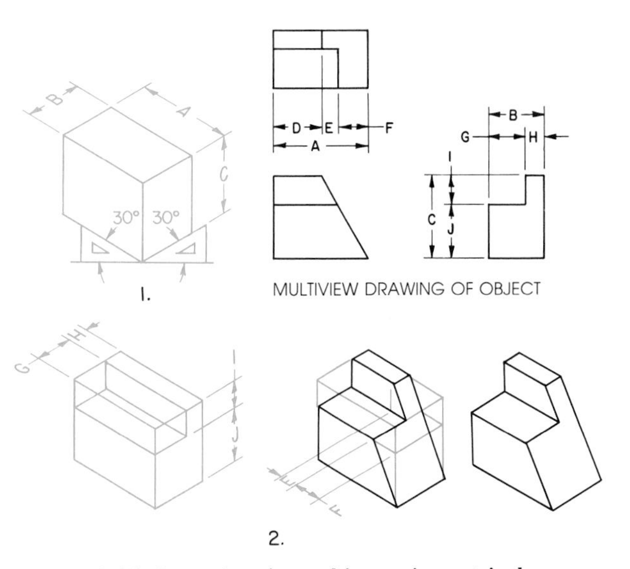

Fig. 8-20. Procedure for making an isometric drawing: (1) Draw the axes and overall dimensions. (2) Mark off distances and darken lines.

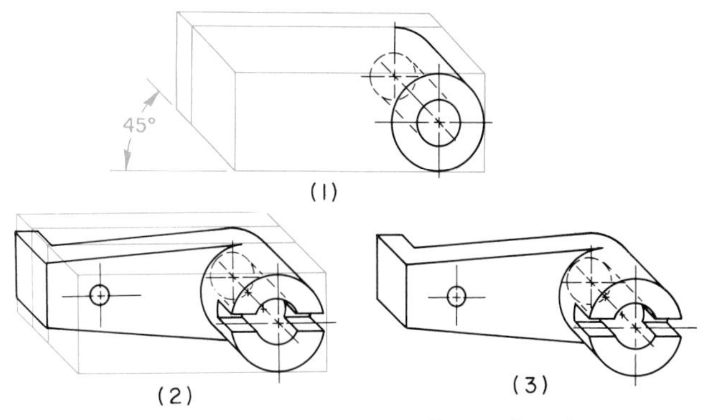

Fig. 8-21. Oblique drawings may be used to show curved or irregularly shaped surfaces.

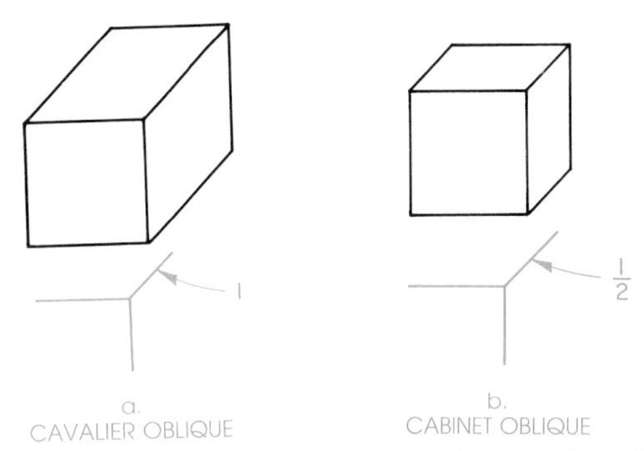

Fig. 8-22. Cavalier oblique shows the true size of the object, but cabinet oblique looks more natural.

Perspective Drawings

Drawings that resemble the way something would look in real life are called **perspective drawings**. The object appears three-dimensional. Receding parallel lines appear to come together in the distance. Fig. 8-23. Perspectives are not often used by drafters. They are primarily found in technical illustration work. However, one field where drafters do use perspectives is architecture. Drawings are prepared that show how buildings or other structures would look when completed. Sometimes, architectural perspectives are done in color.

For perspectives, four factors are of importance.

- *The position of the viewer, relative to the drawn object.* The viewer's position determines what surfaces and lines are visible and how they are shown. Fig. 8-24.
- *The points on the object (usually corners) from which lines can be drawn to the vanishing point.* The vanishing point is that point where the lines meet in the distance.
- *The position of the vanishing point.* The vanishing point is always located somewhere on the horizon.
- *The complexity of the object and its details.*

There are three types of perspective drawings: one-point, two-point, and three-point. Of these, the one-point is most commonly used in industry. Two-point perspectives are often used in architectural and product design illustrations. Three-point perspectives are limited to the field of commercial art.

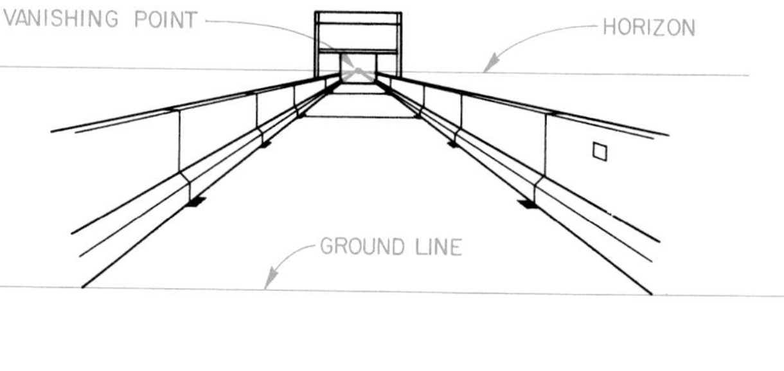

Fig. 8-23. In real life, parallel lines appear to meet in the distance. In perspective drawings, these parallel lines are drawn so that they meet at the vanishing point. This gives the drawings a realistic look.

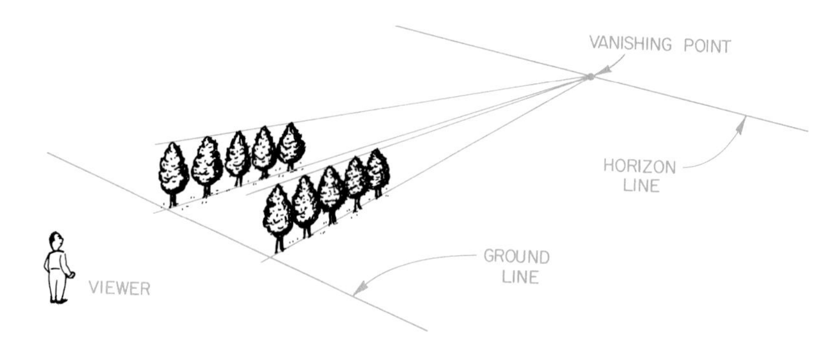

Fig. 8-24. The points of reference for perspective drawings include the viewer's position, the ground line, the horizon line, and the vanishing point.

TECHNOLINKS

GASPARD MONGE: CRACK SHOT WITH A CANNONBALL

During the eighteenth century, wars were fought using cannons. A well-placed cannonball could put a hole in a building or wall. A misplaced cannonball wasn't much use to anybody. Accuracy was important.

On level ground it was not difficult to figure the angle at which the cannon had to be aimed. On an incline, however, things got complicated. In the heat of battle, time-consuming calculations had to be made. In the meantime the enemy could move or even attack! What was needed was a quick method to determine the angle needed for sending a cannonball to its target. Along came Gaspard Monge.

In 1765, at the age of nineteen, Monge entered the artillery school at Mézières, France, where he was to invent a new kind of applied geometry. While at Mézières, Monge came up with a procedure using drawings that plotted the flight of a missile. (In this sense, a *missile* is any object that is thrown or projected in order to hit something.) Not only was Monge's procedure accurate, it was also fast.

When the military leaders of France realized the importance of Monge's invention, they quickly put it to use with artillery weapons. Later it came to be called descriptive geometry. Descriptive geometry is the basis for the technical drawings you are reading about in this chapter.

Monge also applied his theories to other scientific, mathematical, and industrial problems. He soon became a professor of mathematics and physics. Under Napoleon he was Minister of the Navy and was sent on scientific missions. In 1794 he founded the Ecole Polytechnique in Paris, where descriptive geometry and drawing became an important part of the education of military engineering students.

The fact we are still using Monge's theories today is testimony to his importance. It might even be said that, thanks to Gaspard Monge, technical drawing started off with a bang.

A simple one-point perspective drawing uses one vanishing point. Fig. 8-25. To make a one-point perspective, drafters first draw the horizon line and vanishing point. Then the front view of the object is sketched in. Lines from the corners of the object are projected to the vanishing point. Next, drafters estimate the depth of the object and complete the drawing.

Drawing a two-point perspective is more difficult. Fig. 8-26. The object is located so that the vertical lines and axis are parallel and do not meet at a vanishing point. All other lines are drawn to one of two vanishing points.

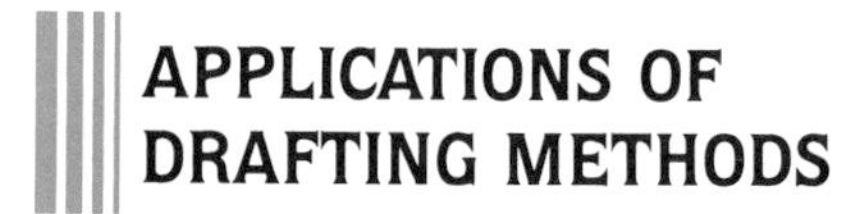

APPLICATIONS OF DRAFTING METHODS

The basic principles and methods presented in this chapter are used in many industries and technical fields. Each specialized field makes use of graphic symbols and presentations known as **conventions**. These conventions provide information about a part without having to draw it as it would appear or write notes about its makeup. Shown in Fig. 8-27 are examples of different drafting conventions.

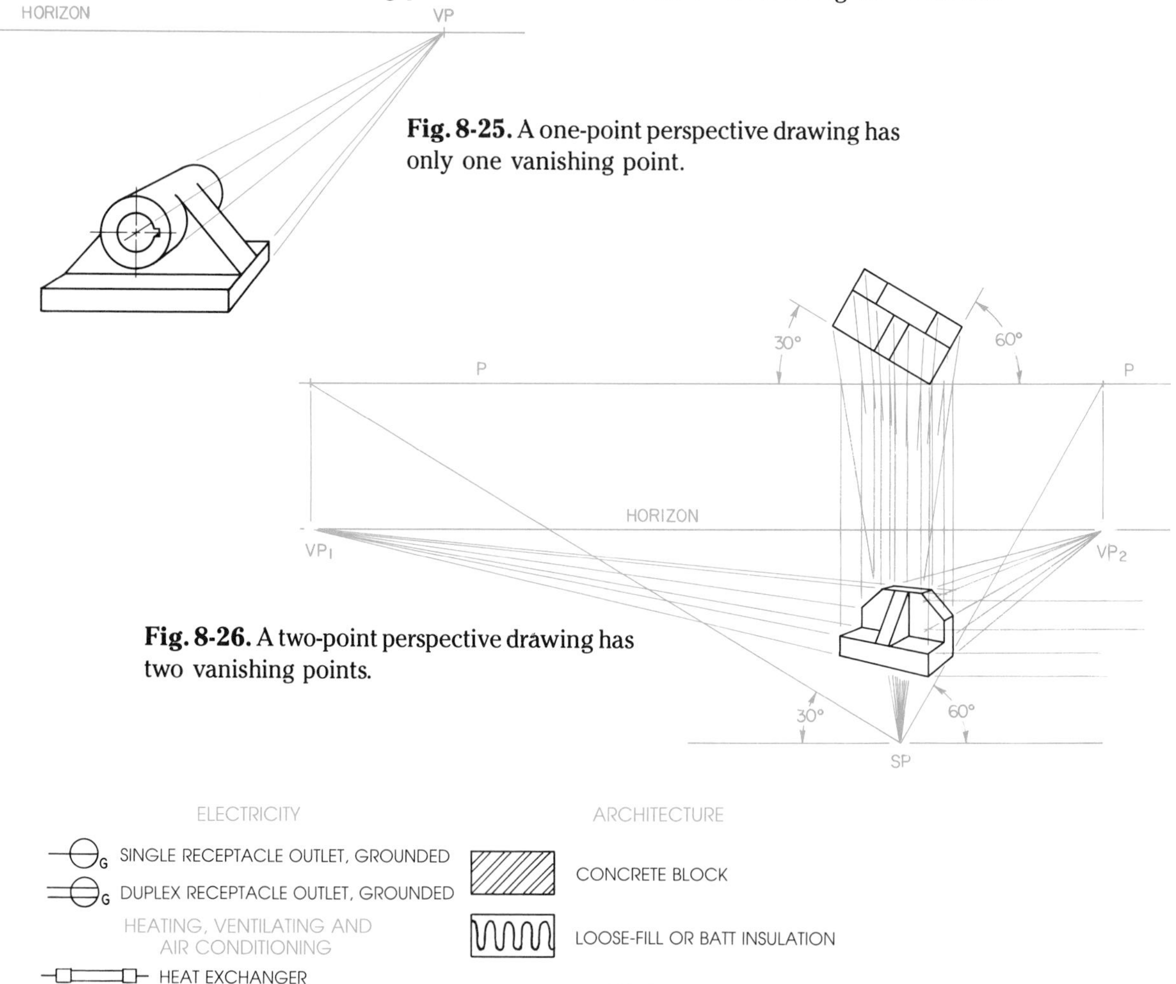

Fig. 8-25. A one-point perspective drawing has only one vanishing point.

Fig. 8-26. A two-point perspective drawing has two vanishing points.

Fig. 8-27. Here are a few examples of drafting conventions used in various fields.

Following are discussions of how drafting methods apply to different technical fields.

Mechanical Engineering

Of all the different fields, mechanical engineering probably makes the widest use of technical drawings. All of the drawings and methods covered in this chapter are commonly used for tool and die design, jigs and fixtures, machines, manufacturing processes, and mechanical and industrial engineering. Differences are usually a result of individual company requirements.

Mechanical drawings can be divided into three major groups: layout drawings, detail drawings, and assembly and subassembly drawings.

Layout drawings are made from sketches prepared by engineers and designers. Their purpose is to show how a product or part fits or operates under working conditions. Unlike sketches, layouts show the product in full size so that engineers can study size, proportion, and relationship to other parts. Fig. 8-28 shows a layout drawing of a product fully assembled. How parts fit together and their true sizes and proportions are all evident. Since most layout drawings are full size, it is not uncommon for them to measure as much as 48 × 96 inches.

Detail drawings are the most common type made by drafters. Fig. 8-29. Once a layout has been made, a detail drawing is created for all parts. The drawing gives all the information needed to make the part.

Information about the part's geometric shape, size, and specifications are included. Some standardized parts, such as screws and bolts, do not have to be detailed. Since they are readily available from suppliers, they are simply noted in a parts list.

Assembly and subassembly drawings are made after all parts have been detailed or noted. Assembly drawings are used to show how parts fit or are assembled together. A subassembly

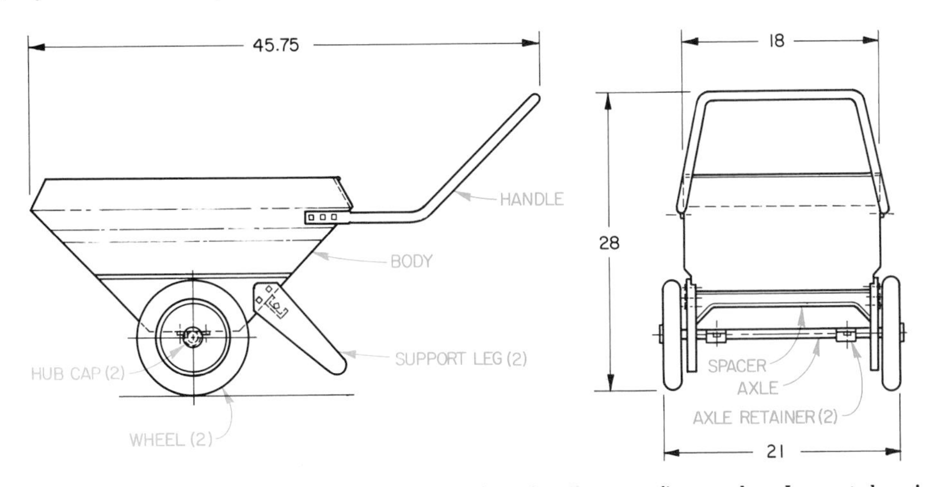

Fig. 8-28. This layout drawing shows how the parts of a wheelbarrow fit together. Layout drawings used in industry are usually full size.

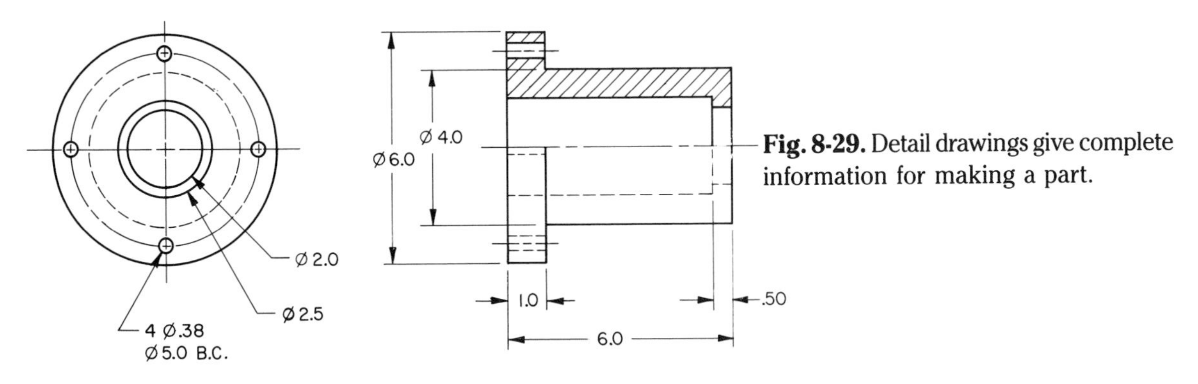

Fig. 8-29. Detail drawings give complete information for making a part.

drawing is similar except that it shows only one section of the total product. Fig. 8-30. Its primary purpose is to further explain the construction of a complex part of the total assembly. Subassemblies are only used to improve the understanding of the general assembly drawing.

Electrical and Electronics Industries

Though technical drawings are used in this area, general drafting practices do not meet all the needs of the electrical and electronics industries. Most electrical drawings are referred to as **diagrams**. These diagrams make use of a system of symbols that represent electrical components such as resistors, transistors, diodes, and capacitors.

Electrical drafters must have a solid knowledge of electronics and electrical theory. They have to understand what they are drawing and identify any design errors. Drawings that drafters are expected to draw are:

- *Block and logic diagrams.* These are used to show how an electrical system or product is supposed to work. They do not give the information needed to make the system. They are primarily used in design work.

  Block diagrams show the flow of electrical power in a system. Fig. 8-31A. Logic diagrams show the operation of a circuit made up of logic gates. The output of a logic gate is a high or a low voltage level, depending upon whether the inputs are high or low, or what type of gate is used. For example, an AND gate must have a high level on both inputs to produce a high level on the output. Fig. 8-31B.

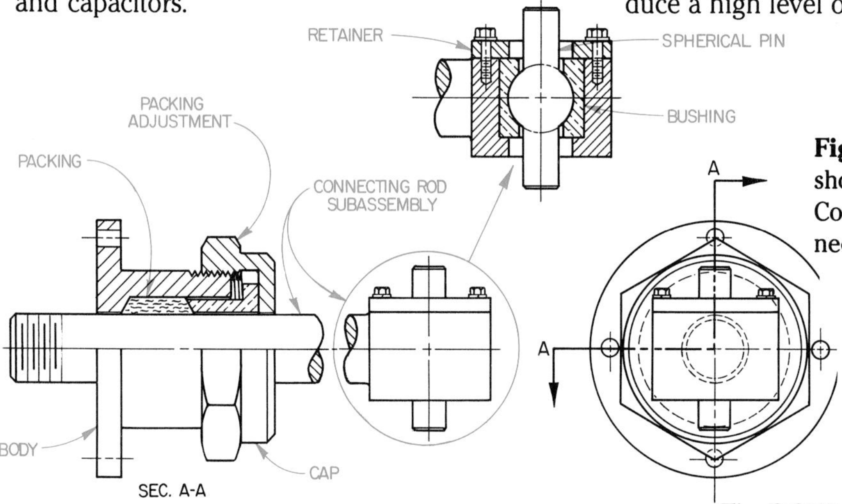

Fig. 8-30. Assembly drawings show how parts fit together. Complex products may also need subassembly drawings.

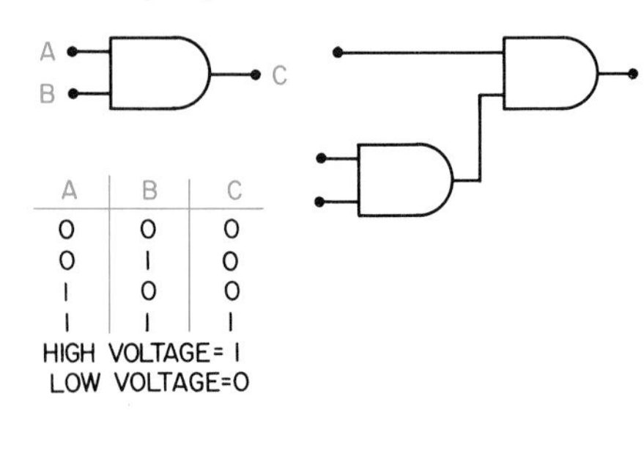

A	B	C
O	O	O
O	I	O
I	O	O
I	I	I

Fig. 8-31B. Shown here are an AND logic gate and its truth table. A truth table shows every possible output of a logic gate. Also shown is a logic diagram. The logic diagram details the signal flow and control of the logic gates.

Fig. 8-31A. This is a block diagram of a digital clock.

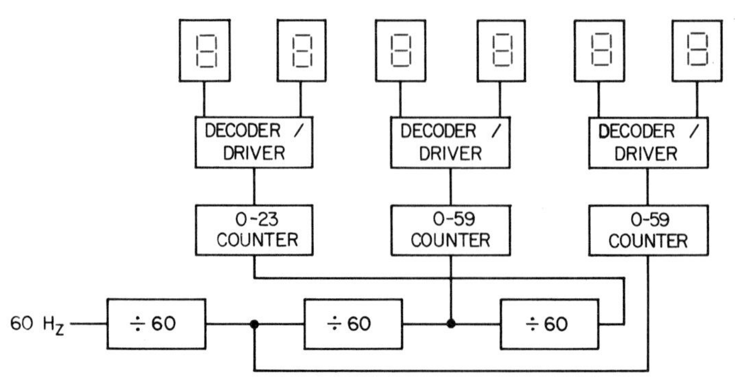

Logic circuits are used in electronic equipment, such as computers. (See Chapter 4.)

- *Wiring diagrams.* These show how electrical products are wired together. There are two categories of wiring diagrams. The first is used in architectural and building drawings. They show the relationships among switches, plugs, and appliances. Fig. 8-32.

  The second type is called a one-line electrical diagram. Fig. 8-33. These are used to show how parts are to be connected or "wired" in a product.
- *Schematic diagrams.* Schematics illustrate how the different electrical and electronic parts in a device are connected to each other as well as to the power sources. Electrical parts appear as symbols. Fig. 8-34.
- *Printed circuit drawings.* Printed circuits replace hand wiring. They consist of foil conductors on an insulating surface, or board. The drawings for printed circuits are made much larger than their finished size. They are photographically reduced and then etched ("printed") on the board. Fig. 8-35.

Fig. 8-32. This is a wiring diagram for the mezzanine of a church.

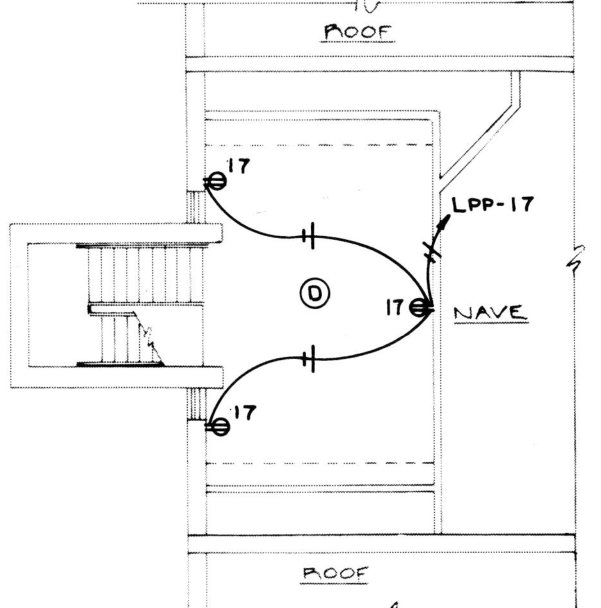

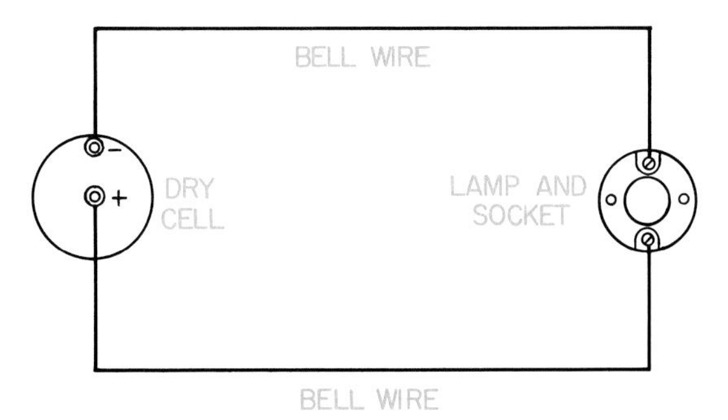

Fig. 8-33. This one-line wiring diagram shows the electrical circuit for a lamp.

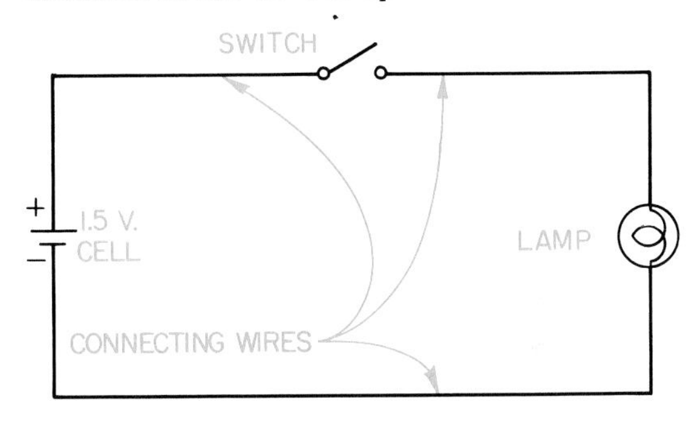

Fig. 8-34. A schematic diagram uses symbols to show how parts are connected. Compare this to the wiring diagram in Fig. 8-33.

Fig. 8-35. Shown here are a drawing for a printed circuit and the finished product. Electronic components will be attached to the printed circuit board.

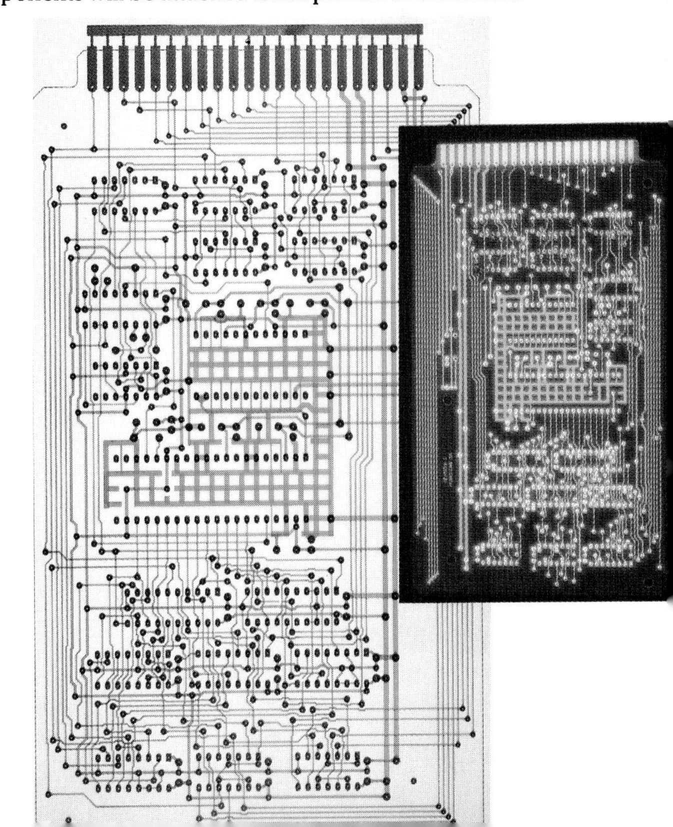

Architecture

The field of architecture and building design makes frequent use of drawings. Drafters who work in this field must be able to prepare several different types of drawings. For large projects, such as office and commercial buildings, architectural firms often hire specialists from various areas to help. Specialists may handle plumbing drawings; electrical systems plans; heating, ventilation, and air-conditioning plans; site and topographic drawings; and structural drawings.

A set of plans for a building can be broken down into four major categories:

- *Foundation plans.* These are used to show how the weight of a structure will be spread out and supported. Structural foundation plans are made by engineering firms and show all the details of construction.

  The architectural foundation plan is perhaps more familiar. It is drawn to show the entire foundation system as part of the building. Basement plans are separate from a foundation plan but are often drawn as part of it.
- *Floor plans.* **Floor plans** are the basis for all architectural drawings. They show the inside of the building viewed from above. Many features may be drawn onto a floor plan. These include electrical, plumbing, heating, cooling, and ventilation details. Fig. 8-36.

Fig. 8-36. This drawing shows part of the floor plan for a church.

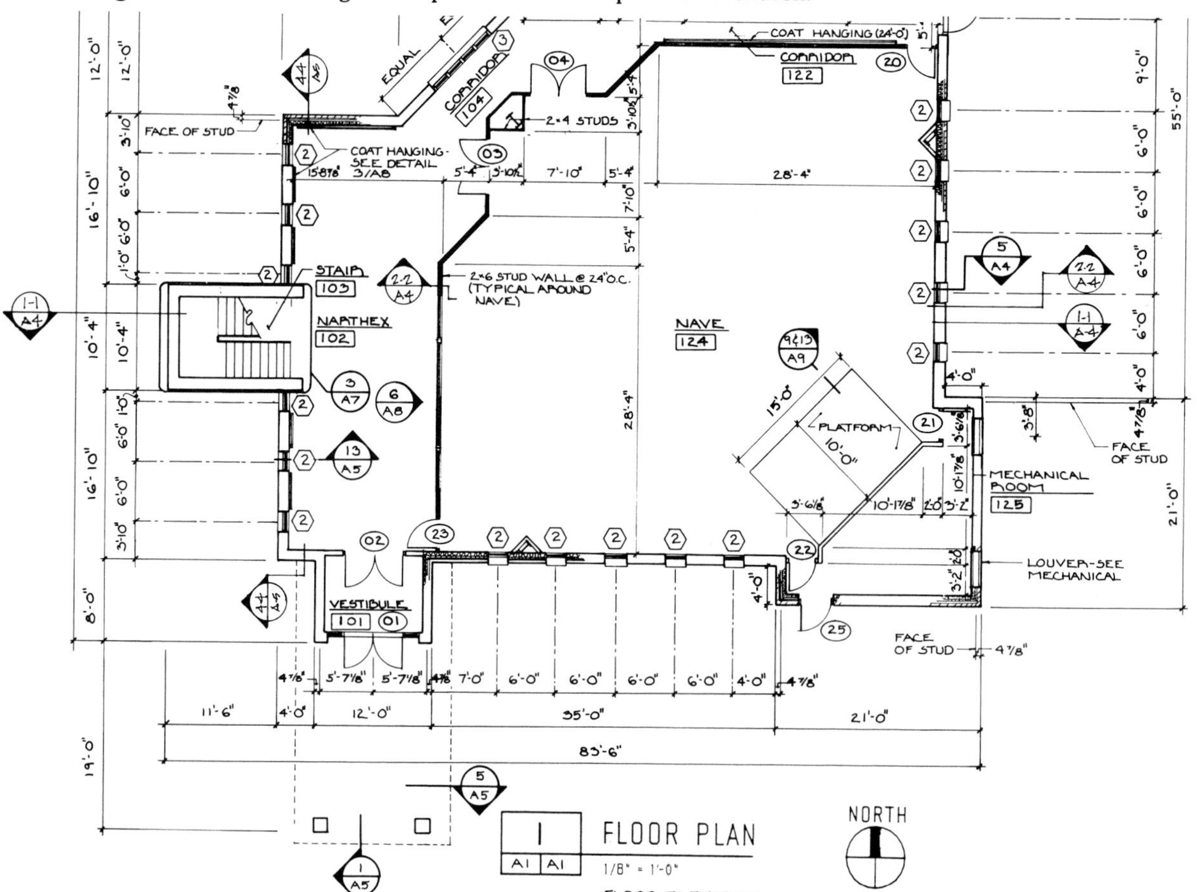

- *Elevations.* These are drawings that show how a building looks from the front, rear, and sides. There are two types of elevation drawings: interior and exterior.

  Interior elevations show construction or design features of the inside of the building. Fig. 8-37. They include items such as cabinets, storage units, or furnaces. Exterior elevations show how the building looks from the outside. Fig. 8-38. Elevations commonly include four views: north, south, east, and west.

- *Detail drawings.* These illustrate details or specifications for construction. They show the parts of a building so that contractors and tradespeople will know how the work should be done.

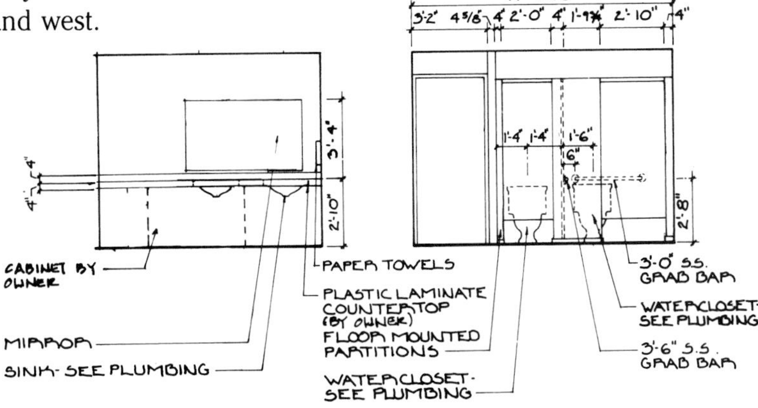

Fig. 8-37. This interior elevation of a bathroom shows cabinets, fixtures, and other features.

Fig. 8-38. Exterior elevations show how a building looks from the outside.

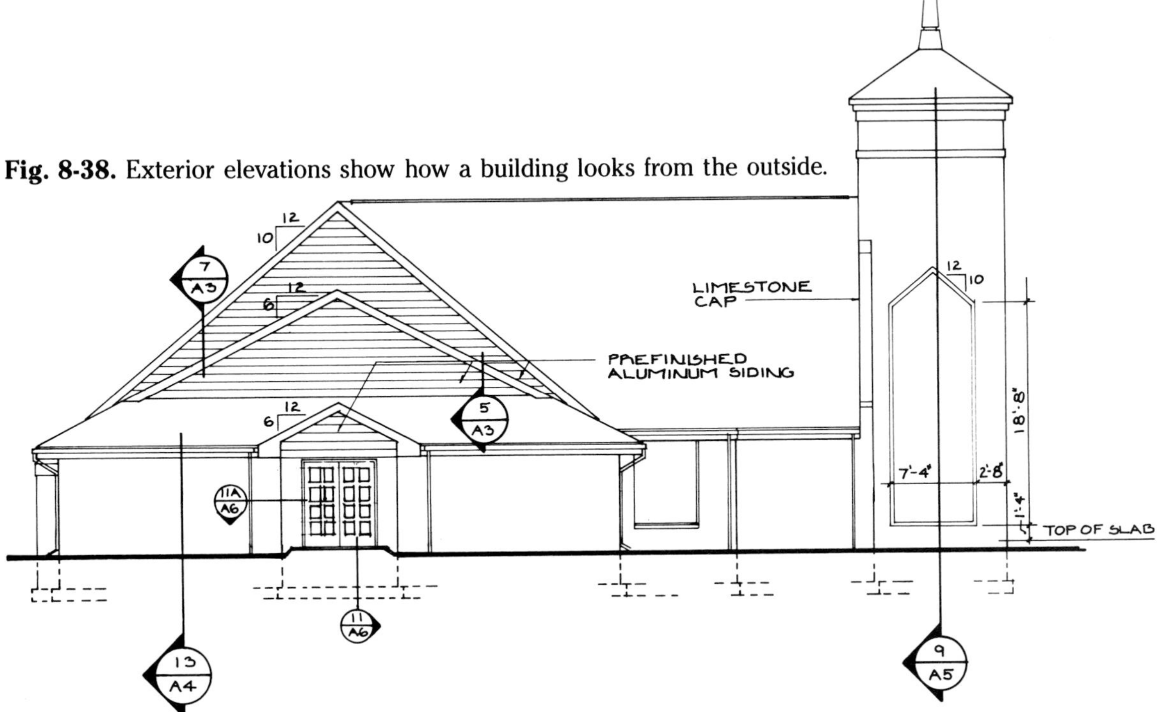

One of the most common types of detail drawings is the wall section. Fig. 8-39. It shows how the wall is to be constructed and includes the type of materials used. Other detail drawings are done for windows, doors, roofs, and ceilings.

Civil and Structural Engineering

Civil and structural engineering are concerned with the design and building of structures that can handle different weights and stresses. Drawings are needed for bridges, industrial plants, mechanical systems, and heavy-duty equipment. Fig. 8-40. They include information about building materials such as timber, reinforced and prestressed concrete, masonry, and steel.

There are two types of structural drawings: general design drawings and shop drawings. General design drawings are used to show the arrangement of a structure and how its members are assembled. Shop drawings are made for parts or steel members that are made in the shop rather than at the construction site.

Most civil and structural engineering projects require several types of drawings.

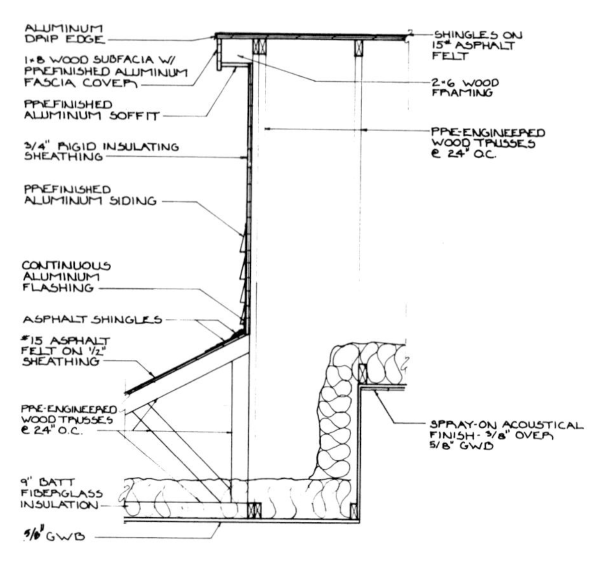

Fig. 8-39. A wall section shows how the wall is to be constructed. Note the many different materials used.

Fig. 8-40. This drawing shows the installation of a double girder crane in an industrial plant. Such cranes are used to lift and move heavy loads.

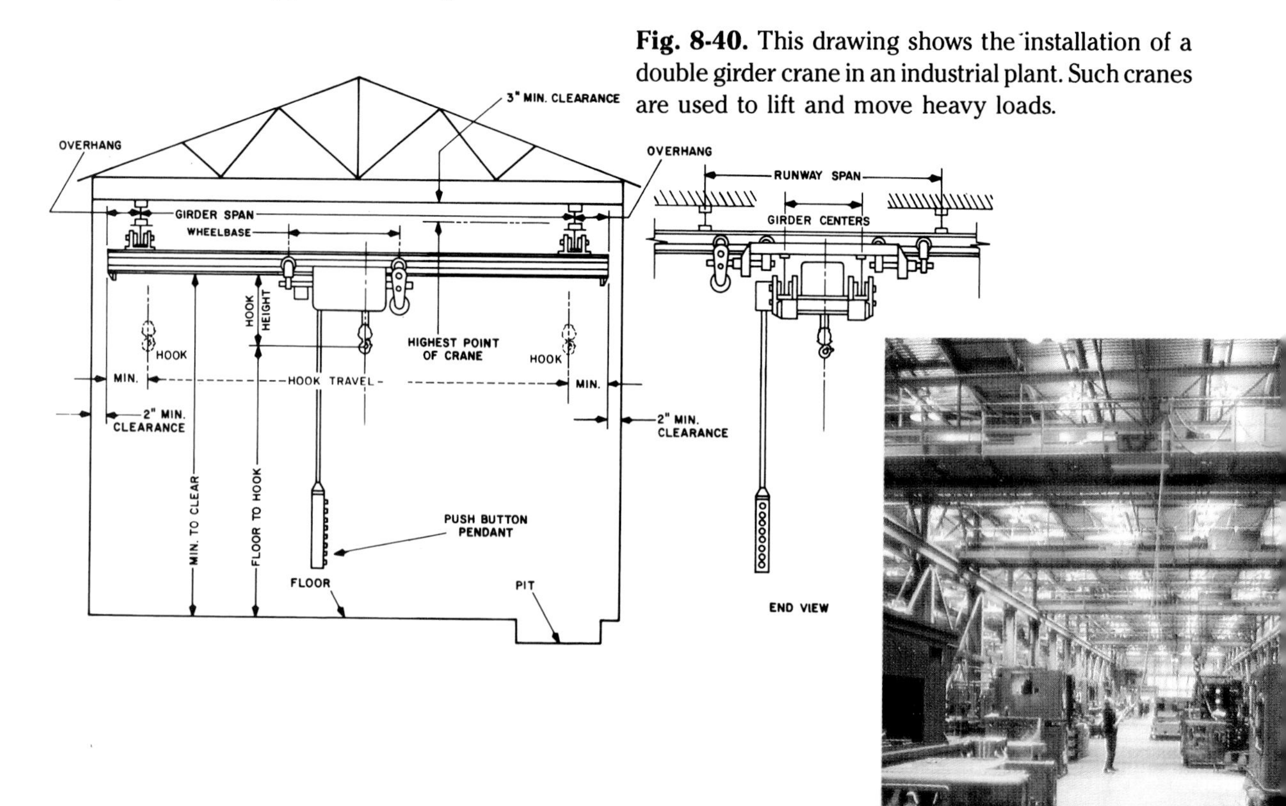

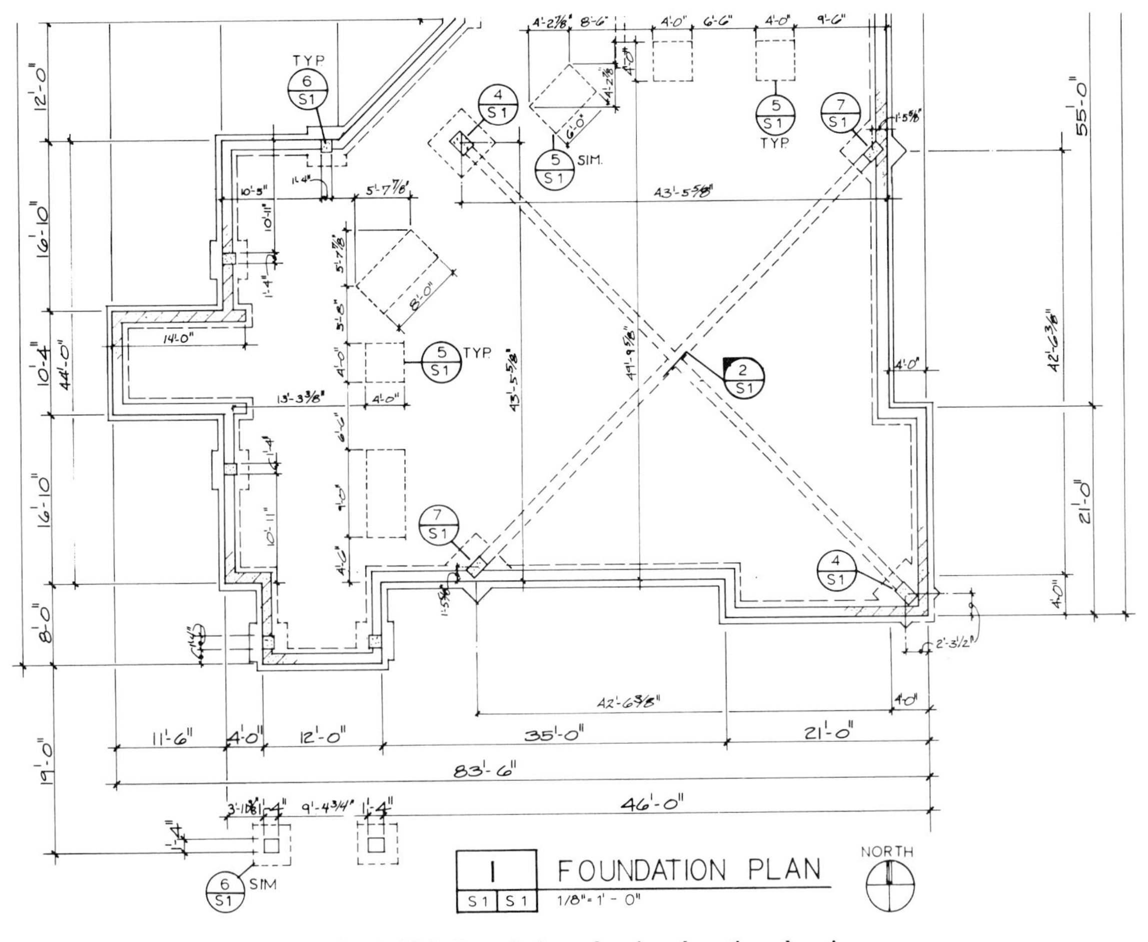

Fig. 8-41. This foundation plan is a location drawing.

Location drawings are simple drawings of a structure and the locations of its parts, such as footings and columns. Fig. 8-41. Since most projects require several location drawings, they are presented in order of construction or assembly.

Detail drawings are drawn on separate sheets or included on location drawings. They are used as a source of additional information about a structural part. Fig. 8-42.

Fig. 8-42. This detail of the foundation plan shows how the beams and columns connect.

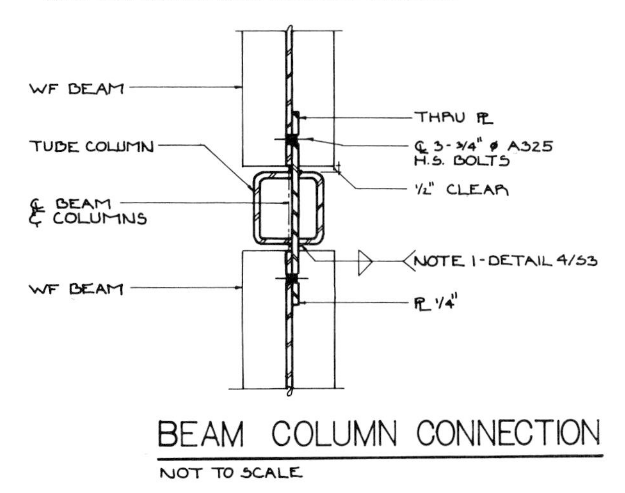

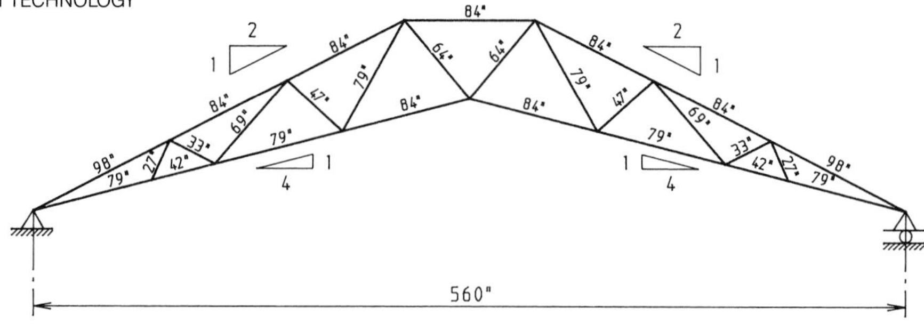

Fig. 8-43. A diagrammatic structural drawing shows distances between points.

Diagrammatic drawings, used in design work, represent the location of and distances between reference points. Reference points are typically where the structure's members are connected. Fig. 8-43.

Topographic Mapping

Drawing the surface features in an area or region of the earth is known as **topographic mapping**. Topographic maps show natural features, such as hills and lakes, as well as structures and boundaries. Information is gathered by surveying. Then topographic maps are drawn by drafters and map makers who use both drafting and photographic methods. Because of the nature of these drawings, topographic drafters use a combination of technical drawing and artistic techniques.

Topographic drawings are used in many technical and engineering fields:

- *Surveying.* This is the process of determining and locating forms and boundaries of a tract of land by measurement. Accurate surveys provide the information needed to prepare topographic drawings.
- *Engineering.* All engineering and architectural projects require several types of topographical maps. For example, civil engineers may use hydrographic charts. These give information about bodies of water. They show features such as navigational aids, underwater hazards, lights, and harbor facilities.
- *Geodesy.* This field locates the exact positions of features along the earth's surface. Perhaps the most common maps produced by geodesy are those showing the geological features of a region.

Topographic information can be presented in several different ways. Contour drawings are used to show the elevation, or height, of land forms. A contour is a continuous line representing a certain elevation. The elevation is usually given in feet above or below sea level. Contour lines show features such as hills, mounds, summits, valleys, and hollows. Fig. 8-44.

Profiles are drawn to show the rise and fall of land. Profiles are used in the design and construction of highways, utility systems, railroads, sidewalks, and waterways. Fig. 8-45.

Fig. 8-44. A contour map shows the shape and height of the land. Each line represents a different elevation.

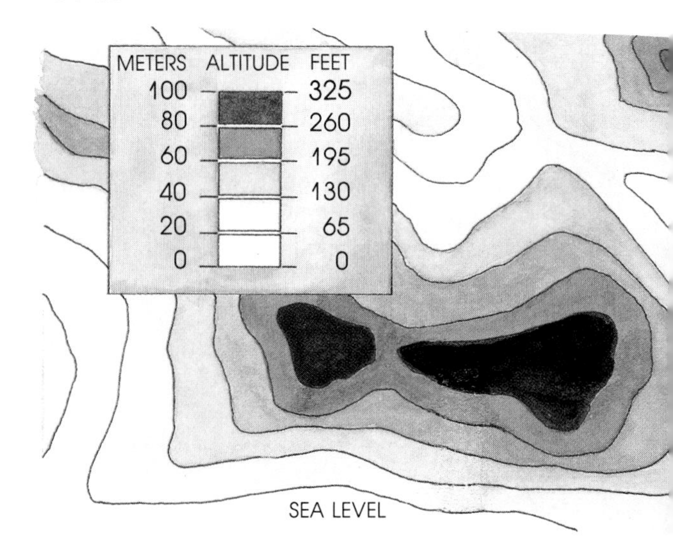

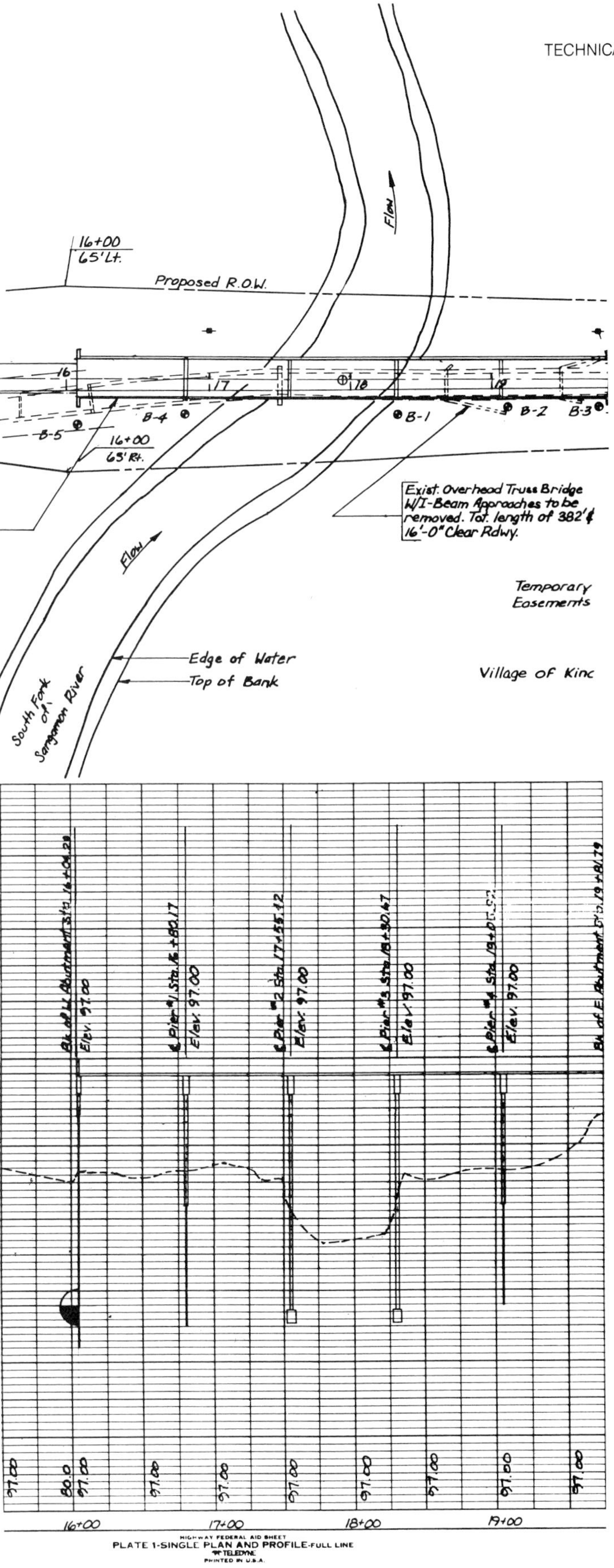

Fig. 8-45. A profile map looks as if someone had made a vertical cut through the land, revealing its elevations. The lower part of this drawing is a profile map.

Photogrammetry uses aerial photography as the source for surveying information. Fig. 8-46. Specialized photogrammetry includes radar-grammetry which uses radar as a measuring tool. Hologrammetry uses holographs to measure surface characteristics.

Fig. 8-46. Surveyors may obtain information from aerial photographs.

Fig. 8-47.

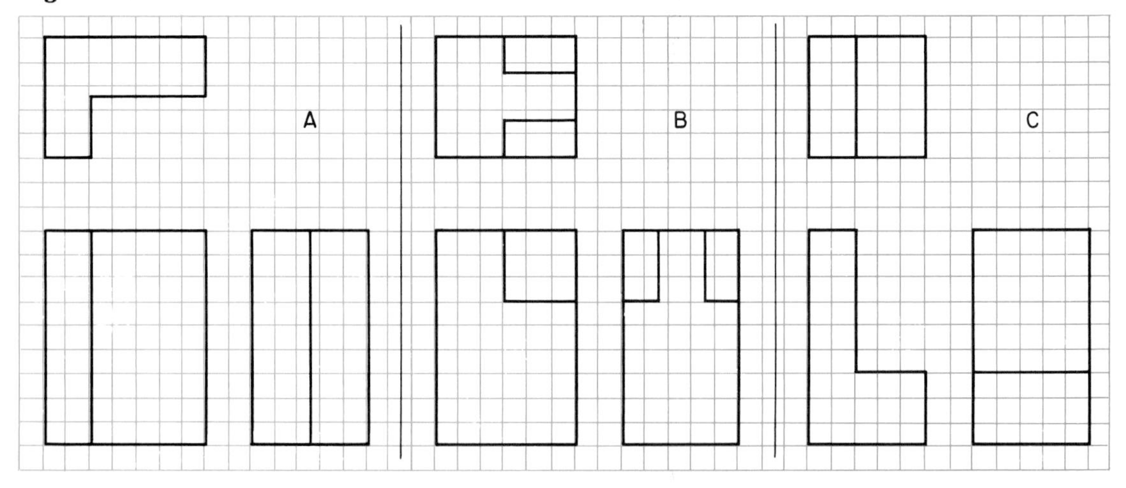

CHAPTER 8 REVIEW

Review Questions

1. What is the difference between a sketch and an illustration?
2. What are multiview drawings used for?
3. Describe orthographic projection.
4. Describe auxiliary and sectional views.
5. Explain the difference between an isometric and an oblique drawing.
6. What four factors are of importance when making a perspective drawing?
7. What is the purpose of detail drawings?
8. Name three types of electrical or electronic drawings and tell their primary use.
9. Tell the differences and similarities between architectural and structural drawings. Give examples of each.
10. Name three ways in which topographic drawings are used.

Activities

1. Select an object and draw a sketch of it using the methods described on pages 162-163. You may use graph paper to draw the object to scale.
2. Make an isometric drawing for the objects shown in Fig. 8-47.
3. Prepare a multiview drawing of one of the objects illustrated in Fig. 8-48. Show at least three views.
4. Obtain five examples of technical illustrations used to sell products. Try to determine what media were used. Prepare a display of the illustrations.
5. Find examples of architectural, electrical, and topographic drawings and identify and label the symbols unique to each.

Fig. 8-48.

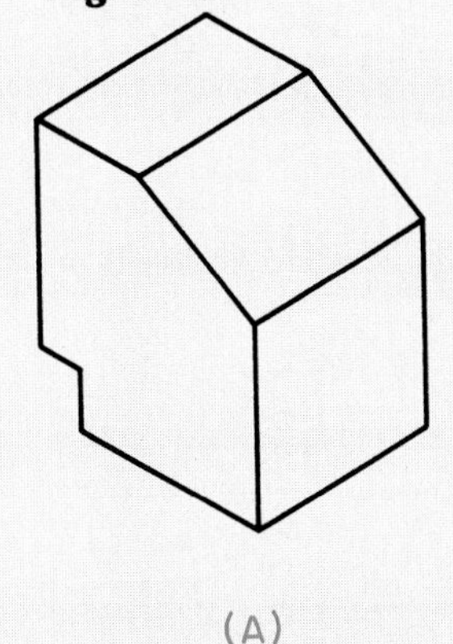

(A)

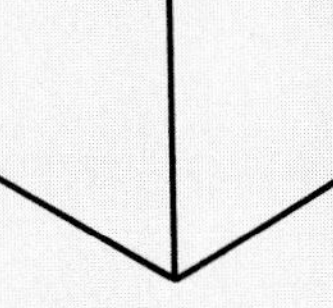

(B)

(C)

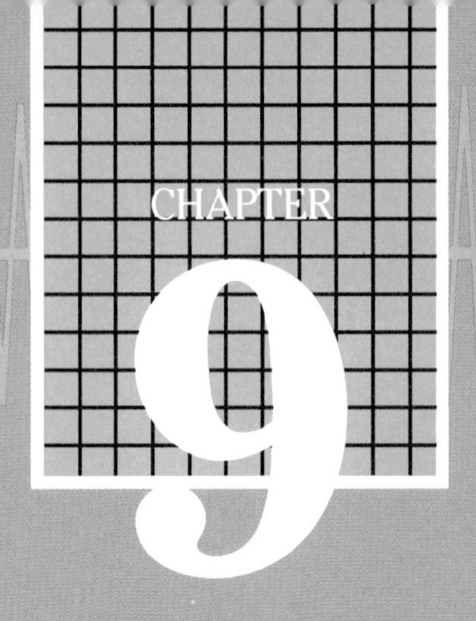

CHAPTER 9

Computer-Aided Design

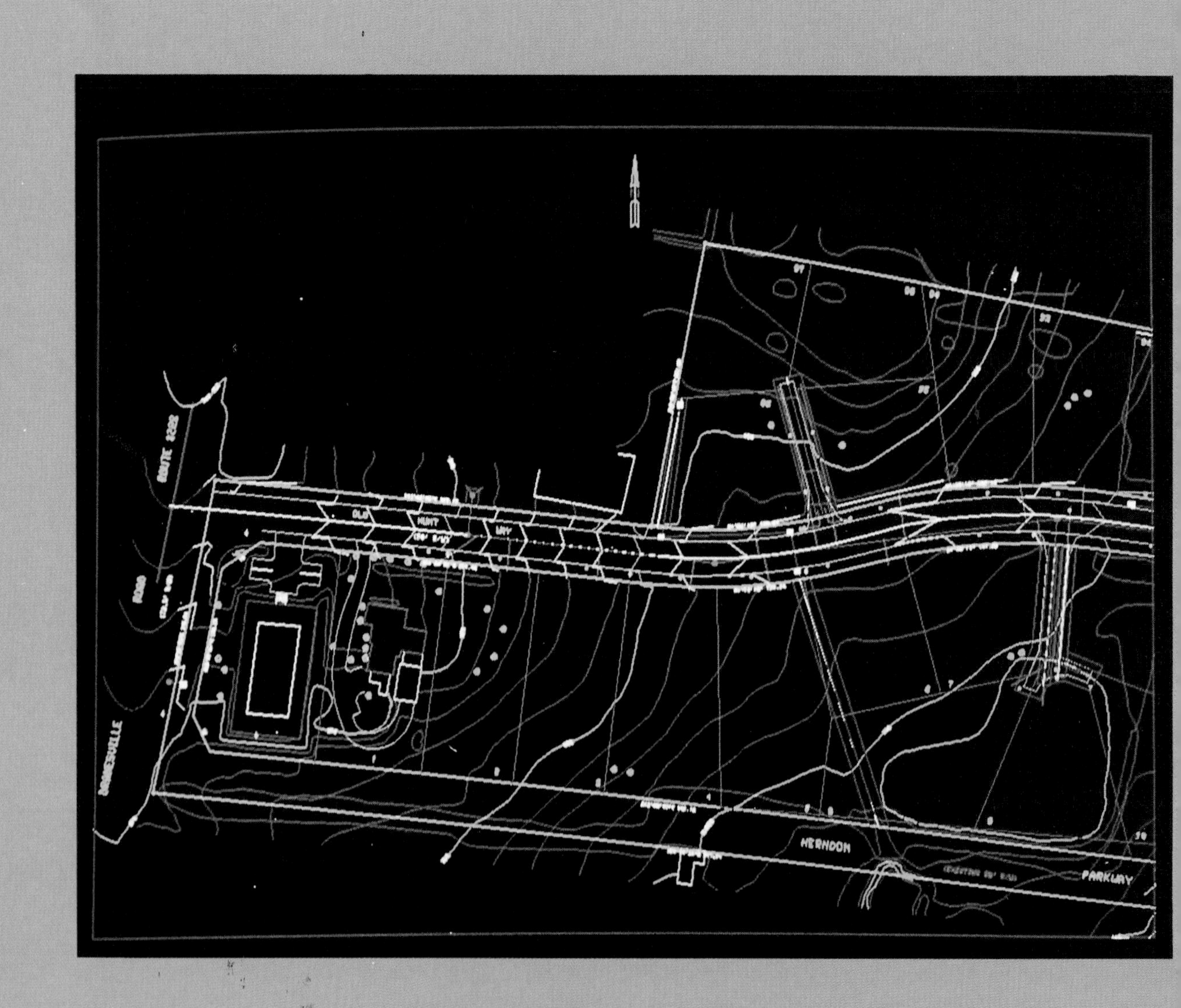

In the technical design field more significant changes have occurred since 1965 than in the previous 200 years! Advanced technology has brought industrial design and drafting into the computer age. Today, not only do engineers, designers, and drafters have to understand and apply various drawing techniques and practices, but they must also be able to work with the computers that have brought about all these changes.

To use computers effectively, professional and technical workers must understand what computers and computer systems can do. Many students make the mistake of thinking, "Since the computer can do so many things, I don't have to become a good drafter." This is wrong. Students *must* have the traditional skills, knowledge, and judgment demanded. Otherwise, they cannot tell the computer what to do or understand what it is producing.

Terms to Learn

layers
parametric design
solid model
stress analysis
symbol libraries
surface model
wireframe model

As you read and study this chapter, you will find answers to questions such as:

- How does CAD drafting differ from traditional drafting?
- What are the basic components of a CAD system?
- What special functions, such as stress analysis, will a CAD system perform?

HOW CAD SYSTEMS WORK

The term "CAD" was originally used for "computer-aided drafting." It was then expanded to "CADD" or "computer-aided drafting and design." Today, CAD is used to mean either or both.

CAD systems may be used for many types of drafting. The drawing is displayed on the computer monitor as the drafter creates it. The drafter interacts with the machine to make additions or changes. Fig. 9-1.

The way CAD systems work depends on the software. Commands and procedures differ, but most programs enable drafters to create drawings in two (or three) dimensions, edit (change) them, store them, and plot (print) them. The CAD programs also have features that make drawing faster and easier. You'll learn about some of these in this chapter.

What Makes CAD Different

CAD differs from traditional drafting in several ways. In traditional drafting, being able to letter and draw correct line weights is essential. Drafters must also be able to use tools such as compasses effectively. For CAD, these skills are not as important. The system automatically produces lines and letters more uniformly than a human drafter.

Advantages of CAD systems include speed, accuracy, neatness, consistency, and ease of making changes. Most original drawings cannot be created any faster using CAD than drawing by hand. However, revisions can be made faster, and both the original and the revised drawing can be kept.

CAD systems can also perform repetitive tasks that are time-consuming for drafters. For instance, a drawing of a standard screw may be pulled from a "library" disk and used, saving the drafter drawing time.

Fig. 9-1. This drafter is using the keyboard and a mouse to make changes to the drawing on the computer monitor.

In general, CAD systems enable drafters to:

- Draw lines as needed at any location and of any size or type.
- Draw circles and arcs of any size.
- Move a drawing or copy it to another area of the screen.
- Set the scale for a drawing and maintain that scale for details.
- Figure distances and do automatic dimensioning.
- Check dimensioning accuracy.
- Make a symmetrical design by creating a mirror image of what has been drawn.
- Turn drawings that have been done separately into "overlays" that can be combined.
- Automatically section or crosshatch a part.
- Insert previously made drawings into new drawings.
- Save and/or plot drawings.

Hardware

A basic CAD system for microcomputers includes the computer plus various input, output, and storage devices.

As you know, input devices are used to put information into the computer. For CAD, these usually include a keyboard and either a digitizing tablet (with a stylus or puck) or a mouse. Fig. 9-2.

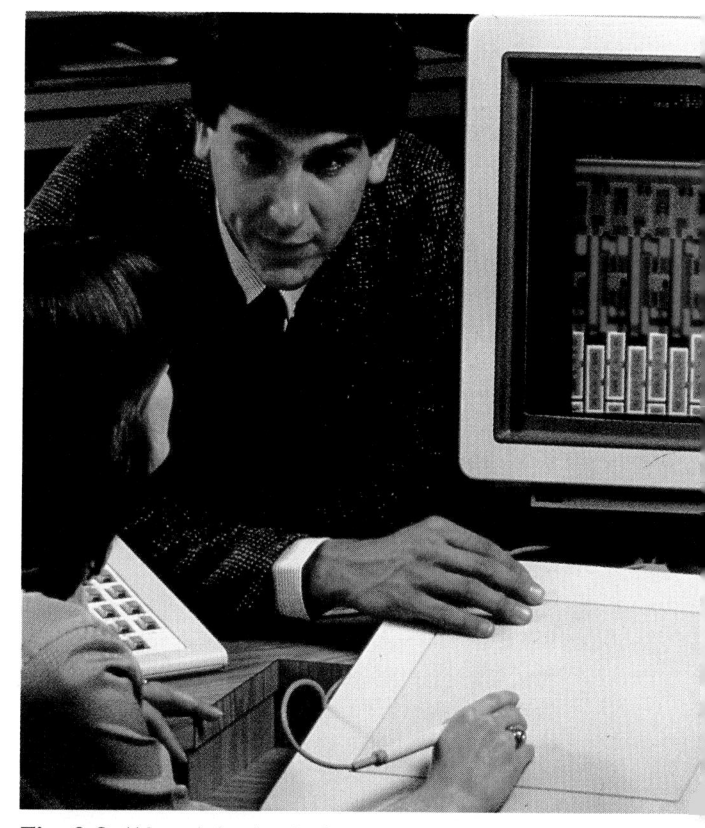

Fig. 9-2. (Above) As the drafter moves the stylus over the surface of the digitizing tablet, a drawing appears on the monitor. (Below) A mouse works in a similar way. Both the stylus and mouse are also used to select items from menus, as the drafter in the picture below is doing.

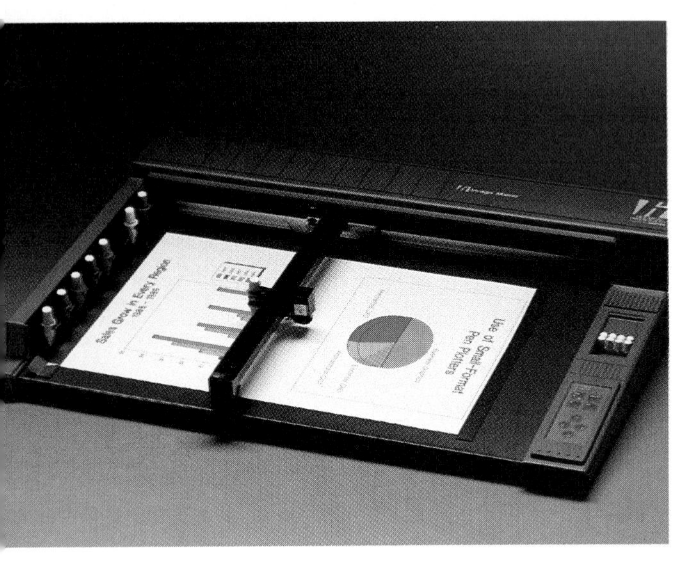

Fig. 9-3. Flatbed plotters (above) usually fit on a desktop. Drum plotters (below) stand on the floor.

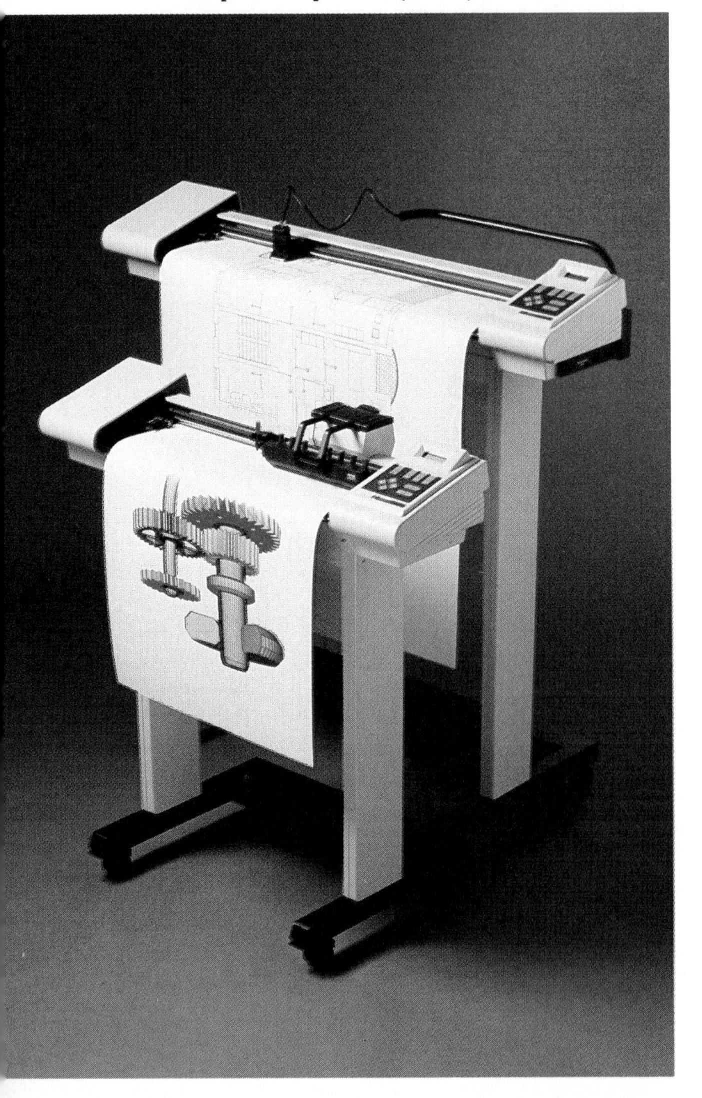

Output devices include plotters, which come in two basic designs, and printers. Flatbed plotters are primarily used for graphs, charts, and small drawings. On a flatbed plotter, the paper (or other substrate) remains stationary, and the pen moves. Drum plotters are used for larger drawings. On drum plotters, both the paper and pen move. Fig. 9-3. Most plotters can be ordered with different colored pens for multi-color plots. Pens can also be added later, as needed.

Printers can also generate images. Dot matrix and ink-jet printers do not produce high-quality technical drawings but are useful for rough copies. Laser printers, on the other hand, can generate highly finished drawings. The main disadvantage of printers is that they cannot produce the large drawings needed in many technical fields. Plotters, on the other hand, can produce large drawings.

Storage devices store the work and information. The user can retrieve it as needed. For CAD, floppy and/or hard disks are commonly used. For added security, some companies make backup copies of their drawings on magnetic tape or optical disks.

Software

The software is, of course, the program or set of instructions that tells the CAD system what to do. Different programs may have different features. Also, not all software works with the same hardware. Software must be designed for a specific computer system.

Drafting with CAD Systems

To make a drawing on a CAD system, the user inputs commands and other information. For example, to draw an arc, the user may first select "Arc" from a screen menu. The computer then asks the user to indicate where the arc should begin. The user responds by selecting a location with the pointing device (stylus, mouse, or puck). Selecting the endpoint of the arc then completes it. Fig. 9-4. The arc may then be moved, copied, rotated, or manipulated in other ways without ever needing to be redrawn.

Collections of pre-drawn symbols, called **symbol libraries**, save drafters time on details. Fig. 9-5. For example, a certain type and size of bolt may be selected from a library menu. Then the computer is told where to place the bolt. It is then automatically drawn in the correct spot.

Another function, semi-automatic dimensioning, is done by first indicating the part. The CAD program then figures the angle or distances, draws dimension and extension lines, and enters the measurements on the drawing. However, the user must specify the size of the lettering, amount of space between lines, sizes of arrowheads, and so on.

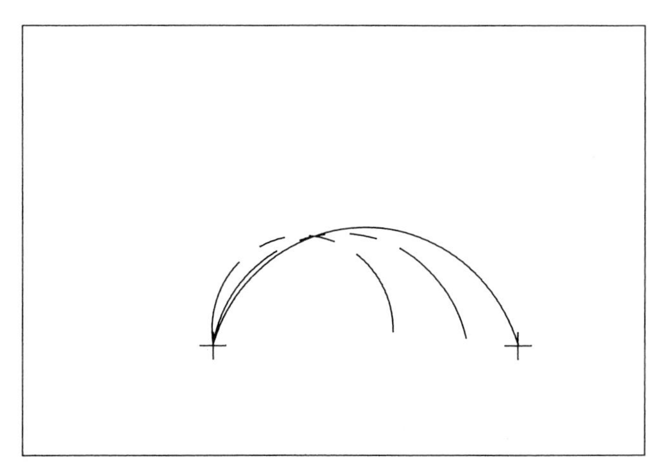

Fig. 9-4. One way to draw an arc is to select the arc's start and end points. The dashed lines represent the phantom arcs that form on the computer monitor as the pointing device is moved. Once the end point of the arc is picked, the arc stops moving. Only the final arc remains on the screen.

Fig. 9-5. This symbol library contains a variety of fasteners.

INCH SQUARE, ROUND, & HEX BOLTS
HEX
HEAVY HEX
SQUARE
SLOTTED CAP SCREWS
FINISH
BOLT
CAP SCREW
SCREW
STRUCT
BOLT
BOLT
ASKEW
FLAT CSK.
ROUND
FILLISTER
COUNTERSUNK
NECK
EYEBOLT
ROUND
T-HEAD
STEP
FLAT
114 DEG
SLOTTED
FIN
SHORT SQR
RIBBED
SQUARE
REGULAR
SHOULDER

Most CAD systems can also produce **layers**. These are similar to the transparent overlays used in manual drawings. Different parts of a drawing can be placed on different layers. The layers can be displayed and plotted in different colors. This feature is especially useful for very complex drawings. Fig. 9-6.

APPLICATIONS OF COMPUTER-AIDED DESIGN

Throughout American industry and business, CAD systems are being used to make high quality products and deliver better services. Many people, from architects to yacht designers, use CAD to replace or supplement manual drafting. But CAD systems can do more than draw. The following sections describe other applications of CAD.

Parametrics

One of the most common uses of CAD systems is in designing families of parts; that is, similar parts that can be applied to different situations. Common examples include fasteners, couplings, gaskets, doors, and windows.

The basic characteristics for each family of parts are stored in the computer. When a new part must be designed, the "family" drawing is called up from storage. The computer asks the user for input—size, material, etc. Using this information, the computer designs the new part. This process is called **parametric design**.

Stress Analysis

When engineers design dams, bridges, large buildings, and other structures, they need to determine whether the structure will be strong enough. CAD aids in structural design with **stress analysis**. A model of the structure is created on the computer. Then the model is analyzed and results are interpreted. Analysis considers the type and location of stresses and strains

Fig. 9-6. This drawing has layers, each with a different color. Each layer can be displayed or hidden as needed, and the user can also choose which layers to plot.

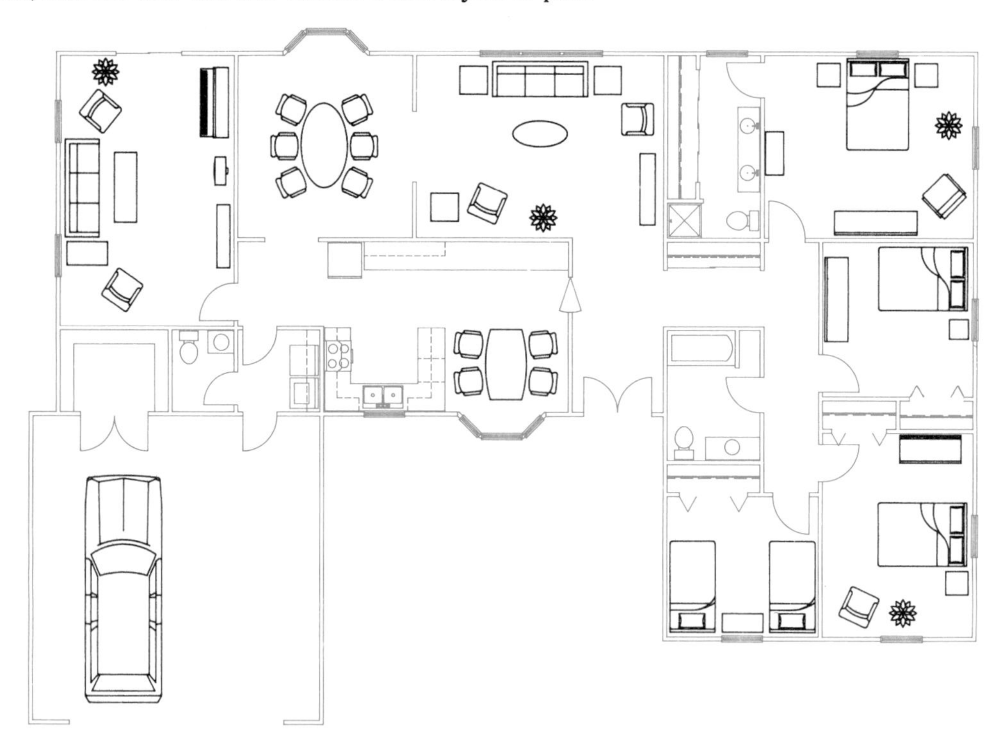

and the ability of a material to resist wear and corrosion. This type of analysis often results in some form of design change. Fig. 9-7.

Computerized analysis saves time because computers can figure mathematical problems rapidly. It also cuts back on the need for building prototypes and then testing them.

SCIENCE FACTS

Stress and Strain

People aren't the only ones who suffer from stress. It also happens to buildings, bridges, machines—in fact, all things are subject to stress. Of course, for nonliving things, *stress* has a different meaning. It is the force created inside a material by forces acting on it from the outside.

Different forces produce different kinds of stress. For example, tensile stress results when forces tend to elongate (stretch) a material. Compressive stress is the opposite; it results from forces that squeeze or crush. Shearing stress can cause a material to separate, like layers of a cake sliding apart.

Sometimes a material changes size or shape as a result of stress. The change is called strain. The greater the stress, the greater the strain. Finally, the material breaks apart. The type of material, its size, the strength of the forces acting on it, and other factors all affect the material's response to stress. Engineers do stress analysis to determine whether the structure or product they are designing will withstand the stresses it will undergo during use.

Modeling

As with manual drafting, most of the drawings done with CAD are two-dimensional. The illusion of depth can also be created with any CAD system. This includes axonometric, oblique, and perspective drawings. However, with some CAD systems, true 3-D drawings can be made. On a true 3-D system, the user can move the object, called a model, along X, Y, and Z axes. The user can also change viewpoints and "travel around" the model while it remains in place.

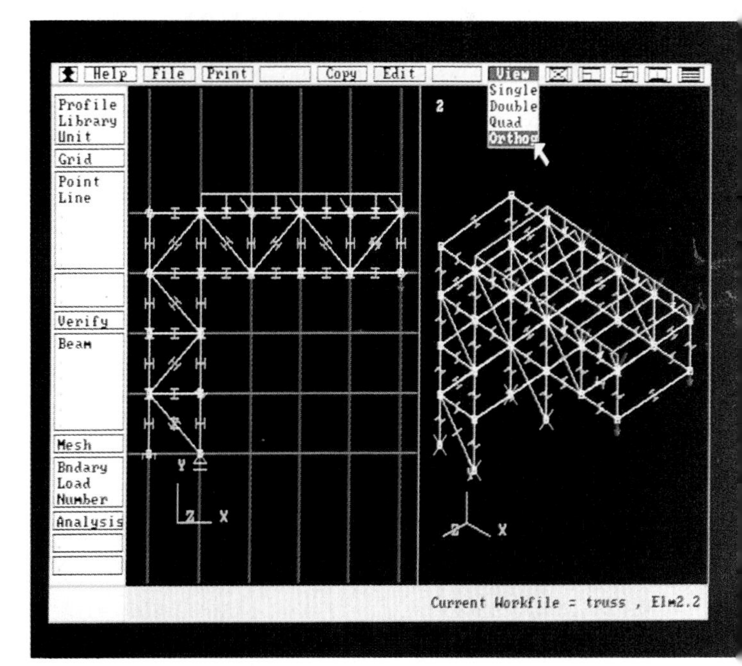

Fig. 9-7. Computerized stress analysis can tell whether a structure will withstand the forces acting on it.

Three types of 3-D models can be made. A **wireframe model** appears as if it were made of bent wires. Fig. 9-8. Any hidden lines can be removed to give the drawing a more realistic look.

A **surface model** is like a wireframe model with a skin over it. Fig. 9-9. By adding shading and color, a drafter can make the object appear solid, although mathematically it is not.

A true **solid model** behaves as a real-life solid object. Fig. 9-10. It contains information about physical properties, such as density, volume, and weight. This type of model is used for stress analysis. True 3-D models cannot be plotted on paper. They must be machined or sculpted.

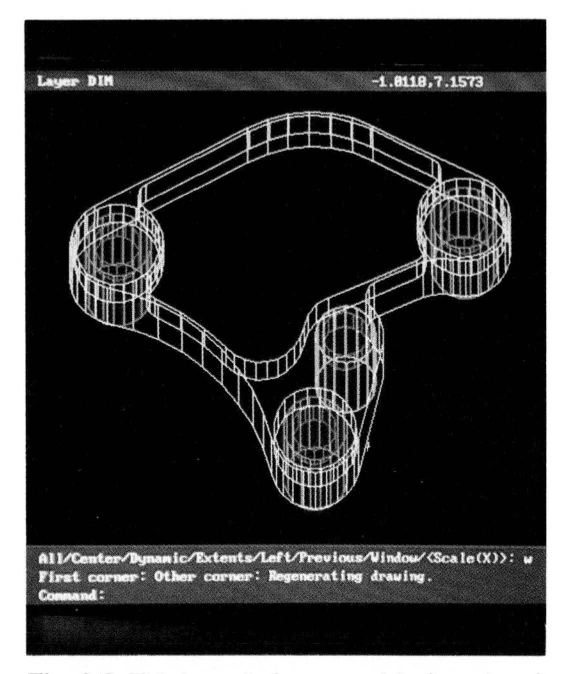

Fig. 9-8. This is a wireframe model of an aircraft hinge.

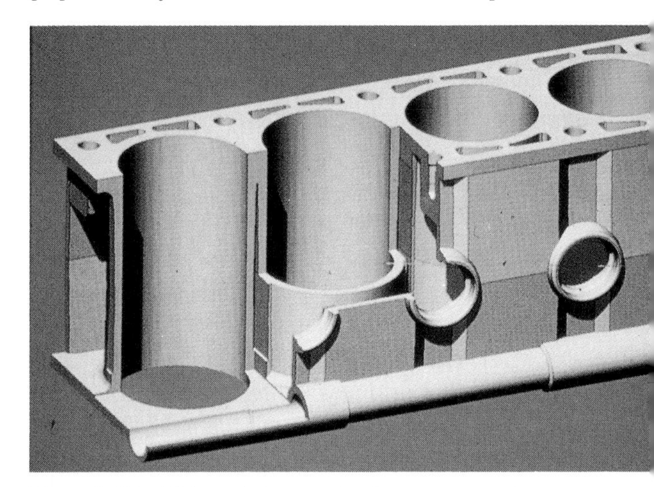

Fig. 9-10. This solid model shows part of an automobile engine.

Fig. 9-9. A surface model looks more realistic than a wireframe model.

CAD and CAM

Until recently, CAD was used primarily for drafting. CAM (computer-aided manufacturing) was viewed as little more than a way to run machine tools with a computer. Even today, CAD/CAM is not used to its full potential—that is, to bridge the gap between product design and production. The goal of CAD/CAM is to design a part on a CAD system, then send the drawing electronically to a machine that produces the part.

Creating the drawing is similar to other drawing on CAD. However, CAM requirements are kept in mind. Then the drawing is sent to the CAM system. CAM converts it into instructions for a computer-controlled machine. Fig. 9-11.

New developments in both hardware and software are improving the CAD/CAM picture. For example:

- Solid modeling software is giving engineers an alternative to wireframe drawings.
- Improvement of hardware now makes it possible to do complex tasks at local workstations, rather than at remote mainframe computer centers. These new high-powered workstations are capable of performing operations such as solid modeling much faster.
- Local computer networks are speeding data transfer within and between plants. Direct links can now be made among engineering areas such as design, analysis, manufacturing, and testing.
- Graphics standardization is making designs more universal. Standards such as the Initial Graphics Exchange Standard (IGES) and General Motor's Manufacturing Automation Protocol (MAP) are gaining wider acceptance among software producers and users.

CAD/CAM systems, such as the Integrated Computer-Aided Engineering and Manufacturing (ICEM) system produced by Control Data, combine 3-D design, 2-D drafting, modeling, and machine programming into a single software package. Fig. 9-12.

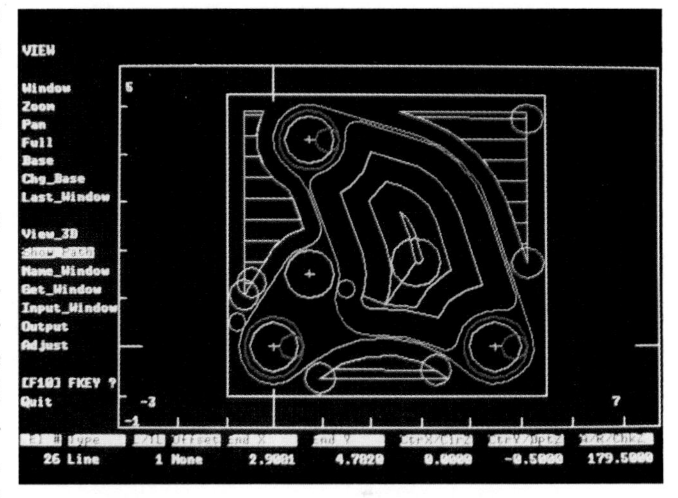

Fig. 9-11. In CAD/CAM the part is designed on a CAD system, loaded into a CAM system, and manufactured by a computer numerically controlled (CNC) tool. Shown on the CAM system here is the aircraft hinge from Fig. 9-8.

Fig. 9-12. Here are graphics produced by the ICEM system. Shown (left to right) are a solid model of an automobile engine, a finite element mesh derived from the solid model and used for structural analysis, and a surface model used to produce numerical control tapes.

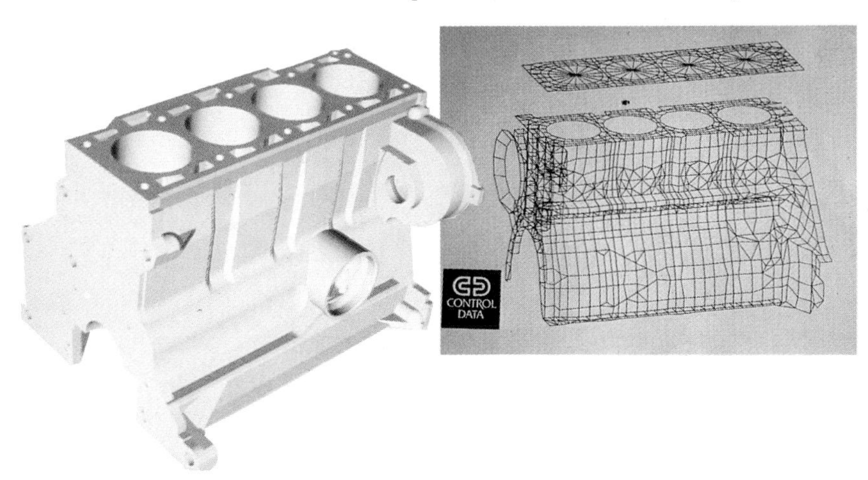

Combining CAD with Non-Drafting Software

CAD can also be used with other software for preparing bills of materials and spreadsheets. Not only does the system format the text, it carries out calculations and shows the effects of any changes.

For desktop publishing, CAD and word processing software allow users to prepare camera-ready text and illustrations. These can then be used for newsletters, annual reports, and company presentations. Fig. 9-13. (See also Chapter 13).

Database files can also be coupled with CAD. For example, a large building, like a school, contains hundreds of electrical fixtures. Information on these fixtures can be kept in a database and later retrieved as needed.

Map-Making

A single map can show only a limited amount of information. However, by combining a map with computer data, the map's usefulness is increased. This is especially important to state and local governments, who must keep up-to-date records. Using CAD, information can be keyed to a base map having different layers.

Fig. 9-13. This newsletter is an example of how CAD drawings can be combined with text and other graphics.

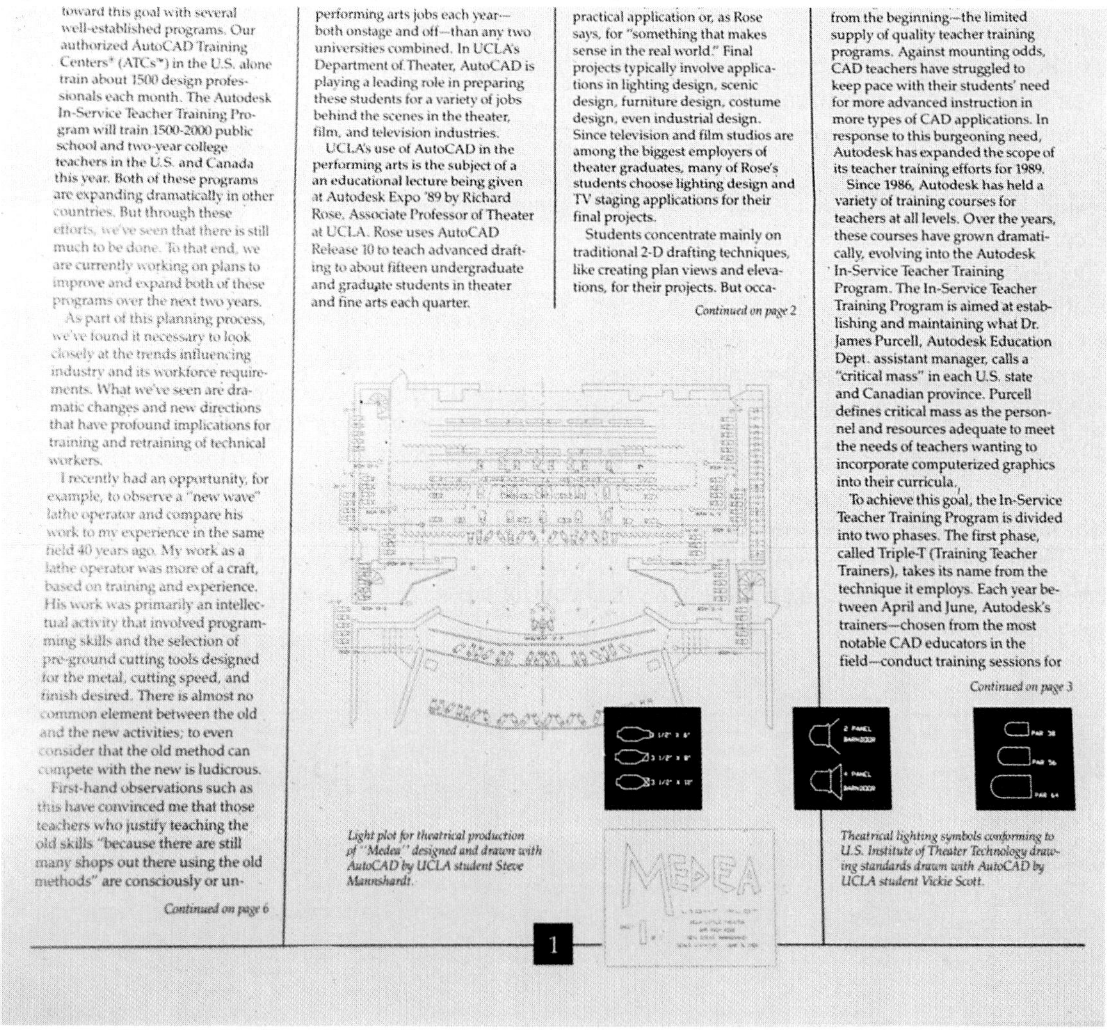

toward this goal with several well-established programs. Our authorized AutoCAD Training Centers* (ATCs™) in the U.S. alone train about 1500 design professionals each month. The Autodesk In-Service Teacher Training Program will train 1500-2000 public school and two-year college teachers in the U.S. and Canada this year. Both of these programs are expanding dramatically in other countries. But through these efforts, we've seen that there is still much to be done. To that end, we are currently working on plans to improve and expand both of these programs over the next two years.

As part of this planning process, we've found it necessary to look closely at the trends influencing industry and its workforce requirements. What we've seen are dramatic changes and new directions that have profound implications for training and retraining of technical workers.

I recently had an opportunity, for example, to observe a "new wave" lathe operator and compare his work to my experience in the same field 40 years ago. My work as a lathe operator was more of a craft, based on training and experience. His work was primarily an intellectual activity that involved programming skills and the selection of pre-ground cutting tools designed for the metal, cutting speed, and finish desired. There is almost no common element between the old and the new activities; to even consider that the old method can compete with the new is ludicrous.

First-hand observations such as this have convinced me that those teachers who justify teaching the old skills "because there are still many shops out there using the old methods" are consciously or un-

Continued on page 6

performing arts jobs each year—both onstage and off—than any two universities combined. In UCLA's Department of Theater, AutoCAD is playing a leading role in preparing these students for a variety of jobs behind the scenes in the theater, film, and television industries.

UCLA's use of AutoCAD in the performing arts is the subject of a an educational lecture being given at Autodesk Expo '89 by Richard Rose, Associate Professor of Theater at UCLA. Rose uses AutoCAD Release 10 to teach advanced drafting to about fifteen undergraduate and graduate students in theater and fine arts each quarter.

practical application or, as Rose says, for "something that makes sense in the real world." Final projects typically involve applications in lighting design, scenic design, furniture design, costume design, even industrial design. Since television and film studios are among the biggest employers of theater graduates, many of Rose's students choose lighting design and TV staging applications for their final projects.

Students concentrate mainly on traditional 2-D drafting techniques, like creating plan views and elevations, for their projects. But occa-

Continued on page 2

from the beginning—the limited supply of quality teacher training programs. Against mounting odds, CAD teachers have struggled to keep pace with their students' need for more advanced instruction in more types of CAD applications. In response to this burgeoning need, Autodesk has expanded the scope of its teacher training efforts for 1989.

Since 1986, Autodesk has held a variety of training courses for teachers at all levels. Over the years, these courses have grown dramatically, evolving into the Autodesk In-Service Teacher Training Program. The In-Service Teacher Training Program is aimed at establishing and maintaining what Dr. James Purcell, Autodesk Education Dept. assistant manager, calls a "critical mass" in each U.S. state and Canadian province. Purcell defines critical mass as the personnel and resources adequate to meet the needs of teachers wanting to incorporate computerized graphics into their curricula.

To achieve this goal, the In-Service Teacher Training Program is divided into two phases. The first phase, called Triple-T (Training Teacher Trainers), takes its name from the technique it employs. Each year between April and June, Autodesk's trainers—chosen from the most notable CAD educators in the field—conduct training sessions for

Continued on page 3

Light plot for theatrical production of "Medea" designed and drawn with AutoCAD by UCLA student Steve Mannshardt.

Theatrical lighting symbols conforming to U.S. Institute of Theater Technology drawing standards drawn with AutoCAD by UCLA student Vickie Scott.

MEDEA

1

For example, a city street map could be used to show locations of street lights, mailboxes, or sewer lines. Population growth, soil types, and stream beds can also be indicated. Fig. 9-14.

Several software packages now available help design maps in many colors and patterns. Other software allows calculating sizes of surface features, such as lakes.

Fig. 9-14. This city map was produced with a CAD system.

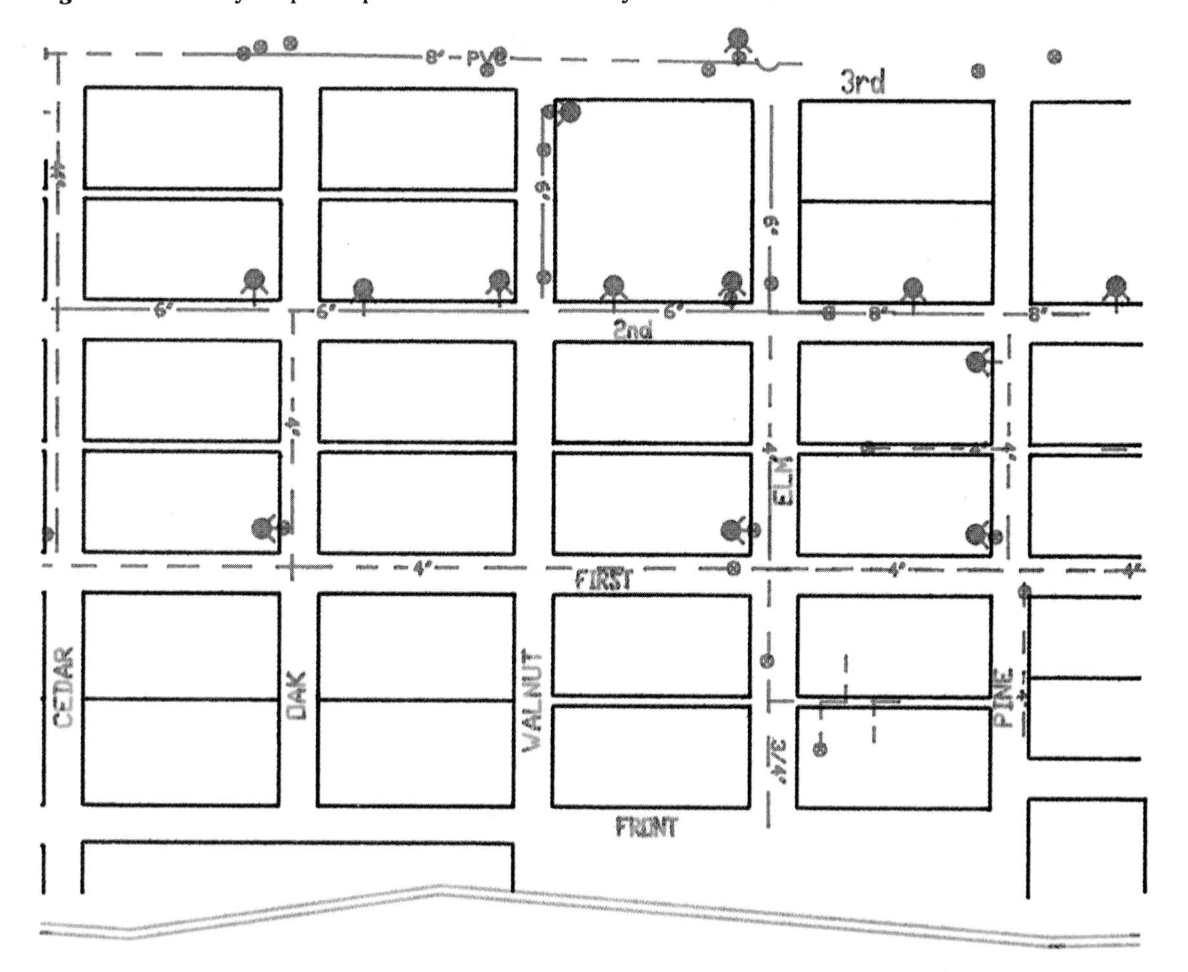

SIMULATORS GIVE FLIERS AN EDGE

Landing an attack plane on an aircraft carrier at night can be scary. The wrong decision can cause a serious accident. Today, Navy pilots' first night landings are safer than they used to be. That's because pilots in the armed forces have already practiced the landing several hundred times in a flight simulator. *Flight simulators* are machines that reproduce what a pilot sees in the plane's cockpit.

The first visual simulator was invented in 1929 by Edwin A. Link to teach flying with instruments. It used a motion picture to simulate what the pilot saw. Since 1975, more flight simulators have been linked to computer generated imagery (CGI) systems. These systems reproduce images of continually changing terrain as it would look out the windows of an aircraft. The landscapes are shaded, three-dimensional images. Large areas, moving targets, various times of the day (sun positions), and different weather conditions, such as haze and fog, can all be simulated. At first, CGI images looked more like cartoons. In recent years, they have become more lifelike.

CGI systems operate with a database. As the simulator "flies" over an area, information from the database continually updates the image with new information.

Next, a "geometric processor" makes mathematical calculations that project three-dimensional models of the information onto the display screen.

Within the last few years, advancements have added greater realism to simulators. General Electric has developed a procedure that adds textures to three-dimensional objects. Honeywell's Systems and Research Center has developed computer generated synthetic imagery (CGSI), using photography. CGSI stores photographs digitally on video-disks. The photographs are then the basis for the simulated scenery.

CHAPTER 9 REVIEW

Review Questions

1. How does CAD drafting differ from traditional drafting?
2. Name three functions a CAD system can perform that would aid a drafter in routine tasks.
3. What input and output devices are used for CAD?
4. How does a flatbed plotter differ from a drum plotter?
5. What is a symbol library and how is it used?
6. What are layers and how are they used?
7. How does CAD aid in stress analysis?
8. How does a surface model differ from a solid model?
9. How is CAD used for creating bills of materials?
10. How can local governments use CAD for record keeping?

Activities

1. Research how computers are used to design special effects for films like "Star Wars." Report your findings to the class.
2. Contact several advertising firms in your local area. Ask if they use computers for designing graphics. Find one that does and ask if you can visit the graphics department. Report on your visit to the class.
3. If your school has a CAD system, use it to create the drawing in Fig. 9-15.
4. If your school has a CAD system, use it to design a simple object such as a doghouse, picture frame, or child's sandbox.
5. Check magazines for information on new developments in CAD. Write a report on your findings. Good sources include *CADENCE*, *CADalyst*, and *MicroCAD News*.

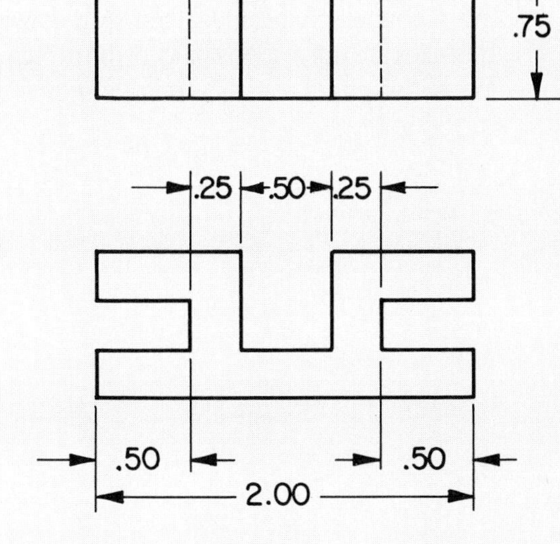

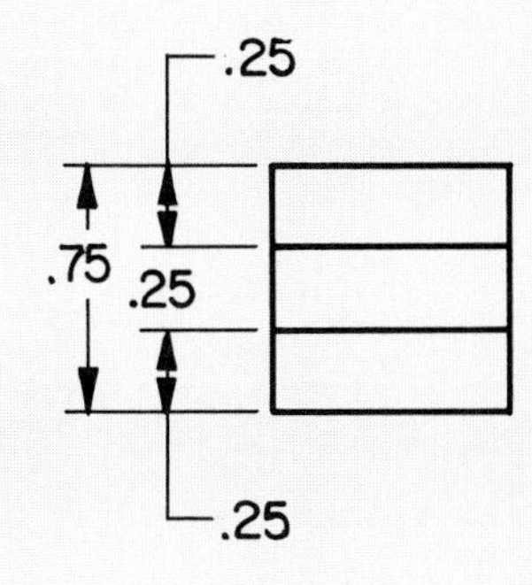

Fig. 9-15.

Careers

Drafter

The nature of a technical design department depends upon the company. The smaller the business, the fewer career levels and the broader the responsibilities for each worker. On the other hand, those working in large companies may find their duties and the climb up the "ladder" more restricted.

Regardless of the type of training and education, newly graduated drafters must start at or near the bottom of the career ladder. Those who have more education, however, usually advance at a much faster rate.

Specific titles and number of career levels vary from one firm to another. However, most drafters fall into these categories:

- The tracer, or first level drafter, is one with little or no on-the-job experience and a minimum of training. To help a tracer develop basic drawing skills and efficiency, she or he is responsible for drawing plans from prints by tracing them.
- The junior drafter's is the beginning position found in many small and medium-sized departments. These drafters trace drawings and make prints. They are expected to be able to make minor corrections and changes on existing drawings.
- Detailers are the backbone of an engineering and design department. They are responsible for making detail drawings of all parts used for planning and production purposes. These individuals have good drawing skills and are competent in reading and understanding all kinds of drawings.
- A layout drafter is skilled as a detailer. Layout drafters must also be able to produce layout drawings from design sketches.
- The checker is the next level in an engineering department. Checkers review all drawings, including revised drawings, details, and layouts, to make sure that they are correct. Checkers also make revisions in layout and detail drawings.
- Designers are at the upper end of the technical design profession. These individuals develop new products and devices and manage all drafting personnel serving under them. Any drawing mistakes are the responsibility of the designer. Many designers are also trained engineers.

Education

Anyone interested in a career in drafting and design will have a better chance of securing a position if he or she has completed a formal drafting program. Because of the complex nature of today's technology, more companies want drafters who have well-rounded backgrounds.

Programs in drafting and design are offered at vocational and technical high schools, postsecondary vocational schools, two-year colleges, and four-year colleges and universities. Less formal training programs are also available through on-the-job training, apprenticeships, and in the Armed Forces.

Two-year degree programs concentrate on drafting, illustration, and design principles. Four-year degree programs give a broader background in related science and math courses.

For individuals who are interested in highly-specialized work, advanced degree programs are offered by some universities. Topics such as engineering graphics, graphical mathematics, computer graphics, and fourth-dimensional descriptive geometry are covered.

For More Information

To learn more about drafting as a career, contact:

Industrial Designer's Institute
441 Madison Avenue
New York, NY 10022

Society of Illustrators
128 East 63rd Street
New York, NY 10021

Bureau of Apprenticeship and Training
U.S. Department of Labor
Washington, D.C. 20036

Aerospace Industries Association of America
1725 DeSales Street, N.W.
Washington, D.C. 20036

Correlations

 *Language Arts*

1. Prepare a one-page written description of a common household tool or appliance that has five or more parts. In your description, give information about size, parts, materials, how it is put together, and any special features.
2. Notes placed on drawings often contain abbreviations. Find out what the following abbreviations mean. (Most textbooks will have this information.) On a separate piece of paper, write your answers.

AL	CDRILL	DIA	MATL	STL
AMP	CHAM	FAO	MAX	SUR
APPROX	CL	GSKT	MIN	TAN
AUX	CRS	HEX	OD	THK
BTU	CSK	ID	REQD	TOL
CBORE	CTR	KWY	SQ	TYP

Science

1. When drawings that have been made on paper are exposed to sunlight over a long period of time, the paper turns yellow, becomes brittle, and tears easily. Research why this happens. Make a display for the class showing the effect of ultraviolet light on cellulose fibers.
2. Explain what happens when an ozalid print is made. Find out what the ammonia-sensitive coating on the print paper is made of. What is the chemical reaction that occurs when the coating is exposed to ammonia vapors?

Math

1. Most drawings are made to scale. The drafter needs to convert the actual size of an object to its scale size. Suppose you were drawing a cylinder whose actual length was 8″ and whose actual diameter was 2″. At the following scales, what would be the cylinder's length and diameter on a drawing?
   Half size (1:2)
   Quarter size (1:4)
   1:10
   2:1
2. Tolerancing is sometimes given in terms of percentages. From the percentage, the drafter has to calculate the tolerances. Give the actual dimensions for the following:
   7.50 ±0.02% tolerance
   4.55 −0.0045% tolerance
   6.000 +1.25% tolerance
   0.500 ±0.185% tolerance
   1.855 ±2.05% tolerance

Social Studies

1. Make a simple floor plan of your school showing concentration of different age groups, subject areas, or some other factor. If possible, make the drawing on a CAD system.
2. Drawings have often played a role in espionage. (It's easier to carry away the plans for a new weapon or aircraft than it is to take the item itself.) Find out who Alfred Frauenknecht was and how he helped the Israelis obtain the plans for Mirage fighter planes in the 1970s.
3. In manual drafting, three-dimensional real-world objects are drawn in two dimensions. Can you imagine a world in which there are only two dimensions? Edwin Abbott, a scholar in Victorian England, did. He called this world *Flatland*. Find out more about Flatland and report to the class.

Basic Activities

Basic Activity #1: Sketching

For this activity, select three small, simple items used at school or in your home, such as a pencil holder, a tape dispenser, and a notepad.

Materials and Equipment
scale
medium-soft pencil
graph paper with 4 divisions to the inch
objects to be drawn

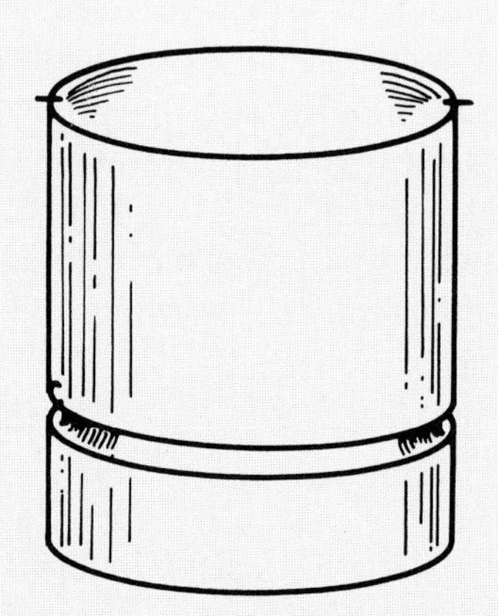

Procedure

1. Using the scale, measure the objects to the nearest ¼ inch.
2. Sketch the objects full size. Show as many views (top, front, side) as are needed to fully describe the shape of the objects.

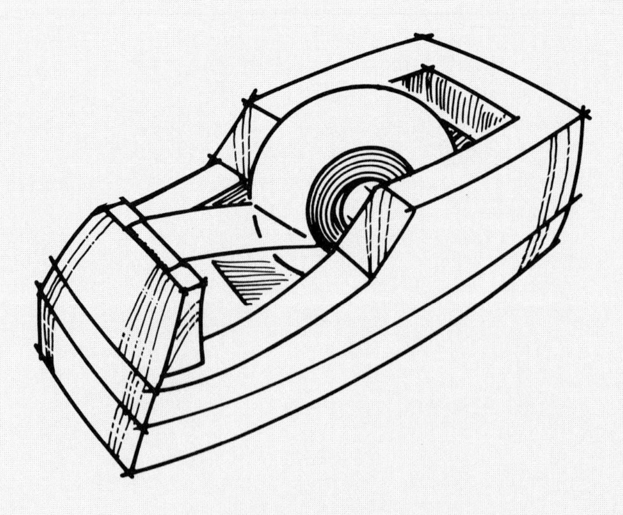

Basic Activity #2: Two-View Drawing

Do a two-view drawing of the cylinder illustrated in Fig. III-1.

Materials and Equipment
drafting table
paper
pencils
parallel straightedge
triangles
scale

Procedure

1. Using the given dimensions, lay out the drawing on a sheet of drawing paper.
2. Darken and fully dimension the drawing.

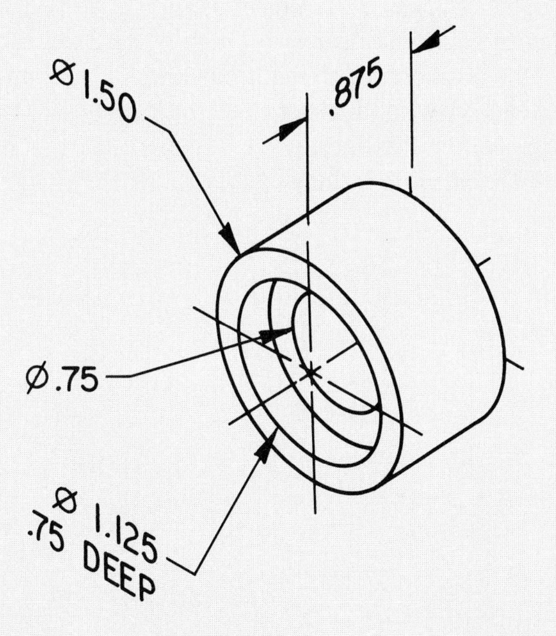

Fig. III-1.

Basic Activity #3: Multiview Drawing

Make a multiview drawing of a simple hand tool.

Materials and Equipment
drafting table
paper
pencils
parallel straightedge
triangles
scale
object to be drawn

Procedure
1. Using the scale, measure the object and note all dimensions to the nearest 1/64 inch.
2. Lay out the positions for all required views.
3. Draw in each view lightly. After checking for accuracy, darken the lines.
4. Add dimensions to the drawing.

Basic Activity #4: Detail Drawings of the Hand Tool

Using the same hand tool drawn in Activity #1, make detail drawings of its parts.

Materials and Equipment
drafting table
paper
pencils
parallel straightedge
triangles
metric scale
object to be drawn

Procedure
1. Take accurate measurements of each part, using a metric scale this time. Measure to the nearest millimeter.
2. Make a detail drawing for each part. Make sure that you have all the required views.
3. Dimension all the drawings.

Intermediate Activities

Intermediate Activity #1: Inked Drawing

In this activity, you will prepare an inked drawing of the product in Fig. III-2. Save the drawing for later use.

Materials and Equipment

drafting table
paper
pencils and drawing pens
parallel straightedge
triangles
scale

Procedure

1. Lay out the drawing based on the sizes shown.
2. Fully dimension the drawing, using *decimals*, not fractions.
3. Prepare a finished, inked drawing.

Intermediate Activity #2: Creating a Drawing on a CAD System

Make the drawing shown in Fig. III-2, using a CAD system.

Materials and Equipment

CAD system, including plotter

Procedure

1. Lay out the drawing based upon the given dimensions.
2. Fully dimension the drawing.
3. Compare the result with the drawing you did for Intermediate Activity #1.

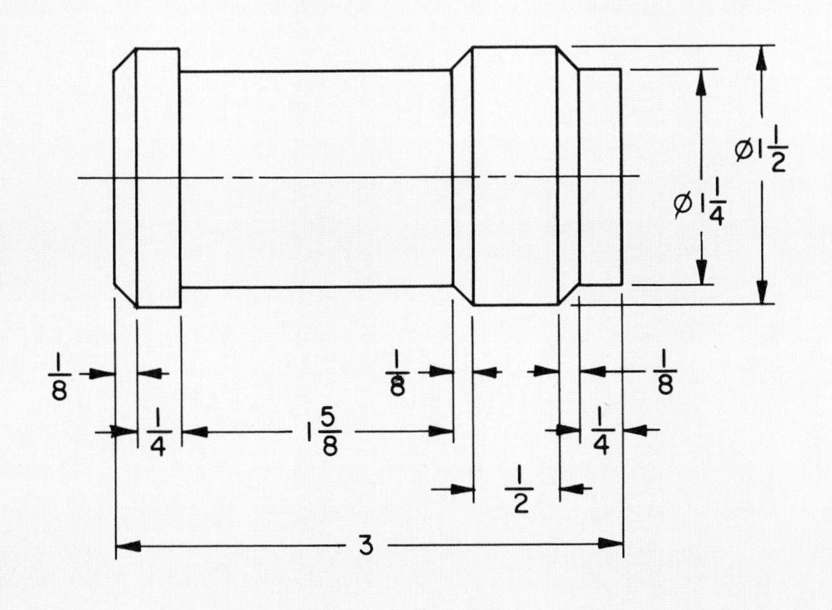

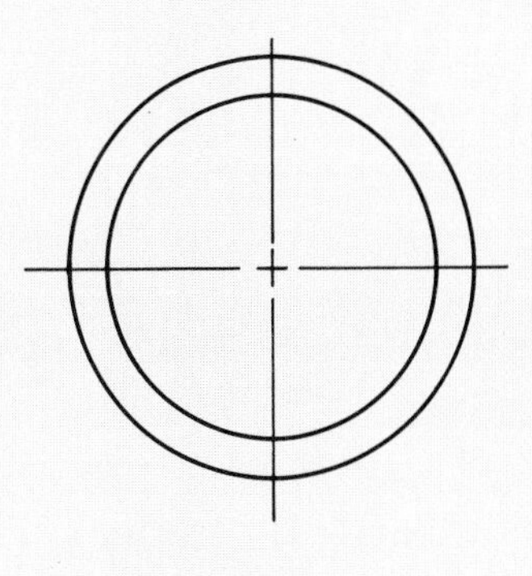

Fig. III-2.

Intermediate Activity #3: Section View

Draw a section view of the hand tool you used for Basic Activity #3.

Materials and Equipment
drafting table
paper
pencils
parallel straightedge
triangles
scale
object to be drawn

Procedure
1. Select a section view of the tool that will describe it more clearly.
2. Draw the view lightly and check for accuracy. Darken the drawing.
3. Add decimal dimensions.
4. Prepare a note indicating the material from which the tool is made.

Intermediate Activity #4: Assembly Drawing for a Product

Select a common household item that has three or more parts and prepare an assembly drawing.

Materials and Equipment
drafting table
paper
pencils
parallel straightedge
triangles
scale
object to be drawn

Procedure
1. Draw the item as it would appear fully assembled.
2. Identify each part and prepare a parts list. Place specifications on the drawing.

Advanced Activities

Advanced Activity #1: Designing a Container

Suppose a fast-food chain has asked you to design a new container for its hamburgers. The container must meet the following specifications.

- It must be large enough to hold a quarter-pound hamburger.
- It must have a lid so that the contents will not fall out.
- It must not let grease or moisture soak through.
- It must be biodegradable.

Materials and Equipment
drafting table
paper
pencils
parallel straightedge
triangles
scale
or
CAD system

Procedure

1. You may find it helpful to begin by sketching your ideas on graph paper. Then select the best one and prepare instrument drawings.
2. Be sure to include dimensions to show the sizes of all parts and notes to indicate what material is to be used.

Advanced Activity #2: Designing a Habitat

Design a habitat for a small pet, such as a hamster or a bird.

Materials and Equipment
drafting table
paper
pencils
parallel straightedge
triangles
scale
or
CAD system

Procedure

1. Determine the design specifications. What kind of animal will live in the habitat? What are that animal's requirements? For example, a bird would need perches of various diameters. What about food and water dishes? How can you make the habitat easy to clean?
2. Sketch your ideas on graph paper.
3. Make working drawings of your best design.
4. Prepare a technical illustration to show how the finished product would look.

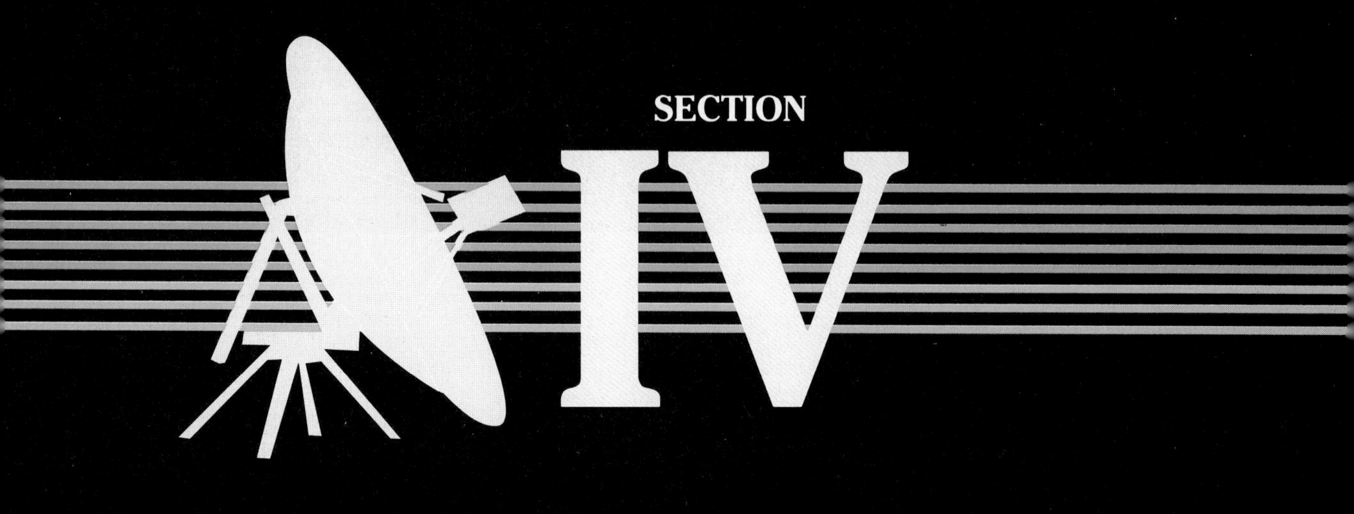

SECTION

IV

Optic Systems

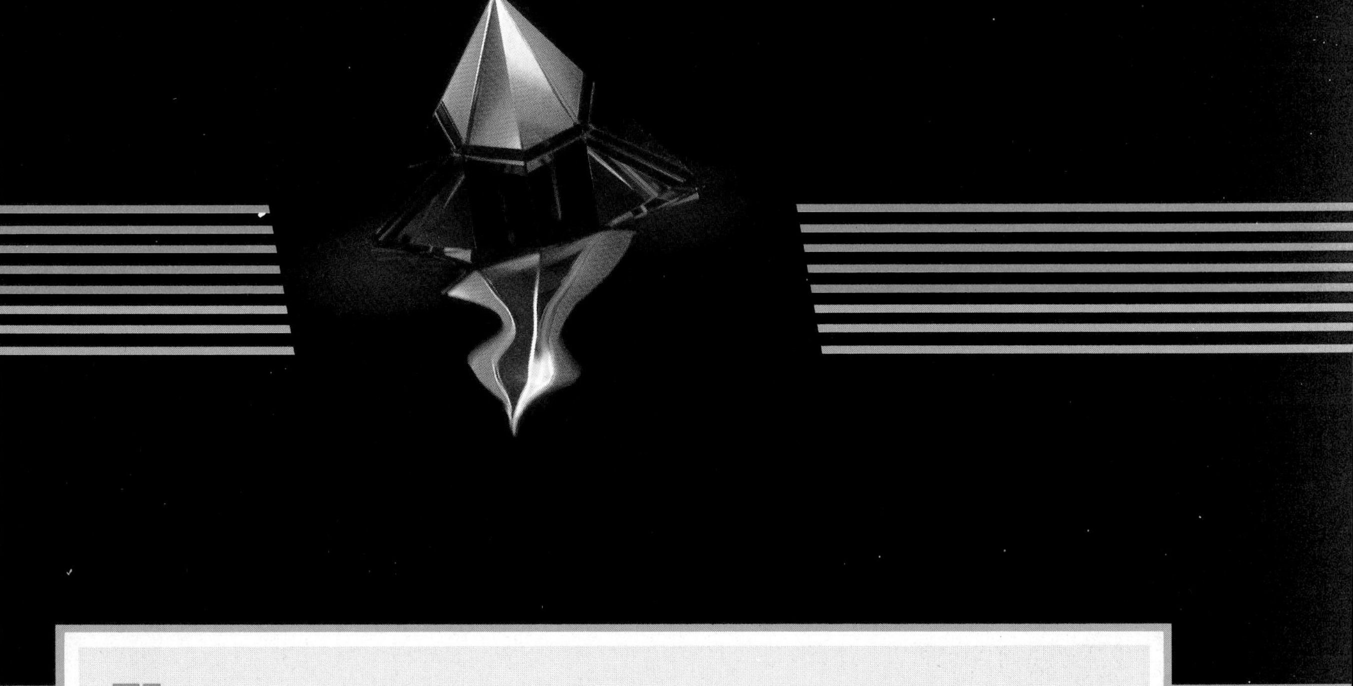

Your eye is the most fundamental of all optic systems . . . and the most complex. It records color, focuses instantly on objects both near and far, and takes in a wide area. Even with computers working on the task for years, no one has yet invented an optic system that even begins to equal the human eye. It is self-cleaning, adjusts quickly to different lighting conditions, has a built-in "lens cover" (the eyelid), changes direction in a flash, sees amazing detail, and lasts a lifetime. Can you name a camera that can do all that?

Like the human eye, optic systems focus and record light. Human-made optic systems have been around for centuries. The camera obscura was a darkened room with a hole in one wall (or the roof) that acted as a lens. Light reflected off objects outside was focused by the hole onto a flat surface inside the room, as shown in the drawing at left. In the 1500s, the camera obscura was used by Renaissance artists to assist them with their paintings. The room was made smaller and smaller until it became a portable box.

In 1727, Johann Schulze discovered that exposure to light would darken a silver nitrate solution. This discovery, combined with the camera obscura concept, led to the invention of photography. For more than a hundred years, the camera has been a popular and useful optic system.

In this section, you'll learn about several kinds of optic systems. You'll study, in detail, the most common application of all—photography. Through photography, you will gain a better understanding of how optic systems in general work.

Principles of Optic Systems

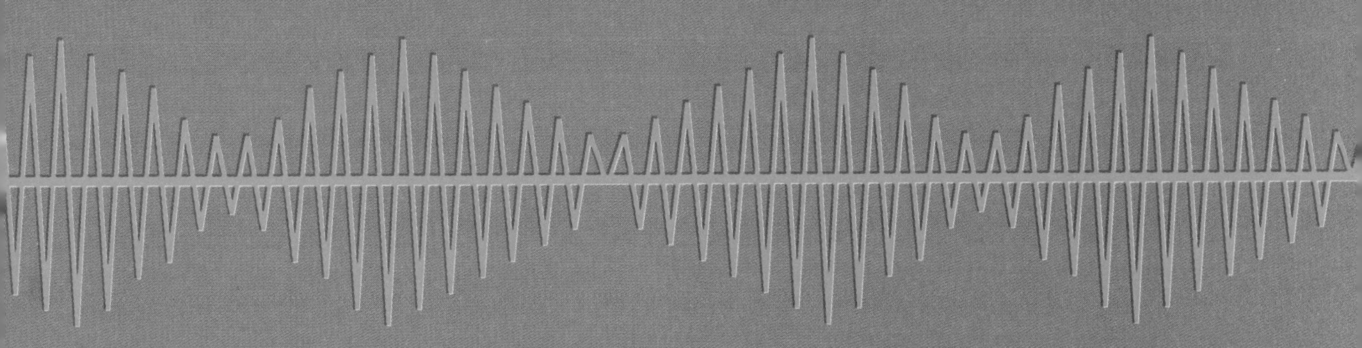

Optic systems use light to record an image or other type of information. To have an optic system you need a light source, a lens to focus the light, and a way to record the image. For example, both the human eye and a camera are optic systems. Our eyes take in the light all around us and our brains tell us what we see. Cameras take in light and record what they "see" on film.

Today, optic systems are being used to create three-dimensional images and to send voices over long distances. In this chapter you will learn about light, the lenses used to focus light, and several different optic systems.

Terms to Learn

additive primary colors
amplitude
focal point
frequency
holography
laser
lens
negative
optical fibers
optic system
photon
polarized
refracted
subtractive primary colors
visible spectrum

As you read and study this chapter, you will find answers to questions such as:

- What is light "made" of?
- What are the properties of light?
- Where do the colors we see come from?
- How does the human eye work?
- How are laser beams created?
- How are holographic images produced?
- What makes fiber optics so useful?

LIGHT

Light is a form of energy. Although light is all around us, flooding our universe, it is not easy for us to see how light works. For instance, did you know that light can turn a corner or that color is present in all light?

How Light "Happens"

Although light has been studied for centuries, scientists were unable to agree on what it consisted of until about 75 years ago. Even today, light still presents mysteries.

As you know from your science classes, all things are made of atoms. We imagine that atoms look like tiny solar systems, as in Fig. 10-1. At the center is a nucleus surrounded by electrons whirling in different orbits. These electrons do not always stay in the same orbits. They may leap from one orbit to another. When an electron leaps, there is a change in its energy. When it loses energy, the energy leaves in the form of an extremely tiny particle called a **photon**. These escaping photons make up what we see as light. Fig. 10-2.

Light, however, does not always act like a system of tiny particles. Sometimes it acts like a system of waves. For instance, light bends around corners the way waves do. Therefore neither explanation of what light is can be considered complete. It seems to show both particle and wave properties, depending on the situation.

Fig. 10-1. An atom consists of a nucleus surrounded by electrons. The electrons orbit the nucleus.

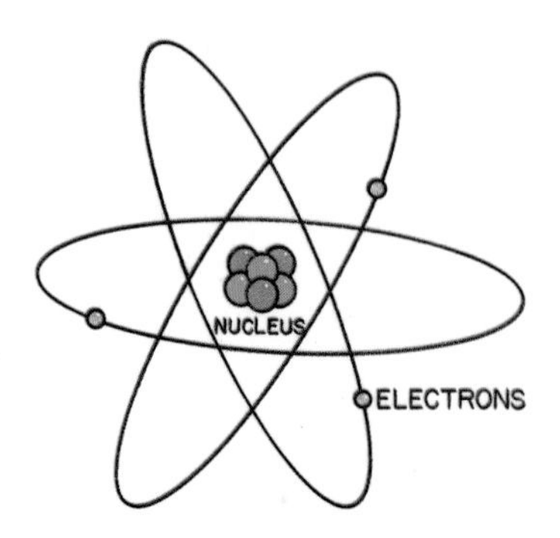

Properties of Light

Light possesses several properties. It can be measured in different ways. Its patterns of travel can be changed. It consists of many colors, and different colors of light have different temperatures.

Measuring Light

Light waves have frequency, amplitude, length, speed, and direction. Fig. 10-3.

- **Frequency** is the number of waves that pass through a given point in one second. Fig. 10-3. About 600 trillion visible light waves strike our eyes every second. Some frequencies are so high we cannot see the light produced. Ultraviolet waves are an example.
- **Amplitude** measures the intensity (strength) of light. A wave's amplitude is its height. Amplitude is measured from the top of a wave to its midpoint. Fig. 10-3.
- The length of a light wave is measured from a point on one wave to the same point on the next. Fig. 10-3. The length of visible light waves is about 0.00005 centimeters.
- The speed of light always remains the same—186,000 miles per second.
- Light waves radiate outward in every direction. (Light is a form of radiant energy.) Unless something interferes with light waves, they travel in a straight line.

Fig. 10-2. When the electron at point (A) is excited by an electric charge, it leaps to another orbit (B). When it returns to its normal orbit (C), it releases a photon.

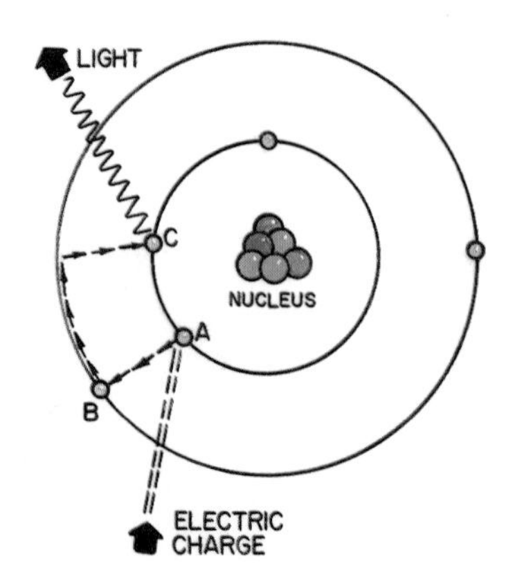

TECHNOLINKS

QUANTUM MECHANICS—HOW WE CAME TO UNDERSTAND LIGHT

Quantum mechanics is the field of physics that describes the nature of both the atom and light. Although it was not recognized as a separate field until the 1900s, the roots of quantum mechanics go back to 1666. This was when the English scientist, Sir Isaac Newton, proposed that light consisted of tiny particles. About the same time, a Dutch physicist, Christian Huygens, suggested that light was made of waves. Who was right? Scientists argued about the two theories for over 100 years. Gradually, the field of quantum mechanics took shape.

Between 1800 and 1864, the idea that light was made of waves gained in popularity. Scientists demonstrated that light beams could cancel each other under certain conditions. This was how waves in water acted. They also proved mathematically that vibrating electrical charges sent waves of energy traveling through space. Light waves were supposed to be caused by electrical charges in the atom. Therefore light must travel in waves, right?

However, in 1900 Max Planck, a German physicist, swung the argument in favor of particles when he announced that energy came in little packets he called *quanta*. Quanta were later renamed photons, which is what we call them today, but it was from "quanta" that quantum mechanics got its name. The old arguments began anew.

Then in 1905, the work of another German physicist, Albert Einstein, finally established that light consisted of energy particles having wave properties. Thus, both theories were right. The field of quantum mechanics

expanded rapidly. New scientists developed new ideas.

A Danish physicist, Niels Bohr, proposed the idea of the atom's electron structure in 1913. He also showed how atoms give off light. Louis de Broglie of France, Erwin Schrödinger of Austria, and Werner Heisenberg of Germany all developed forms of quantum mechanics. Soon all this knowledge was combined into what we know as modern quantum mechanics.

The study of quantum mechanics contributed greatly to the development of such important devices as microchips and lasers. Lasers, for example, were possible because scientists understood how light waves travel. Ordinary light waves move outward in all directions. Laser light waves move in step with one another. Because of the knowledge gained from quantum mechanics, scientists were able to theorize about lasers and eventually make them.

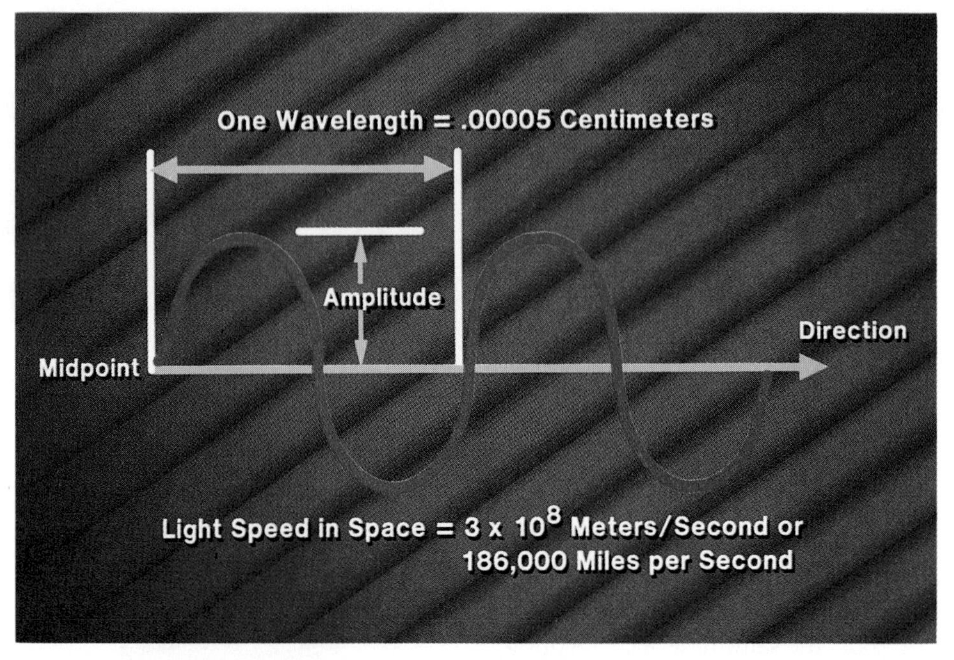

Fig. 10-3. Properties of light waves include frequency, amplitude, length, speed, and direction.

Patterns of Travel

Light passes (is transmitted) through clear objects, such as glass. When light waves bounce off an object, they are said to be reflected. Fig. 10-4. Some objects, such as those painted black,

Fig. 10-4. When light hits matter, it is either transmitted, reflected, absorbed, or refracted.

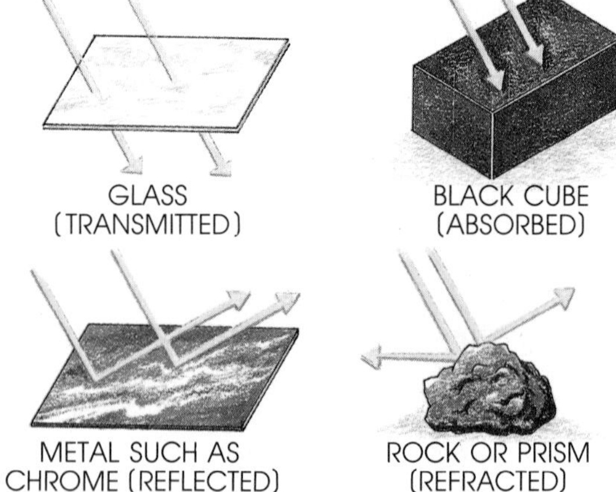

SCIENCE FACTS

Light—The Fastest Way to Travel

Light is the fastest thing in the universe. It travels at 186,000 miles per second. A beam of light travels from the earth to the moon in 1¼ *seconds*. It took the Apollo astronauts 4 *days* to make the same trip using rocket power.

absorb light. Light waves can also be **refracted**, or bent. The optical illusion in Figure 10-5 is an example of the refraction of light. The water in the glass bends the light so that when the image reaches your eyes the straw, too, appears bent.

Light can also be refracted to produce a rainbow. White light is a combination of many colors. Each has a different wavelength. When light passes through a prism, the different wavelengths bend in different amounts. Fig. 10-6. As the colors leave the prism, they leave separately.

When light is **polarized** its travel is limited to a single plane. This is done by sending the light through a filter. Fig. 10-7. The filter allows only those waves traveling in the right plane to pass through. Such filters, which cut down the glare of light, are used in cameras and sunglasses.

Fig. 10-5. Refraction of light makes the straw look bent.

LIGHT RAY

PRISM

Fig. 10-6. When white light passes through a prism, it is separated into its component colors.

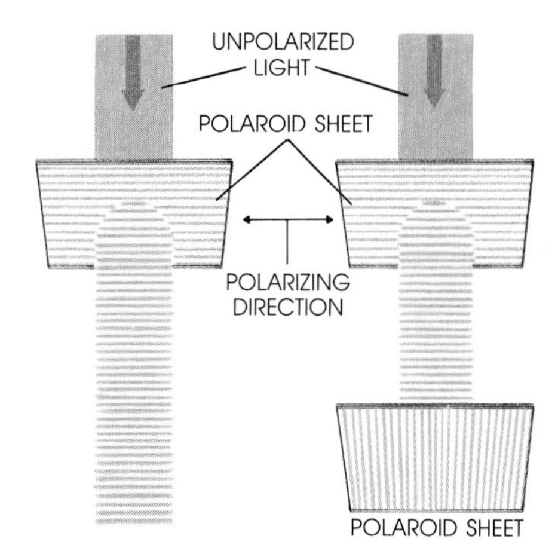

Fig. 10-7. Polarizing filters screen out all waves except those on a certain plane.

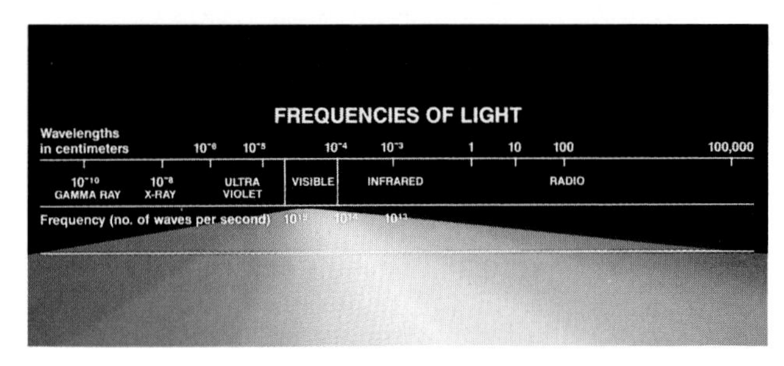

Fig. 10-8. Light can have many frequencies, but we see only a few of them (the visible spectrum). Within the visible spectrum, we see the frequencies as colors.

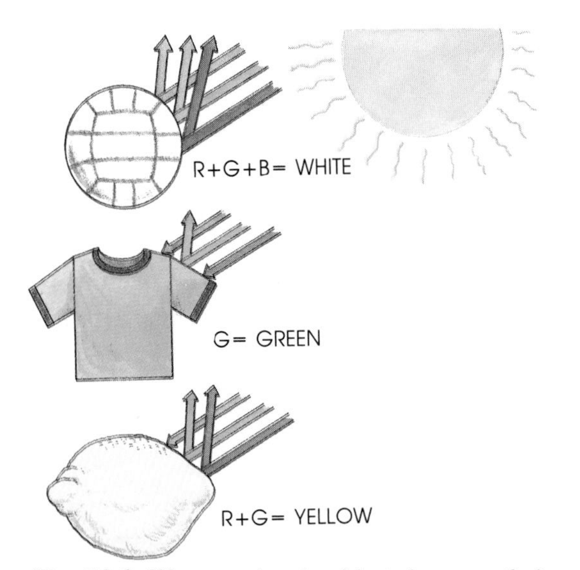

Fig. 10-9. We see colors in objects because their surfaces reflect certain wavelengths of light.

Color

White light is composed of many colors. We see white light when all colors of light are present at the same time. Sunlight is white light. When sunlight passes through a prism or through droplets of water, it is broken into the major colors of red, orange, yellow, green, blue, indigo, and violet. These colors are called the **visible spectrum**.

The different colors within white light have their own frequencies. Fig. 10-8. Red light has a frequency of about 460 trillion waves per second. Violet light has about 710 trillion waves per second. All other visible colors fall between those two.

The objects that fill our world get their colors from light. If light itself did not contain many colors, the world would be gray. All objects reflect some of the colors within light and absorb others. Fig. 10-9. The reflected colors are the ones we see. A leaf of lettuce, for example, reflects the wavelengths we see as green and absorbs all the others. Thus, the lettuce looks green. Objects that reflect all light appear white. Those that absorb all light are black.

Color has hue, intensity, and value. Hue is the name of a color. Red, green, and blue are the primary hues of light. Intensity is the brightness or dullness of a hue. If an object reflects only yellow light, the yellow will be very bright, because pure hues have the highest intensity. Value depends on the amount of light reflected. Not all hues have the same value. Pink is a light value of red; burgundy is a dark value.

Red, green, and blue are called the **additive primary colors** of light. This is because, when added together in different combinations, red, green, and blue can create all other colors. They are used to produce the different colors you see on a color television set. Look very closely at the screen of a color TV and you will see tiny red, green, and blue dots or stripes.

SCIENCE FACTS

Color Vision in Animals

Do animals see the same colorful world we see? In most cases, no. In general, animals that are diurnal (active in daylight) and have good vision can see color. Most monkeys and apes, for example, have color vision. However, cats see the world in black, white, and gray. Cats also are nocturnal (active at night). Dogs have rather poor vision, and they are color-blind. How do you think an animal's ability to survive in the wild might be affected by color vision?

When equal portions of red, green, and blue light are combined, white light results. Fig. 10-10. Notice that each overlapping of colors also creates a new color. These new colors are called **subtractive primary colors**.

Temperature

The balance of colors in light changes from one kind of light source to another. For example, some "white" light sources look blue while others look more yellow. This is because different light sources have different temperatures. The unit of measurement for the temperature of light is the kelvin. Figure 10-11 shows the kelvin temperature rating for different natural and artificial light sources.

Some colors look different under different light sources. This happens because the different sources produce wavelengths of different frequencies that are absorbed at different rates. In the printing industry a standard 5000 K light is used when color reproductions are checked to see if they match the original photograph. Fig. 10-12.

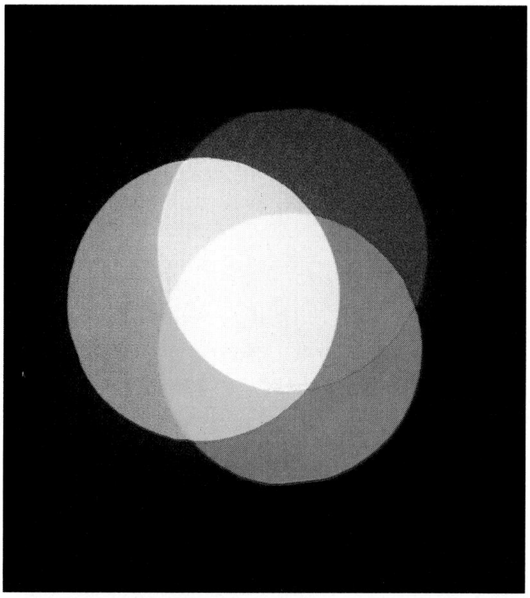

Fig. 10-10. White light is made up of equal parts of red, green, and blue light.

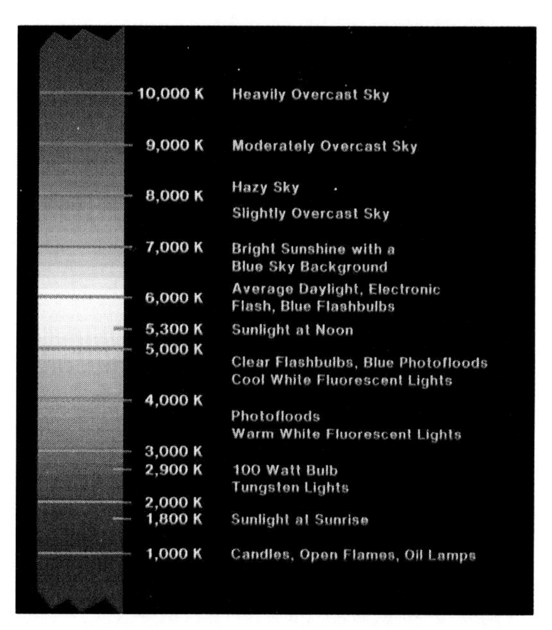

Fig. 10-11. Each light source projects a different color and temperature.

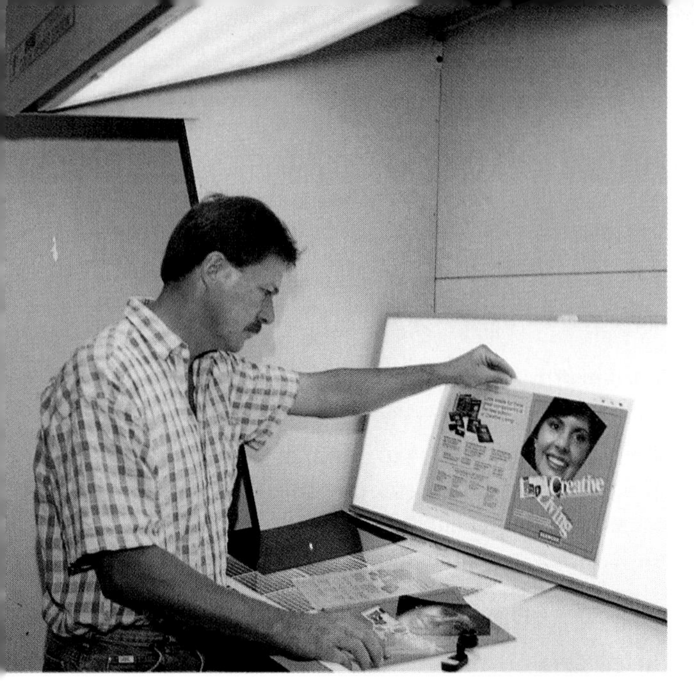

Fig. 10-12. Printed copy is compared to the original inside a 5000 K color viewing booth.

SCIENCE FACTS

The Kelvin Scale

The Kelvin scale is used by scientists to measure temperature. A degree on the Kelvin scale is the same "size" as a degree on the Celsius scale. The difference is in the scales' starting points. A temperature of 1°C (Celsius) is one degree above the freezing point of water. A temperature of 1 K (kelvin) is one degree above absolute zero.

Absolute zero is the lowest temperature. It can be expressed as −273.15°C, −459.67°F, or 0 K. In theory, all atomic movement would stop at absolute zero. However, it's impossible to cool a substance down to absolute zero. To do so, you would need something even colder, and nothing is colder than absolute zero. However, scientists have come close: within .000022 of a degree.

LENSES

A **lens** is a piece of transparent material that is used to focus light. A lens has two opposite surfaces. Both may be curved or one may be flat. Light and a lens together form two major parts of an optic system.

Water, glass, plastic, and even air will refract light, and all act as lenses. By refracting light the lens can help us see more clearly. It can bring distant objects closer or project images onto a screen or film.

Depending upon the shape of the lens, the light is either concentrated on one point or made to spread out.

Convex lenses are thicker in the center and thinner at the edges. A convex lens concentrates the light. When light is concentrated, the point where the rays meet is called the **focal point**. Fig. 10-13. Light is bent more going through the rounded outer edges of the lens than through the center.

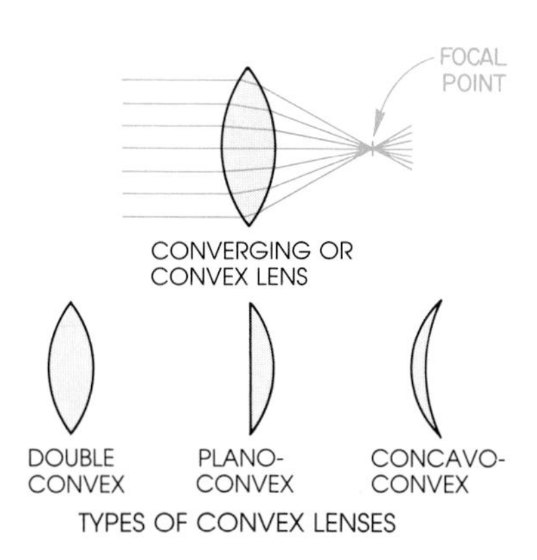

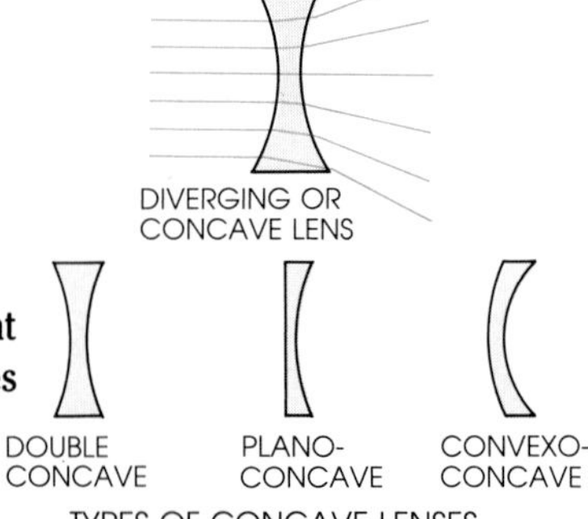

Fig. 10-13. Convex lenses cause light waves to come together. Concave lenses make them spread out.

Concave lenses are thinner in the center and thicker at the edges. Concave lenses spread out the light rays. The light is spread farther when it travels through the outer edges of the convex lens than through the center.

Images can become distorted when they pass through sharply curved lenses. An example is a photograph taken through a "fisheye," or a wide-angle lens. Fig. 10-14.

Lenses may be combined with one another to form the lens system used in cameras, enlargers, and other optic devices. Fig. 10-15. When concave and convex lenses are combined, they are called a compound lens. Fig. 10-16. This combination produces a high-quality image.

Fig. 10-14. A fisheye lens creates a distorted image. Photographers use it for special effects.

Fig. 10-15. Many devices use lenses to aid in focusing images either to our eyes or to film.

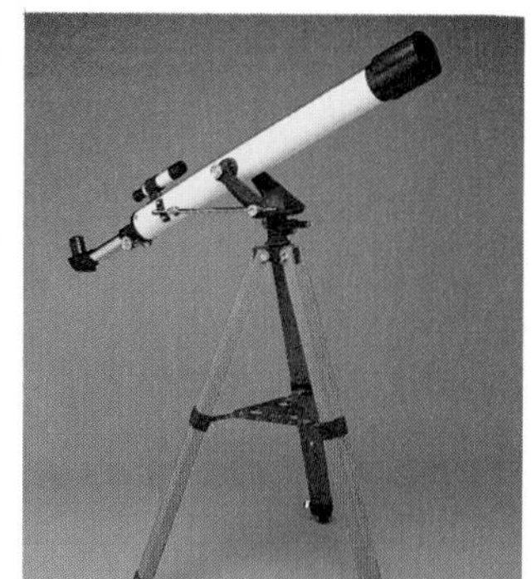

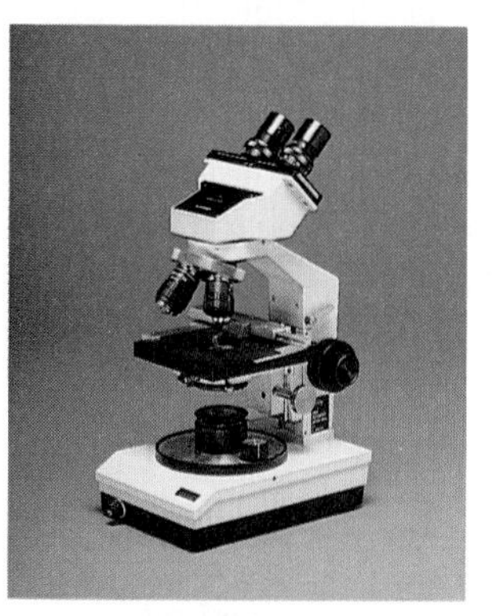

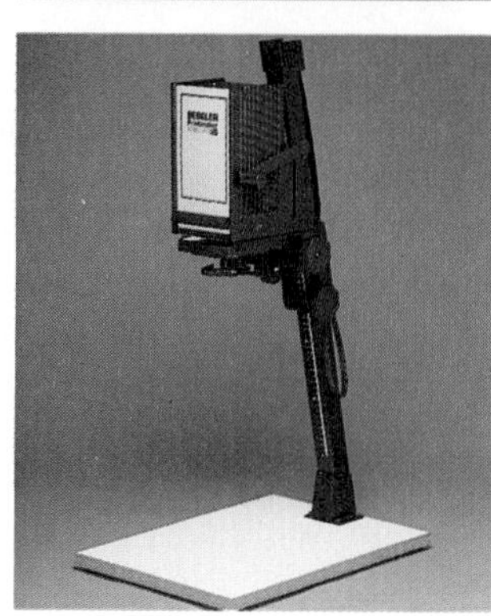

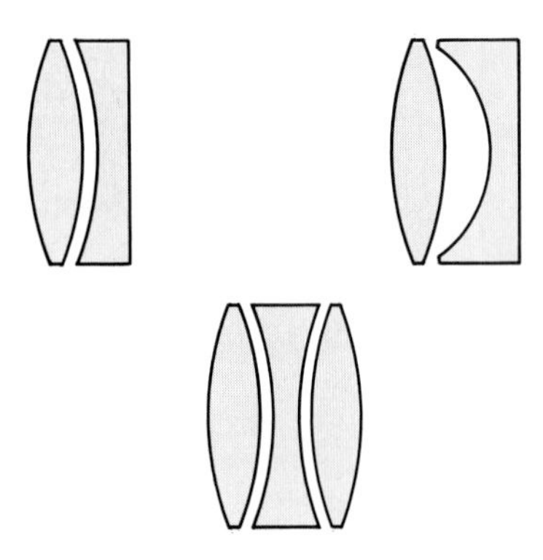

Fig. 10-16. A compound lens includes lens elements that are concave and/or convex.

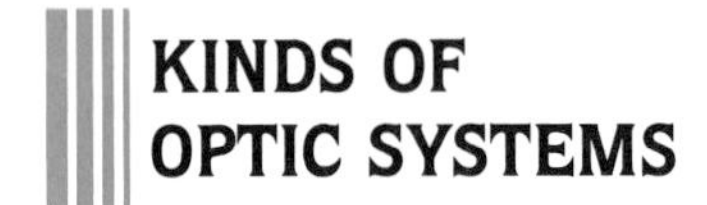

KINDS OF OPTIC SYSTEMS

The human eye is an optic system; so is a camera. To have an optic system you need a light source, a lens to focus the light, and a way to record the image.

The Human Eye

The human eye acts as our optic system. It receives light and transmits to our brain pieces of visual information.

Parts of the Eye

The eye is made up of seven major parts. Fig. 10-17. The cornea is a clear, curved tissue located at the front of the eye. It lets in the light rays and helps focus them on the back of the eye. The iris is the colored part of the eye. It looks like a flat doughnut, and its opening can be made larger or smaller. The opening is called the pupil, and it controls the amount of light allowed into the eye. You have probably noticed your pupils changing when you go from a bright room to a dim one, and vice versa.

The lens is a clear, bean-shaped tissue that can become thicker or thinner to adjust for viewing close up or far away. The lens also helps to focus the light rays on the back of the eye. At the back of the eyeball is the retina. The retina is a light-sensitive coating. Within the retina is the macula, an oval area responsible for color and detail vision. The center of the macula is the fovea, which is the point of clearest vision.

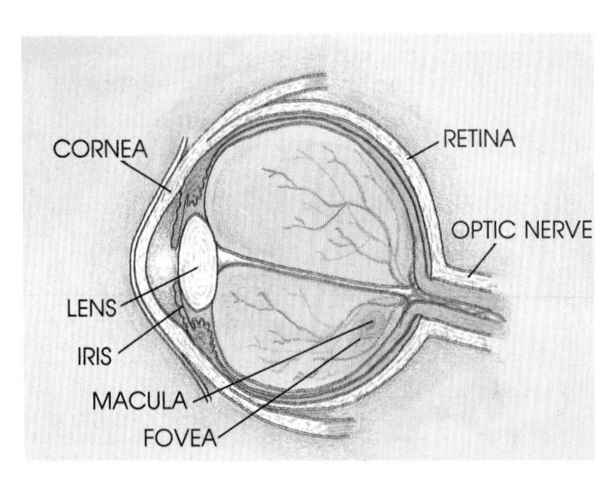

Fig. 10-17. Each part of the eye plays a role in receiving light waves or transmitting them to the brain.

How the Eye Works

The retina receives the light rays focused by the cornea and lens. Behind the retina is the optic nerve. The light rays are transformed by the optic nerve into electrical impulses and sent to the brain. The brain translates these impulses into the images we see. The eye and brain can see all wavelengths in the visible spectrum at the same time.

The retina contains two kinds of light-sensitive cells: rods and cones. Rods register amounts of light. The rods are color blind. If you had only rods in your retinas you could see, but everything would be in shades of gray. The cones are color sensitive. Some record red, some green, and some blue light waves. More than six million cones cover the surface of the retina. The fovea contains 50,000 cones in an area smaller than a square millimeter.

People who can't see colors normally are considered to be color deficient, or color "blind." Fig. 10-18. Color deficiencies occur when cones are absent or not fully formed. Most color deficient people confuse the colors red and green. These colors appear as shades of gray. The rods are functioning but the cones are not. Can you read the numbers in Figure 10-19?

Fig. 10-18. If you were totally color blind, colors would look like shades of gray to you.

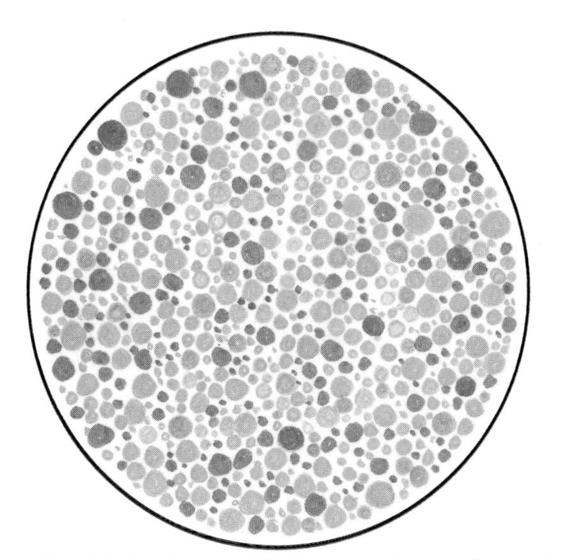

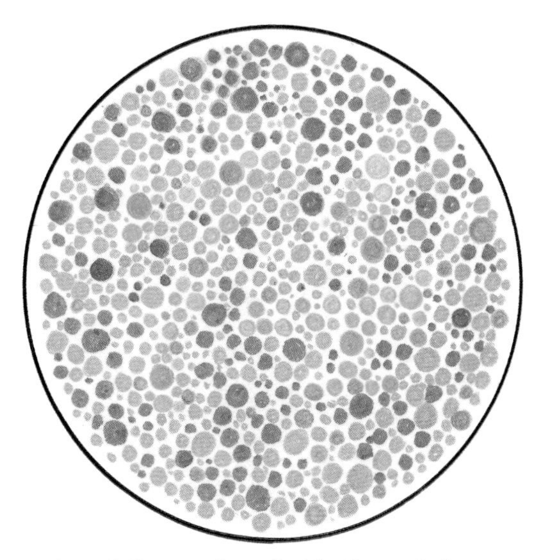

Fig. 10-19. People whose eyesight is color deficient cannot read the numbers inside these circles.

The Camera

A camera is a simple kind of optic system. Fig. 10-20. The light source is the scene being photographed. Light reflected from this scene enters an opening at the front of the camera. Behind this opening is a lens. The lens focuses the light on the film at the back of the camera. The film is a light-sensitive material, and it records the image.

The "box" of the camera is light-tight, so no unwanted light can enter. The size of the opening at the front of the camera can be controlled. This in turn controls the amount of light that strikes the film. The camera has a curtain-like shutter that covers the opening when a picture is not being taken.

The coating on film undergoes a chemical change when light hits it. Exposed film is processed with chemicals to form negatives. In a **negative** the light areas of the scene look dark and the dark areas look light. Fig. 10-21. To make a print, the negative is placed over a light-sensitive paper. Then light is projected through the negative. Where the negative is light, or clear, a lot of light strikes the paper. These areas turn dark on the print. Where the negative is dark, less light strikes the paper. These areas remain light on the print. Since light is reflected from objects in varying amounts, photographs have a range of gray, or middle, tones.

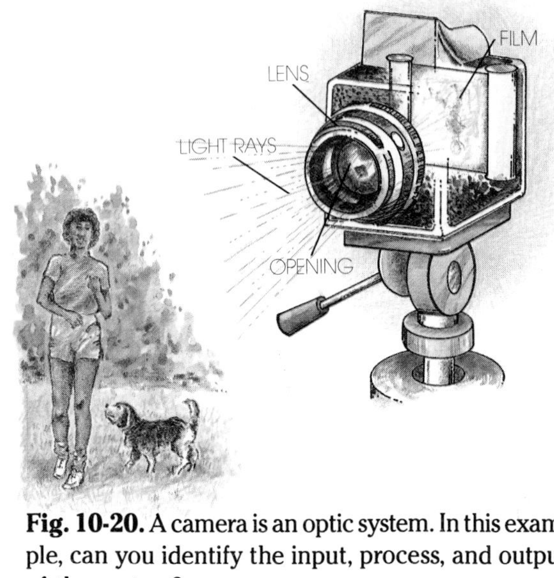

Fig. 10-20. A camera is an optic system. In this example, can you identify the input, process, and output of the system?

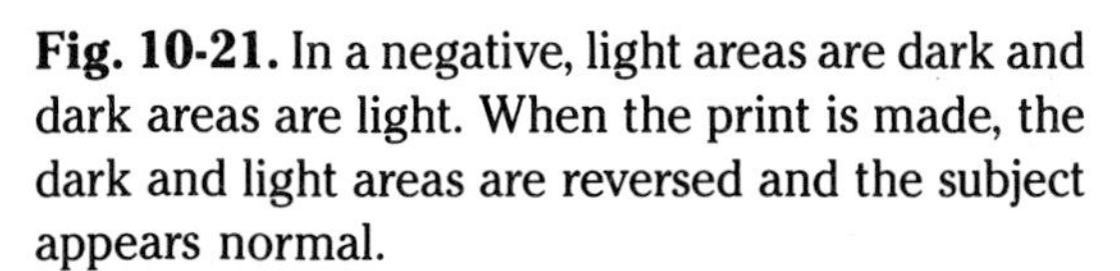

Fig. 10-21. In a negative, light areas are dark and dark areas are light. When the print is made, the dark and light areas are reversed and the subject appears normal.

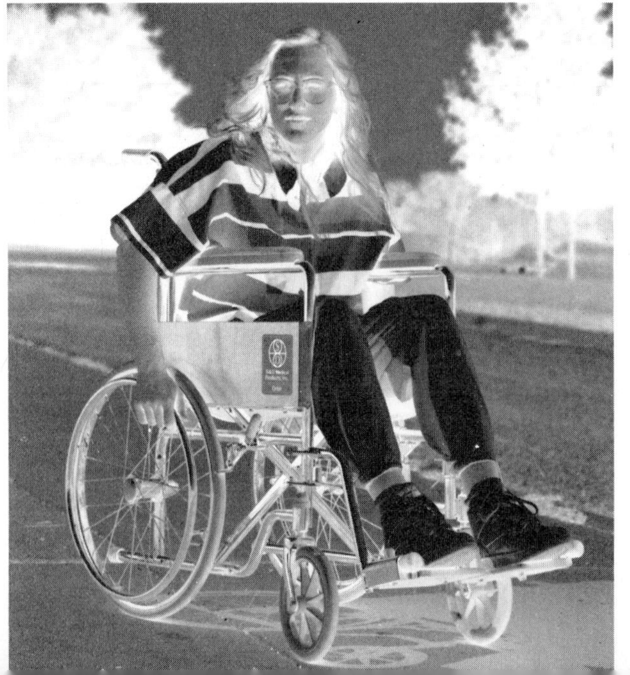

HEALTH & SAFETY

CARING FOR YOUR EYES

Because your eyes are so important, there is a variety of trained professionals available to care for them. Do you know the differences among the following?

Ophthalmologist. An ophthalmologist is a medical doctor and surgeon who is licensed to treat any eye problems. Ophthalmologists may prescribe glasses or medication. They may also perform eye surgery.

Optometrist. An optometrist can test people's eyes for defects and prescribe glasses. However, an optometrist is not a medical doctor.

Optician. An optician makes or sells glasses according to a prescription from an ophthalmologist or optometrist.

Have regular eye checkups. If you have any of the signs or symptoms below, tell your family, the school nurse, or your teacher. You may need the help of an eye specialist.

- blurry vision not helped by glasses
- double vision
- dimming or sudden loss of vision
- red eye or eye pain
- loss of peripheral (side) vision
- halos around lights
- crossed eyes
- wandering eye
- difference in eye size
- twitching eye
- flashes of light
- new spots or shadows in the eye
- discharge, crusting, or chronic, heavy tearfulness
- swelling of an eye
- bulging of an eye
- diabetes

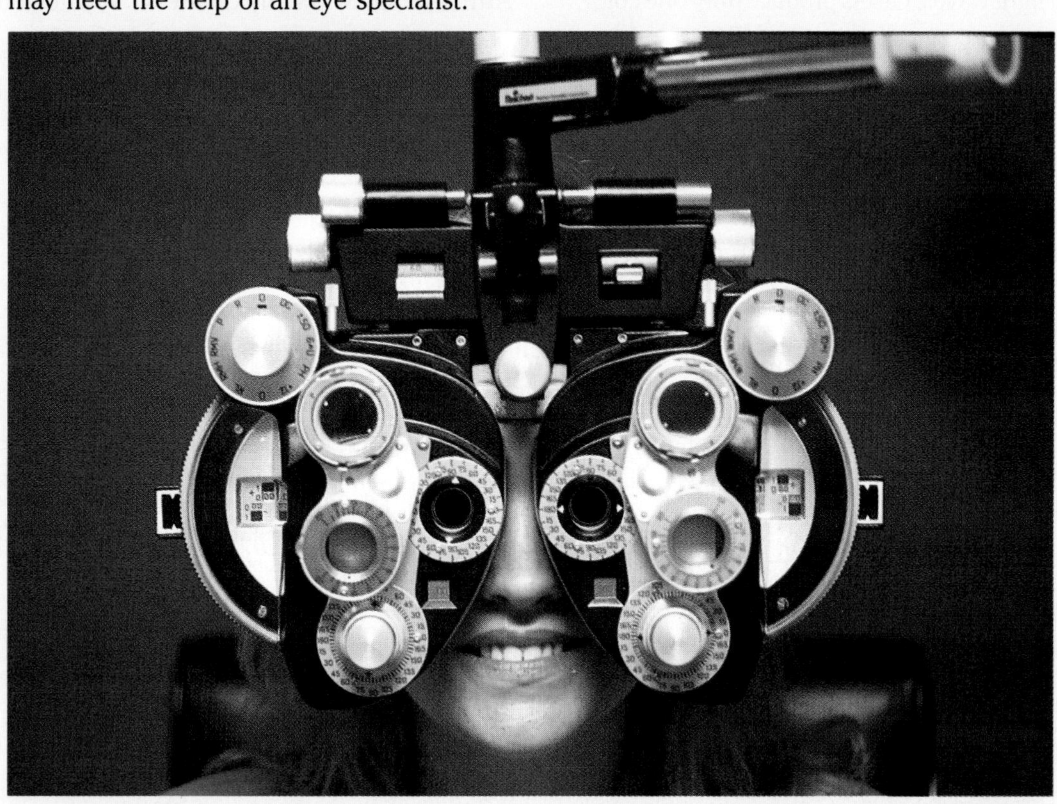

The first permanent photograph was produced in 1826. Sixty-three years later, the first simple box camera was introduced. Soon photography had become very popular, and it remains so today. Instant (Polaroid®) pictures, movies, and medical X-rays are all developments resulting from that first photograph.

Chapters 11 and 12 discuss the camera and photography in detail.

Lasers

As you already know, light radiates in all directions. That's why when you turn on a light in the middle of a room, it brightens the whole room. As you move farther away, the light gets dimmer. This is because the light waves have so many different frequencies that they bump into and scatter one another.

What happens when light waves are not allowed to scatter but are made to travel in only one direction? They become very powerful and can travel over great distances without dimming. A **laser** (Light Amplification by Stimulated Emission of Radiation) is a narrow beam of such parallel light waves. Lasers produce only one color, or wavelength, of light. These waves travel "in step" with each other, like marching soldiers.

How a Laser Works

Lasers may be made from crystal, gas, chemicals, dyes, or semiconductors, but they all work in basically the same way. Fig. 10-22. In a ruby laser, for example, light from a photographic flash is beamed onto a ruby crystal inside a tube. (The tube resembles a fluorescent light bulb.) The light hits the atoms in the crystal and excites their electrons so that they give off a large number of photons. The light waves created by the photons are all of the same length. Some of the light is reflected back and forth between the mirror-like ends of the crystal, and this excites even more electrons. Finally a shutter is opened at one end of the laser unit, and the pulsing parallel light waves burst out as a powerful laser beam.

Crystal lasers send brilliant flashes of light while gas lasers produce a continuous beam of light. Each type of laser has many uses in communication technology. Laser beams can be used to "carry" information much as radio waves do. The idea of a laser was first proposed by Arthur L. Schawlow and Charles H. Townes, but

Fig. 10-22. Lasers produce light of a single wavelength.

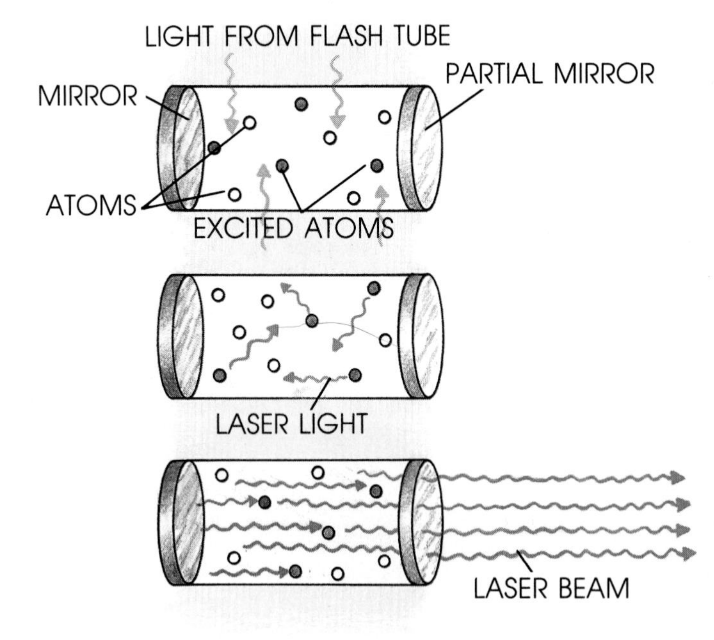

A flash tube sends intense light through the crystal. The light excites atoms in the crystal.

Excited atoms radiate light. Some of this light passes out the sides of the crystal, but part travels along the crystal's axis as laser light. The light is amplified many times as mirrors reflect it back and forth.

Some of the light passes through the partial mirror as a laser beam.

Fig. 10-23. The first laser emitted bursts of ruby red light.

Fig. 10-24. These photos show a single hologram viewed from above, straight on, and from the left. Note how the image changes. (*Space Lover's Series, Untitled No. 3,* Douglas E. Tyler, copyright 1980 by the artist.)

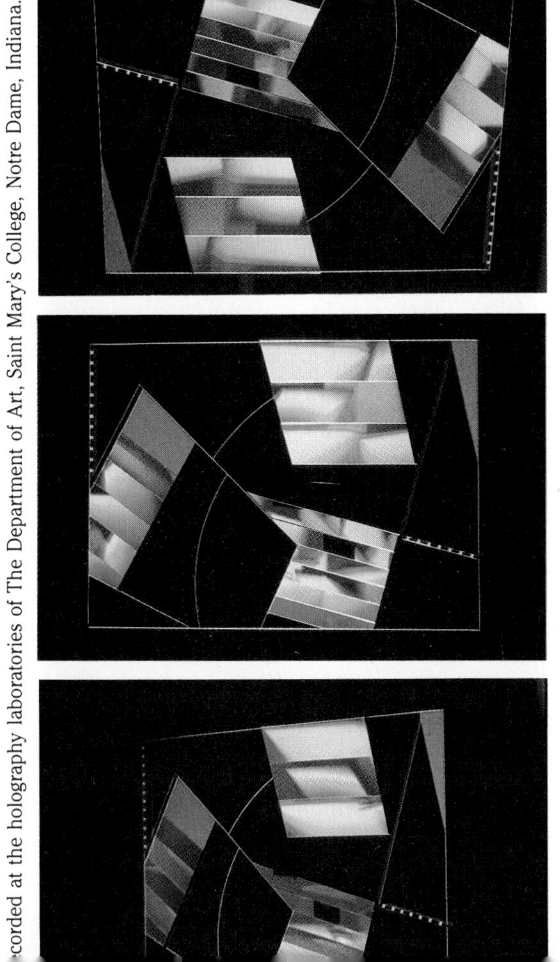

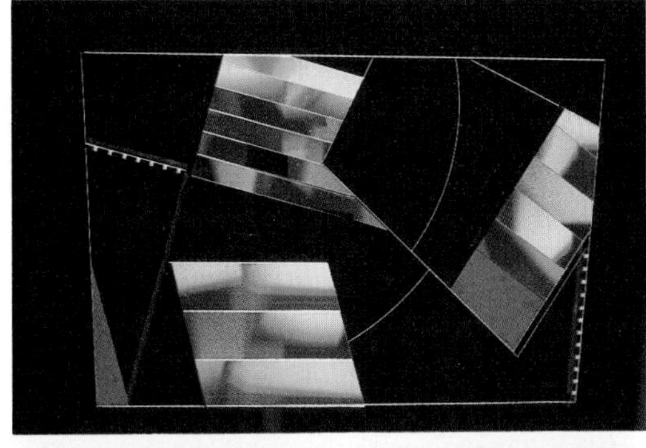

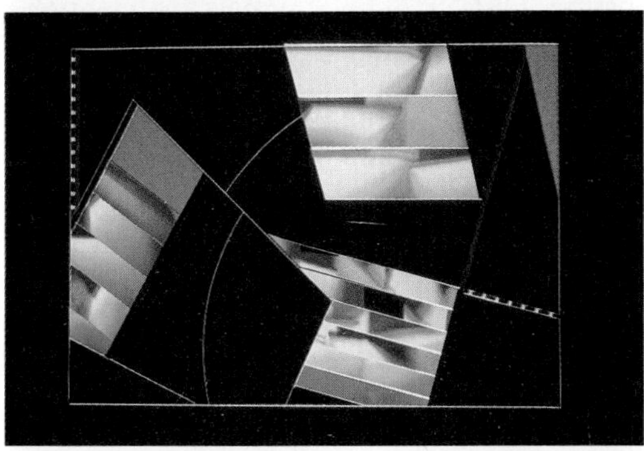

corded at the holography laboratories of The Department of Art, Saint Mary's College, Notre Dame, Indiana.

it was Theodore H. Maiman who built the first ruby laser in 1960. Fig. 10-23. Maiman's laser was developed at Hughes Aircraft Company and radiated bursts of red light. A year later, Ali Javan operated a continuous laser that used neon and helium gases.

Holography

Holography is the use of lasers to record realistic images of three-dimensional objects. The image, too, is three-dimensional, and the viewer can see the back as well as the front and sides. Fig. 10-24.

Holography is done with lasers because they produce parallel light waves of only one frequency. In one type of holography, the light from the laser is split into two beams. Fig. 10-25. One beam, called the object beam, shines on the object and is reflected onto a photographic plate. The other beam, called the reference beam, is reflected from a mirror and falls on the same plate at an angle. The two wave patterns overlap and interfere with one another. The pattern of this interference is recorded on the plate.

The image produced is called a transmission hologram. When this hologram is shown, light is directed at the processed plate from the side opposite the viewer. Part of the light creates a three-dimensional, colorful image of the original object. Another part of the light forms an image similar to an ordinary two-dimensional photograph.

Fig. 10-25. In making a transmission hologram, two beams of light transmit the interference pattern onto the photographic plate.

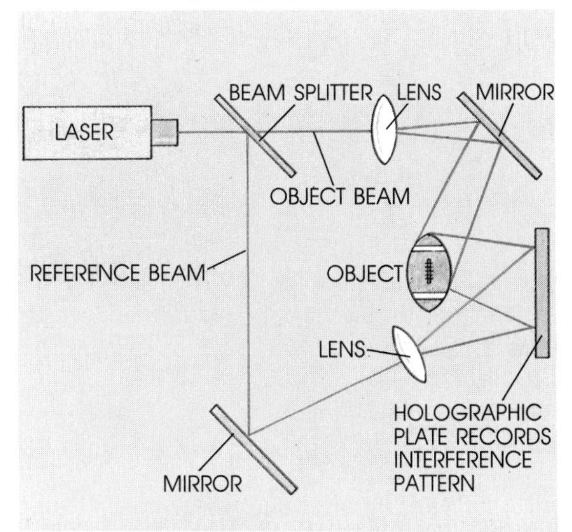

A second holographic method produces reflection holograms. It uses a laser beam shining first onto a mirror and then through a concave lens. Fig. 10-26. Next, the light passes through the photographic plate and then reaches the real object. The light striking the object is reflected back to the plate. The interference takes place inside the photographic emulsion (layer of chemicals) on the plate. To view the image, a white light is directed onto the surface at the same angle as the original laser reference beam. The light source is on the same side of the plate as the viewer.

Although there are other types of holograms, they are created using variations of these same methods.

Optical Fibers

Light can be sent through transparent fibers of glass. Fig. 10-27. These **optical fibers** can carry the light over a few inches or for more than 100 miles. They can also bend it around corners. The light is used to send messages. Optical fiber cable no more than three-quarters of an inch in diameter can carry as many as 40,000 phone calls at one time.

Optical fibers are about the size of a human hair. There are two kinds: single-mode and multimode. Single-mode fibers require special lasers as their light source. Their cores are very small and will accept light only along their center. Single-mode fibers must be precisely connected.

Multi-mode fibers have larger cores and can accept light from a variety of sources. They are cheaper than single-mode but cannot be used over long distances.

How Optical Fibers Work

The fibers consist of a pure core of glass surrounded by a cladding (covering) material. Fig. 10-28. An electronic signal activates a light source, which pulses on and off in a kind of code. The code is what relays the message. The light enters one end of the fiber. As it travels through the core, the light is kept inside by the cladding. At the other end, the light is received by some kind of sensor. When the light signal reaches the sensor, the sensor converts it back into an electronic signal. This signal is changed back to sound, computer data, visual images, or other information.

As discussed earlier, light waves have frequencies. The higher the frequency, the more signals can be carried. The frequencies of light waves are much higher than that of radio waves, for instance. (Both light waves and radio waves are part of the electromagnetic spectrum.) Light waves can carry 200,000 times more signals than can radio waves. In addition, the signal is less likely to need boosting, or repeating, along its journey. Ordinary electrical signals require constant boosting. Boosting often causes noise or other interference. However, light signals travel almost four times the same distance before they require boosting. Thus, fiber optic signals are "cleaner."

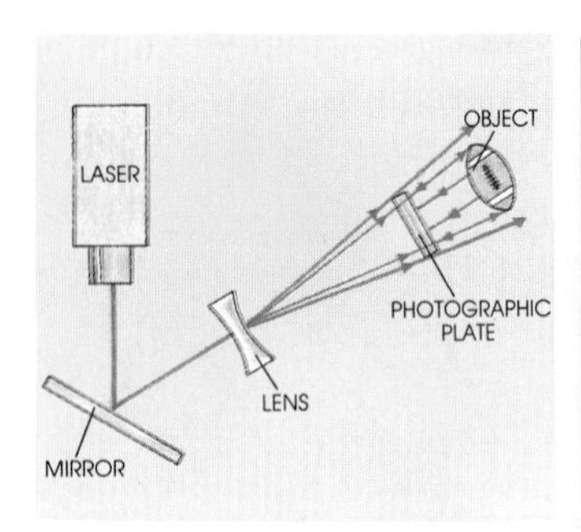

Fig. 10-26. To make a reflection hologram, the laser light beams are reflected off the object, and the interference pattern is created within the plate.

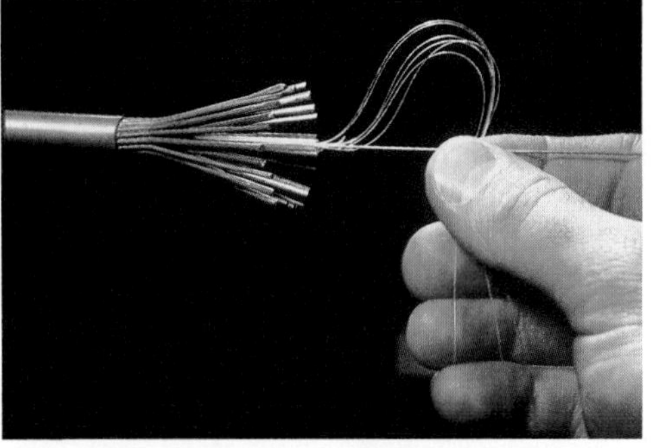

Fig. 10-27. Optical fibers are specially made thin strands of glass that transmit light.

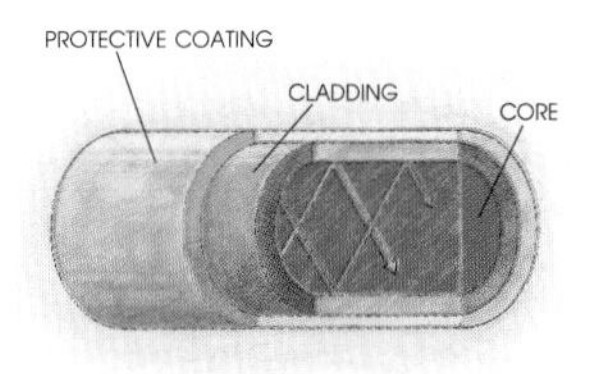

Fig. 10-28. Optical fiber has a core of glass covered by a reflective cladding and an outer protective layer.

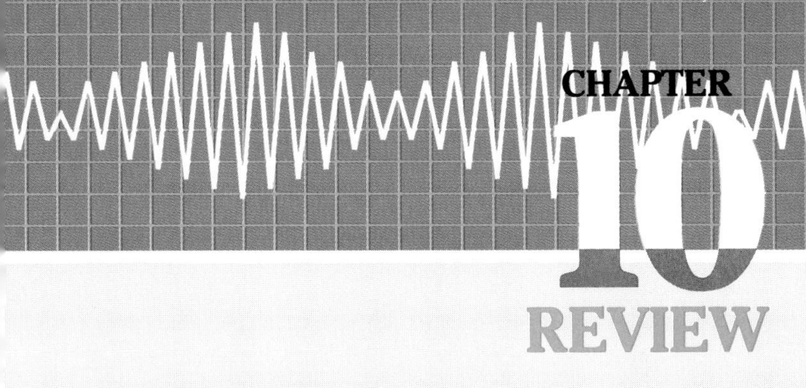

CHAPTER 10 REVIEW

Review Questions

1. Explain what an optic system is and give two examples.
2. What occurs when sunlight passes through a prism?
3. Describe the way light acts and travels.
4. Name the additive primary colors. How are subtractive colors created?
5. If an object looks blue, what causes us to see it as that color?
6. Name the seven major parts of the human eye.
7. What are the two primary lens shapes? What happens when light passes through each?
8. What is a laser and how is a laser beam created?
9. Describe holography.
10. What is optical fiber?

Activities

1. For one day, keep a list of all the optic systems you notice around you. You may find them in magazines, in your home, in stores, etc.
2. Send a beam of white light through a prism and observe the visible spectrum. Place a second prism where the "rainbow" occurs. Describe what happens to the rainbow of light that passes through the second prism.
3. Select three different types of indoor light based on color (fluorescent, blue-white, yellow, or pink). Collect a variety of different-colored items. View each item under the different lights. (Remember to turn off all other lights around you when you try this. Avoid windows where sunlight may enter.) Which objects seem to change color when viewed under the lights? Make a chart to show the differences.
4. Interview a local eye doctor and learn how eye examinations are made. Ask the doctor to demonstrate the equipment used to determine how well you focus, the pressure levels in your eye, or your field of vision.
5. Obtain a variety of lenses (glasses, a magnifying glass, slide projector lens, camera lens, telescope, microscope, binoculars, etc.). Look at the same object through each lens. Describe how the object appears in each case. What causes any differences?

CHAPTER 11

Photography: Equipment and Methods

As with any optic system, the most important tool in photography is light. All cameras either control or respond to light. The word photography comes from the Greek words "photos" and "graphe," meaning "to carve from light." Without light, there is no photograph.

You read about light and optic systems in Chapter 10. This chapter discusses some basics about photography and photographic equipment. You will learn the anatomy of a camera and how lenses and film work.

Terms to Learn

contrast
developer
emulsion
enlargers
field of view
filters
fixer
focal length
light meter
panchromatic films
shutter
single lens reflex camera
stop bath
telephoto lenses
wide-angle lenses

As you read and study this chapter, you will find answers to questions such as:

- What is the difference between a rangefinder and an SLR camera?
- What accessories are useful for taking good photographs?
- What does an enlarger do?
- How does color photography differ from black and white?
- What takes place during film processing?

THE CAMERA

The camera is a common example of an optic system. As you learned in the last chapter, an optic system includes a light source, a means of controlling the light, and some way to record the image. Cameras are used to control and record light on film.

All cameras have a body, a lens, and a shutter. Fig. 11-1. The body is a light-tight container that holds the film on which the image is to be recorded. The only way light can get into this box is through the lens. The lens, as you learned in the last chapter, focuses light. In the case of the camera, it focuses the light onto the film. The **shutter** of the camera opens and closes the lens opening. Light can enter the camera only when the shutter is open.

In its simplest form, a camera may be just a box with a hole in one end and a place for film in the other. The hole directs light onto the film and thus acts as a lens. The length of time that light enters the camera is controlled by a piece of cardboard taped over the hole. The cardboard serves as a shutter. This type of camera is known as a pinhole camera. Fig. 11-2. As primitive as this camera is, it can be used to create some interesting photographs.

Today there are many types of cameras on the market. Some are simple "point and shoot" models; others are quite complex. The following describes the basic features of rangefinder, twin lens reflex, view, and single lens reflex cameras.

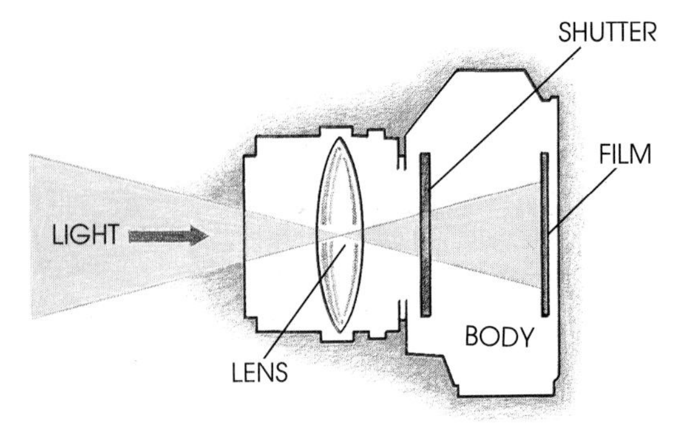

Fig. 11-1. The basic parts of a camera include a light-tight body, a lens to focus light onto the film, and a shutter that opens and closes.

Fig. 11-2. The simple camera above is a light-tight container with a pinhole at one end. The other end is removable so that film can be mounted inside. (Basic Activity Number 2 at the end of Section IV tells how to make such a camera.) Shown below is a more elaborate pinhole camera.

Rangefinder Camera

Rangefinder cameras have two lenses. Fig. 11-3. The photographer looks through the simple lens of the viewfinder to view the subject of the photograph. A separate, more complex lens focuses the image on the film. Because the subject is viewed through one lens and photographed through another, the resulting picture may not be what the photographer saw. Fig. 11-4. This happens often when taking photographs at close range.

Some inexpensive rangefinder cameras take good photographs. These have simple lenses that are permanently attached to the camera. Some are small and use film in small sizes. They are popular for snapshots, since they are often lightweight and easy to use.

More expensive rangefinder cameras may have interchangeable lenses, automatic focus and exposure control, built-in flash, and even power winding of the film. Fig. 11-5. They can be electronic marvels. Since they use 35mm film (described later in this chapter), they can produce very good photographs in a wide range of situations. These rangefinder cameras have become popular with consumers because they are so easy to use.

Fig. 11-3. A rangefinder camera has separate lenses for viewing the subject and taking the picture.

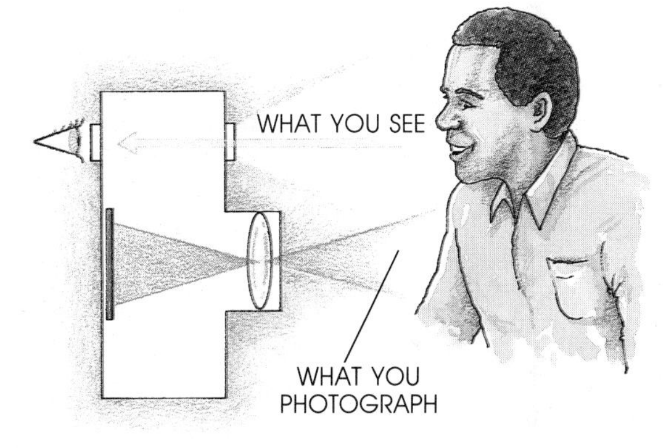

Fig. 11-4. What you see through the viewfinder may not be exactly what appears in the photo.

Fig. 11-5. A modern rangefinder camera has many sophisticated features, such as automatic focusing. Such cameras are called autofocus cameras.

Twin Lens Reflex Camera

Like the rangefinder, the twin lens reflex camera has two lenses. Fig. 11-6. One is for viewing the subject, and the other is for focusing the light onto the film. The viewfinder is on top of the camera and the photographer looks down into it. The term "reflex" means there is a mirror inside the camera that *reflects* the image of the subject up to the photographer's eye. Fig. 11-7.

Twin lens reflex cameras are larger than rangefinder cameras. They use a larger size film and are capable of taking excellent photographs. They are, however, heavy and expensive. Therefore, they are generally used only by professional and serious amateur photographers.

Fig. 11-6. Twin lens reflex (TLR) cameras have two lenses. The upper lens is for viewing the subject. The lower one is used to take the picture.

View Camera

View cameras are the biggest of all. Fig. 11-8. They look like an accordion because the lens is attached to the back of the camera with a collapsible bellows. Because of their size, view cameras are generally mounted on a three-legged stand called a tripod.

Unlike the smaller cameras, the view camera uses individual sheets of film. The sheets may measure 4 × 5 inches, 8 × 10 inches, or even larger. Film this large results in photographs with sharp detail. When negatives are enlarged to make a photograph, the film grain is also enlarged. This makes the photograph appear "grainy," or a little fuzzy. Because the negatives

Fig. 11-7. A fixed mirror inside the TLR camera reflects the image up to the photographer's eye.

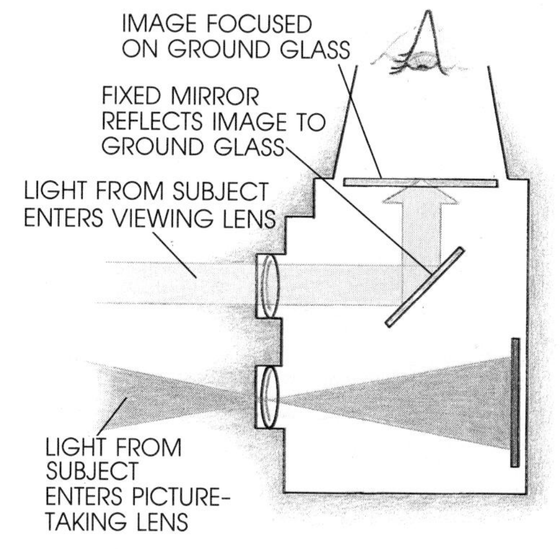

Fig. 11-8. View cameras use fairly large sheets of film to take high quality photographs. Both the lens and the film back may be tilted to change the perspective of the photograph.

produced by a view camera are large to begin with, they need not be enlarged as much to make the photograph. Therefore the photographs taken with a view camera have very fine detail. That's one advantage of the view camera.

Another advantage is its ability to produce sharp focus and increased depth in a photograph. This is because the lens and film holder may be tilted to change perspective. The photographer has a lot of control.

Single Lens Reflex (SLR) Camera

The 35mm **single lens reflex camera** has become the workhorse for both professional and serious amateur photographers. Fig. 11-9. These cameras are easily carried and are capable of excellent photographs. In addition, they come in a wide range of models with a huge assortment of accessories.

Single lens reflex (SLR) means that both viewing and picture taking are done through a single lens. There is a mirror inside the camera which reflects the image to a five-sided prism. Fig. 11-10. The prism reflects the image to the viewfinder. The mirror flips up out of the way when the shutter opens. The primary advantage of using the main lens to view is that what you see is what you get. What you see through the viewfinder is what appears in the photograph you take.

In general, 35mm SLR cameras are made with interchangeable lenses. You can quickly remove one lens from the camera body and replace it with another. Using different lenses allows you to photograph objects very close up or very far away.

Fig. 11-9. Single lens reflex (SLR) cameras allow you to photograph exactly what you see. The same lens is used for viewing and for taking the picture.

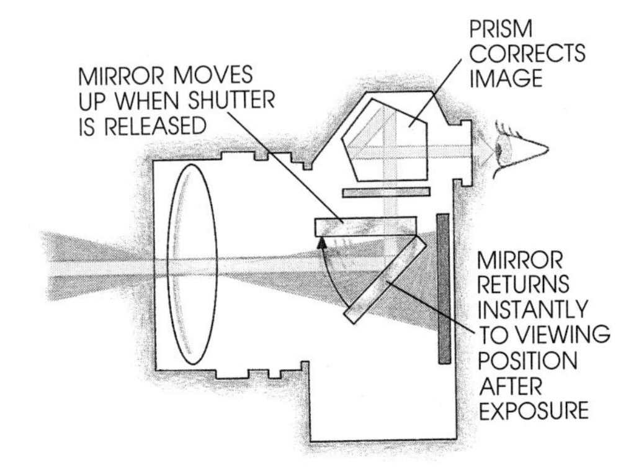

Fig. 11-10. The mirror flips out of the way when the shutter opens.

The 35mm SLR cameras may be manual, automatic, or both. For the manual type the exposure must be set each time a picture is taken. Exposures are set automatically on the automatic versions. Some 35mm cameras with the automatic feature may also be set manually. See Fig. 11-11 for the parts of a manual SLR.

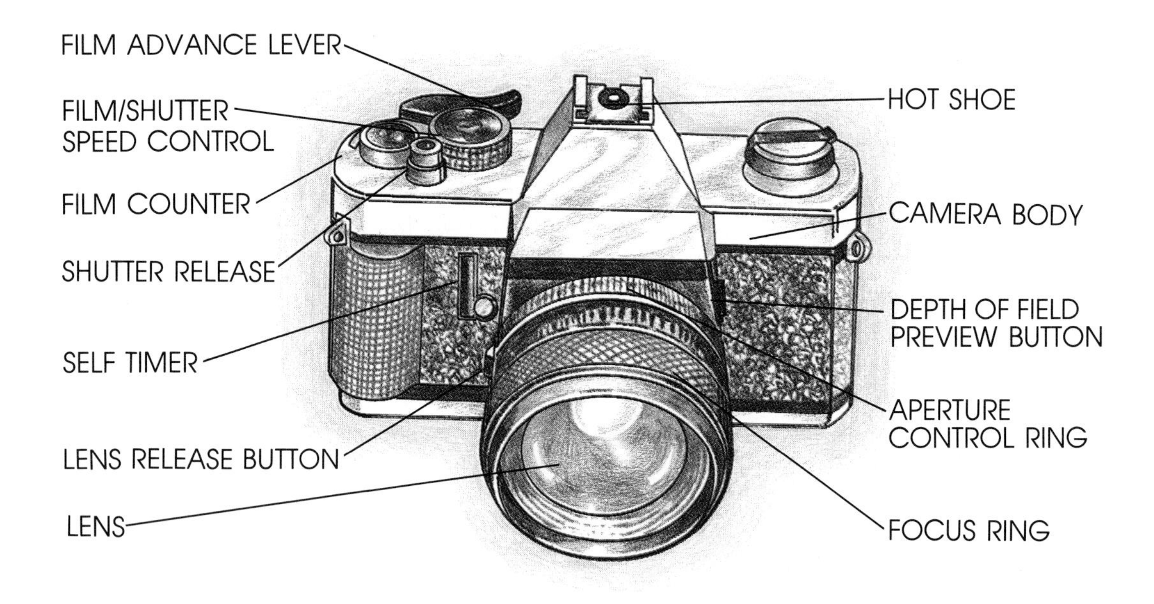

Fig. 11-11. The parts of this manual SLR camera are described on the following page.

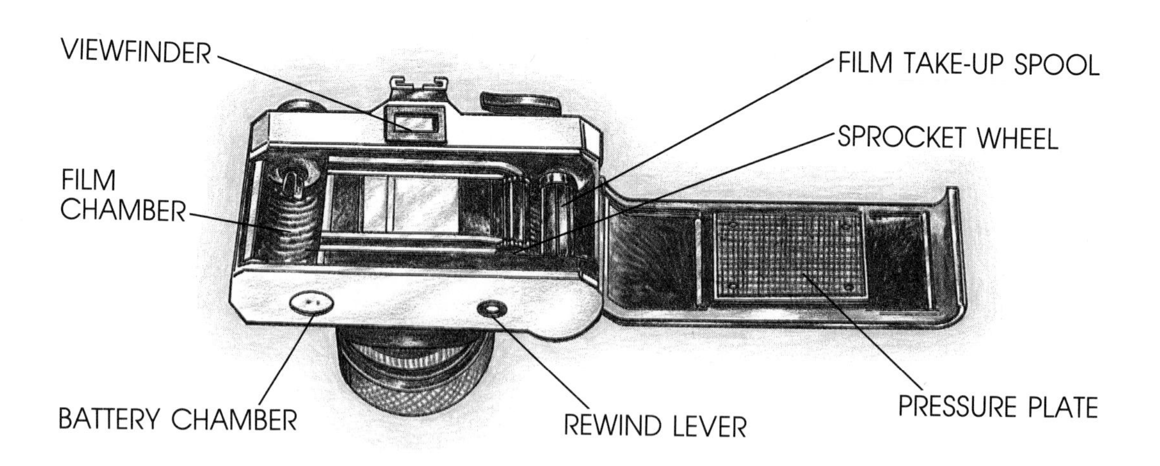

The Basic Parts of a Manual 35mm SLR Camera:

- **Camera Body:** This may be either metal or plastic. The back opens so the film may be loaded.
- **Lens:** The lens shown is an ordinary one. It may be replaced with special lenses.
- **Aperture Control Ring:** This ring is rotated to change the size of the aperture (lens opening) and control the amount of light entering the camera.
- **Focus Ring:** This is rotated back and forth to focus on the subject. Markings on this ring indicate how much of the subject will be in focus.
- **Film/Shutter Speed Control:** This dial may be rotated to change the shutter speed. It may also be lifted up and turned to adjust the built-in light meter for the proper film speed.
- **Film Advance Lever:** Pushing this lever moves the film forward.
- **Film Counter:** This indicates how many exposures have been made so far.
- **Shutter Release:** Pressing this button opens and closes the shutter.
- **Rewind Lever:** This is turned to rewind the film into its own container. Lifting up on the lever opens the back of the camera so the film may be removed.
- **Self Timer:** When the timer is set, the camera automatically takes a picture after a short delay. Photographers sometimes use this feature to take their own pictures. They set the timer and then place themselves in front of the camera.
- **Lens Release Button:** Pressing this button releases the lens so that it can be replaced with a different lens.
- **Depth of Field Preview Button:** By pushing this button, the photographer may see how much of the subject will actually be in focus.
- **Hot Shoe:** A flash attachment is mounted on the camera here.
- **Viewfinder:** This is what the photographer looks through.
- **Battery Chamber:** A small battery is placed under this cap. It supplies the power for the light meter.
- **Film Chamber:** Film is inserted here.
- **Sprocket Wheel:** The sprockets on this wheel match holes in the film. When the film advance lever is pushed, this wheel turns and moves the film.
- **Film Take-Up Spool:** As film advances through the camera, it is wound onto this spool.
- **Pressure Plate:** This plate holds the film in the proper position during exposure.

There is also a 2¼″ SLR camera. It is much like the 35mm SLR, except that it is larger. Fig. 11-12. The large negatives it produces result in photos with sharp detail. This makes it popular among professional photographers. However, the 2¼″ SLR generally costs more than most amateur photographers wish to invest.

Fig. 11-12. The 2¼″ SLR camera is used by many professional photographers.

DIGITAL PHOTOGRAPHY

Conventional color photography begins with cameras that record color images on light-sensitive film. The film is processed to create a negative. Light is projected through this negative onto photographic paper and processed to make a print.

Digital cameras don't use film. Instead, they store the image as digital data inside the camera. These data may then be downloaded directly to a computer's hard drive. From there, they can be sent to a printer or edited.

Like conventional cameras, digital cameras use glass lenses to capture and direct the light that reflects off the subject. However, the similarity ends there. In a conventional camera, this light is focused onto film, which is used to create the final photograph. In a digital camera, the lens directs the incoming light to a charge-coupled device (CCD) array, like the one shown on this page. The CCD is coated with red, green, and blue filter material, which allows it to convert the incoming light into electrical signals that represent the red, green, and blue components of the original subject. These analog electrical signals are then converted into digital data and compressed.

The compression process removes unnecessary information. The resulting file size is much smaller than the original file. Compression allows the camera to store more images and decreases the time needed to transfer these digital images to a computer. Once the digital files are downloaded to a desktop computer, they can be edited with software such as Adobe Photoshop™. For example, the image may easily be cropped (trimmed), resized, color-corrected, or changed in almost any way with the editing software.

Digital photographs—images originating with digital cameras—are used in many different ways. Newspaper and magazine photographers send these files over telephone lines and/or via satellite to their publishers, who may then quickly combine them with digital text provided by a writer. Since film processing isn't necessary, the publisher saves time.

Similarly, digital images may be imported directly into a digital multimedia presentation. This saves considerable time over the alternative—scanning a conventional photograph.

Digital images are standard fare on the World Wide Web. Images taken with digital cameras may be converted into the GIF file format, allowing them to be displayed on the World Wide Web.

Speed is only one of the advantages of digital imaging. Digital images do not degrade over time or from one generation to the next as they do in conventional formats. In contrast, film negatives and photographs fade over time, and duplicating them results in a loss of detail.

There are many benefits of digital imaging over conventional photography. As all of our media continue to shift from analog to digital formats, digital cameras will become commonplace.

LENSES

Perhaps the most important part of a camera is the lens. The better the lens, the better the camera. Although lenses can be made of plastic, the more expensive cameras usually have glass lenses.

Many cameras are designed so that lenses may be changed very easily. Some types screw onto the camera, and others are twisted on and off.

Lenses differ in focal length. Fig. 11-13. The **focal length** of a lens is the distance between the center of the lens and the film, when the lens is focused to infinity. The longer the focal length, the more a lens magnifies what it sees. However, it also sees a smaller area, or **field of view**. The shorter the focal length, the smaller the subject appears; but a greater field of view is visible.

Fig. 11-13. The focal length of a lens is measured in millimeters. It is marked on the lens housing.

Types of Lenses

Many lenses have fixed focal lengths. These lenses are either normal, short, or long. Fig. 11-14. A normal lens sees much as the human eye does. It has a focal length that is about the same as the length of a diagonal line drawn across a negative made with the same camera. For example, a normal lens for a 35mm camera usually has a focal length of 50mm.

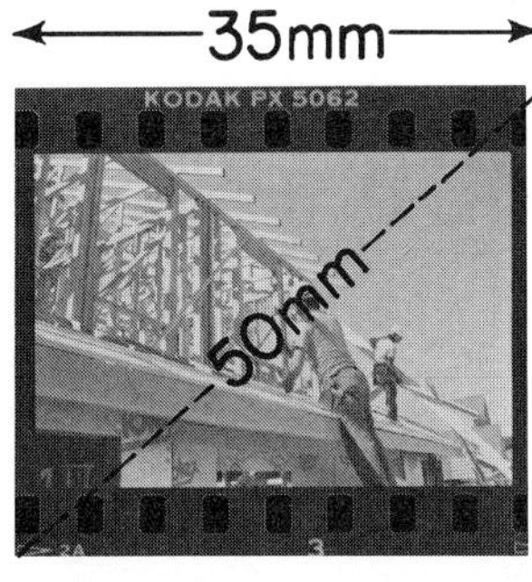

Fig. 11-14. Shown below are the types of lenses a professional photographer or a serious amateur might have. (Above) A normal lens has a focal length about equal to the diagonal of the film negative.

Short lenses, as their name suggests, have a focal length shorter than that of a normal lens. They tend to have a wide field of view. Thus they are known as **wide angle lenses**. They are good to use when you cannot stand far enough away from the subject to photograph all of it with a normal lens. A special purpose short lens is the "fisheye" lens. This lens has a 180 degree field of view. This is even wider than the human eye's field of view. However, it produces a distorted image. Fig. 11-15.

Long lenses, sometimes referred to as **telephoto lenses**, make faraway objects appear closer. They enlarge subjects that are far away from the camera the same way binoculars do. At the same time, though, they limit the field of view. Long lenses are very useful for photographing subjects from a distance. For example, if you wished to photograph the action in a football game, a long lens might be a good choice. Fig. 11-15.

Unlike lenses with fixed focal lengths, zoom lenses allow the photographer to change the focal length as needed. By twisting the outside of the lens, the photographer can change the focal length from short to long. A subject that is far away "zooms" into focus, which is why these are known as zoom lenses.

Special lenses are also available for photographing very small objects at very close range. These close-focus, or macro, lenses allow you to photograph something as small as an ant and make it seem larger than a football.

Extension tubes are hollow tubes that fit between the camera body and the lens. They can also be used for close focus photography, but cost much less than a macro lens.

Fig. 11-15. All these photographs were taken from the same spot. The view at top left was taken with a wide angle lens. Compare it with the one at top right, which was taken with a normal lens. A telephoto lens was used for the photo at bottom left, and a fisheye lens was used for the bottom right view.

ACCESSORIES

Lenses are the most important pieces of equipment, but photographers use many other things to take good pictures. Fig. 11-16.

Light Meters

You can't take a photograph without light. It may be natural sunlight or artificial light, but either way, you have to have it.

Many cameras have a built-in **light meter** to measure the amount of light present. This allows photographers to select the proper aperture (lens opening) and shutter speed. For more accurate readings, photographers use hand-held light meters. Incident-light meters measure the amount of light falling on the subject. Reflected-light meters measure the amount of light reflected off the subject.

Lighting Equipment

Common lighting equipment is shown in Fig. 11-17. Artificial lighting equipment is of two types: continuous or flash. Continuous lights are like the lights in a room. You turn them on and they stay on. Because the lights remain on, you can use a light meter to help you determine the best camera settings.

Continuous lighting may either be flood lighting or spot lighting. Flood lights project light over a large area. Spot lights direct a narrow beam of light at a small area.

Electronic flash, or strobe, lighting illuminates the subject only at the exact instant the picture is taken. The simplest strobe lights mount directly on the hot shoe of the camera. They are synchronized to give a burst of light when the shutter opens. More expensive types may be mounted on floor stands.

Fig. 11-16. These are some of the accessories photographers use.

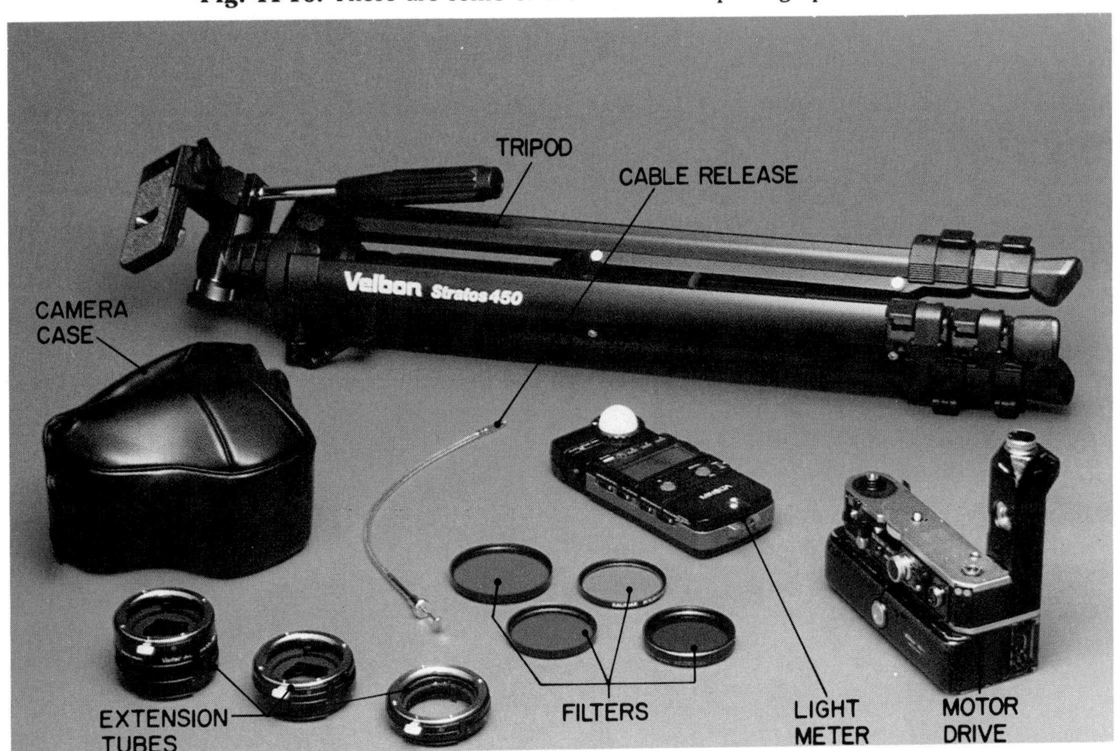

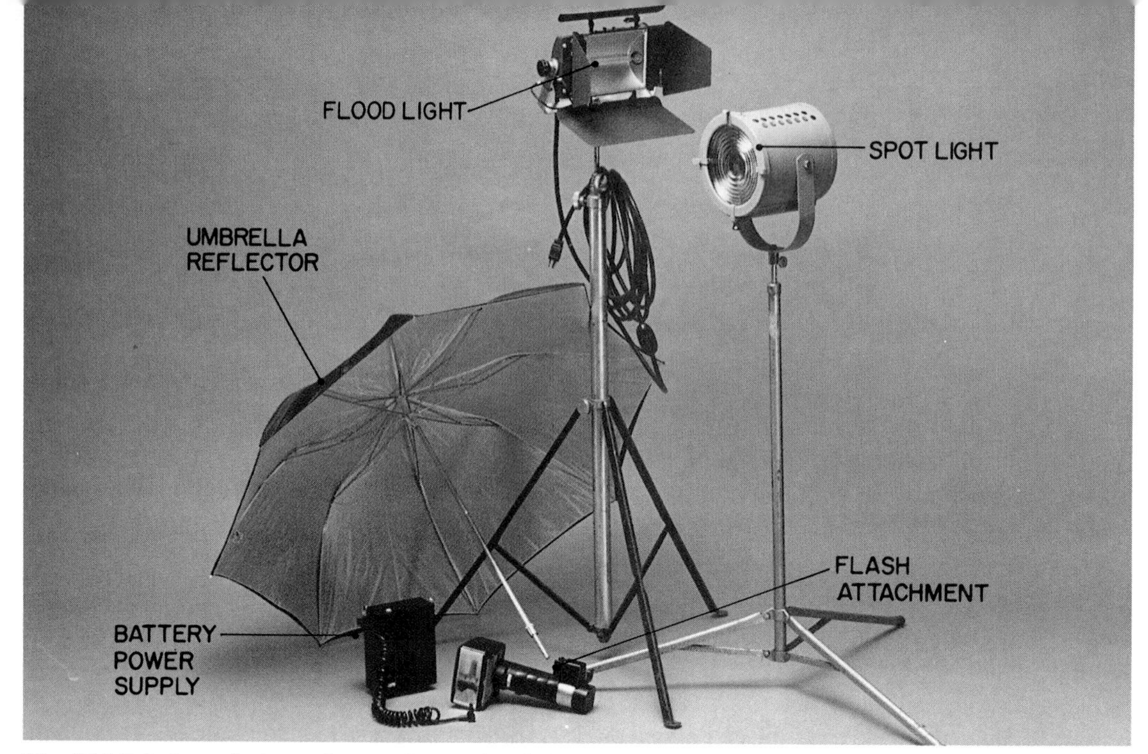

Fig. 11-17. Indoor photography usually requires lighting equipment. The room lights alone are not enough.

Photographers cannot always count on having a source of electricity close by. Portable battery packs allow them to use lighting equipment anywhere.

Lights, by themselves, are not always enough. To control the "look" of the light, photographers use special reflectors. The umbrella reflector is used to diffuse (spread) light over a wide area. It provides even illumination and a softer appearance.

Camera Equipment

Filters

Sometimes photographers attach filters to the camera lens. **Filters** screen out certain wavelengths of light. There are three basic types: contrast, special effects, and color correction filters. In black and white photography, contrast filters are used to make some areas of the photograph stand out. For example, used with black and white film, a yellow contrast filter will make white clouds more visible in a blue sky.

Special effects filters are used to create a wide range of interesting images, such as sunbursts or mirror images. Color correction filters are used with color film to correct for the type of lighting used. For example, if a film designed for natural (sun) light is used indoors, a color correction filter can improve the results.

A skylight filter is a good filter to leave on the lens at all times. It is clear and helps to screen out ultraviolet light, which can produce a haze on the photograph. At the same time, it protects the lens from scratches and dirt.

Film Winders

Power winders that automatically advance the film after each exposure are popular with professional photographers. The film is always ready for the next picture. Power winders are handy when taking action photographs, such as at a sporting event. This is because several photographs can be taken each second.

User's Guide to Technology

BUYING A CAMERA

You can spend a few dollars or a few thousand dollars on a new camera. Which will it be? In addition to cost, there are a lot of things to consider.

Decide what you plan to use it for most. Will it be for travel pictures? Do you want to become a creative black-and-white photographer? Perhaps you want to take pictures for the yearbook?

If you think you will just be taking casual snapshots, start with a viewfinder camera. (Remember, a viewfinder camera has one lens for viewing and another lens that takes the picture.) In general, the smaller the film size, the lower the picture quality. Thus, a 35mm viewfinder camera will probably produce better photos than the smaller 110 format camera. The strengths of a 110 format camera are low price and compact size, not high quality.

There is a wide range of 35mm viewfinder cameras on the market. The inexpensive types are okay for travel snapshots, but that's about it. They will not produce quality photographs under a variety of conditions.

The more expensive 35mm viewfinder cameras have features that make them an excellent choice for many beginning photographers. Features to look for include automatic focus, automatic exposure, zoom lens, self-timer, film winder, and even interchangeable lens capability.

Automatic focus and exposure will allow you to take decent photographs with the "point and click" method. A zoom lens will let you take pictures of subjects at some distance. The self-timer delays the shutter release, so the photographer may move into the picture before it is taken. A film winder automatically advances the film after each exposure. This lets you take more photos in

shorter time than the manual advance on many cameras. The interchangeable lens feature allows you to take all different types of photographs . . . as long as you bought the extra lenses.

If you plan to get serious about photography, a 35mm single lens reflex (SLR) camera is a good place to start. This is the "workhorse" for most professional photographers. Most 35mm SLR cameras have interchangeable lenses, and you can get dozens of accessories for them. In short, you can expand this type of "system" to meet your future needs, even if you get very serious about your photography. Here again, the more features and accessories you add, the more flexibility you have (and the more expensive the camera).

Whatever type of camera you choose, make sure it's easy for you to hold. Everyone's hands are different sizes, and it's important that you are comfortable with the camera and its controls. Be sure to try out several different brands in several camera stores before you buy one.

A good warranty is important, and you should ask about repairs. If something goes wrong, do you have to send the camera back to the manufacturer, or can it be fixed locally?

You should shop around for a good price. Keep in mind that cameras are very rarely sold for the "suggested retail" price. They almost always sell for much less, so don't assume you are getting a great deal until you have compared prices at different stores.

Cable Release

A cable release is a handy tool that attaches to the shutter release of the camera. If the camera is on a stand, the photographer can take a picture without even touching it. This helps prevent jiggling the camera when using slow shutter speeds. At slow speeds any movement will blur the picture.

Camera Supports

A tripod is a three-legged stand that holds the camera while the photograph is being taken. Tripods are collapsible, so they may be easily moved from place to place. One-legged stands (monopods) are also available.

Camera Case

A camera case protects a camera from normal wear and tear. It may be as simple as a leather or plastic cover that snaps in place around the camera, or it may be as big and sturdy as a suitcase. Serious amateur photographers generally carry their equipment in a soft camera bag. Professional photographers often carry large cases full of equipment.

FILM AND PHOTOGRAPHIC PAPERS

In the early days of photography, photographers made their own films and papers by coating glass or paper with mixtures of light-sensitive chemicals. Today, we can buy a wide range of photographic papers and plastic films, as well as the chemicals to develop them. Different combinations of film, paper, and chemicals allow the photographer to create an amazing range of effects.

How Film Works

Black and white film is made of four layers. Fig. 11-18(a). The **emulsion** layer contains tiny light-sensitive silver halide crystals. These crystals are suspended in a gelatin, much the way fruit is suspended in a gelatin dessert. The emulsion side of the film is always the lighter side.

The type and size of the silver crystals determine the characteristics of the film. When light strikes the crystals, their chemistry changes. When the film is processed, those crystals struck by light turn into black metallic silver. The others are washed away. The remaining (black) crystals form the image on the negative.

During film manufacture, a durable coating is applied to the emulsion. This coating protects the emulsion from scratches.

Beneath the emulsion is a thin sheet of clear plastic, called the base layer. Very stable plastics, such as polyester or polystyrene, are generally used for the base. It is important that the base not expand or shrink much with temperature or humidity changes.

The back of the film is coated with a dark antihalation layer. This layer absorbs light and prevents it from reflecting back through the emulsion layer and creating a second "halo-like" image on the film.

Different types of film emulsions have different properties. The characteristics of black and white film are: light sensitivity, color sensitivity, contrast, and grain.

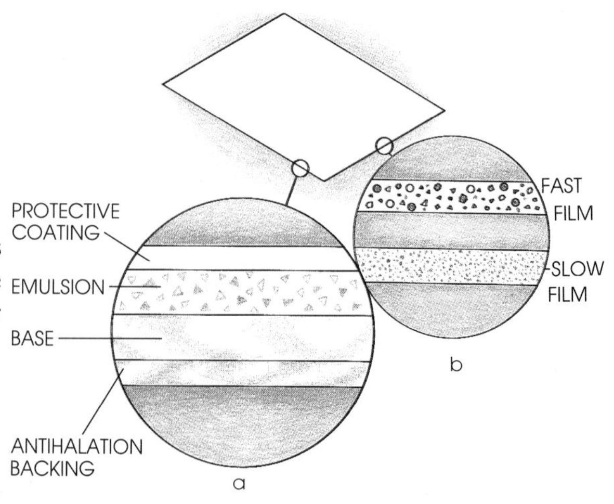

Fig. 11-18. The cross section (a) shows the four layers in a piece of black and white film. In general, the larger the silver crystals in the emulsion (b), the faster the film speed.

Light Sensitivity

Light sensitivity refers to the speed with which the film reacts to light. A "fast" film requires relatively little light for proper exposure. This is because it has larger crystals of silver in its emulsion layer. Fig. 11-18(b). Fast films are useful in low light.

Light sensitivity is referred to as film "speed" and is indicated with an ISO number. (ISO stands for International Standards Organization. Common ISO numbers for black and white films are 1000, 400, 125, and 32. An older designation, ASA, is also sometimes used.) The higher the number, the greater the light sensitivity of the film. For example, ISO 32 film requires far more light than ISO 1000 for proper exposure.

Color Sensitivity

Color sensitivity refers to the fact that different films are sensitive to different colors. Fig. 11-19. **Panchromatic films** are sensitive to nearly all visible light, especially blue light. Most panchromatic films produce a wide range of gray tones. Therefore they are used for most black and white photography.

Orthochromatic films are sensitive to nearly all visible light except red. They may therefore be handled under red "safe" light without danger of exposure.

Infrared films are sensitive to nearly all visible light, plus invisible infrared wavelengths. Blue-sensitive films are only sensitive to blue light.

SCIENCE FACTS

Infrared Film

On the electromagnetic spectrum, infrared radiation occupies the area between visible light and microwaves. We can't see infrared radiation, but we can feel it as heat. Sixty percent of the radiation we get from the sun is in the infrared wavelengths.

Infrared radiation can penetrate haze and mist better than visible light waves can. Therefore film that is sensitive to infrared radiation can be used to take pictures in haze or mist. Things that are warm give off infrared radiation. Infrared film can photograph these, even in darkness.

You've probably seen satellite photographs of the earth in which forests and other vegetation appear red instead of green. These photographs were taken with "false color" infrared film. In these pictures, anything giving off infrared radiation appears red. Scientists use these photos to study many things, such as patterns of plant growth.

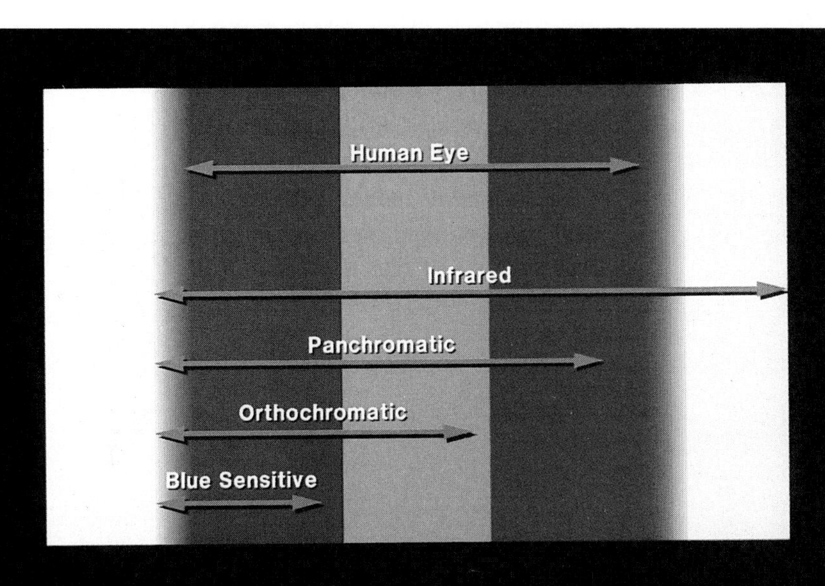

Fig. 11-19. Here you can see a comparison of the color sensitivity of various films with that of the human eye.

Fig. 11-20. Compare the high contrast photo at left with the regular photo at right. What differences do you see?

Fig. 11-21. When a photo is enlarged, grain becomes more noticeable.

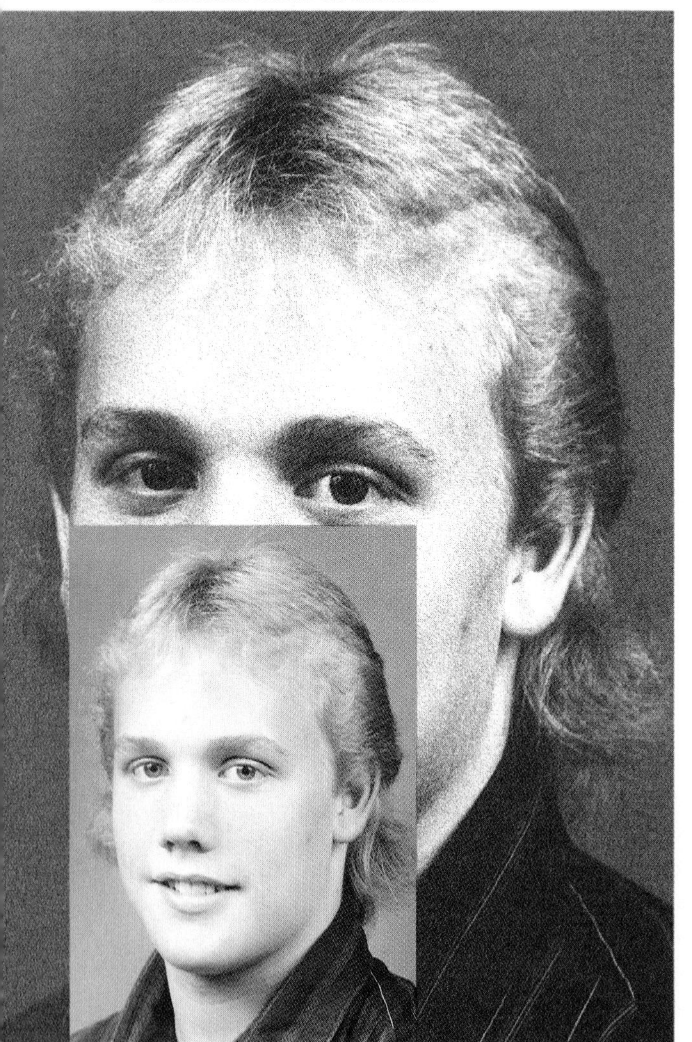

Contrast

Contrast is the amount of difference between light and dark areas in a photograph. "High-contrast" means there is a lot of difference between the light and dark areas and there are few grays in between. The amount of contrast in a photograph depends on the method of development as well as the type of film. Fig. 11-20.

Grain

When film is developed, a grain pattern appears in the silver crystals. If the grain is fine enough, it is hard to see in the photograph. Some films, however, have very coarse grain that becomes visible in enlargements. Fig. 11-21.

In general, the greater the light sensitivity of the film (higher ISO number) the more visible grain. This is because the silver crystals in these faster films are larger.

Usually, photographers try to make photographs that show little, if any, grain. Once in a while, however, grain is exaggerated for effect.

Film Formats and Sizes

Panchromatic films are packaged in a number of different ways. You may buy them in cartridges, rolls, or sheets. The cartridges are designed for quick loading.

Roll films require a little more effort to load but are less expensive than cartridge types. Serious photographers often "bulk load" their own rolls. That is, they buy 100-foot rolls of film and wind it onto individual small rolls themselves. This saves quite a bit of money.

Sheet films are also available. These are generally used in the larger view cameras.

Since cameras come in different sizes there must be films in different sizes to fit them. Numbers such as 110, 116, 120, 126, 127, 135, 220, 616, 620, and 828 are used to designate different sizes of film. The safest thing to do is to refer to the manual that comes with a camera when purchasing film.

Photographic Papers

Photographic papers are used to make prints. Different papers create different feelings and effects. Black and white papers are identified by their weight, surface texture, contrast range, surface coating, and tone. Fig. 11-22.

The weight of the paper refers to its thickness. Photographic papers may be single-weight, medium-weight, or double-weight.

Many surface textures are available. Glossy papers have a smooth, glossy surface. These are commonly used when the photograph is to be reproduced in a book or magazine. Glossy surfaces reproduce better than do other surfaces. Matte papers have a dull finish. They are a good choice if the photograph is to be displayed. Papers with pebbly or fabric-like surfaces can create interesting effects.

Just as films are available in different contrasts, photographic papers are made to produce different contrasts. Fig. 11-23. A numbering system indicates the degree of contrast. For example, a #1 contrast paper has low contrast. Photographs made with this paper appear "soft." A #5 paper, on the other hand, is a high-contrast paper with few gray tones.

Some papers are capable of producing a range of different contrasts. These variable contrast papers require the use of contrast filters during processing. The higher the filter number, the more contrast produced.

Resin-coated papers have a plastic coating on their surface. This helps to keep chemicals from soaking into the paper during the processing stage. However, they generally do not produce as wide a range of tones as do uncoated papers.

Color Film

Much of what you have already studied in this chapter also holds true for color film. However, there are some basic differences. Color films are of two types: negative and positive. Color negative films produce color prints, while the positive films result in color slides.

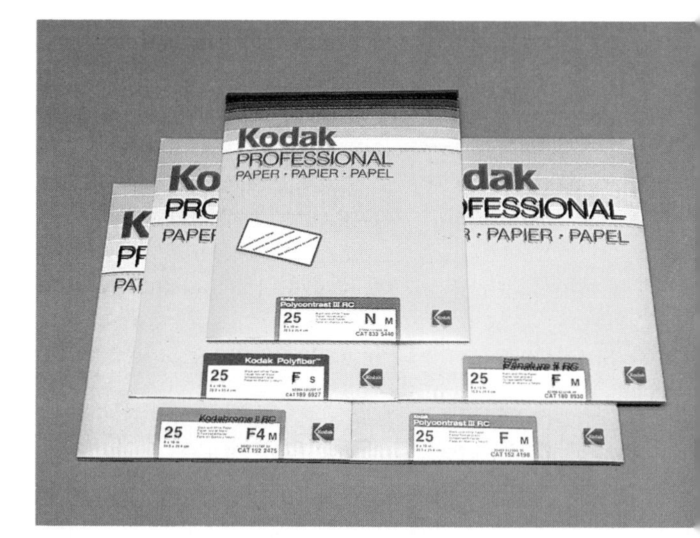

Fig. 11-22. Photographic papers come in many varieties. Serious photographers often keep several kinds on hand.

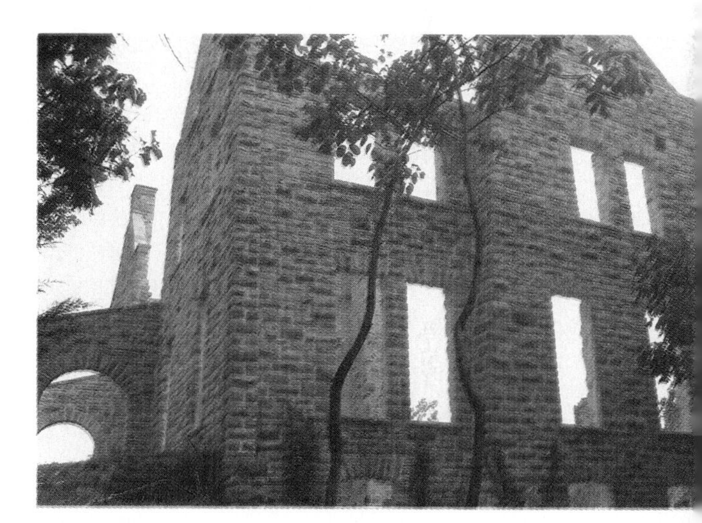

Fig. 11-23. The choice of paper can affect the amount of contrast in the print.

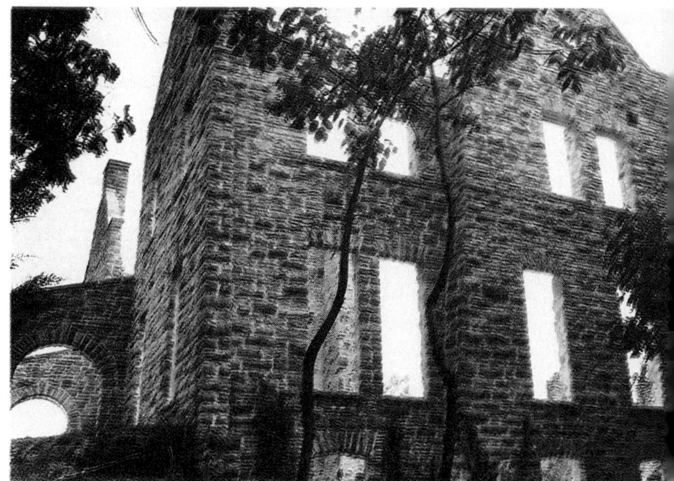

Color negative film consists of a clear plastic base coated with three emulsion layers. Fig. 11-24. Each layer is sensitive to one of the additive primary colors described in Chapter 10: red, green, or blue. When you take a picture of a red car, the red light reflected from the car exposes the red layer. When the film is processed, this layer is changed to cyan (its subtractive complement) by a cyan dye. In the same way, a green car would be magenta on the negative, and a blue car would be yellow.

Color Photographic Papers

Color photographic papers work much the same way as color film. They have three light-sensitive layers: red, green, and blue.

Consider the red car described above. Remember, it produced a cyan area in the film negative. Light passing through this cyan area exposes the blue and green layers of the photographic paper (cyan really is a combination of blue and green). When the paper is developed, the exposed blue layer is converted to its subtractive complementary color (yellow). At the same time, the exposed areas in the green layer are changed to magenta (the subtractive complement of green). When white light passes through these yellow and magenta layers and reflects off the paper to our eyes, it looks red.

Color Transparencies

Color transparencies, commonly called slides, work very much like color negatives. The difference is in the processing. When slide film is processed, it is first changed into a black and white negative. Then, subtractive dyes create a positive image in the subtractive colors. When you look at the slide, you are really looking through three subtractive filters. Since these colors vary in density across the slide, they combine to reproduce all colors in the visible spectrum.

SCIENCE FACTS

Tungsten Light

Tungsten light is light from ordinary lightbulbs. The lamps in your home produce this kind of light. Photographers call it tungsten light because the filament inside the bulb is made of the metal tungsten.

The light produced by tungsten bulbs contains more yellow and red than blue. Therefore film used under tungsten light should be sensitive to blue. Otherwise, the film will not record enough blue light and the photograph will have a yellow or orange tone.

Color Balance

The sun produces a different color of white light than do tungsten or fluorescent lightbulbs. Because of this, color films must be balanced for the light source being used. You may purchase films balanced for either daylight (sun) or tungsten light. Using a daylight film indoors with tungsten lighting produces a photograph that looks too yellow or orange. Either of these films used under fluorescent lights results in an off-color photograph. To counter this color shift, color filters must be used when the picture is taken.

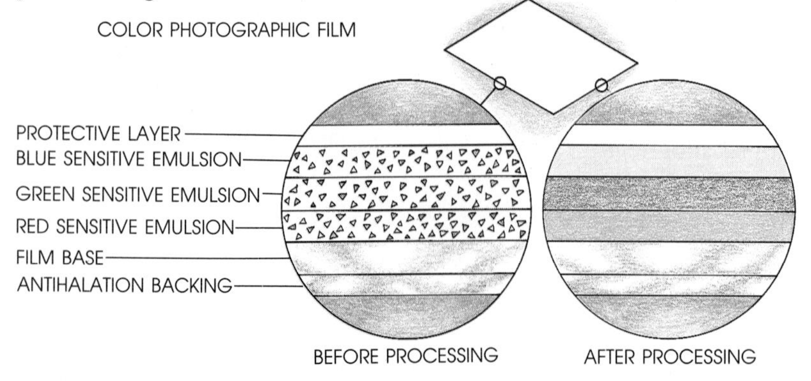

Fig. 11-24. Color photographic films have three emulsion layers which are dyed yellow, magenta, and cyan during the film processing stage.

PROCESSING AND PRINTING

Turning film into prints requires several procedures. First, the exposed film is developed. The resulting negative is transferred to light-sensitive paper. Then the paper is developed. The following describes the materials and equipment needed. In Chapter 12, you'll learn the step-by-step procedures.

Processing Chemicals

Whether developing film or photographic paper, the chemicals are roughly the same: developer, stop bath, fixer, and water wash.

The **developer** causes the exposed silver crystals to turn black. Fig. 11-25. The **stop bath** is a solution of 18 percent acetic acid. It halts the developing process by neutralizing the developer. The **fixer** makes the image permanent and removes any unexposed silver crystals. Finally, the water wash removes all fixer and excess chemicals that might otherwise ruin the film or paper after it dries.

During processing, the film is bathed in this order: developer, stop bath, fixer, and then water wash. The order must not be changed, and the process must be controlled with time, temperature, and agitation (movement).

Most chemicals need a certain amount of time to work. How much is needed depends on the type of chemical and type of film. The developing time affects density and contrast. If the film is overdeveloped (left in the developer too long), the exposed areas will become too dark. A print made from such a negative will look "washed out." Underdeveloping the film will result in prints that are too dark.

Fig. 11-25. Silver particles that were exposed turn black when developed. The unexposed silver is removed by the fixer.

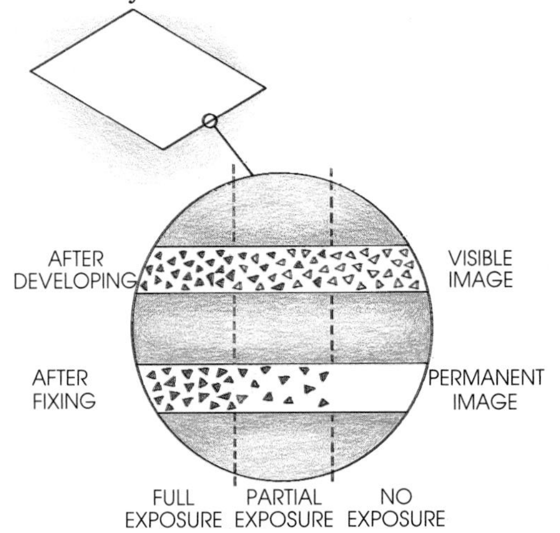

Ideally, all chemicals used for processing should be the same temperature. It is especially important for the developer to be at the right temperature. If it is too warm, the film may quickly become overdeveloped. If it's too cool, development will be slow. Check the manufacturer's directions for best temperature.

Agitation causes the chemicals to wash over the film's surface. This keeps the film in contact with fresh chemicals and prevents air bubbles from forming. Agitation is accomplished by rocking the developing tray or tapping and moving the developing tank in a circular motion.

Film and paper developers are manufactured to produce different levels of contrast and grain. By selecting a fine-grain film and fine-grain developer, a photographer can produce negatives that show very little grain. Similarly, a high-contrast film developer might be selected if the photographer was trying to produce a photograph with a lot of contrast. Sometimes a low-contrast paper developer is used to soften the effect of a high-contrast negative. What is used depends on the result desired.

Negatives that are underexposed or underdeveloped appear very light. These do not produce good photographs, as there is not enough contrast. A chemical intensifier can be used to build up the overall density of the negative. This improves the appearance of any photographs made. The opposite effect, lightening a dark negative, may be done with a chemical known as a reducer.

Toners are chemicals used to change the color of photographic paper. The most common is sepia toner, which causes the paper to turn a pale brown color. Sepia toner gives a brand new photograph an antique look. Toners are available in many different colors.

Equipment

Processing photographic films requires developing tanks. Fig. 11-26. A developing tank is a light-tight container that holds one or more developing reels. In a darkroom, film is loaded onto the reel and then put inside the tank. Once the film is inside the tank, developing may take place in normal room light.

Developing tanks may be of plastic or stainless steel. The plastic type are a little easier to use, but the metal type last longer if well cared for. Another advantage of the metal type is their ability to conduct heat. Chemicals poured into a metal tank may be warmed or cooled more easily.

A basic developing tank holds only one reel. However, larger tanks hold two or more reels at a time. Plastic reels are often adjustable for different film sizes. This is handy if you have several rolls of film to develop at one time. Metal reels are not adjustable.

Fig. 11-26. Film is wound on a reel and placed in the developing tank.

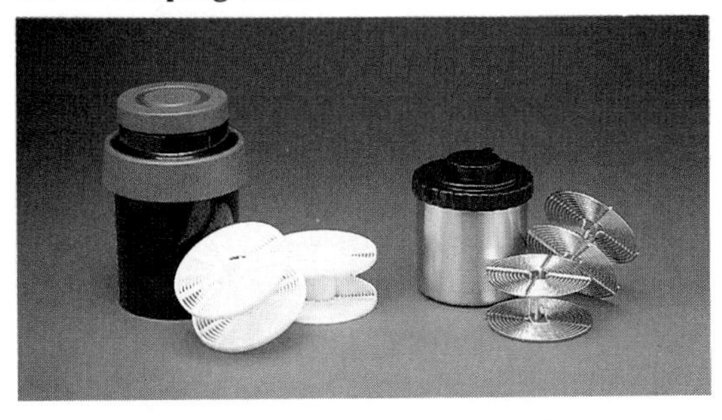

Fig. 11-27. A photographic enlarger is used to make prints that are bigger than the negatives.

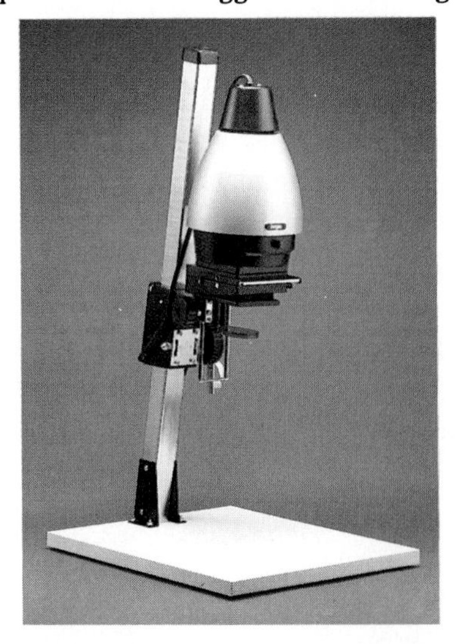

Film negatives are generally much smaller than the photographs they produce. **Enlargers** are used to "blow up" these negatives. Light is projected through the negative while it is held at a distance from the photographic paper.

The basic parts of an enlarger are the head, negative carrier, lens, base, and support. Fig. 11-27. Inside the head are a special lightbulb and a lens that distributes light evenly across the negative. The negative carrier holds the negative in place.

An enlarging lens is attached to the bottom of the head. It focuses the image onto the photographic paper. There are different size negative carriers and lenses for different size negatives.

The head may be moved up and down along the support. The farther it is from the paper, the larger the image produced.

An easel is generally used to hold the photographic paper in the proper position. Some are adjustable, while others have openings of different sizes for different size photographs. The frame of the easel produces the white border on a photograph because it blocks light, and the paper is not exposed.

Photographic papers are generally processed in developing trays. Fig. 11-28. Ideally, these trays should be just slightly larger than the paper being developed. That way, few chemicals are wasted. At least four trays are needed for black and white photography. These hold developer, stop bath, fixer, and a water wash.

The steps followed during printmaking will be discussed in Chapter 12.

Fig. 11-28. Plastic or stainless steel developing trays are used for processing photographic papers.

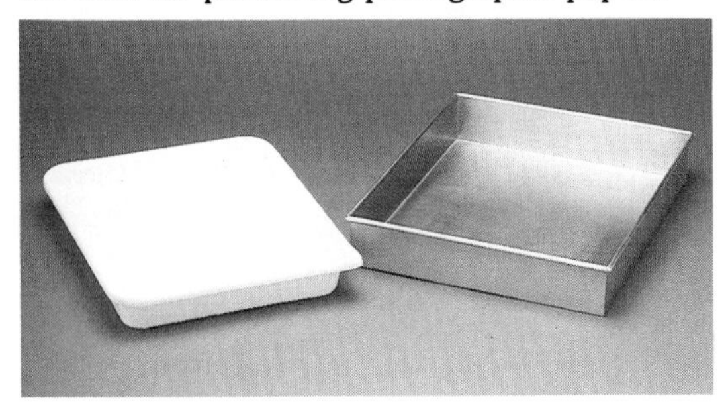

CHAPTER 11 REVIEW

Review Questions

1. Describe and tell the purpose of the following: viewing mechanism, lens, shutter.
2. Why does the viewfinder on a rangefinder camera cause problems when viewing the subject?
3. What is the purpose of the mirror in a single lens reflex camera?
4. What is focal length? How does it affect a lens?
5. What is the purpose of a light meter?
6. How does electronic flash differ from continuous lighting?
7. Describe the structure of black and white film. How does it differ from that of color film?
8. Define film speed, contrast, and grain.
9. Tell the purposes of developer, stop bath, and fixer.
10. How are time, temperature, and agitation important during processing?

Activities

1. With a manual to guide you, sit down with a 35mm SLR camera and identify each of the basic parts. Without any film in the camera, make all of the adjustments/settings.
2. Conduct research in your school library on some aspect of the history of photography. Write a report on your findings.
3. Experiment with lighting a model for a photograph. Use different sources of light set up in different positions. Observe the results of these changing set-ups, and take notes on the effects they create.
4. Practice loading a roll of exposed film onto a developing reel. When you get good at it, try doing it behind your back or with your hands inside a cloth bag. This will prepare you for working in a darkroom.
5. Locate photographs (or reproductions of photographs in magazines) that show a) high contrast, b) low contrast, and c) coarse grain. Label them and make a display for the class.

Applications of Photography

Photography is used in a wide variety of applications. Commercial photographers create photographs for advertising and illustration. Scientists use photography to document their experiments and communicate their ideas to other scientists. Photojournalists document newsworthy events, while artists create photographs for personal expression.

For most people, applying the principles of photography means taking their own pictures. They may be simple snapshots or high quality photographs made with complex equipment. Some people also like to process their own film and make their own prints.

As simple as photography can be, it is also a science and an art that can take years to master. Every single photograph is an opportunity to learn something new.

Terms to Learn

aperture
burning-in
composition
contact prints
depth of field
dodging
exposure control
f-stops
inverse square law
projection printing
rule of thirds
safelights
spot retouching
step test

As you read and study this chapter, you will find answers to questions such as:

- How do photographers think about their subjects before taking a picture?
- How do lighting and camera settings affect a photo?
- What are the steps in processing film?
- How can mistakes in negatives be corrected?
- How are prints made?

TAKING PICTURES

Taking pictures involves more than just clicking the shutter button. It means understanding composition and knowing how to handle the camera and its adjustments.

Composition

Have you ever seen photographers make a little "window" at arm's length with their hands? Sometimes this is done to get a clearer idea of the composition. **Composition** is the way in which all the elements in a photograph are arranged. Every time photographers take a photograph, they are thinking about its composition. Good photographs rarely happen by chance. They are more likely to be well thought out in advance.

The Subject

Interesting photographs have one thing in common: they have an obvious subject. When you look at them, your eye is drawn to what the photographer wanted you to see. It may be a person, animal, or object. Whatever the subject, it was decided on *before* the photograph was taken.

Except in rare cases, photographers focus on the subject. The subject should look sharp and clear.

When it comes to placing the subject within the picture, photographers use the **rule of thirds**. The rule of thirds divides the viewing area into thirds both vertically and horizontally. Many amateurs locate the subject in the center. However, when the subject is located at a spot where the lines intersect, it creates a more pleasing composition. Fig. 12-1.

Fig. 12-1. Imagine a view divided into three parts horizontally and another three parts vertically. Composition is more interesting if the subject is located at the intersections of these divisions (below) rather than in the exact center (above).

Because photography is an art, photographs can benefit from the elements of art and principles of design discussed in Chapter 13. For instance, diagonal lines are more interesting than horizontal or vertical lines. Diagonals tend to lead the eye into a picture. Fig. 12-2. They may be used to lead the eye directly to the subject of the photograph.

In general, the larger the subject in relation to the field of view, the better. As photographs are enlarged, they become grainy and lose detail. If the subject fills the field of view, the photo does not have to be enlarged as much. The result will be clearer. Fig. 12-3.

In getting a close-up of the subject, care must be taken not to crop off important parts. For example, photographers usually do not want to cut off part of the subject's head. Arms and legs may be less important to the picture and are more often cropped.

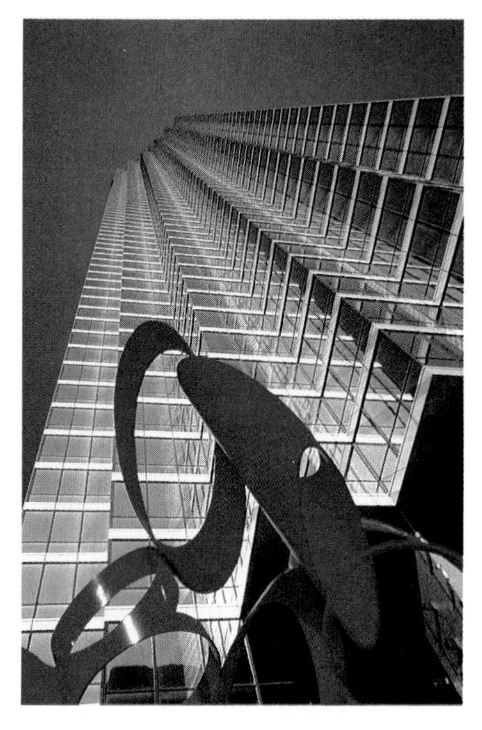

Fig. 12-2. The unusual angle of this photograph makes it more interesting.

Fig. 12-3. A close-up of the subject makes the photograph more interesting.

Fig. 12-4A. Background objects can be used to frame a subject.

Sometimes other elements in the photograph can create a natural frame around the subject. The trunk and branches of a tree, for example, can frame a subject standing beneath the tree. Fig. 12-4A.

Background

It is easy to concentrate so much on the subject that the background is neglected. This is a common mistake among beginners. Too often the background detracts from the subject. For example, a person photographed in front of a busy background can look as if objects were "growing" out of his or her head. Fig. 12-4B.

A simple solution is to use plain backgrounds or those having consistent texture. Studio photographers often hang a roll of paper behind a model to create a solid background. Shadows in the background can also detract. This can be remedied with lighting as discussed on page 255.

Timing

Many photographs require split-second timing. A baby changes facial expression many times in one minute. Sporting events often have exciting moments that last only a second or two. To capture these pictures, a photographer must be ready to act instantly. This means composition and camera adjustments must be determined in advance as much as possible.

Fig. 12-4B. Busy backgrounds detract from a subject. It's better to use a simple background that lets the subject be the center of attention.

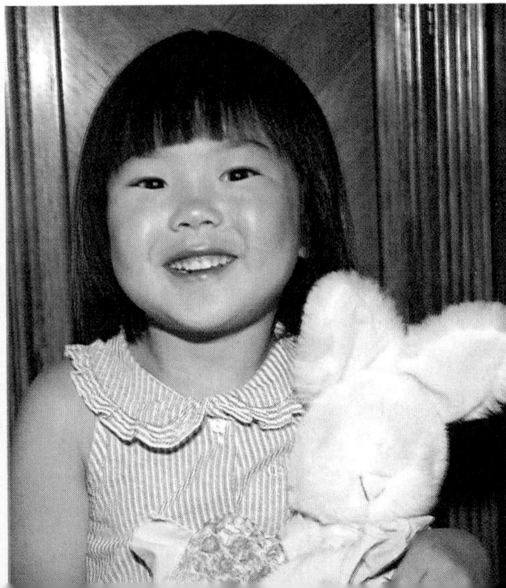

Using the Camera

Snapshot photography requires little or no skill. The "point and click" method produces snapshots that are suitable for many people's needs. For higher quality results, learning additional techniques is a must.

Holding the Camera

Random movement of the camera while taking a photograph results in a blurred image. To avoid out-of-focus pictures, the camera must be supported properly.

A 35mm SLR camera should be held in the palm of the left hand. The thumb, index, and middle fingers of the left hand are then used to adjust the f/stop and focus ring. Fig. 12-5. The right hand is then free to adjust the shutter speed and press the shutter release. With a little practice, this technique becomes second nature.

For shutter speeds slower than 1/60 of a second, a tripod should be used. Most people cannot hold the camera steady enough to avoid blurring at these slow speeds. If a tripod is not available, the camera and/or body can be rested against some other support such as a wall. A cable release also makes it easier to take pictures at slow speeds without moving the camera. (See page 242.)

Lighting

As discussed in Chapter 11, lighting is the most important concern when taking photographs. Using existing lighting (either natural or artificial) often results in an average photograph. By giving thought to the lighting, better results can be obtained. For example, light coming through a window can be used as a side or back light.

When photographing subjects outdoors, it is generally a good idea to keep the sun behind the camera or to the side. This limits glare that can appear in the picture and ruin it. On the other hand, dramatic results can be achieved by silhouetting a subject. The subject is dark and without visible detail against a bright light. This is done by locating the subject between the sun and the camera at sunrise or sunset. Fig. 12-6. When creating the silhouette effect, care must be taken to hide all or part of the sun from the camera lens or the scene will be overexposed. Also, damage to the eyes can occur if the sun is viewed directly.

Effective indoor lighting is easiest to achieve in a photographic studio. In the studio, a photographer can have a wide range of lighting equipment on hand at all times: flood lights, spot lights, umbrellas, a power supply, backdrops, and so forth. These can be positioned around

Fig. 12-5. This is the correct way to hold a camera.

Fig. 12-6. Silhouetting the subject can produce dramatic photographs.

the studio to create different lighting effects. Fig. 12-7.

When away from the studio, photographers often carry lighting equipment with them. The most portable of all lighting equipment are flash or strobe lights. These may be mounted on the camera or held by hand or on a stand. With practice, flash attachments can be used for a range of different lighting effects. Many may be angled to bounce the light off a wall or ceiling to soften the effect. Covering them with a handkerchief also softens the light.

At ten feet away, a light is only one fourth as bright as it is at five feet. The same light at twenty feet is only 1/16 as bright. This is the **inverse square law** of physics. The reduction in light must be taken into account when using a flash. From three feet away, the flash may wash out the subject. From the back row at a concert, it is useless. Many flash attachments have sensing devices that judge how far they are from the subject and provide more or less light as needed. Nevertheless, flash photography is tricky and requires practice for consistent results.

Regardless of the type of lighting, care must be taken with shadows. Shadows can produce dramatic effects. They can also hide important details or produce large unwanted dark areas.

Fig. 12-7. Different lighting was used to achieve these results: top left — front lighting; top right — side lighting; bottom left — main and fill lighting; bottom right — main, fill, and accent lighting.

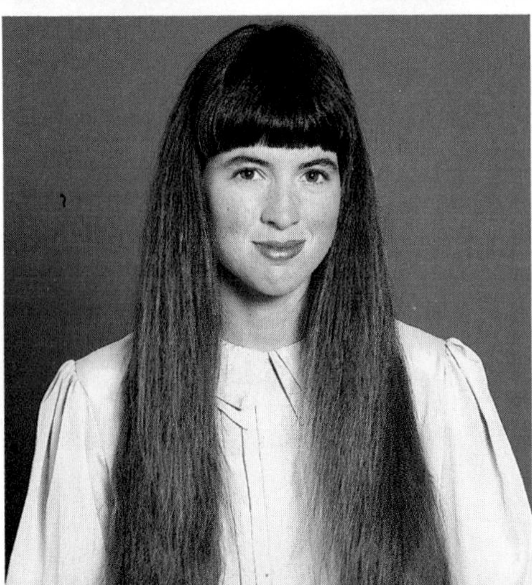

Camera Settings

Light by itself is not enough for a good picture. It must be the right amount of light. This is known as **exposure control**.

On some cameras, exposure control is automatic. There are also cameras on which the settings are built in at the factory and cannot be changed. However, many cameras, such as the 35mm SLR camera, do allow the user to control the exposure. This is done by adjusting the shutter speed and the size of the **aperture** (lens opening). The right combination can produce some interesting effects.

A typical 35mm SLR camera will have shutter speeds from 1 second to 1/1000 of a second. It will also have a "B" setting that allows the shutter to be held open manually for any amount of time.

Shutter speeds are designed so that each one is half or twice as long as the next: 1, 1/2, 1/4, 1/8, 1/16, 1/30, 1/60, 1/125, 1/250, 1/500, and 1/1000 of a second. If a shutter speed is adjusted to two speeds "faster," only 1/4 as much light is let into the camera. Three shutter speeds faster lets in 1/8 as much light, and so forth.

The aperture is controlled by a diaphragm. The opening created by the diaphragm may be adjusted with the aperture ring on the lens. The different size openings created this way are called **f-stops**. Like the shutter speeds, each f-stop lets in half or twice as much light as the next. Typical f-stop numbers on a 35mm SLR camera are f/22, f/16, f/11, f/8, f/5.6, f/4, f/2.8 and f/1.8. The larger the f-stop number, the smaller the aperture. The f-stop numbers are the result of dividing the focal length of the lens by the diameter of the aperture ($f = F/D$).

Since both the f-stop and shutter speed are set up to let in half or twice as much light from one to the next, it is possible to create the same exposure with different settings. For example, using the following settings, the same exposure is obtained.

Shutter Speeds	**f-stops**
1/30	f/11
1/60	f/8
1/125	f/5.6
1/250	f/4
1/500	f/2.8

If the shutter is set to remain open longer, then the aperture is made smaller, or vice versa. (Remember, the larger the f-stop number, the smaller the opening.)

Even though different settings can produce the same exposures, the pictures may look different. The reason for this is that different exposure settings create different depths of field and record motion differently. **Depth of field** refers to how much of the field of view, from front to back, remains in focus. A small aperture, f/22 for instance, creates a large depth of field — that is, objects close to the camera as well as objects far away are in focus. Fig. 12-8. Large apertures, on the other hand, have a short depth of field. If a subject in the foreground is photographed with a large aperture, the background will be out of focus. This is often done to draw attention to the subject. When taking pictures with large apertures, care must be taken to keep the subject in focus.

Fig. 12-8. The photo at right has greater depth of field. All the items in the picture are in focus.

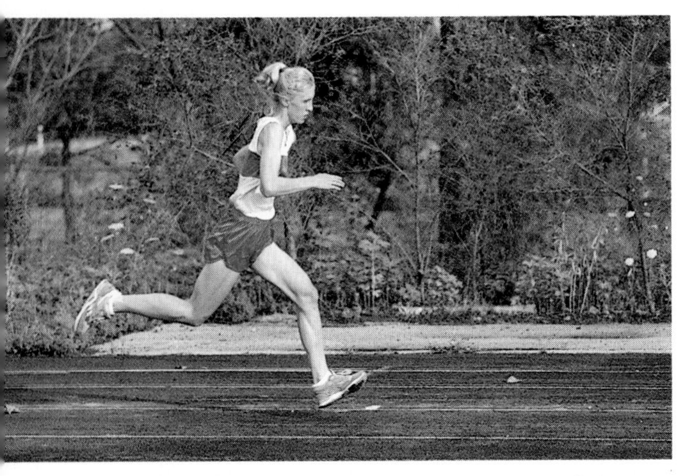

Fig. 12-9. (Above) Fast shutter speed freezes motion. (Below) A slow shutter speed captures only a blur.

Fig. 12-10. According to this light meter, the aperture should be set at f/4 for correct exposure.

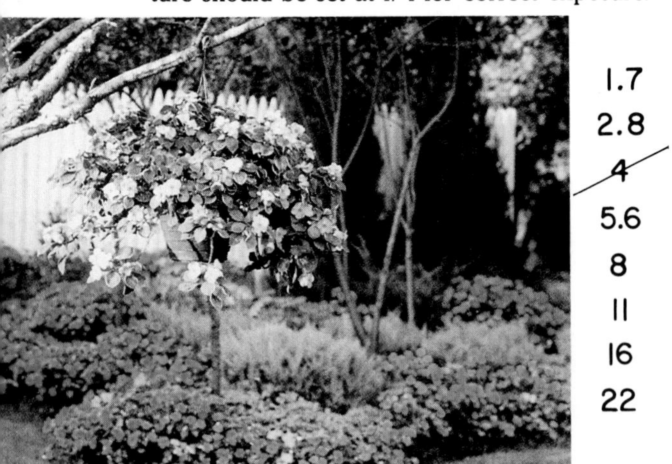

Shutter speed is important when the subject is moving. Fast shutter speeds can make moving objects appear to stand still. A runner photographed at 1/1000 of a second can be "frozen" in the photograph. On the other hand, the same runner can be caught in a blur of motion using a ½ second shutter speed. Fig. 12-9.

As you can see, exposure control may be used by a photographer to control the appearance of the photograph. It is not simply a matter of getting the correct amount of light on the film. It is also a matter of selecting a shutter speed/aperture combination that will produce the desired effect. This comes with lots of practice and careful observation.

Another adjustment that affects exposure is determined by the film's ability to react to light. It is possible to adjust the ISO (ASA) setting on the camera. This tells the camera's built-in light meter how to react based on the film being used. ISO (ASA) adjustments are usually indicated on the camera by number, such as 400, 125, and 32. As film is loaded, the setting for that type of film should be made. Some cameras do this automatically.

Reading the Light Meter

Most SLRs have a built-in light meter. The meter is usually read through the viewfinder. Fig. 12-10. Some meters simply indicate with a **+** or **–** if more or less light is needed. Others suggest an f-stop setting that will give proper exposure.

Occasionally a meter can be misleading. This occurs when a dark subject is against a bright background. The meter will read the bright background, and the resulting photograph will be a silhouette of the subject. To avoid this, a meter reading should be taken close to the subject and camera adjustments made. Then when the photograph is actually taken at the desired distance, the result will be better.

HEALTH & SAFETY

DARKROOM SAFETY

Although developing photographs is not a dangerous activity, there are certain safety precautions you should take.

- Some people have skin reactions to photographic chemicals. To be safe, wear gloves. Use tongs to lift negatives or photos from developing trays.
- Wear safety glasses when mixing chemicals to avoid getting the chemicals in your eyes. If this happens, flush your eyes with cold water.
- Chemicals are poisonous. Don't bring them in contact with food or beverages.
- Be sure the darkroom has sufficient ventilation so fumes won't accumulate.
- Read the manufacturer's directions carefully before mixing and using chemicals.
- Store supplies in such a way that they don't create fire hazards.
- Keep walkways clear to prevent tripping or injuries.

FILM PROCESSING

Since panchromatic film is sensitive to nearly all visible light, processing must take place in total darkness. Film developing requires seven different steps. Fig. 12-11.

Step 1: Loading the film into the tank. Loading the film into the developing tank must be done in complete darkness. A totally dark room or a loading bag may be used. The trick is to get the roll of film onto the developing reel while in total darkness. Practicing this with the reel and a scrap of film in room light is the best way to learn. When it can be done without looking, it can be tried in total darkness.

Step 2: Developing the film. The purpose of the developer is to turn the exposed silver crystals into black metallic silver to create the image. This is a chemical reaction. Careful control of time, temperature, and agitation is important to produce good negatives. The manufacturer supplies a chart to determine the proper developer, temperature, and time required. Utensils must be clean, measurement of developer must be cor-

Fig. 12-11. These seven steps turn a roll of exposed film into usable negatives.

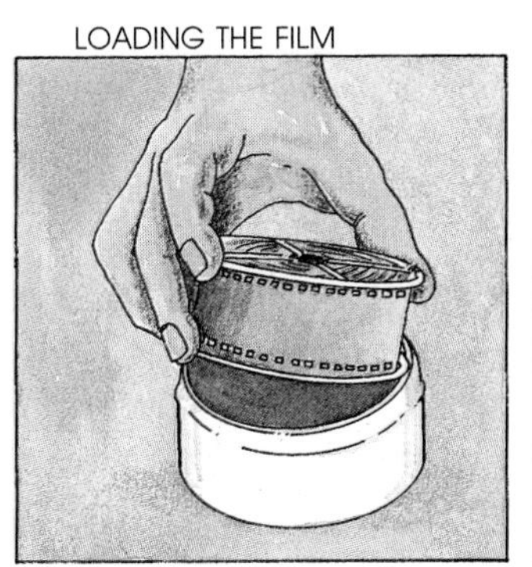

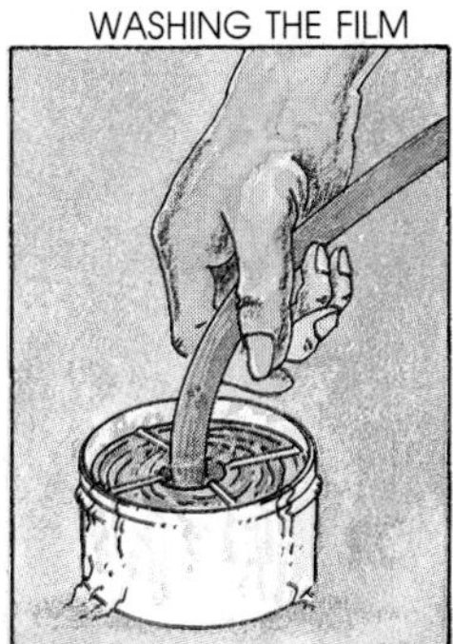

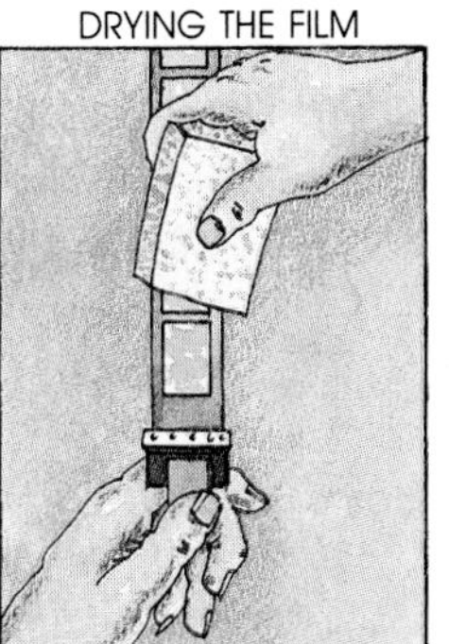

rect, and it must be the right temperature. This is usually between 68 and 72 degrees.

The exact time should be noted and developer poured (without stopping) into the developing tank. Tapping the developing tank several times on a counter top removes air bubbles from the film. Then the tank should be agitated by turning it upside down or moving it with a circular motion. Agitation should be repeated once every 30 seconds throughout the developing stage. Careful and consistent agitation is important in achieving good quality negatives.

Step 3: Stopping development. At the end of the developing time, the developer should be discarded and the tank filled with stop bath at room temperature. The stop bath is a mild acid solution that neutralizes the developer. This causes developing to stop. The reaction takes only about 10 seconds. The tank should be constantly agitated during the stop bath stage.

Step 4: Fixing the image. Next, the tank should be filled with fixer at room temperature and agitated continuously for several minutes. The purpose of the fixer is to remove all of the unexposed silver halide from the film. Unexposed silver has a milky white appearance. After several minutes of fixing, the tank should be opened and the film checked to be sure all of the unexposed silver has been removed. If the film is not clear, the lid is put back on the tank and fixing continues for another minute or two. Fixer may be reused until it no longer clears film within several minutes.

Step 5: Washing the film. The image is now permanent, but there are still chemicals on the film. Unless they are removed, the film will discolor as it dries. To accomplish this, the tank with the reel still in it should be placed under running water. Cold water must be allowed to run over the film for five to ten minutes.

Step 6: Wetting the film. When water dries on the negatives, it can leave water spots that would ruin them. To prevent this, a wetting agent may be used. The wetting agent causes the water to flow from the film more evenly, reducing spots. While the film is still on the reel, the reel should be dipped in a wetting agent. The excess is then shaken off.

Step 7: Drying the film. Negatives should be carefully hung to dry in a completely dust-free place. If any dust particles come to rest on the negatives while they dry, they will be ruined. Once dried, the processed film can be cut into strips of five negatives and stored in a film file.

MAKING PRINTS

Projection printing is making prints by projecting light through the negative onto a sheet of photographic paper. The farther the negative is from the paper, the larger the resulting print. All of this is done in a darkroom, under filtered **safelights**. Safelights have coatings that filter out certain colors of light. The remaining light does not expose photographic paper.

The following steps are required in print making. Fig. 12-12.

Step 1: Cleaning the negative. Negatives used for enlarging should be perfectly clean. Even the smallest piece of dust will be enlarged and appear on the print. A soft brush or compressed air, shot out of a can, is used to remove dust.

Step 2: Setting up the enlarger. The room lights should be turned off and the safelights turned on. The negative is then carefully placed with the emulsion-side (curled side) facing down in the negative carrier of the enlarger. With the enlarger bulb on, the enlarger head is moved up and down until the desired size print is projected on the easel. Adjustable easels allow the photographer to crop unwanted portions of the image. The easel is moved from side to side or back and forth. Then the image is brought into focus by eye and/or with a device known

Fig. 12-12. These are the steps for making a print.

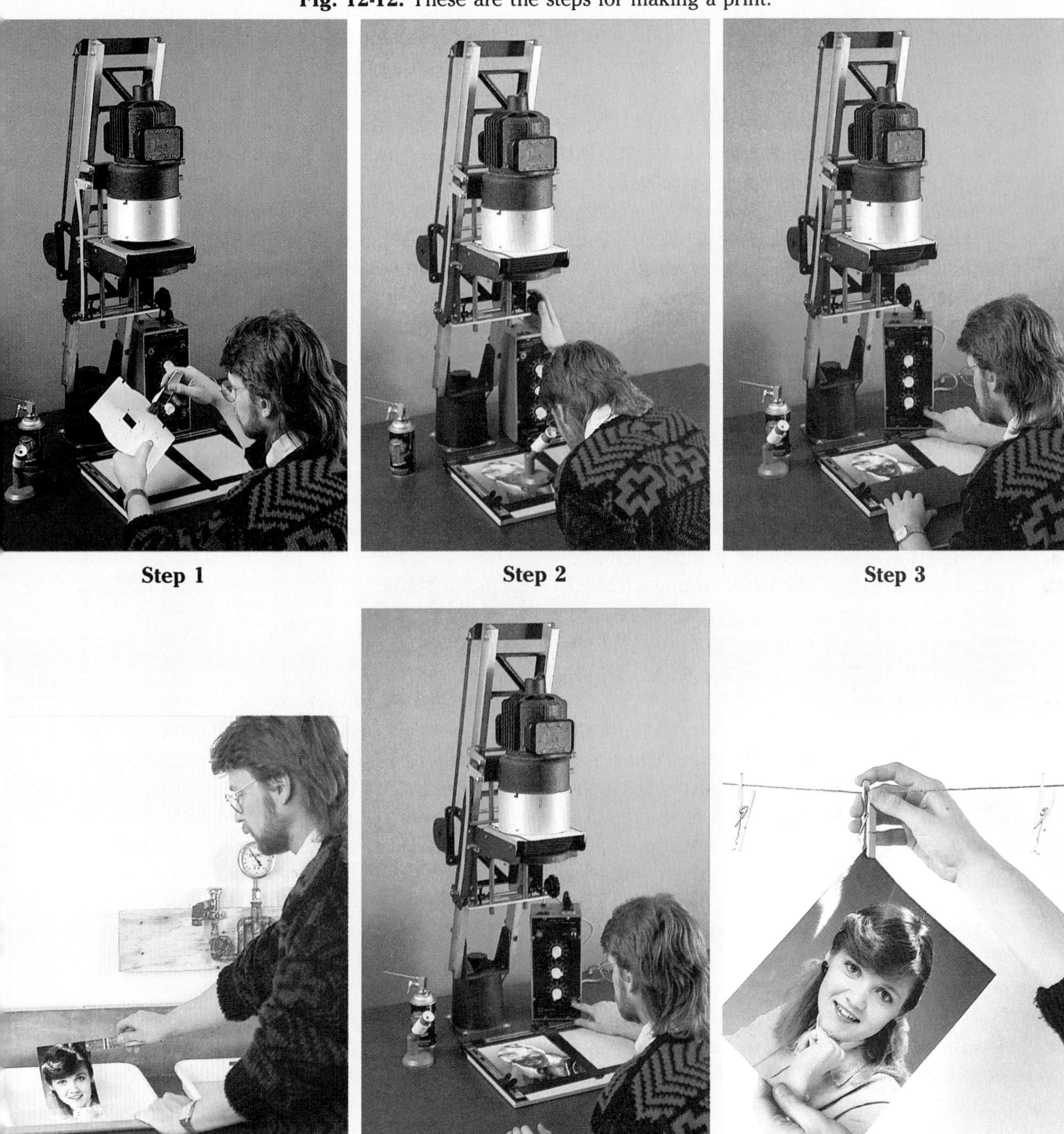

as a focus finder. Focusing is done with the enlarging lens wide open. The more light, the easier it is to focus.

Step 3: Exposing the step test. A test must be done to determine the proper amount of light, or exposure, to use. Since every negative is different, each requires different exposure. Also, as the size of the photograph is changed, the amount of light required changes.

The size of the lens aperture is selected with the f/stop ring. Aperture size varies with the density (darkness) of the negative. Light negatives require a smaller aperture than dark negatives. Beginners use a trial-and-error approach. Experienced photographers develop a sense of how bright the projected light should appear.

A **step test** is made to save time and photographic paper. A strip of unexposed photographic paper is placed on the easel. All but the edge is covered to protect it from light projected through the negative. A brief exposure is made (perhaps 3 seconds). Then the covering is moved over a bit so more paper shows. A second exposure is made. This is repeated perhaps four more times, producing exposures of 3, 6, 9, 12, and 15 seconds.

Step 4: Processing the step test print. All photographic papers should be processed at the time and temperature recommended by the manufacturer. This is important to achieve consistent quality.

Processing is done in four steps: developer, stop bath, fixer, and water wash. The process is the same as for film except in two ways: The developer is a print (rather than film) developer. Processing is done under a safelight, so the photographer may watch what happens.

As with film developing, cleanliness is critical. It is important not to touch the photographic paper with wet hands, as this produces fingerprints on the photograph. Agitation should be constant in the developer. Proper fixing and washing prevent discoloration of the finished photograph.

Step 5: Printing the photograph. The processed step test should display five different exposure densities. Under white light, the exposure time that produced the best overall density is selected. If they are all too light or too dark, the aperture setting on the enlarger lens must be changed and another step test made. The final photograph is exposed for the time indicated by the best test results and processed exactly the same way.

Step 6: Drying the photographic print. The simplest method of drying prints is to hang them in a dust-free place. Drying may be speeded up by removing excess moisture with a squeegee and/or blotting the print with special blotting paper. Print dryers with heated surfaces on which the print is placed may also be used.

SCIENCE FACTS

Instant Photography

A child's eagerness to see vacation pictures led to the development of instant photography. In the late 1940s, Dr. Edwin H. Land was a research scientist working for Polaroid Corporation. After a family vacation, his daughter wondered why it had to take so long before they could see their vacation pictures. Dr. Land decided to find a way to speed up the process, and in 1947 he succeeded.

Instant photography uses the same processes as regular photography. The difference is that the developer, fixer, and print material are already present in the film pack. They are activated after exposure takes place. There is no need for tanks, trays, enlargers, or darkrooms.

Contact Prints

It can be hard to imagine how a picture will look from a negative. The 35mm negatives are especially difficult since they are so small. To make it easier to tell just what is on a roll of film, a **contact print** is made. Fig. 12-13. Sometimes called a proof sheet, the contact print allows photographers to preview the pictures. Then only the best are selected for prints.

A sheet of photographic paper is held in contact with the negatives and exposed. A light from an enlarger or a contact printing light will both work. The negatives are held in place by a clean sheet of glass. It is a good idea to leave the negatives in the clear film file during the exposure to prevent damage. The contact print is then processed just like any other photograph.

Correcting Mistakes

When negatives don't turn out as well as they should, there are a number of things the photographer can do to improve the print.

Fig. 12-13. Contact prints allow photographers to preview pictures and select the best ones for making enlarged prints.

Contrast Filters

One of the most common problems beginning photographers face is working with negatives that lack contrast. In other words, there is not a wide range of tone between light and dark areas. This usually happens because of improper film exposure or development.

Contrast may be improved by using a variable contrast paper and a high contrast filter during the enlarging process. The filter is held in place below the enlarger lens while the step test and print are made. A low contrast filter may be used, if there is too much contrast. Fig. 12-14.

Burning-in and Dodging

Another common problem occurs when an otherwise good negative has an area that appears too light or dark. If the area is too light, it may be corrected by **burning-in**. Burning-in is the process of adding more light to a particular area during enlarging. This process darkens the area on the print and improves the image detail.

First, the entire print is given the proper exposure. Then a mask is created with the hands or paper over the area that is correctly exposed. Then an additional exposure is made. Fig. 12-15. The mask must be moved slightly during exposure to soften the edges of the too-light area. The additional exposure darkens the area that was too light.

The opposite effect, lightening a dark area to improve detail, may be achieved by **dodging**. A small piece of paper in the shape of the dark area is taped to a thin piece of wire. The paper is held in place under the enlarging lens during part of the normal exposure. Fig. 12-16. Jiggling it slightly softens the edges. The result is a lighter area on the photograph.

Spot Retouching

Even under the best conditions, dust spots can appear on a photograph. In some cases, these can be painted over with a gray spotting solution. Fig. 12-17. **Spot retouching** requires a great deal of skill developed by long hours of practice.

Fig. 12-14. If the negative has poor contrast, use variable contrast paper and a high contrast filter to produce the print.

Fig. 12-15. Burning-in darkens areas that are too light. In these two prints, note the difference in the lapels.

Fig. 12-16. Dark areas can be lightened by dodging. In this example, dodging improved the detail in the duck's head.

Fig. 12-17. When skillfully done, spot retouching can greatly improve a photograph.

User's Guide to 

PUTTING PHOTOGRAPHS ON DISPLAY

The appearance of a photograph may be enhanced by the way in which it is displayed. Typically, a photo is mounted (attached) to a piece of matboard. (Matboard is a type of cardboard made for use in framing. It comes in many colors and can be bought at art supply stores.) Another piece of matboard, with the center cut out to allow the picture to show through, is placed on top of the photo. This assembly is then installed in a frame, with or without glass. A faster, less expensive method is to simply mount the photo on matboard. Whether mounted or framed, a few basic tricks can help a photo look its best.

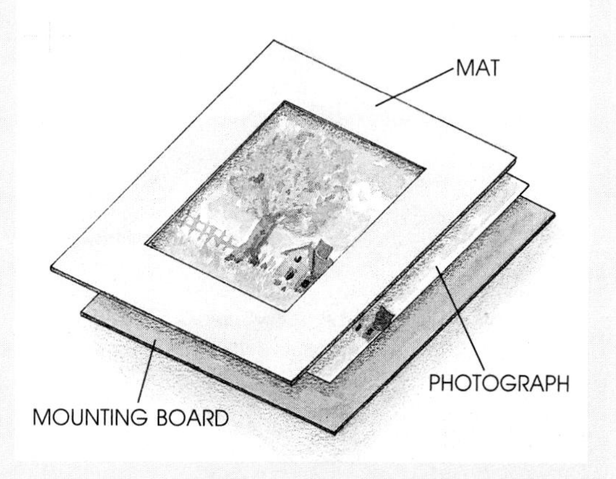

Neutral colors, such as gray, black, or white, should be used when framing or mounting black and white photographs. Other colors will overpower the gray tones of the photograph. For a color photo, choose materials that will coordinate with the colors in the photo. Remember, people's eyes should be drawn to the photo, not to its framing.

Mounting the photograph on the board may be done with dry mount tissue or spray adhesive. Dry mount tissue is a thin sheet of paper that has a waxy coating on both sides. The tissue is sandwiched between the photograph and mount board. Heating this "sandwich" in a dry mount press permanently attaches the photograph to the board. Spray adhesives are a little easier to use, but are not as permanent.

There should be an ample border around the photograph. An 8″ × 10″ print, for example, should have a mat that extends for at least 1½″ all the way around. It is a good idea to have a slightly wider extension on the bottom than on the sides and top.

There are many interesting frames from which to choose. A new "frameless" frame—just brackets and clips—is perfect for mounted photos. Simple gold or silver frames are also good choices.

Arranging several photos on a wall is sometimes difficult. The wall could be full of nail holes before the best arrangement is found. To avoid this, first arrange the photos on a sheet of brown wrapping paper or newspapers taped together. The size of the paper should be the same as the wall space for the photos. Move the photos around until you are pleased with the way they look. Then draw around them with a felt-tip marker. Remove the photos and lightly tape the paper to the wall. Drive nails into the wall through the paper in the appropriate spots. Gently tear the paper away. Hang the photographs.

OTHER APPLICATIONS

The uses of photography include more than pictures for personal enjoyment. New applications are constantly being developed, such as those in the entertainment field.

Motion Pictures

Motion pictures are made by photographing moving objects with film that passes through the camera at 24 frames per second. Each frame records a different position of the object or person in motion. When these frames are run through a projector at the same 24-frames-per-second rate, they recreate the original motion. Even though there are "jumps" in the movement, the human eye cannot see them at this speed.

Animation

The same motion picture principle is used in film animation. Instead of showing the motions of actors, drawings are used. The most celebrated animator of all time, Walt Disney, created color images on clear cellulose film. Each "cel" showed the character in a slightly different position. Twenty-four were needed for each second of animation. After thousands of these cels were created, they were individually photographed. When the film was run at 24 frames per second, the images seemed to move.

Animation is now seldom created by artists drawing individual cels. That is now considered too time-consuming and expensive. Today, animation models and computers are used. The models can be placed in any position. They are set up in a scene and photographed. Then they are moved slightly and photographed again. This is done over and over to produce the final film.

Computers have made a new kind of animation possible. Fig. 12-18. Computer graphic images are first created using a video digitizer or standard computer illustration tools. Generally, the operator creates the beginning and ending position of a particular character/motion. This tells the computer what it must work with. Then the software automatically creates the different images in between. This process is known as "tweening."

Fig. 12-18. The film *Tin Toy* combined a number of computer animation techniques to tell the story of a wind-up toy's encounter with a baby. It was the first computer animated film to win an Oscar.

THE MANY USES OF PHOTOGRAPHY

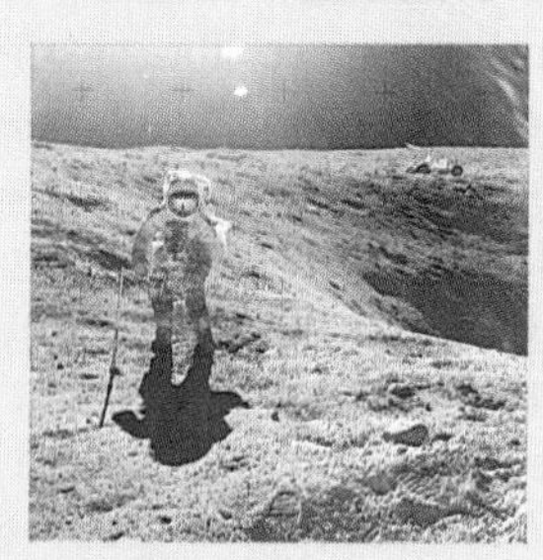

Most of us think of photographs in terms of memories. They help us remember the vacations we took or the people we care about. Sometimes they're a means of keeping personal records. When you were a small child, your parents probably used dozens of rolls of film to capture for all time your first smiles, your first steps, your first Halloween costume. However, photography has many other applications that we don't always think about.

Photographs can take us to places we've never been. How many people will ever visit the mountains of Tibet? Yet almost everyone has seen pictures of that or other faraway countries. Photos also take us to places we can't go. Cameras can go beneath the sea to levels too deep for divers. Most of us will never go to the moon, but we've seen photographs.

Photos show us things too small to be seen with the naked eye, such as strands of DNA, the molecules that determine genetic characteristics. We even have photographs of the streaks that subatomic particles make as they travel through certain substances.

Some uses for photography are so common we almost forget they're there. Take advertising, for instance. Almost every product made is photographed so it can be shown to a prospective buyer. Newspapers use photographs to give us a look at what happened that day. Such photographs can often relay information better than the written word.

In a world that moves very fast, photographs help us look at and learn about things we couldn't catch up with otherwise. Take the motion of a bee's wing as it flies, or the arc a player's bat makes as he swings at a baseball. We can see these things in detail because of photographs.

Almost everyone's photograph appears on a drivers' license or other type of identification. A photograph can assure others of who we are better than our names or most other methods of identification.

Doctors use photography to take pictures, such as X-rays, of the insides of our bodies. Special heat-sensitive films help doctors spot cancer and other diseases.

For certain artists, photographs provide the means for expressing themselves. Ansel Adams, for example, took pictures of Yosemite National Park that made him famous. Other artists, like Edward Steichen, have become well-known for their photographs of famous people.

Since the invention of the camera, photographs have also been used to capture, and sometimes influence, history. Some photographs have come to symbolize the events they recorded. Nearly everyone has seen the picture of Neil Armstrong setting foot on the moon. The people of the future will know about our lives because they too will be able to look at pictures and see what it was like in the "old days."

CHAPTER 12 REVIEW

Review Questions

1. What is composition and how can it improve a photograph?
2. When should a tripod be used to avoid blurring a picture?
3. Describe three ways lighting can be altered to improve a photograph.
4. What is depth of field and how is it achieved?
5. When might you use a shutter speed of 1/1000 of a second?
6. When can a light meter be misleading?
7. Name the seven steps in film processing.
8. What is a contact print and what is it used for?
9. What is "burning-in" and how does it differ from "dodging"?
10. How is film animation created today?

Activities

1. A "photogram" is created when photographic paper is exposed while objects rest on it. After processing, shapes appear on the paper that match the objects. Make several photograms. Use some transparent as well as opaque objects. Describe how results were achieved to the class.
2. Find examples of snapshots or photographs from magazines and newspapers that demonstrate the following principles:
   a. the rule of thirds
   b. use of diagonal lines or some other element of art
   c. framing
   d. stop action
   e. a plain background
3. Plan a portrait that can be shot on location. Select a location suited to the person chosen. For example, if the person enjoys sports it might be at a stadium. Sketch a lighting plan using existing light.
4. Under your teacher's supervision, try to make your own photographic emulsion. Do research on the materials needed. Coat the emulsion on paper or some other suitable material. Try making a photogram (see Basic Activity #1) using the emulsion.
5. Try to create your own animated motion picture. Draw an object or scene on the right half of a 3″ × 5″ index card. Redraw the object or scene, changing it slightly, on a second card. Repeat this until you have a dozen or more cards. Then, flip through the scenes with your thumb, as if you were shuffling a deck of cards. Describe your efforts to the class.

Careers

Industrial Photographer

Many different businesses and industries employ their own photographers. This is because photography is such a good way of recording information.

Industrial photographers are given assignments that include portraits of company personnel, pictures of products and processes, visual materials for presentations and displays, and photographs for brochures, catalogs, and other promotional materials.

Technical photographers make visual records of experiments done for scientific research. These are then used in publications and research reports. They become part of the data collected by the researcher.

Industrial training materials often require photographs and slides. These too might be produced by an industrial photographer. Photographs are also used by businesses and industry to document their work.

Education

Business and industry generally require a photographer to have a college degree. This is important for two reasons. First, industrial photographers are called upon to produce a wide variety of photographs. It is difficult to learn all about these without formal training. Second, industrial photographers must work with many different kinds of people and in many different situations. They must be able to communicate, schedule, plan, and adapt. College helps prepare them for such duties.

A number of two- and four-year colleges and universities have programs in photography. For a comprehensive listing of such programs, write for Kodak Publication T-17 (address below).

Related Careers

Photojournalism
Commercial photographer
Free-lance photographer
Portrait photographer

For More Information

The following sources provide information about education and careers:

Kodak Publication T-17, *College Instruction in Photography*
Eastman Kodak Company
343 State Street
Rochester, NY 14650

Photographic Art and Science Foundation
111 Stratford
Des Plaines, IL 60016

Accrediting Council on Education in Journalism and Mass Communication
School of Journalism
University of Kansas
Lawrence, KS 66045

Correlations

 *Language Arts*

You've probably heard the old saying "A picture is worth a thousand words." Select a photograph from a newpaper or magazine that you think says a lot all by itself. Write a paragraph about what it communicates to you.

Science

1. Our eyes often play tricks on us. Some of these tricks have to do with how we see color. We may even see color where there is none. For instance, narrow black lines placed close together will look like colored lines if you look at them long enough. Try this experiment first devised by Englishman Charles E. Benham (1860-1929). Draw the disk shown in Fig. A with black ink on white cardboard. Cut out the disk. Stick a pin or toothpick through the center. Hold one end of the pin and spin the disk while looking at it. Spin first in one direction and then the other. What do you see? Do the colors change depending upon the direction of spin?
2. Try this experiment. Find two objects, one bright red and one bright green. Next, obtain an ordinary desk or table lamp (such as a gooseneck) that can be focused on the objects. Cover the light with colored cellophane paper, first in blue, then in red, then in green. Look at the red and green objects under each color of light. How do their own colors change?

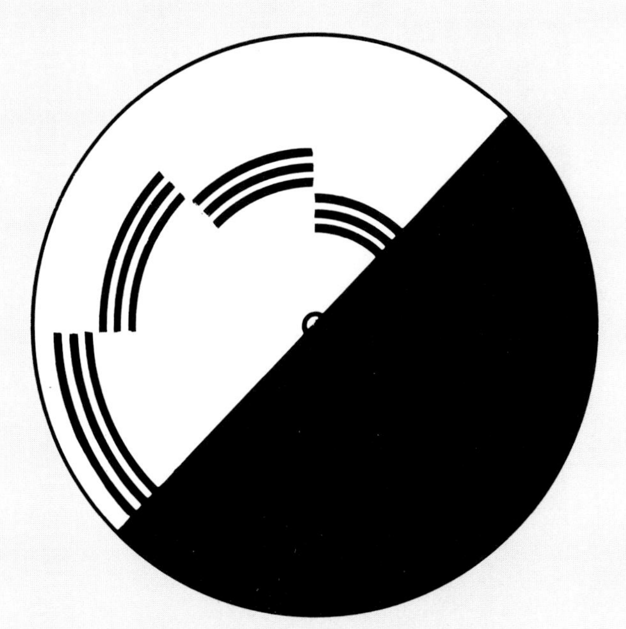

Fig. A.

3. Another interesting effect of color is that the same color may look different on different backgrounds. Look at the squares in Fig. B. The same colors are shown next to one another. How do the background colors seem to affect them?

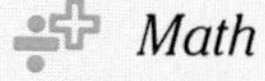

Math

The f-stop of a lens is determined by dividing the focal length of the lens by the diameter of the aperture ($f = F \div D$); F is the focal length of the lens, and D is the diameter of the aperture (lens opening).

Figure the effective f-stops of the following (round your answer off to the nearest tenth)

F = 50 mm and D = 28 mm
F = 50 mm and D = 9 mm
F = 80 mm and D = 5 mm
F = 200 mm and D = 25 mm

Social Studies

Photographs are an excellent way of documenting history. Photographs can be the best record we have of a particular event.

In the special collections section of your local library locate old photographs of the area in which you live. If the library can't help you, try the local historical society. Select several photographs from a certain period and study them. What do they tell you about that particular time and place? Present your findings to the class.

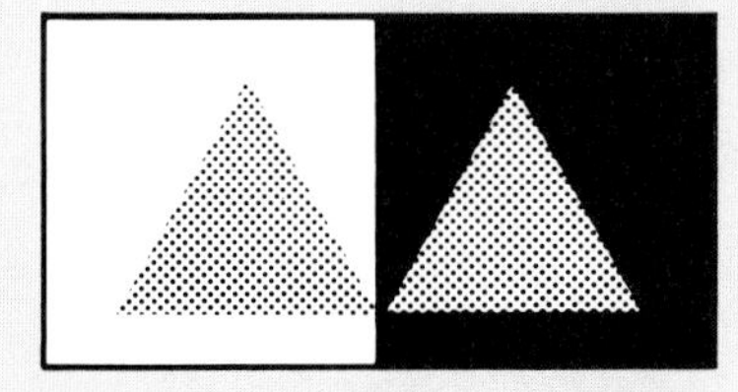

Fig. B.

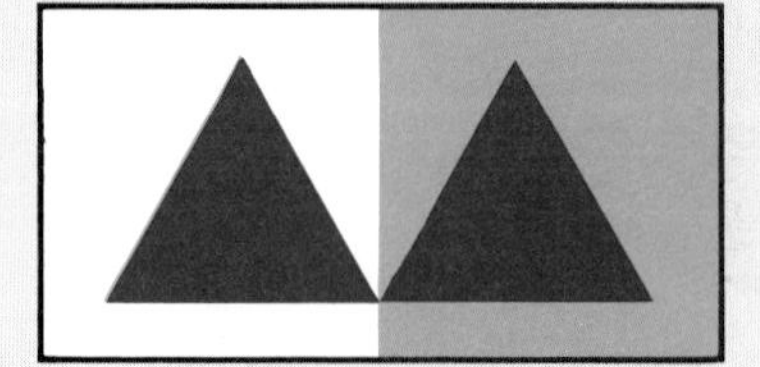

Basic Activities

Basic Activity #1: Photogram

The simplest of all photographic images to produce is the photogram. A photogram resembles a white drawing on a black background. It is created by placing objects directly on light-sensitive paper and exposing it to white light. When processed, the paper turns black everywhere except where the objects blocked the light. Some interesting photograms may be made with a little planning.

Materials and Equipment

photographic paper
developer
stop bath
fixer
trays
clip for drying

Procedure

1. Plan the image you would like to create. Keep in mind that objects that allow some light to pass through, such as certain plastics, create more interesting effects. Solid objects simply create shapes.
2. In a dark or safelit room, place the objects on a sheet of photographic paper.
3. Turn on the ordinary lights in the room for a few seconds.
4. Process the photographic paper in a tray of developer for one minute.
5. Stop the development in a tray of stop bath.
6. Fix the image by placing the paper in a tray of fixer for several minutes.
7. If the photogram appears too light or dark overall, adjust the white light exposure time and repeat the process.
8. Wash the photogram thoroughly in cold running water for several minutes.
9. Dry the photogram.

Basic Activity #2: Pinhole Photography

The most basic of all cameras is the pinhole camera. It is simply a box with a tiny hole in one end. Film is placed in the other end, and an exposure is made through the pinhole onto the film.

Materials and Equipment

small box, such as a round oatmeal box
clear tape
black tape
small sheet of orthochromatic film

Procedure

1. Construct a pinhole camera, following the general guidelines shown in Fig. IV-1.
2. In a safelit room, open the back of the camera. Attach the orthochromatic film, emulsion side up, to the inside of the camera back with a little bit of clear tape. Do not tape over the area to be exposed.
3. Shut the camera. Be sure the camera back is light tight. You may need to tape it shut with the black tape.
4. Cover the pinhole with a piece of black tape.
5. Under normal light, arrange your subject for the photograph. Be sure you have some way of holding the pinhole camera perfectly still during the exposure. For example, you could prop it on a table.
6. Without jiggling the camera, open the pinhole "lens" by peeling back the tape. The exposure time will depend upon how bright the light is, the size of the lens opening, and the type of film. If your subject is in bright sunlight and you are using fast film, expose it for 1 second.
7. Replace the tape over the lens.
8. Turn off the lights. Remove the film in total darkness. Store it in a light-tight bag or container. Reload the camera with another piece of film.
9. Expose this film for twice as much time as the last piece.
10. Repeat steps 7 through 9 until you have made a range of exposures of your subject.

Fig. IV-1.

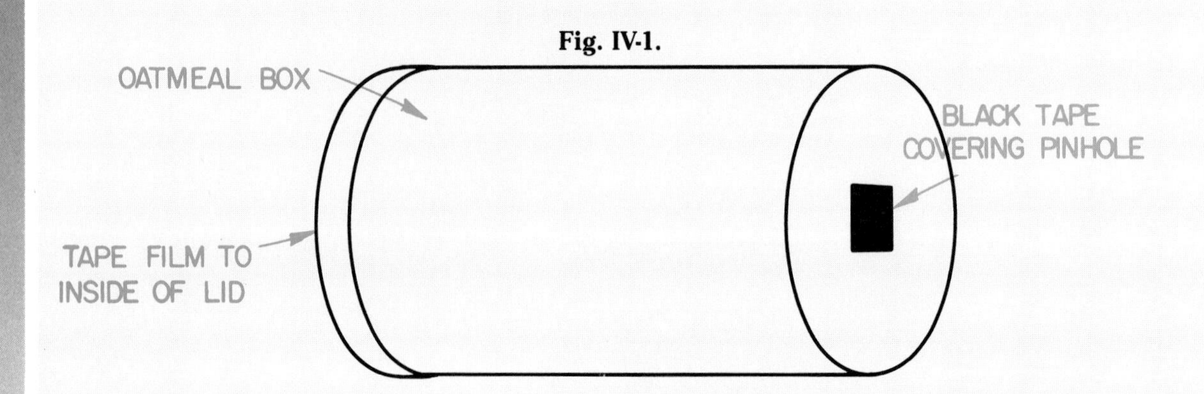

Basic Activity #3: Film Processing

Develop the pinhole photographs to create photographic negatives from which prints can be made.

Materials and Equipment

continuous tone film developer
stop bath
fixer
developing trays
clip for drying

Procedure

1. Set up the developing, stop, and fixing trays in a safelit darkroom.
2. Remove the film from the camera and place it in the developing tray.
3. Agitate the developer throughout the process. The time required will depend upon the film and developer being used. See the manufacturer's instructions. Experiment to determine the best developing time.
4. When the image is properly developed, note the developing time. Place the film in the stop bath for 10 seconds. Remove it.
5. Place the film in the fixer and agitate continuously for twice the time it takes the film to clear.
6. Wash the film in cold running water for five minutes.
7. Dry the negative.

Basic Activity #4: Contact Printing

The negative developed in Basic Activity #3 should be large enough for a contact print.

Materials and Equipment

continuous tone photographic paper
developer
stop bath
fixer
clip for drying

Procedure

1. In a safelit darkroom, place the emulsion (dull) side of the negative in contact with a sheet of the photographic paper.
2. Place a sheet of clean glass over the negative.
3. Expose through the negative with white light. (You may need to do an exposure step test first to determine the best exposure time.)
4. Process the photographic paper at the time and temperature recommended by the manufacturer.
5. Dry the pinhole photograph.

Intermediate Activities

Intermediate Activity #1: Composition

The basics of composition can be learned with almost any camera. With care and a little planning, you can take snapshots that demonstrate understanding of composition.

Materials and Equipment
camera
film

Procedure
1. Following the instructions that came with the camera, load the film.
2. Take three pictures that illustrate the "rule of thirds," which is discussed in Chapter 12.
3. Take three exposures that illustrate the use of diagonals to create interest. Continue to take pictures that illustrate the rules of good composition, as described in Chapter 12.
4. Have the roll of film processed commercially.
5. Using your photos, prepare a display with captions illustrating different composition techniques.

Intermediate Activity #2: 35mm Copy Stand Photography

Amateur and professional photographers alike find the 35mm camera very useful. In this activity, you will have an opportunity to use one. A copy stand is a stand used when making photographic copies of printed material or other photos.

Materials and Equipment
35mm SLR camera
copy stand
24-exposure roll of black and white film
8 magazine pictures

Procedure
1. Load the film into the camera. Mount the camera on the copy stand. Be sure the camera is adjusted for the proper film speed.
2. Place one of the magazine pictures on the copy stand.
3. Move the camera up or down on the copy stand so that the picture more or less fills the entire field of view. Focus the camera.
4. Adjust the shutter speed and f-stop for a normal exposure as indicated by the light meter.
5. Make an exposure. Record the frame number, f-stop, and shutter speed on a sheet of paper. This will be your exposure log.

6. Open the aperture one full f-stop. For example, you might go from f/11 to f/8. Make a second exposure. Record this in the exposure log.
7. Make a third exposure one full f-stop down from the first exposure. In our example, this would be f/16. Record this data in the exposure log.
8. Repeat steps 2-7 for each of the other magazine pictures.
9. Rewind and remove the film from the camera.

Intermediate Activity #3: Processing a Roll of Black and White Film

In this activity, you will produce a roll of film negatives by processing the film exposed in Intermediate Activity #2.

Materials and Equipment

several feet of unwanted 35mm film
the roll of 35mm film you used for the previous activity
bottle opener
developing tank and reel
continuous tone film developer
stop bath
fixer
wetting agent
trays
clips for drying

Procedure

1. Practice loading the unwanted film onto the developing reel until you can do it without looking.
2. In total darkness, remove the good film from its metal container using a bottle opener. Load it onto the developing reel.
3. Place the reel into the developing tank and secure the light-tight cover.
4. Measure out the proper amount of developer and check its temperature.
5. Using the manufacturer's chart, select the appropriate time and temperature for developing.
6. Using a clock with a second hand as a guide, pour the developer into the tank. Immediately tap the tank on the countertop and agitate with a circular motion for five seconds. Every 30 seconds, agitate with a circular motion for 5 seconds. When the developing time is up, pour out the developer.
7. Immediately pour stop bath into the developing tank and agitate for 10 seconds. Return the stop bath to its container.
8. Pour fixer into the developing tank. Agitate for several minutes. Remove the cover of the tank and check to see that the film no longer has a milky white appearance. If it is still white, cover the tank and continue fixing until it appears clear in the nonimage areas.
9. Wash the film in cold running water for 10 minutes.

10. Dip the reel of film into a tank filled with a wetting agent. Squeegee excess water from the film with your index and middle fingers.
11. Hang the film to dry in a dust-free place with a weight (such as a clothespin) at the bottom. The weight prevents the film from curling.
12. Store the negatives in a clear plastic film file to protect them from scratches and dust.

Intermediate Activity #4: Projection Printing

In this activity you will select and enlarge one or more of the negatives produced in Intermediate Activity #3.

Materials and Equipment

35mm negatives from Activity #3
photographic paper
developer
stop bath
fixer
enlarger
easel
trays
clips for drying

Procedure

1. Make contact prints of the negatives by following the same procedures outlined in Basic Activity #4.
2. Study the contact prints. Select a negative with good contrast for projection printing.
3. Dust off the chosen negative with a soft brush or compressed air.
4. Place the negative emulsion (dull) side down in the negative carrier of the enlarger.
5. Adjust the height of the enlarger head for the desired enlargement size. Focus the image with a wide-open f-stop.
6. Close down the f-stop two or three stops. Cover all but a strip of photographic paper in the easel. Expose that strip for 3-5 seconds. Move the cover an inch or two and make another 3-5 second exposure. Repeat to create five or six exposure steps.
7. Process this test strip at the time and temperature recommended by the paper manufacturer.
8. Select the best exposure time. Repeat steps 6-7 at a different f-stop if necessary to get a properly exposed area on the test strip.
9. Make the projection print at the selected exposure time and process it.
10. Dry the completed photograph.

Advanced Activities

Advanced Activity #1: Color Slides

Color slides produce excellent color and may be produced in much the same way as black and white negatives. Of course, the chemistry used is much different.

Materials and Equipment

35mm camera
E-6 slide film
developing tank
E-6 processing kit
bottle opener
slide mounts

Procedure

1. Select a theme or a model for a photo session, such as a school event. Using the 35mm camera, shoot the roll of E-6 slide film, being careful that all pictures relate to your theme or subject.
2. Carefully read the instructions that come with the E-6 slide processing kit. Prepare all chemistry as directed.
3. In the dark, load the slide film into the developing tank.
4. Process the slides according to the instructions that come with the kit.
5. When the slide film is dry, cut it into individual slides. Secure these in slide mounts.
6. Select the best slides from the group and show them to your class.

Advanced Activity #2: Digital Documentary

Digital photographs are increasingly used in place of conventional photographs. They may be quickly imported into desktop publications or multimedia productions. In this activity, you will plan and create a series of digital images that could be used to document the work being done in your communication technology class.

Materials and Equipment

digital camera
computer
digital editing software

Procedure

1. Design (plan) your digital images well ahead of time. Identify what you think the most effective shots might be. Consider the lighting, props, models, etc. If you are going to shoot equipment, plan to include one or more people in the picture who will be using the equipment. You'll be much more successful if you have models (fellow students) pose for these shots than if you attempt to take unrehearsed shots.
2. Carefully review the manual that came with the digital camera. Practice with it on still (nonliving) subjects before taking pictures of people doing work in the lab. Experiment with room lighting, flash, etc.
3. Carefully set up each shot. Look through the lens and be sure both the foreground and the background look good before taking the shot. Make a series of exposures with the digital camera according to your plan.
4. Download the images to a computer.
5. Edit them with digital image editing software to create the best possible images.
6. Print out "thumbnail" size images on a color printer, or use a black and white laser printer.
7. Develop a plan for using these images in a digital slide show, a World Wide Web page, or a desktop publication that promotes the work being done in your communication technology laboratory.

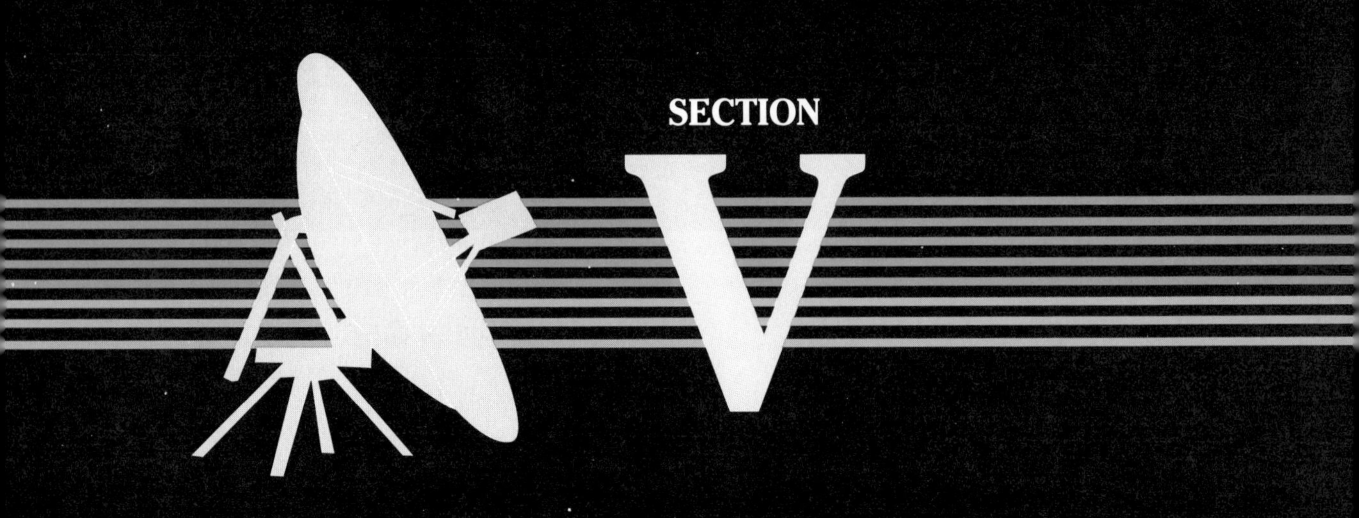

SECTION V

Graphic Production Systems

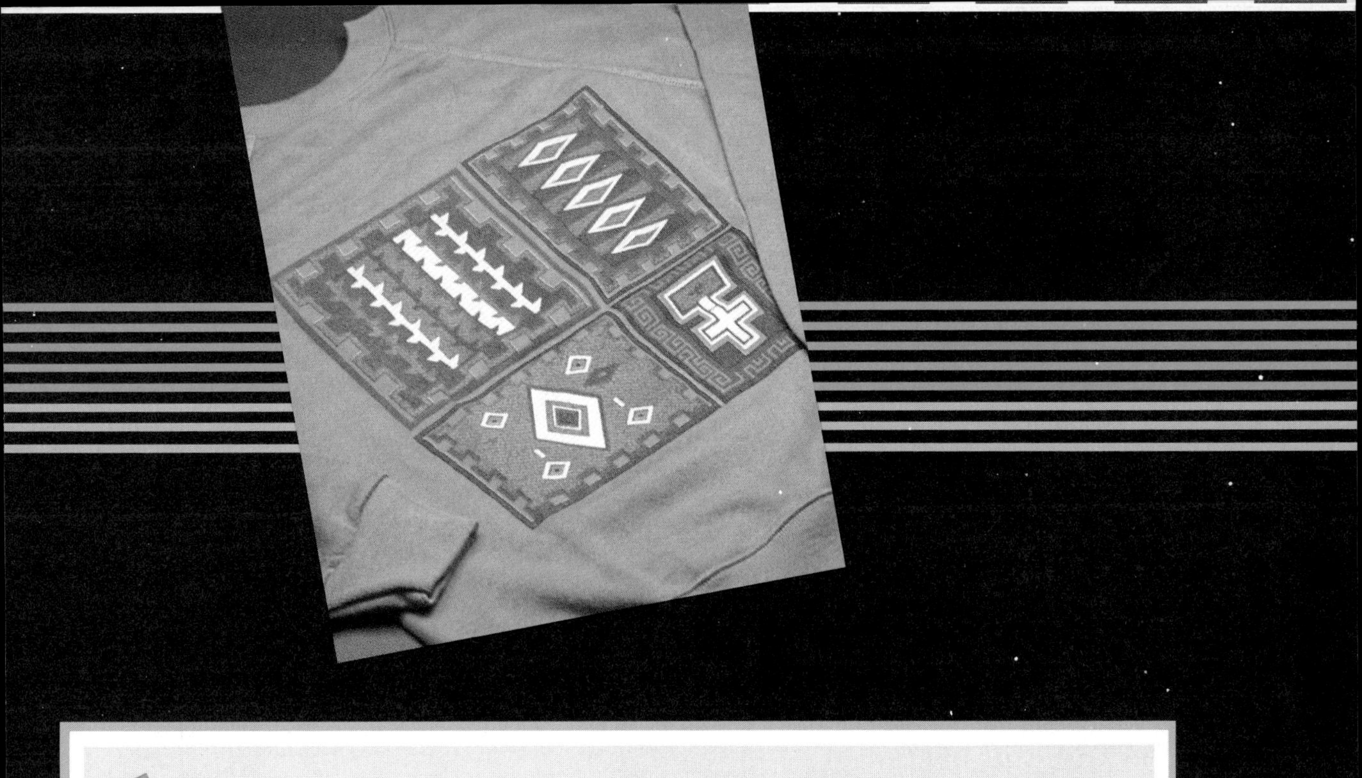

Although there are a variety of graphic production methods, common processes are used by all of them to create, develop, and reproduce graphic messages. These processes make up the system of graphic production, which includes the following phases:

1. *Message Design.* Typewritten copy looks plain. Graphic designers are called upon to add "pizzazz" to the message with interesting type styles and illustrations.
2. *Message Composition.* After the message has been designed, all of the text and artwork must be produced in final form. This process is known as composition. In the composition process, words are set in type and illustrations are prepared in such a way that they will be suitable for printing.
3. *Message Assembly.* At this point, things are coming together from all directions—typeset words from the typesetter, photographs from the photographer, and drawings from the artist. In the message assembly process, the various elements are arranged according to the planned design.
4. *Film Conversion.* Assembled graphic elements are usually converted to film negatives or positives using photographic techniques. Some computer-based systems are able to skip this stage of the process.
5. *Film Assembly.* Often, a graphic production job will require many different film negatives. The film assembly stage brings these negatives together. The message then appears as it should when it is printed.
6. *Message Transfer.* Finally, the message must be transferred to (printed on) paper, plastic, metal, or some other material. There are different transfer methods for different sorts of jobs. For example, you wouldn't use the same method to transfer a message to a T-shirt as you would to include it in a magazine.
7. *Product Conversion.* After the message has been printed, many jobs require additional work, such as cutting a printed sheet to final size.

In this section, you will learn about the seven phases of graphic production. You'll see how printing—one of the oldest forms of communication—is being changed by new methods and new technology.

CHAPTER 13

Message Design, Composition, and Assembly

When you read a book, the cover on a record album, or the wrapper on a candy bar, you are receiving a graphic message. How well you understand what you read may depend on how well the message was designed. The design may even affect whether you want to read the message at all! It may also affect whether you want to buy one product instead of another.

The content—words and illustrations—of a message helps determine its design. The design, in turn, influences the choice of words and illustrations. In this chapter, you will read how design, words, and illustrations are put together to create a graphic message.

Terms to Learn

color system
composition
continuous-tone images
copyfitting
design elements
desktop publishing systems
fonts
graphic designer
line art
mechanical
pica
point
principles of design
substrate
typefaces

As you read and study this chapter, you will find answers to questions such as:

- What guidelines should be followed when designing graphic messages?
- What materials and techniques help graphic designers do their work?
- What kinds of illustrations are used for graphic messages?
- What must be done to convert typewritten copy to the form generally seen in books and magazines?
- How do words and illustrations come together on the page?
- How have computers changed the way graphic messages are created and produced?

DESIGNING A GRAPHIC MESSAGE

Why does a book or magazine page look the way it does? Notice how some words are bigger than others. Color may be used. Some pages may contain a great deal of information while others have more empty space. A book or some other printed communication looks the way it does because a **graphic designer** planned every detail, right down to the location of the page numbers!

The graphic designer is part of a team that also includes writers, artists, and photographers. Often, the work of writing and illustrating begins even before a design is created. Editors coordinate the work of the designer, writers, artists, and photographers to make sure the final product turns out as planned.

Principles of Design

To assist them in their work, graphic designers follow guidelines as to what looks good on a printed page. Included among these guidelines, or **principles of design**, are rhythm, balance, proportion, variety, emphasis, and harmony.

Rhythm

Rhythm is repetition. In music, rhythm is created by repeating a beat. In graphic design, rhythm occurs when a certain element is repeated. For example, look at Fig. 13-1. What is repeated in this picture? Rhythm can add movement to a design. It looks like something is happening!

Balance

When an acrobat balances on a tightrope, his or her weight is positioned in such a way that the person remains steady. The same is true in graphic design. All the elements are placed in a way that gives an impression of steadiness.

There are two kinds of balance. Formal balance is achieved when a line drawn through the center of the design would create two halves that are similar to one another, or symmetrical. Fig. 13-2(a) shows an example of formal balance.

Informal balance, on the other hand, is more subtle. It is a balance of objects that may look different but that have equal weight to the eye. In a photograph, for example, a large building might be balanced against several smaller buildings. What elements are balanced against one another in Fig. 13-2(b)?

Fig. 13-1. Rhythm occurs when an element is repeated. How does rhythm help make this design effective?

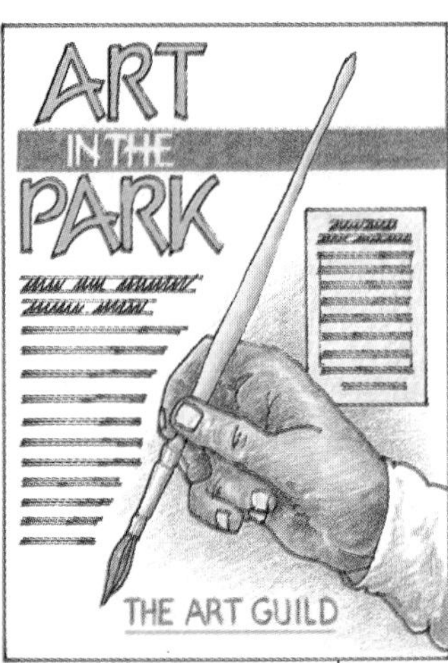

Fig. 13-2. Balance may be formal or informal. Which type of balance would you use for a wedding invitation?

Proportion

"I wish these shoes weren't so tight!"

"There's too much salt on this hamburger!"

Both of these complaints are about problems of proportion. Proportion has to do with the size relationship of one part to another. By itself, the size of an object has little meaning. Only by comparing it to something else can we say it's too big or too small. In graphic design the proportions of one element to another should be correct. Although some elements may be larger or smaller than others, the effect should be pleasing. Fig. 13-3.

Fig. 13-3. Do the parts of these designs seem to belong together? The proportion in the design at the left is wrong. Something seems to be missing. In the design at the right, the parts are in better proportion.

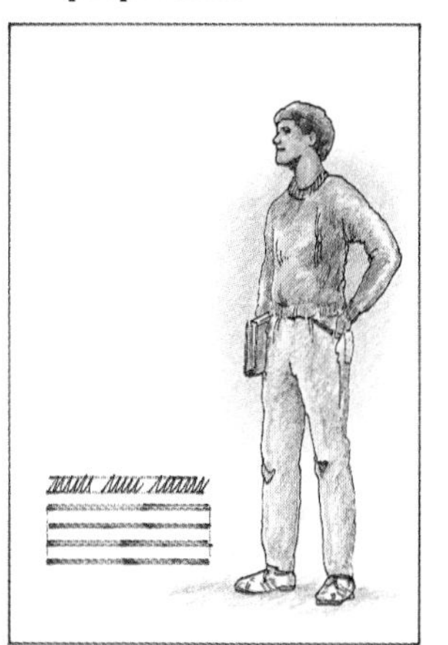

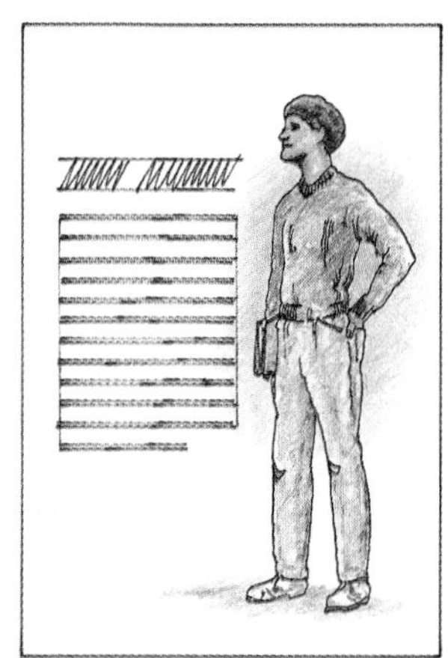

Variety

Variety is difference. Without variety, life would be boring because everything would be the same. Variety in graphic design may add interest and excitement. What adds variety to the design in Fig. 13-4?

Emphasis

Graphic designers are constantly thinking of ways to attract your eye to the most important part of the message design. One of the ways this is done is by emphasis, making one element stand out.

Emphasis may be achieved in several ways, such as with size or color. Fig. 13-5. The designer of this book, for example, used darker type to emphasize the most important terms. Isn't it easier to spot those words than the others around them?

Fig. 13-4. How many different typefaces are used in this design?

Fig. 13-5. What is the first thing you notice in this design?

Harmony

It is important that the various elements of a message design work well together. When this happens, the design has harmony. For example, a fancy style of type might not look good next to a very plain style of type. When all elements of the design look as if they belong together, harmony has been achieved. Look at Figs. 13-4 and 13-5. Do you think these designs have harmony?

Design Elements

Design principles provide some general guidelines for the design of graphic messages. To achieve a desired effect, designers rely on different **design elements**. The basic elements include line, shape, form, space, color, texture, and shades of dark or light. For example, an illustration made with dots will look very different from one done with lines or with bold shapes. Fig. 13-6. By changing the texture, tone, or space in an illustration, different results may be achieved also. Color is a popular design element.

Materials and Techniques

In addition to knowing about design elements and principles, a graphic designer must be familiar with the different materials and techniques that can be used. For this discussion we can mention only a few general categories. They include type, color systems, substrates, measurement, sketches, and computer design.

SCIENCE FACTS

Optical Illusions

We don't always see things as they really are. Such visual mistakes are called optical illusions. Look at the drawings below.

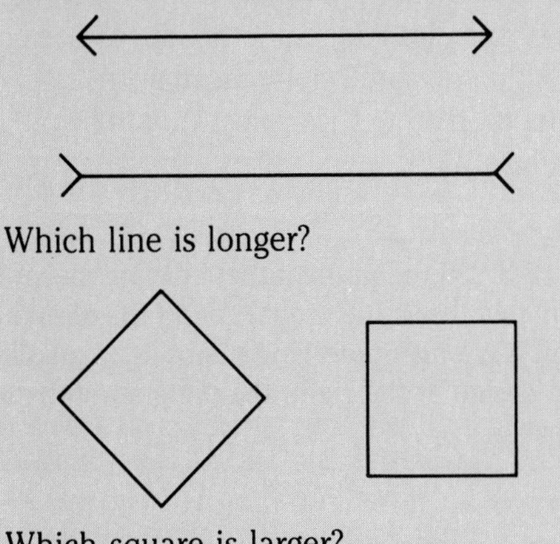

Which line is longer?

Which square is larger?

Actually, both lines are the same length, and both squares are the same size. These are two common optical illusions. They demonstrate that the placement of an object or the things around it can influence its apparent size. Designers often make use of optical illusions to achieve a particular effect.

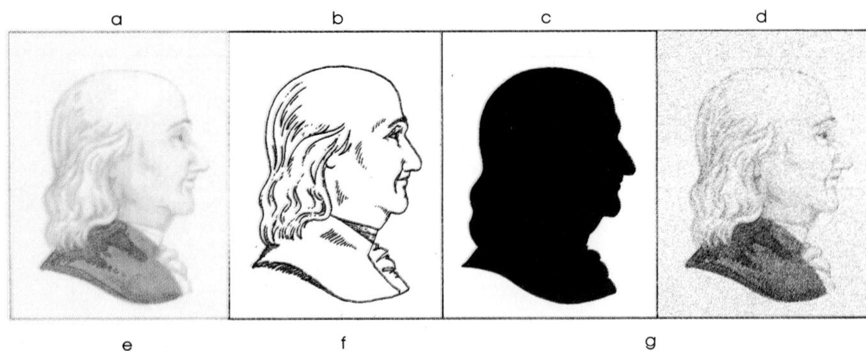

Fig. 13-6. The different elements of design have changed the appearance of this illustration: (a) this halftone version is made of dots; (b) lines add shading here; (c) a silhouette relies on shape; (d) texture has been added; (e) a screen tint creates tone; (f) color makes this version more lifelike; (g) on the left space has been cropped, and on the right it has been added.

TECHNOLINKS

FROM GUTENBERG TO THE LASER BEAM: 500 YEARS OF SETTING TYPE

For centuries, type was set by hand. The Chinese probably invented block printing. Characters were carved on wooden blocks and inked.

Johannes Gutenberg, a metalsmith in Mainz, Germany, invented the first movable type around AD 1450. Gutenberg cast letters and other characters by hand from an alloy of lead, tin, and antimony. These letters were then arranged in a hand-held device to form the words needed for a particular job. After printing, each letter was put back into the type case to be used for future jobs.

In 1886, another German, Ottmar Mergenthaler, invented a machine that would revolutionize the printing industry. The machine was called the Linotype, and it set type automatically. When the operator struck a key on the keyboard for the letter wanted, a tiny brass mold moved into place. When all the words for a single line were assembled, the machine squirted molten lead into the molds. After this "hot type" had cooled, the line of type was removed and the molds stored to be used again.

The Linotype was used through the first half of the 20th century. Then came the typesetter, a machine that created type photographically rather than with hot metal.

The first commercial typesetter, introduced in 1946, operated in ways similar to the Linotype. Instead of a mold, however, a film negative of the letter to be set was used. Light was projected through this negative onto light-sensitive photographic paper.

a R 9 G 2
q w H
R F Z P

A new generation of typesetters kept all of the film negatives of each character or letter on one continuous piece of film. This film spun around while a strobe light, operated by means of the keyboard, exposed the letters onto photographic paper.

The next generation of typesetters used a cathode ray tube, like the one in a television set or computer monitor. Images of the letters to be set were first displayed on the cathode ray tube and then projected through a lens to the photographic paper.

The most modern typesetters use a laser beam and a computer to generate images. In some cases, the laser beam exposes photographic paper. In other cases, the laser printer works much like a photocopy machine, outputting on plain paper. The computer controls the printing functions.

How things have changed in 500 years!

Typography

Look at the different styles and sizes of type used on this page. Using different type adds variety. At the same time the different kinds used must be in harmony with one another.

Type comes in many designs called **typefaces**. Many different typefaces are created by changing certain parts of the characters. See Fig. 13-7 for the parts of a character of type.

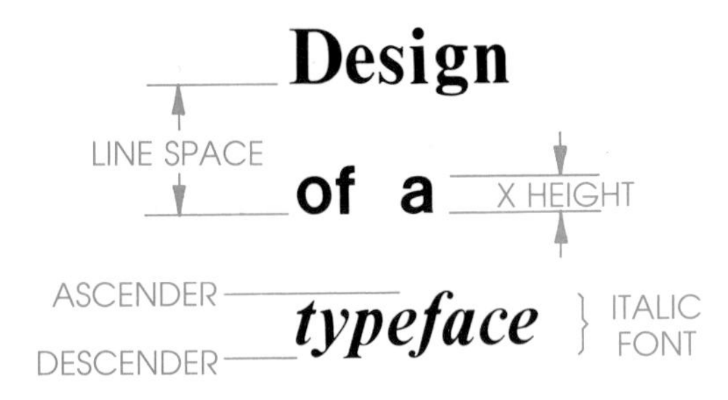

Fig. 13-7. The variety of typographic elements provides designers with many options.

Each typeface is available in an assortment of sizes, thicknesses, and other variations called **fonts**. For example, there are many fonts available in Cheltenham typeface. Fig. 13-8.

Cheltenham Light
Cheltenham Light Italic
Cheltenham Book
Cheltenham Book Italic
Cheltenham Book Condensed
Cheltenham Bold
Cheltenham Bold Italic
Cheltenham Bold Condensed
Cheltenham Ultra

Fig. 13-8. Cheltenham is available in these nine fonts. What differences do you notice among them.

All the thousands of different typefaces are grouped into six major styles: Roman, square serif, sans serif, text, script, and novelty. Fig. 13-9.

1. Technology
2. Technology
3. Technology
4. Technology
5. Technology
6. Technology

Fig. 13-9. There are six major styles of type: (1) roman, (2) square serif, (3) sans serif, (4) text, (5) script, and (6) novelty. Text and script styles should never be set in all capital letters because in that format they are too difficult to read.

Roman. This typestyle was created in early Roman times when words were cut in stone. Roman typefaces are made with a serif, or little tail, on most letters. The width of the strokes used to form a letter varies. Roman typefaces are the easiest to read. They are commonly used in books and magazines, where there are many words.

Square serif. A number of typefaces have square serifs. This sets them apart from other typefaces.

Sans serif. Sans is a French word that means "without." Sans serif typefaces, therefore, are those that have no serifs. While they are more difficult to read than those with serifs, they are less formal. They also lend a clean or "modern" look to the page.

Text. Text typefaces are styled after the work of scribes who copied books by hand during the Middle Ages. They are very detailed and are sometimes used for a formal look. Text typefaces are nearly impossible to read when set in all capitals and should never be used this way.

Script. Script typefaces are those that look like handwriting. Some are rather formal, while others are quite informal. They appear in designs for everything from wedding announcements to greeting cards. As with text typestyles, script styles are difficult to read when set in all capitals.

Novelty. Typographers are always designing new typefaces. Many do not fit neatly into any of the other major styles. These are called novelty typefaces. Novelty typefaces may be dramatic, funny, or even grotesque. They are generally eye-catching because of their unusual designs and can be used to add emphasis.

Measurement

Printers have always worked with a measuring system that differs from those we use every day. The basic unit of this measuring system is the **point**, which is equal to 1/72 of an inch. Because points are such small units, they are very handy for measuring type sizes and spaces between lines. For example, 12-point type may be set on a 14-point line space.

The other primary unit of measure is the **pica**, which is equal to 12 points (or 1/6 of an inch). Picas are used to measure larger sizes, such as the length of a line or column of type. For example, "10 on 12 by 20" means "10 point type set on a 12 point line space over a 20 pica line length."

Points and picas are measured with a typographer's rule. The rule has six different scales on it, including those for points and picas. Fig. 13-10.

The character size scale shows capital letters in various sizes. This clear scale is laid over a line of type to measure its point size. Point size is measured from the top of a capital letter to the bottom of the descender on a lowercase letter. Capital letters are generally about two-thirds the height of the overall point size. For example, a capital "C" in a 30-point font is actually only about 20 points from top to bottom.

The rule (line) width scale may be laid over thin lines to determine their width. The line space scale may be laid over a paragraph of type to measure the space between lines.

There is also an inch scale for handy reference, though inches and ordinary fractions are rarely used when working with type.

Substrates

When we think of printing, we generally think of printing on paper. Graphic messages are actu-

Fig. 13-10. A typographer's rule measures inches, line spacing, points, rule widths, picas, and character size.

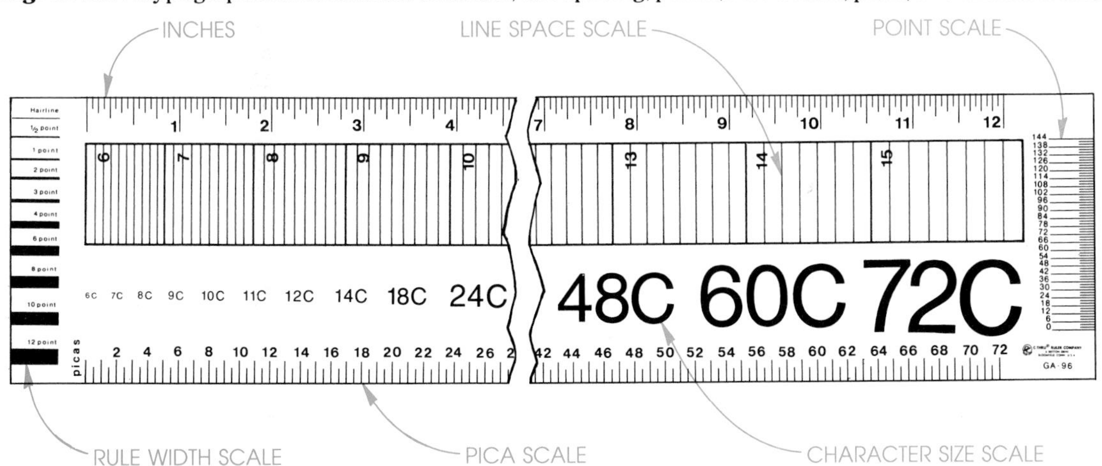

ally printed on a wide range of materials, including paper, plastic, fabric, metal, glass, ceramics, and corrugated cardboard. Any material on which printing is done is called a **substrate**.

Paper, the most commonly used substrate, is available in a wide assortment of sizes, kinds, and finishes. Fig. 13-11. "Finish" refers to the appearance of the paper's surface. For example, a paper's finish might be dull or glossy. Paper may be calendered, which means it has been polished smooth between two hard rollers. Of course, paper may also be purchased in a variety of colors and textures.

Designers must choose paper carefully. A high-quality paper can go a long way toward "dressing up" a design. On the other hand, low-quality paper can ruin an otherwise successful design.

Layout Sketches

After discussing a project with a client, the graphic designer creates a series of sketches. The client, in turn, provides the designer with information and opinions at each stage. In this way, they eventually come to agree upon the final design. Fig. 13-12.

Kind of Paper	Basic Sheet Size	Common Uses
Book	25″ × 38″	Books, Catalogs, Brochures
Writing	17″ × 22″	Stationery, Duplicating
Cover	20″ × 26″	Directory Covers, Booklets
Bristol	22½″ × 28½″	Business Cards, File Folders, Post Cards
Other: Newsprint, Carbonless Transfer, DuPont Tyvek, Onionskin, etc.		

Fig. 13-11. Paper is classified into five categories. The basic sheet size is the standard size for each category. The weight of one ream (500 sheets) of the basic sheet size is known as the basis weight.

Fig. 13-12. Here you can see the graphic design process from thumbnail sketches, to rough layout, to full-color comprehensive layout.

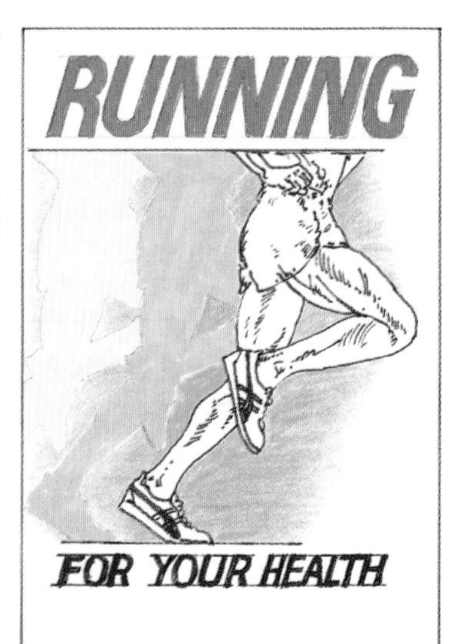

Thumbnail sketches are generally small, quick sketches drawn to try out several ideas and show how something might look. Designers often make a large number of thumbnail sketches and then select the idea they like best for further development. Thumbnails are usually drawn with pencil and include little detail. Any illustrations are shown by blocking in shapes, and type is indicated by simple straight lines.

After a thumbnail sketch is selected, the graphic designer prepares a more accurate drawing known as a rough layout. This "rough" contains all the information a printer needs, such as size and location of the elements, to produce the final product. The rough also gives the client a good idea of how the finished design will look.

Roughs are also generally drawn with pencil, but they are the actual size of the final product. Headlines may be sketched the way they will look when printed. Smaller type is still only indicated with lines. Illustrations may be drawn in some detail.

If a job is complex and/or costly, a client may want to see a full-color version that shows how it will look after it has been printed. This is known as a comprehensive layout. For example, a comprehensive layout for a cereal box might be done in full color and might even be folded into a box shape. By looking over the comprehensive layout, the client can ask for changes before the job is produced, saving both time and money.

Color Systems

Graphic messages often appear in many colors. Color is always used with the principles of design in mind. However, color presents some problems. If you were told to print something a "light red," for example, you would have to select from dozens of colors that could be called light red. Then you would hope that the client agreed with your choice.

To reduce confusion, color systems, such as the PANTONE MATCHING SYSTEM, have been developed. A **color system** is a series of colors that have each been given a number for identification. Fig. 13-13. The colors and their numbers always remain the same. When giving specifications to a printer, the designer lists the color number, such as PANTONE 191C, rather than a description such as light red. Printers, in turn, mix their inks to match the color with that number. The final result is what the designer intended.

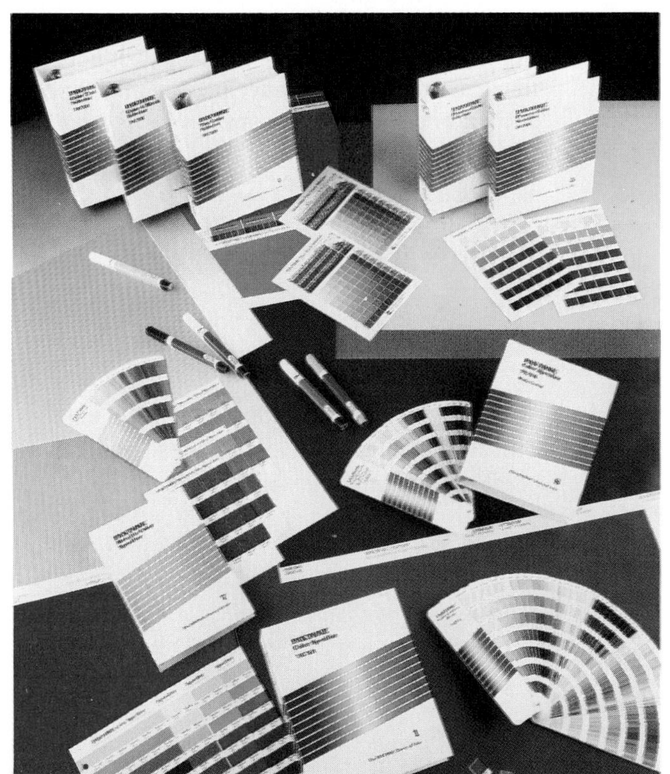

Fig. 13-13. Color systems, such as the Pantone Matching System*, allow graphic designers to work effectively with color.

*Pantone, Inc.'s check-standard trademark for color reproduction and color reproduction materials

Process color reproduction may not match PANTONE®-identified solid color standards. Refer to current PANTONE Color Publications for the accurate color.

Computer Design

As you learned in Section 3, microcomputers are beginning to have an important effect on graphic design. Graphic designers can now sit at a computer and completely design a graphic message without using the materials or manual techniques we have just described.

In many cases, it is now more efficient to design with a computer. The Yellow Pages in the telephone directory, for example, are often laid out entirely by computer. Even simple computer graphic systems make it possible to draw, cut and paste, copy, store, and print designs without ever touching paper or pencil. Fig. 13-14.

COMPOSING THE GRAPHIC MESSAGE

Once a client has approved a design, the creation of the words ("copy") and illustrations can be completed. Converting words and illustrations into a form that will be ready for printing is called **composition**. In the early stages of composition, the words and illustrations are handled separately. Words are set in type, while the illustrations are reproduced photographically. (Desktop publishing is an exception and will be discussed later.)

Fig. 13-14. Computer graphic systems allow the graphic designer to try out many variations of a design. Each version can be stored as is and then edited at the computer to produce a different version. The designer can go back to an earlier version at any time.

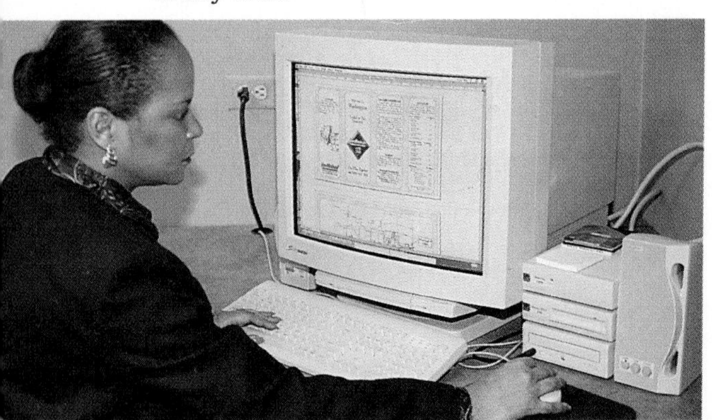

Typesetting is the process of turning written text into the final typefaces. The primary difference between typeset copy and that produced on a typewriter is the greater variety in sizes and styles of type. Another difference is in the spacing. Fig. 13-15. A typewritten "i" takes up the same space on the page as does the typewritten letter "w." Typeset letters, on the other hand, take up only the space they need. A typeset "i" occupies less space than does a typeset "w." Typeset copy is also much sharper because the edges of each letter are cleaner than those made on a typewriter.

Today, setting type is similar to using a typewriter. For example, you have to set the margins, decide on single or double spacing, and locate the page numbers. In addition you must decide among a variety of type sizes, styles, weights (such as regular and bold), and line lengths.

There are usually so many of these decisions to be made that the copy is "marked up" before it is composed in its final form. Marking up means that directions for style, point size, line spacing, line length, and other things are written in the margins.

Fig. 13-15. The words are the same, but notice the difference in appearance between the typewritten and the typeset copy.

```
Speech is civilization itself. The
word, even the most contradictory
word, preserves contact--it is
silence which isolates.

                        Thomas Mann
```

Speech is civilization itself. The word, even the most contradictory word, preserves contact — it is silence which isolates.

Thomas Mann

Copyfitting

Most printed messages are designed to be a certain size. For example, an ad might be planned to fill one 8″ × 11″ page in a magazine. All the words in the ad must fit on that one page, with enough room left for the illustrations.

Making the words fit the available space is called **copyfitting**. To do copyfitting, the writers or designers count how many characters (letters, numbers, and symbols) will fit in one line of type. They multiply this number by the number of lines that will fit on one page. This gives the total number of characters. They compare this number with the actual number of characters in the written copy. In some cases, if the copy is too long, it can be shortened. If it is too short, more copy can be added so that it fills the available space.

Fig. 13-16. These technical pens come in a variety of widths and colors.

Staedtler
MARSMATIC 700

Line Widths

Red	00	(.30mm)
Blue	0	(.35mm)
Green	1	(.45mm)
Yellow	2	(.50mm)
White	2½	(.70mm)
Gray	3	(.80mm)
Black	3½	(1.0mm)

Illustrations

Illustrations may include line art, continuous-tone images, and full-color art. Actual reproduction of illustrations will be covered in Chapter 14.

Line Art

Line art is made up of solid, dark (usually black) lines and shapes drawn on a white surface. Technical pens, which can make solid lines of uniform width, are used for technical illustrations. Fig. 13-16. More relaxed pen and ink techniques are used for freehand illustrations.

Line drawings may also be purchased from a "clip art" company. Clip art is artwork (often bound in a large book) that can be clipped out for use in the buyer's graphic designs. Clip art provides a wide variety of images, many of which have to do with major holidays and seasonal activities, Fig. 13-17.

Fig. 13-17. These pictures are examples of clip art. Designers using computers can buy clip art that is stored on computer disks.

Continuous-Tone Images

In addition to line drawings, artists may create a wide range of continuous-tone artwork. **Continuous-tone images** are composed of varying shades of gray, such as those drawn with charcoal, airbrush, or pencil. Black and white photographs are another form of continuous-tone images. Fig. 13-18.

Fig. 13-18. The black and white halftone (left) and the charcoal drawing (right) both contain a range of grays.

Full-Color Artwork

Full-color artwork includes illustrations done with watercolor or oil paint, colored pens and pencils, pastels, or color photography. Color can add a great deal to a graphic message.

Desktop Publishing

In the mid-1980s, microcomputers and laser printers were teamed with each other and called **desktop publishing systems**. Fig. 13-19. Another term for these is electronic publishing systems. Although the quality of the text was not as good as that produced by commercial typesetters, it was good enough for many publications. Those producing company newsletters, for example, found desktop publishing systems helpful. As a result, the systems became popular among many people outside the printing industry.

Desktop publishing systems have been fairly inexpensive. Advances in software have made them easier to operate than the typesetters that were on the market in the 1980s. Unlike most typesetters used in the '80s, desktop publishing systems can also create and print graphics.

With a desktop publishing system, illustrations may be "drawn" by moving a mouse. (See Chapter 5 for a description of computer mice.) As the mouse is moved around, lines and shapes appear on the computer monitor. Once drawn, the image may easily be changed, printed on a laser printer, or saved on a computer disk.

A second method of creating graphics makes use of a device known as a scanner. Scanners reflect light off original artwork, such as a drawing or photograph, and change the reflection into a computer image. Fig. 13-20. The image may then be edited (changed) using the mouse, printed, or stored on computer disk.

The third method available to those with desktop publishing systems is to purchase computer clip art. Computer clip art is sold on a disk.

Fig. 13-19. Desktop publishing systems combine a microcomputer, laser printer, and software, such as Xerox Corporation's Ventura Publisher or Aldus Corporation's PageMaker.

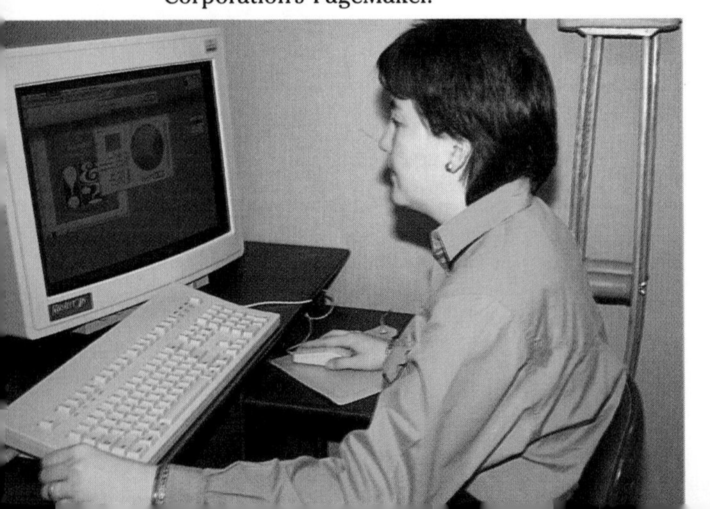

Fig. 13-20. Photographs may be "captured" with a scanner. The resulting halftone image may then be changed electronically to produce different effects.

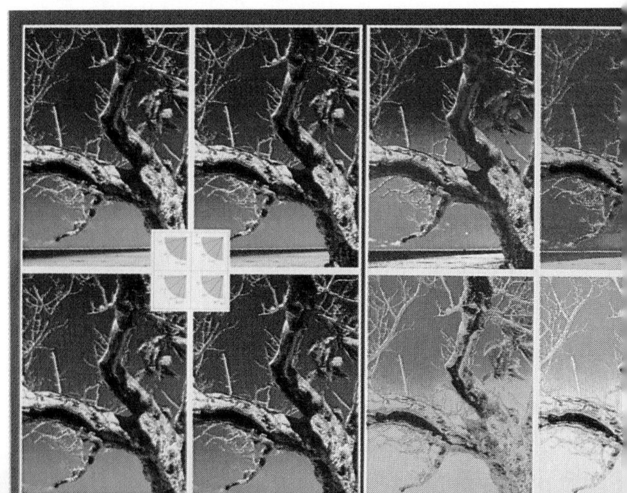

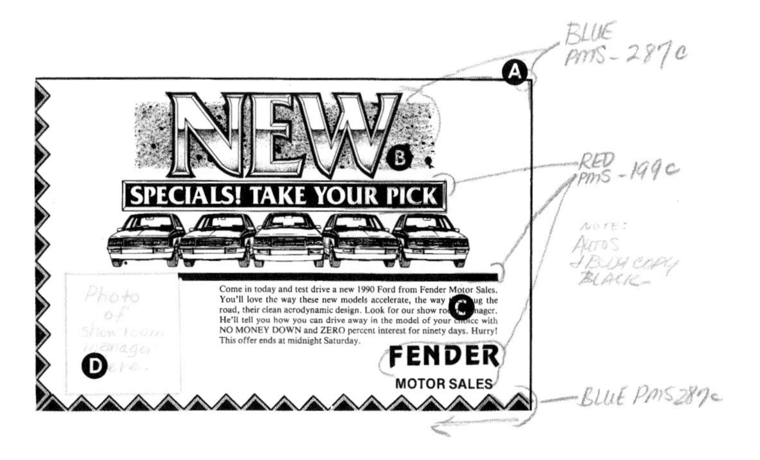

Fig. 13-21. Border tape, clip art, typeset text, a halftone window, and non-reproducible blue lines have all been used to make this mechanical.

ASSEMBLING THE GRAPHIC MESSAGE

After all of the text and illustrations have been composed, these elements must be brought together. This process is known as assembly.

Paste-Up

Traditional assembly is done by pasting the copy and line art in place on a white "board." Some boards have light blue guidelines to indicate margins, column widths, and so on.

Working from the rough layout supplied by the designer, a paste-up artist arranges the typeset copy and the reproductions of line art on the board. When a continuous-tone illustration is to be included on the page, there are two options. For the first, a "window" must be created. This is done by attaching a patch of red paper or plastic to the place where the photograph is to appear. This window must be the exact size and shape as the continuous-tone illustration. For the second option, the artist may attach a prescreened halftone or stat to the board. Halftones are discussed in the next chapter.

The elements are coated on the back with wax. The wax acts as a kind of glue, but it doesn't dry the way glue does. The elements can be moved around until they are in exactly the right place.

Borders, such as those around illustrations or blocks of type, may be added using a pen or tape. Some border tapes are very thin, while others have fancy designs to create decorative effects.

Texture may be added to an area by covering it with tiny dots. Texture patterns are available on adhesive-backed sheets that may be cut and a section lifted off and stuck in place. Common shapes, such as arrowheads, are also available in this form.

After everything is in final position, a rubber roller is used to press everything down and secure it in place. The end result is called a paste-up, or **mechanical**. Fig. 13-21.

Computerized Message Assembly

Just as desktop publishing systems are used to create text and illustrations, they are also capable of assembling the message. Special software allows the operator to pull text and illustrations from different storage disks and assemble them. Rather than working with pieces of paper, wax, and a paste-up board, the operator simply arranges the different graphic elements on the computer monitor.

After assembly, the electronic "page" may be stored on a disk or output to a laser printer. More advanced systems are linked to a typesetter and create a high-quality mechanical without the operator ever having to cut or paste a single image!

DESKTOP PUBLISHING: CHANGING THE NATURE OF THE GRAPHIC COMMUNICATION INDUSTRY

Using computers to generate publications isn't new. Professional printers have been using computers in their work since the late 1960s. Today's desktop computer, however, is *far* more powerful than the large mainframe computers some printers used in the late 1960s. As a result, anyone with a desktop computer has the capability to do work that only printers could do a few decades ago.

Desktop computers allow you to completely lay out a publication on your own. With a conventional paste-up technique, the manuscript is typeset by a compositor and artwork is generated by a graphic artist. Then the job is pasted up by a printer. If you do all of this using a computer page layout program, you'll save time. You will also have the freedom and/or responsibility of making design decisions. Since printers are in business to make money, doing the work yourself also saves you money.

The future of desktop publishing is with telecommunication. Linking computers together in networks allows people in different offices to work on the same publication "on-line." That is, a writer, editor, and designer can all be working on the same publication from different offices. They can even be doing so at the same time — the computer equivalent of a conference telephone call. When everyone is satisfied, telecommunication devices allow them to send the file directly to a commercial printer for output on a very high-quality imagesetting device. (Imagesetters provide 2400 dot-per-inch or better resolution for both text *and* graphics.) The printer, in turn, can print the job and mail it back overnight to the client. Again, this process saves time and money, and it puts more control with the document originators.

There are, however, drawbacks. Since most people are not trained as graphic designers, many desktop publications don't look as good as they should. While they may save time and money, the time spent with the computer layout is time away from other things they might be doing.

Perhaps the biggest impact of desktop publishing has to do with education. Because of desktop publishing, a lot more people now know a lot more about printing. They have a better idea of what it takes to produce a document from beginning to end. That means, even if they choose to let someone else do the work, they can communicate their needs to the printer better than ever before.

In the long run, this means quicker turnaround time for publications and more satisfied clients. Better communication makes it easier on both the printer and client. The end result is better publications.

CHAPTER 13 REVIEW

Review Questions

1. Identify and describe the six principles of design.
2. What are the six major styles of type?
3. Name the elements of design.
4. What type style is used for the text in most textbooks? Why?
5. How is the point size of a typeface determined?
6. Name and describe the three kinds of sketches produced during the graphic design process.
7. Tell the difference between line art and continuous-tone images.
8. List the steps in copyfitting.
9. What is a desktop publishing system?
10. Briefly describe the process of preparing a paste-up.

Activities

1. Photocopy an advertisement that shows formal balance, cut it apart, and redesign it using informal balance.
2. Create a series of thumbnail sketches and a rough layout for an advertisement promoting a student activity.
3. Prepare a color comprehensive for the advertisement created in 2 above.
4. Determine the point size, line length, and line space used for the main text in this book.
5. Create a mechanical for the advertisement designed in Activity 2 above.

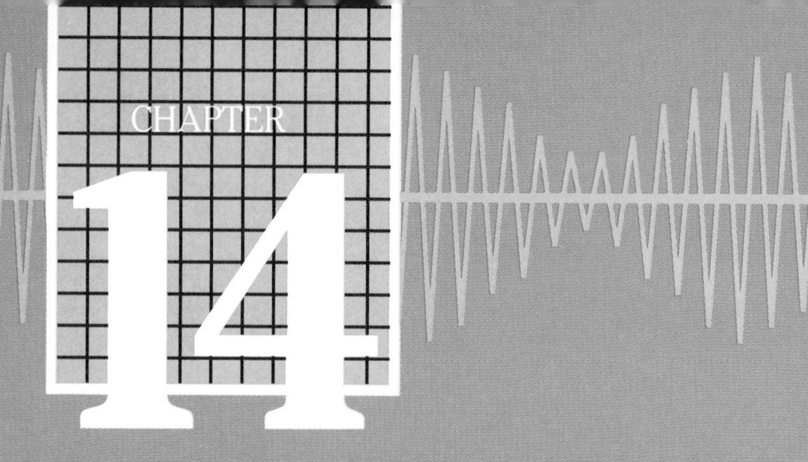

CHAPTER 14

Film Conversion and Assembly

Did you know that a camera was used to prepare the pages of this book? It was so big it filled an entire room. Look closely at the picture on the facing page. Use a magnifying glass if you have to. Do you see the patterns of tiny dots? Both the giant camera and the tiny dots are used to reproduce graphic messages. They are only the biggest and smallest parts of the process that turns graphic messages into images on film. Many other steps take place. Some have been around a long time. Others involve the use of computers and are helping revolutionize the industry.

Terms to Learn

color scanners
color separation
electronic pagination systems
exposure
film conversion
film positive
flat
gray scale
halftone photography
line photography
orthochromatic film
pin registration
process camera
stripping

As you read and study this chapter, you will find answers to questions such as:

- What is the role of film in the graphic production process?
- What are line negatives and how are they produced?
- What is a halftone negative?
- How are color illustrations reproduced?
- How are computers used in the film conversion process today?

FILM CONVERSION

In Chapter 13, you learned about paste-ups, or mechanicals. The next step in graphic production is **film conversion**. In film conversion, graphic components such as mechanicals or continuous-tone art are photographed to produce film negatives or film positives.

The Process Camera

The camera used to create film negatives is called a **process camera**. (Film positives will be discussed later in this chapter.) Process cameras are quite large, often filling one or two small rooms. There are two basic types of process cameras: horizontal and vertical. As their names imply, one is set up parallel to the floor, while the other looks as if it were turned on end. Figs. 14-1 and 14-2.

Parts of the Camera

Both types of process camera have four main parts: copyboard, lights, lens, and vacuum back.

The copyboard is the surface on which the mechanical, or camera-ready copy, is placed. A glass frame over the copyboard holds the copy securely in place. At the other end of the camera, the vacuum back holds the film in place. The lights on the process camera have special bulbs in them which are far more intense than ordinary lightbulbs. Light reflected off the copy passes through the lens, which focuses the light onto the film. The lens has an adjustable lens opening, or aperture. It controls the amount of light that passes through the lens. Behind the lens is the shutter, which controls the length of time the aperture stays open.

Both the copyboard and the lens may be moved back and forth in relation to one another. This movement controls the size of the image on the film. A typical process camera can enlarge an image up to three times or reduce it down to one-third of its original size.

Fig. 14-1. This horizontal process camera will be installed so that the back portion is in the darkroom and the front (with copyboard and lights) is in a lighted room.

Fig. 14-2. A vertical process camera can be used either in the darkroom or in a lighted room. Usually the darkroom is preferred because film loading is easier there.

SCIENCE FACTS

Lamps

Various kinds of lamps may be used for process photography. The most common types are the pulsed-xenon lamps and quartz iodine lamps. Unlike some other lamps, these do not dim with age, and they do not produce harmful fumes. Furthermore, the light from a pulsed-xenon lamp is similar to daylight.

No matter what type of lamp is used, you should never look directly at it. The lamps used in process photography produce very strong light, which can be harmful to the eyes.

Controlling Exposure

The **exposure**, or amount of light reaching the film, must be carefully controlled. This is done by varying the length of time the shutter is open and by controlling the size of the aperture.

On many process cameras, a timer opens the shutter, then closes it again after a given number of seconds. More modern cameras use a light integrator rather than a timer to control exposure. The light integrator measures the amount of light passing through the lens rather than the time the shutter is open. Light integrators are more accurate than timers.

A device known as a diaphragm controls the size of the aperture. The diaphragm may be adjusted to different f-stop settings. F/22, f/16, and f/11 are common f-stops on a process camera. As you learned in Chapter 12, there are two rules to remember about f-stops:

- The smaller the f-stop number, the larger the aperture.
- Each f-stop lets in either half or twice as much light through the lens as the next f-stop. For example, f/16 lets twice as much light through the lens as f/22 and half as much light as f/11. Therefore, a 10-second exposure at f/16 will let in the same amount of light as a 20-second exposure at f/22. Fig. 14-3.

Orthochromatic Film

Orthochromatic film, also known as lith or line film, is a high-contrast film used in the process camera. "High-contrast" means there are no gray areas or middle tones. Unlike films used to make ordinary black and white photographs, orthochromatic film produces only black and clear areas on the negative.

Orthochromatic film is not sensitive to light at the red end of the color spectrum. It is extremely sensitive to light at the blue end of the spectrum. This is an advantage in two ways. First, because the film is not sensitive to red light, it can be handled under red "safe" lights in a darkroom without fear of exposing it. Because it is nearly as sensitive to light blue as it is to white, light blue guidelines drawn on a mechanical come out as black on the negative. In other words, when the negative is printed, the light

Fig. 14-3. In this drawing you can see that although the number is smaller, the opening for f/8 is larger than that for f/22.

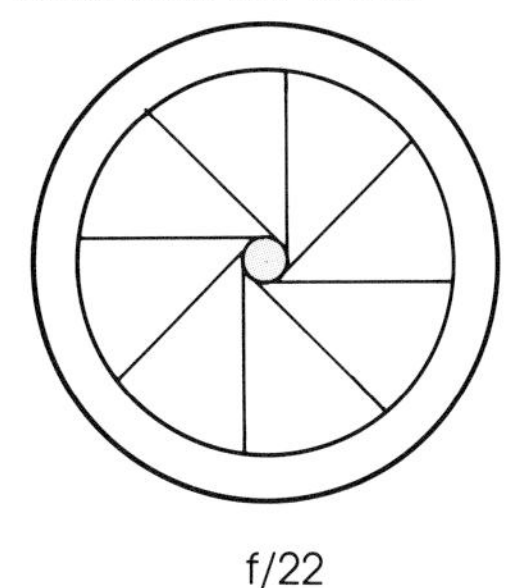
f/22

f/16

f/11

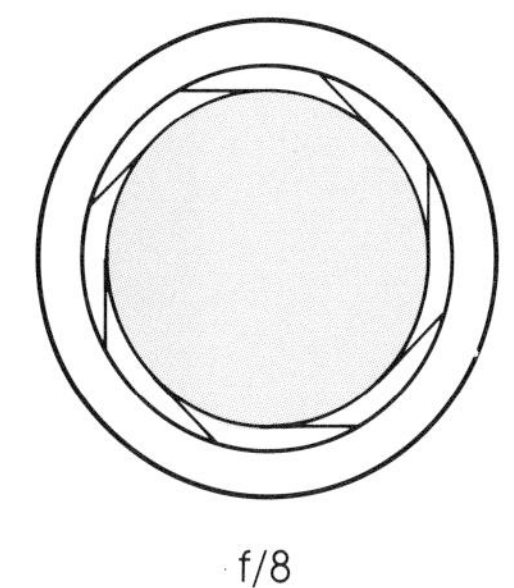
f/8

blue layout lines will not show up at all. That is why "nonreproducing" blue pencil is always used for guidelines on a mechanical.

Orthochromatic film, like other black and white films, is made of four layers. (To review how film is made, see Chapter 11, Fig. 11-18.)

Photographing Black and White Copy

Two kinds of copy are photographed using the process camera: line and continuous-tone.

Line Photography

Line photography is the process of converting line copy, such as the words and line art of a paste-up, to a line negative.

During the exposure of a line negative in the process camera, light reaching the copyboard does one of two things. (1) It may be absorbed by the black (image) areas of the line copy. Or, (2) it may be reflected off the white (non-image) areas. The reflected light from the non-image areas passes through the camera lens, which then focuses this light onto the film. An invisible change occurs where light strikes the film. A latent (hidden) image is created. When the film is processed, these non-image areas turn black. The areas with the hidden images become clear.

It is important to be able to identify the emulsion side of a line negative. There are three ways of doing so. On the emulsion side:

- Any image appears backwards.
- The surface is duller.
- The surface may be scratched with a sharp instrument, such as a razor blade.

To make a line negative, the copy is placed on the copyboard of the process camera. A gray scale is placed next to the copy. A **gray scale** is a strip of special paper, usually divided into twelve density steps. The steps range from white to black. The gray scale is photographed along with the line copy. Later, when the negative is developed, the gray scale serves as a guide in determining how long the negative should be left in the developer. Fig. 14-4.

The lens and copyboard on the camera may be adjusted back and forth for the desired reproduction size. Remember, the process camera can enlarge or reduce images. Usually, these are unchanged. The f-stop is set and the film is loaded. The exposure time is set and the film is exposed.

Fig. 14-4. As the line negative is developed, each step on the gray scale turns black, beginning with Step 1, followed by Step 2, and so on. When Step 4 turns solid black, the negative is properly developed.

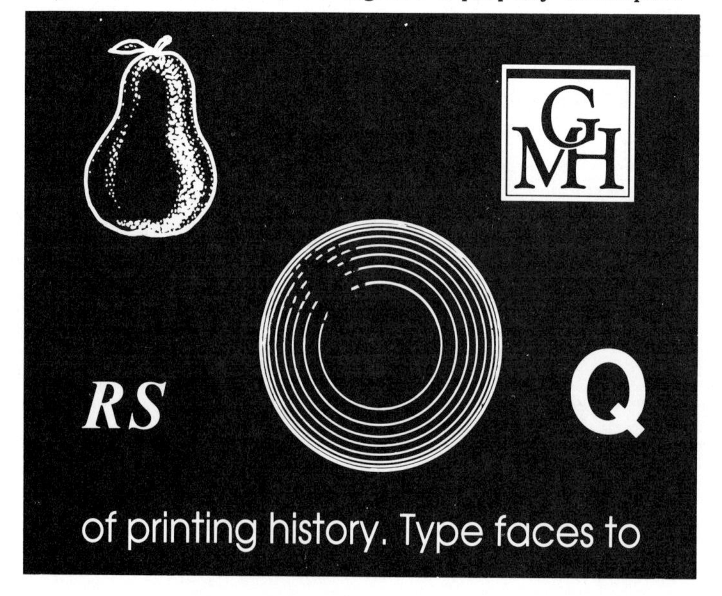

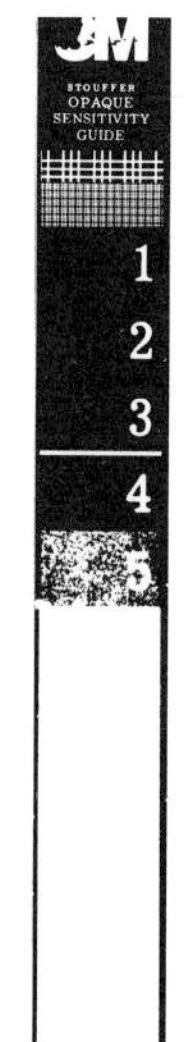

TECHNOLINKS

FIVE CENTURIES OF COMMERCIAL ILLUSTRATION

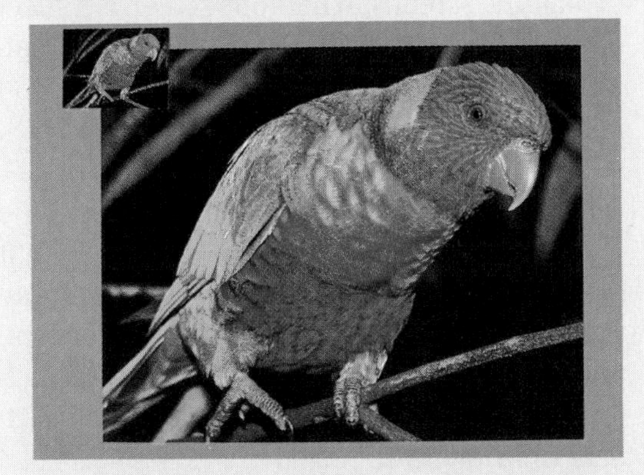

Pictures in books look a lot different today than they did centuries ago. That's because the technology of graphic illustration has changed over the centuries.

Long before the invention of movable metal type (in the mid-1400s), illustrations were reproduced from wooden blocks. These illustrations were made by first carving away the non-image areas from the surface of a flat wooden block. The image area would thus be raised above the non-image areas. Ink could be rolled over these raised surfaces and then transferred to paper with pressure.

Illustrators developed another technique called wood engraving. Wood engravings (or woodcuts) were done on the end grain of the wood, as this is harder than the face grain used for block prints. With this technique, the illustrator would use a chisel-like tool to carve the image area as thin lines. Ink would be pushed down into these grooves and then wiped off the smooth surface of the end grain. With pressure, this image could be transferred to paper.

Metal engravings, made with copper or zinc, worked the same way. With the use of metal, the recessed lines could be produced in a number of different ways. Besides engraving the image, the illustrator could etch it into the metal with nitric acid. The famous Dutch painter, Rembrandt van Rijn, produced many beautiful etchings this way. The paper money used in the United States today is produced from metal engravings.

Beginning in the late 1830s, those experimenting with photography (Daguerre, Niepce, and Talbot, among others) were also trying to devise a way to print their photographs with ink on paper. In the early 1870s, New Yorker John Moss developed a practical way of doing so. By the 1890s, photographic methods were routinely used to make metal printing plates. These were known as photoengravings.

Photoengravings were fine for line drawings, but printing photographs presented another problem. Photographs have differing gray tones throughout, but printing presses only print one color of ink at a time, usually black. A means of representing the gray tones with only black ink had to be developed.

The solution to this problem was the halftone screen, which was developed in the 1880s. This screen was made by etching lines at right angles to each other in glass. Passing light through this screen broke the photographic image into lines or dots. The halftone screen converted light gray areas to small dots and dark gray areas to large dots. Today, most black and white photographs are printed with the aid of halftone screens. Nearly all color photographs are electronically screened using computer software.

Halftone Photography

Black and white photographs contain a wide range of gray tones. Some tones are nearly white and some are nearly black. There is a continuous range of gray tones in between. Unfortunately, none of the different printing methods can reproduce continuous tones on paper. Printers are limited to printing one tone at a time, such as black ink on white paper. To overcome this limitation, printers take advantage of an optical illusion.

You've probably noticed that photographs printed in a newspaper seem to be made up of tiny black dots. Fig. 14-5. If the tiny dots of black have white paper showing between them, the area looks light gray. If the dots are arranged close together and little white paper shows through, the area appears dark gray.

Fig. 14-5. By using the different patterns of dots, the illusion of gray is created in the halftone.

With this illusion in mind, printers change continuous-tone images to patterns of dots. The lightest areas of a photograph, known as the highlights, end up as tiny dots. The darkest areas of the photograph are changed to large dots. This conversion process is known as **halftone photography**.

Halftone negatives are made with the help of halftone screens. A halftone screen is a sheet of clear plastic that is covered with a dot pattern. Generally, there are between 65 and 150 of these dots per inch. The dots are solid black in the center and get lighter toward their edges.

If a lot of light hits any dot on the screen, the light can go through the entire dot. If less light strikes a dot, the light might go through only the dot's outer edges and not the center. The amount of light going through dots on the halftone screen is what produces the different size dots on a halftone negative.

To make a halftone, a photograph is placed on the copyboard of the process camera. Fig. 14-6. A halftone screen is placed directly over the orthochromatic film on the vacuum back. A "main" exposure is then made.

Light reflecting off the whitest areas of the photograph (highlights) usually penetrates the entire dot on the halftone screen. This exposes a large

Fig. 14-6. To make a main (through the lens) exposure, the vacuum back is closed (a). A flash exposure is made through a yellow filter and the back is left open (b). A halftone screen is in place for both exposures.

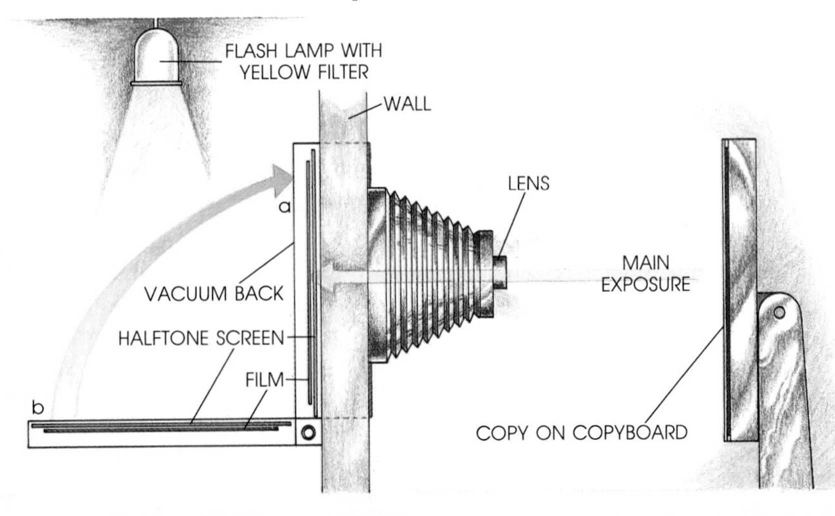

dot on the film behind the screen. Less light is reflected off the darker areas of the photograph (shadows) and light does not penetrate very much of the dot on the halftone screen. This exposes a smaller dot on the film.

Usually the main exposure cannot represent all of the tones of a photograph by itself. In most cases a second exposure, known as a flash, is needed, chiefly to reinforce the dots in the shadow areas of the halftone. The flash exposure is not made through the lens of the camera. Instead, the camera back is opened and the flash is made through a yellow filter and the halftone screen.

The dots produced by the halftone process look like continuous tones when printed. However, they are printed in only one color, usually black. If you look carefully at the black and white photograph appearing below, you will see that it is a halftone. The photograph contains 133 dots per square inch. This number is common for photos in books and magazines. Newspaper halftones contain 65 dots per inch and the dots are easier to see. Fig. 14-7.

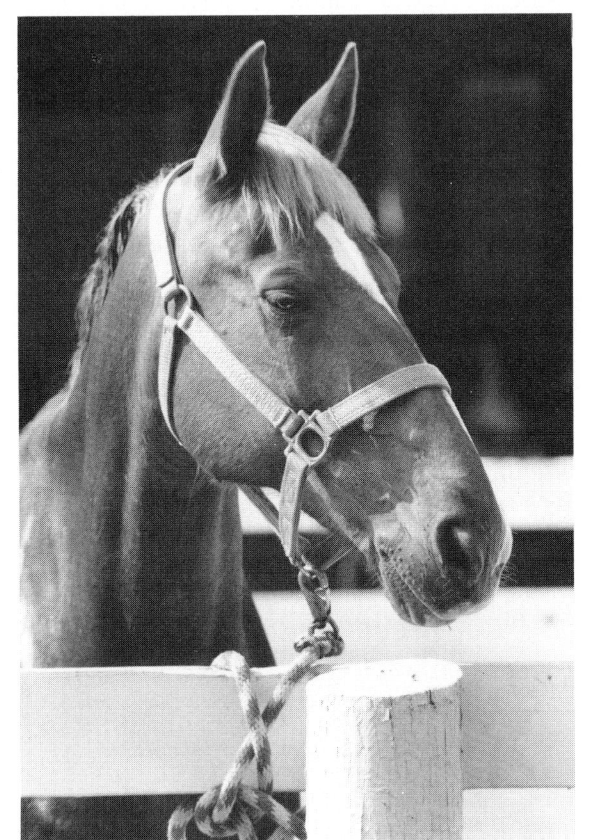

Fig. 14-7. The halftone on the left was made with a 133-line screen. Newspapers generally use 65-line halftones, while 133-line halftones are commonly used in books and magazines.

Processing Orthochromatic Film

Processing orthochromatic film is very much like processing ordinary black and white film. It is done in developing trays in a darkroom. There are four processing steps. They include developing the film, stopping development, fixing the image, and washing the film.

The time for development is recommended by the manufacturer and should be used as a general guideline. However, a gray scale gives a more accurate reading. As developing continues, the steps on the scale turn black, first Step 1, then 2, and so forth. As a general rule of thumb, negatives should be developed until Step 4 on the gray scale turns solid black. This means that all images on the copy that are as light as, or lighter than, Step 4 will be changed to black on the negative.

For a review of the steps in manual film processing, see Chapter 12.

Automatic Film Processing

When a lot of negatives have to be processed, an automatic film processor is generally used. A film processor uses a developer, stop bath, fixer, and a water wash. The operator feeds the film into the processor. The processor automatically moves the film through each of these chemicals. The film remains in each chemical for the correct amount of time. The film exits the processor completely developed and dried.

Film Positives

For certain printing processes, a film positive is needed rather than a negative. A **film positive** is black in the image areas and clear in the non-image areas. It is the opposite of a line negative. The emulsion side of a film positive is shiny, and the images are right-reading rather than backwards.

Film positives are not made with a process camera but with a vacuum frame and light source. Fig. 14-8. The vacuum frame holds a line negative in close contact with a fresh sheet of orthochromatic film. The light source, installed above the frame, provides the necessary exposure. Film positives are processed in the same manner as are line negatives.

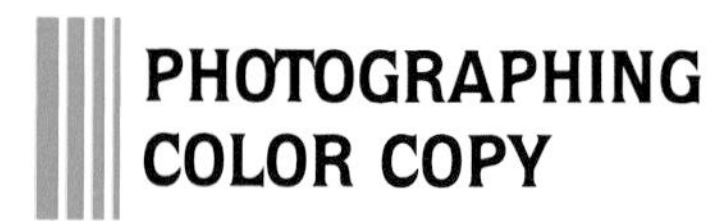

PHOTOGRAPHING COLOR COPY

We say there are seven basic colors in the spectrum: red, orange, yellow, green, blue, indigo, and violet. Each one of these colors can be combined into many additional colors. The next time you're in a hardware store, look at the many paint chip displays. These are only a portion of the colors which are possible.

Did you know that some computer monitors can display more than 1000 different colors? Printers can mix almost all known colors separately and print them one at a time, but this is extremely impractical. Imagine how long it might take to print a picture that shows the color monitor's display.

Color Separation

Just as halftone screens create the illusion of continuous tones on the printed page, a process called **color separation** is used to create the illusion of a range of colors. In the past, a process camera and color filters were used to separate the image of the original color artwork or photograph into a series of four halftones. Each halftone represented a different color to be printed: yellow, magenta (red), cyan (blue), or black. When the four color halftones were printed, one on top of the other, they produced the illusion of continuous-tone color. Fig. 14-9.

Color separation negatives had to be processed using the highest standards. Even then, problems occurred and the color might have been off a bit. Highly skilled workers were able to alter separation negatives to produce a desired end result. All of this took time and was therefore quite expensive.

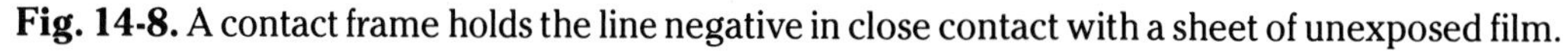

Fig. 14-8. A contact frame holds the line negative in close contact with a sheet of unexposed film.

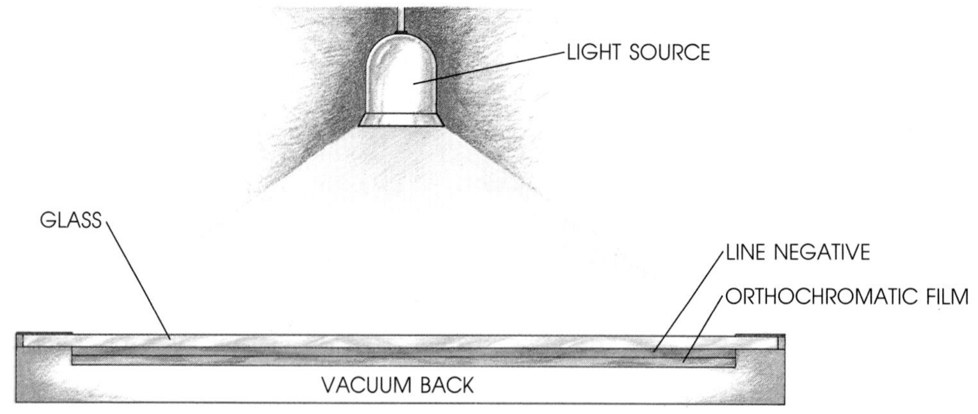

Fig. 14-9. A color photograph can be separated into the four subtractive colors: yellow, magenta, cyan, and black. When printed as halftones, one over the other, these "separations" create the illusion of a full-color image. Note the dot pattern shown at center right.

Color Scanning

Only a small percentage of color separations is still made with a process camera. Today the work is done by computers. **Color scanners**, as they are called, can create quality color separations in a fraction of the time required in the past. Fig. 14-10.

Artwork, color photographs, or color transparencies (slides) are mounted on the scanner and revolved at high speed. A light beam is reflected off or projected through the artwork. The computer separates the light the way color filters do on a process camera. The light is then stored as electronic data. The data are then fed to a unit that exposes the color separation negatives.

The effect of the color scanner on the use of color in printed matter has been tremendous! Not long ago, color was used sparingly. Few color illustrations appeared in a textbook such as this. Now we see color used generously in books, magazines, and even the daily newspaper. By reducing costs, the scanner has made color illustrations more practical.

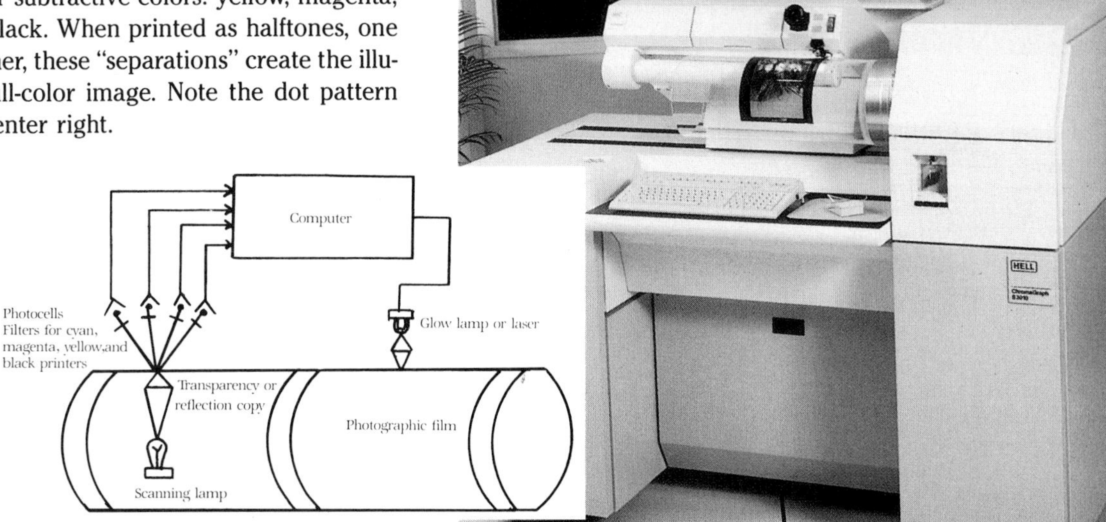

Fig. 14-10. Color scanners electronically separate color artwork and produce color separation negatives.

ELECTRONIC PAGINATION SYSTEMS

In the early 1980s, the Scitex Corporation amazed the printing world with a system that combined computer-created text with color scanning. **Electronic pagination systems** allow a single operator to compose an entire full-color page electronically. Fig. 14-11. In addition, the operator may electronically edit, or change, any element on the screen. Changing the color of a model's eyes from brown to blue, for example, is simply a matter of a few minutes' work at the controls. Almost any change can be made, with all changes instantly visible on the color viewing monitor. Once the scanning and editing have taken place, it is fairly simple to output a complete and final set of color separation negatives.

Although electronic pagination systems are amazing tools, they remain fairly expensive. Therefore, they are used only by the larger companies in the industry. Smaller companies generally send their artwork to a company that makes color separations on a scanner.

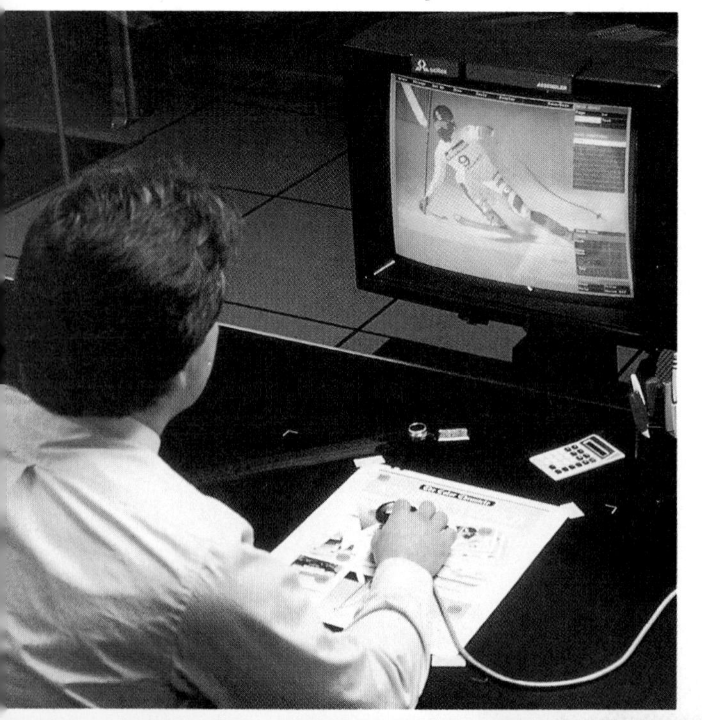

Fig. 14-11. Electronic pagination systems combine color scanners with computers.

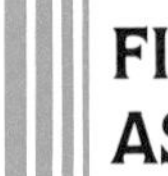

FILM ASSEMBLY

Just as text and illustrations are brought together in a paste-up, very often film negatives must be assembled. For example, halftone negatives must be placed in the windows of the paste-up negatives to make the pages complete. Negatives of entire pages are also assembled. The pages of this book, for example, were not printed separately but grouped together and printed 32 at a time. This is done because presses are made to handle very large sheets of paper. Also, whenever more than one color is printed, there must be negatives representing each color. These negatives must each be placed accurately, one over the other. The process of combining and arranging negatives in this way is known as film assembly.

Assembling negatives is similar to assembling graphic elements on a mechanical. The difference is you cannot paste the negatives on a board because light must be free to pass through them during the next stage of the production process. As a result, the negatives are attached to a thin material that can be cut away from the image areas. This material is known as a masking sheet. The process of attaching the negatives to the masking sheet is called **stripping**. Fig. 14-12. For simple jobs, masking sheets with preprinted

Fig. 14-12. This worker is attaching negatives to a masking sheet.

grids are available. The grids help with correct alignment of the negatives.

Stripping a one-color job begins by taping the negative, right-reading side up, to a light table. Light shining through the negative allows a masking sheet to be positioned accurately over the negative. Small openings are then cut through the masking sheet. Red cellophane opaque tape is used to temporarily hold the masking sheet to the negative. The tape holding the negative to the light table is removed, and the entire assembly is turned over. This allows the stripper to tape the negative to the masking sheet on the back side. The **flat**, as it is now called, is once again turned over, and more of the masking sheet is cut away so light can pass through the image areas. Finally, any tiny scratches or "pinholes" in the negative are blocked out with an opaque material.

Pin Registration

For color assembly, each of the four separation negatives must be stripped to a different flat. Each flat, in turn, must be in perfect alignment, or register, with the others or the colors will blur. Alignment is done by means of **pin registration**. Holes are accurately punched in each of the flats before stripping begins. Metal pins matching the punched holes are attached to the light table. Each of the color separation flats is placed on these pins. Everything is then in perfect register throughout the remaining work to be done.

Pin registration is also used to align screen tints, textures, surprints, and reverses. A screen tint is an area of uniform dots that creates the illusion of a lighter shade. Fig. 14-13. Screen tints are indicated by percentages. A 10% screen tint, for example, is very light while a 90% tint is quite dark.

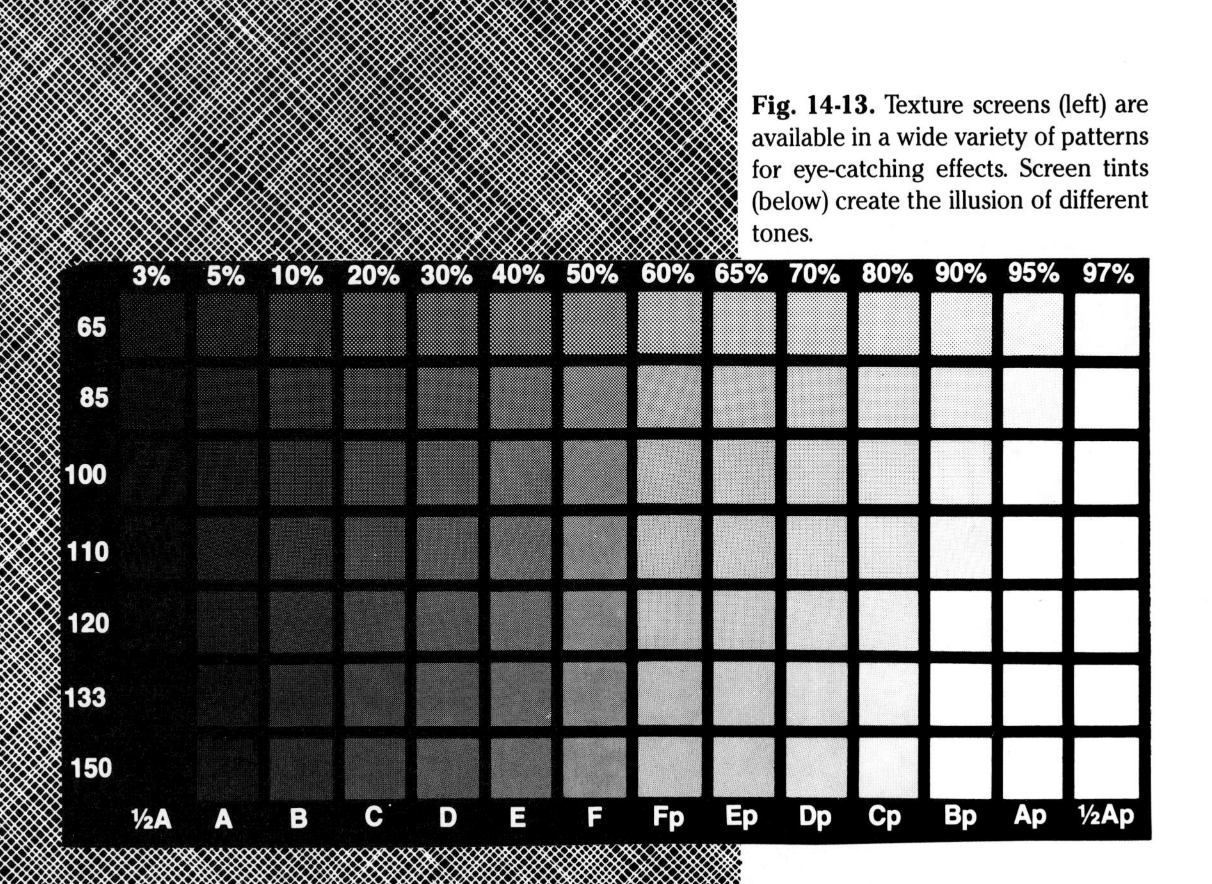

Fig. 14-13. Texture screens (left) are available in a wide variety of patterns for eye-catching effects. Screen tints (below) create the illusion of different tones.

A texture may be created in precisely the same manner. The only difference is that a texture screen, one creating the look of wood grain for instance, is used instead of a tint screen.

Surprint refers to the overlaying of one image (generally text) on top of another (usually a halftone). Fig. 14-14. This is done by aligning the line negative for the text with the other image.

A reverse is the opposite of a surprint. Fig. 14-14. The type is a light color on a dark background, often a halftone. To make a reverse, a film positive is made of the text. The positive is stripped in pin register with the halftone flat.

Automated Film Assembly

For some high-production work, such as for a weekly magazine, automatic stripping equipment is now available. Fig. 14-15. Computer-controlled robot-like devices pick up negatives from a stack, position them, and put them through the next stage of the graphic production process, which is platemaking. You'll read about platemaking in Chapter 15.

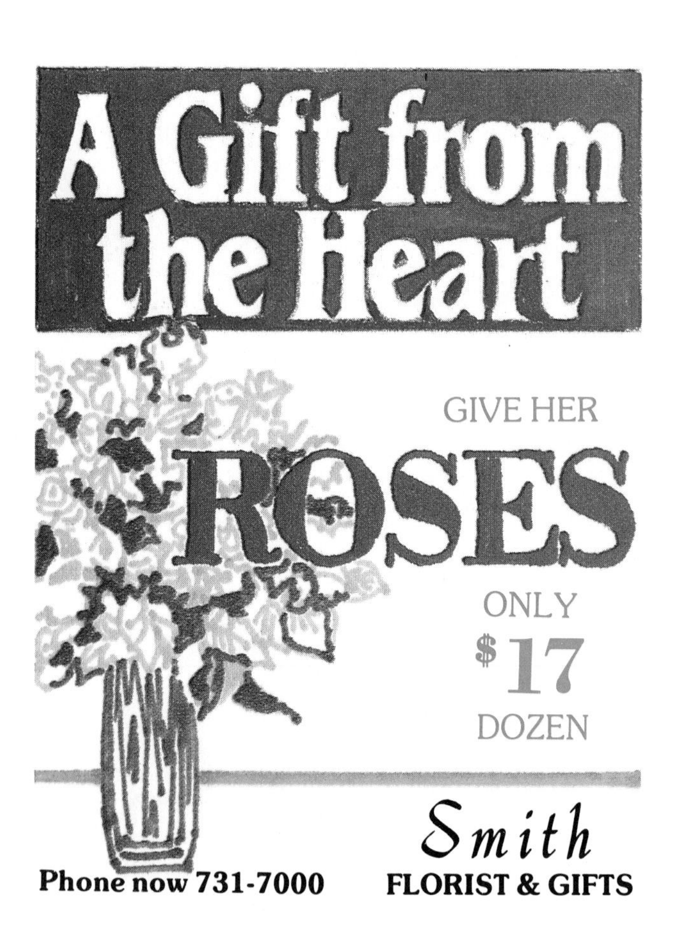

Fig. 14-14. The word "Roses" has been surprinted over the image of the flowers. "A Gift from the Heart" is reversed out of the background.

Fig. 14-15. Automatic stripping systems can expose a stack of negatives in their appropriate positions across a single plate. They bypass the process of making a flat.

CHAPTER 14 REVIEW

Review Questions

1. When photographing copy with a process camera, how is the size of the image changed?
2. Describe the relationship between the f-stop number, the size of the aperture, and the amount of light that passes through the aperture.
3. To which step on a gray scale should a line negative be developed?
4. How does a film positive differ from a negative?
5. What is a halftone? Describe how one is made.
6. Describe the process of color separation.
7. For what purpose is a color scanner used?
8. How have electronic pagination systems changed the film conversion process?
9. Define stripping and tell how it is done.
10. Why is pin registration necessary?

Activities

1. Cover half of a piece of orthographic film while exposing the other half to room light. Process the film and explain the result.
2. Research the chemistry of film and film processing. Write a report on your findings.
3. Take a mirror into a darkroom and turn out all lights. After several minutes, turn on a light and look in the mirror. What has happened to the pupils of your eyes? In what way is your eye like the lens of a process camera? Report on your experiment to the class.
4. Find examples of a screen tint, a screen texture, a surprint, and a reverse. Make a display labeling each.
5. Find out the cost of having a set of color separations made for a 5″ × 7″ color photograph. Report your findings to the class.

Message Transfer and Product Conversion

What do postage stamps, *National Geographic* magazine, and wallpaper have in common? The answer is "gravure," the process by which each of these products was printed.

Printing also helped create the box your breakfast cereal comes in, this textbook, and the circuit boards in your television set. Each of these products was produced by a different printing process. After printing, the products underwent such processes as folding, trimming, or cutting to convert them to usable form.

Printing is often called *message transfer* because the message (words and/or pictures) is transferred from one medium (such as a printing plate) to another (such as paper).

In this chapter, you will be introduced to the major message transfer and product conversion processes. You will begin to understand why certain processes make sense for one product but not for another.

Terms to Learn

binding
die cutting
electrostatic printing
embossing
flexography
gravure
ink jet printing
letterpress printing
offset lithography
offset plate
photopolymer
thermography
screen printing
stencil

As you read and study this chapter, you will find answers to questions such as:

- What are the major message transfer processes and how do they work?
- For what uses is each of these processes best suited?
- How are printed materials converted to finished products?

METHODS OF PRINTING

All printed products are produced by one of the following methods: offset lithography, relief printing, gravure, screen printing, electrostatic printing, ink jet printing, or low-cost duplication. Although the methods differ, the general process by which they transfer the message to the substrate is the same. Generally, the message transfer process involves: (1) Exposing a message carrier through a film assembly and (2) transferring the message from the carrier to the substrate with a printing press.

After printing, products usually undergo some kind of conversion process, such as embossing, folding, binding, or package conversion.

Fig. 15-1. This example of an early lithograph was made by French artist Henri de Toulouse-Lautrec in 1895. It is titled *Le Revue Blanche.*

Fig. 15-2. This drawing illustrates the principle of lithography. Because the printing is done from a plane (flat) surface, lithography is a planographic printing process.

OFFSET LITHOGRAPHY

In 1798, Aloys Senefelder, a playwright living in Munich, Germany, invented a new way of printing. Senefelder drew a picture with a greasy "crayon" on a smooth slab of limestone. When he rolled water and then ink over the picture, two things happened. First, the water was repelled by the greasy crayon but clung to the stone background. Second, the ink clung to the crayon but was repelled by the wet stone. Senefelder transferred the inked picture to paper simply by pressing the paper against the limestone. He called the process "lithography," which means "stone writing." Figs. 15-1 and 15-2.

By the nineteenth century, the process was done by machines. The quality of the printed image was improved by first transferring, or offsetting, it onto a rubber blanket. The improved process became known as **offset lithography**. Today, offset lithography accounts for nearly half of all dollars earned in the printing industry.

Offset Plates

A wide variety of message carriers, or **offset plates**, are used for offset lithography. These include paper, diffusion transfer, aluminum foil, sheet aluminum, and bimetal plates. They vary as to cost and durability. Paper plates, for example, are capable of printing only several hundred impressions before they wear out. Bimetal plates last for more than 500,000 impressions.

Offset platemaking begins with the assembled film, or flat. The flat is placed against the plate and then exposed with a bright light. Next, the plates are processed by hand at a plate processing sink or by machine. Fig. 15-3. In this way an "ink-loving" image area is created on the surface of the plate.

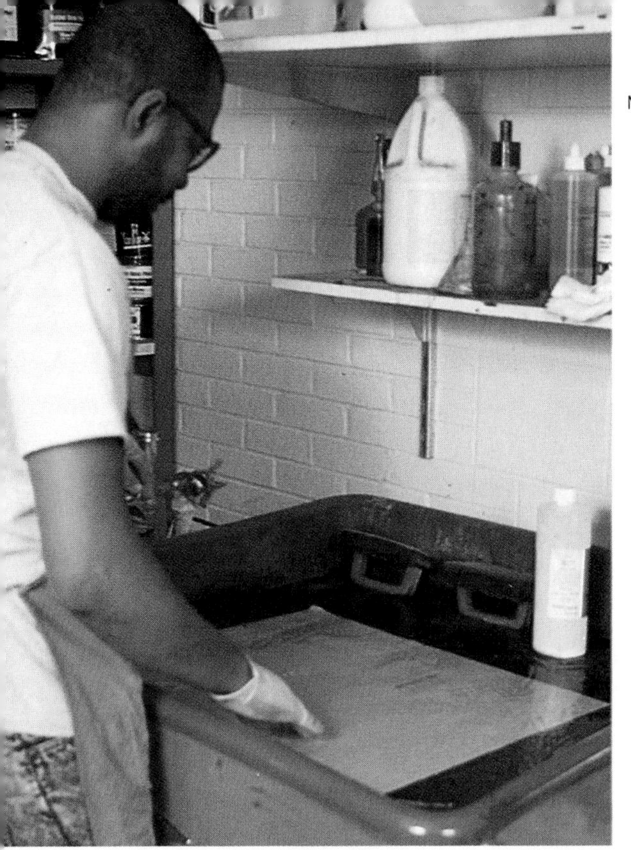

Fig. 15-3. Processing a plate is similar to developing film. The methods vary, depending on the type of plate. Here, for a small job, the processing is done by hand.

Offset Lithographic Message Transfer

An offset press transfers ink and water to the plate. In doing so, it transfers the message from the plate to the substrate. Any offset lithographic press performs these functions: inking, dampening, feeding, registration, printing, and delivery. Figs. 15-4 and 15-5.

Fig. 15-4. The diagram below shows how the inking, dampening, and printing units are set up. A modern web offset press is shown at right.

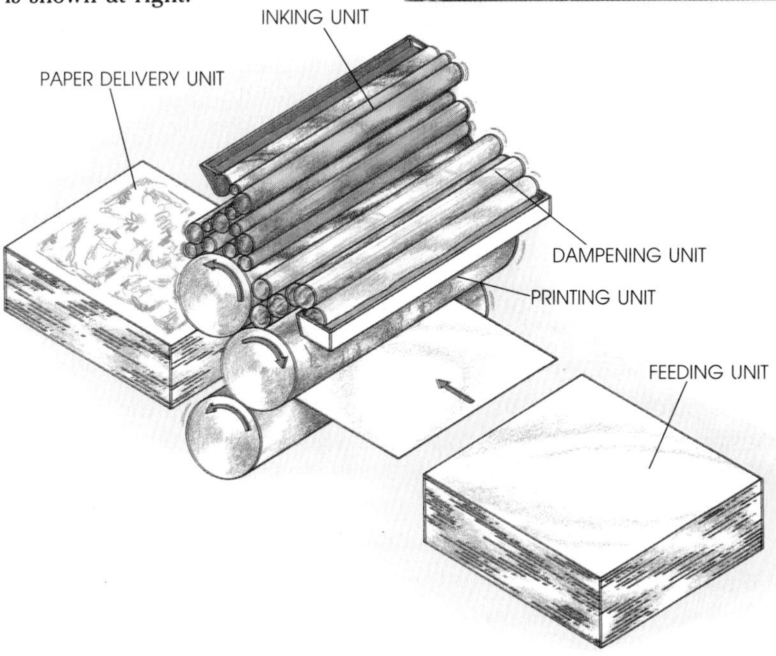

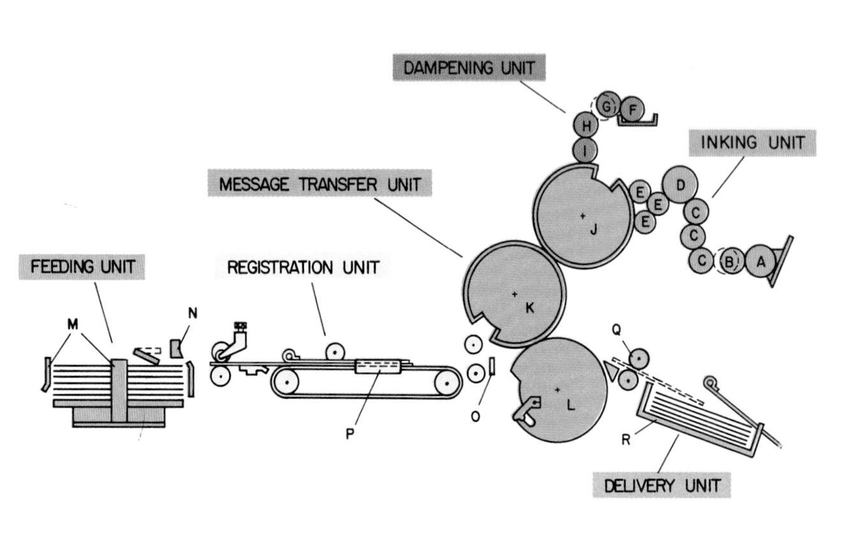

Inking Unit
A Fountain roller
B Ductor roller
C Distribution rollers
D Oscillating distribution rollers
E Form rollers

Dampening Unit
F Fountain roller
G Ductor roller
H Distribution rollers
I Form roller

Message Transfer
J Plate cylinder
K Blanket cylinder
L Impression cylinder

Feeding Unit
M Paper pile guides
N Suction feet

Registration Unit
O Stop fingers
P Side jogger

Delivery Unit
Q Ejector wheels
R Delivery tray

Fig. 15-5. This is a schematic of a sheet-fed, single-color offset press.

Inking

The purpose of inking is to spread a thin film of ink over the image area on the offset plate. The press operator adds ink to the ink fountain. As the press runs, the fountain roller picks it up. The amount of ink that is picked up may be controlled by adjusting the ink keys located on the fountain. As the fountain roller rotates, it carries ink to the ductor roller. The ductor roller moves back and forth between the fountain roller and the distribution rollers. The distribution rollers carry the ink to the form rollers. The form rollers transfer the ink to the image areas of the plate.

Dampening

Dampening is also done by a series of rollers. These rollers carry fountain solution, made mostly of water, to the offset plate. Fountain solution is used to wet the non-image areas. The fountain solution is placed in the water fountain. It is then transferred from the fountain roller to the ductor roller and then to the distribution roller. Finally, the form rollers transfer the fountain solution to the plate, where it clings to the non-image areas.

Feeding

The feeding unit carries paper or another substrate into the press. Feeding is done by a feed table, paper pile guides, blower tubes, and suction feet. The pile guides hold the paper in place on the feed table. Each sheet must be fed into the press in exactly the same spot. Air forced through the blower tubes lifts the top sheet of paper. Then the suction feet lift it and move it forward.

Registration

The paper must be held in precisely the same spot each time so that colors and images align properly when one sheet is printed upon two or more times. Registration is aided by stop fingers that stop the movement of the paper while the side jogger pushes or pulls it sideways. Some small offset presses cannot place the paper accurately and are therefore not well suited to multicolor printing.

Printing

Printing is done by three cylinders that rotate in contact with each other during the printing cycle. They are called the plate cylinder, the blanket cylinder, and the impression cylinder. Some presses combine the plate and impression cylinders.

An offset plate is mounted on the plate cylinder. It picks up ink and fountain solution from the form rollers as it rotates. The image is first transferred onto a rubber offset blanket mounted on the blanket cylinder. The substrate, generally paper, is fed between the blanket cylinder and the impression cylinder. Gripper fingers, located on the impression cylinder, carry the paper through the press. The impression cylinder applies pressure to the back of the paper, transferring the image.

Delivery

After the image has been transferred, the paper is removed from the press and stacked neatly on the delivery table. In chute delivery systems, the paper is simply ejected onto the delivery table with ejector wheels. In chain delivery systems, gripper fingers attached to a chain deliver the printed sheets.

Multi-Color Presses

Some offset presses are designed to print one, two, four, or even seven colors of ink in one pass. Fig. 15-6. This is done by connecting several inking, dampening, and printing units together. Drying devices between each unit quickly dry the ink after each color is laid on. A four-color press, for example, is commonly used to print the yellow, magenta, cyan, and black needed for full-color reproduction.

Uses for Offset Lithography

Offset lithography can be used to print on a wide range of substrates. This book (and nearly all others) was printed by means of offset lithography. So are most newspapers, magazines, brochures, posters, and many other products.

Lithography may be used for as few as 10 or as many as a million copies. For the local printer who prints "quick and dirty" jobs, or the large commercial printer who prints first-rate four-color jobs, offset lithography is considered the single most important printing process.

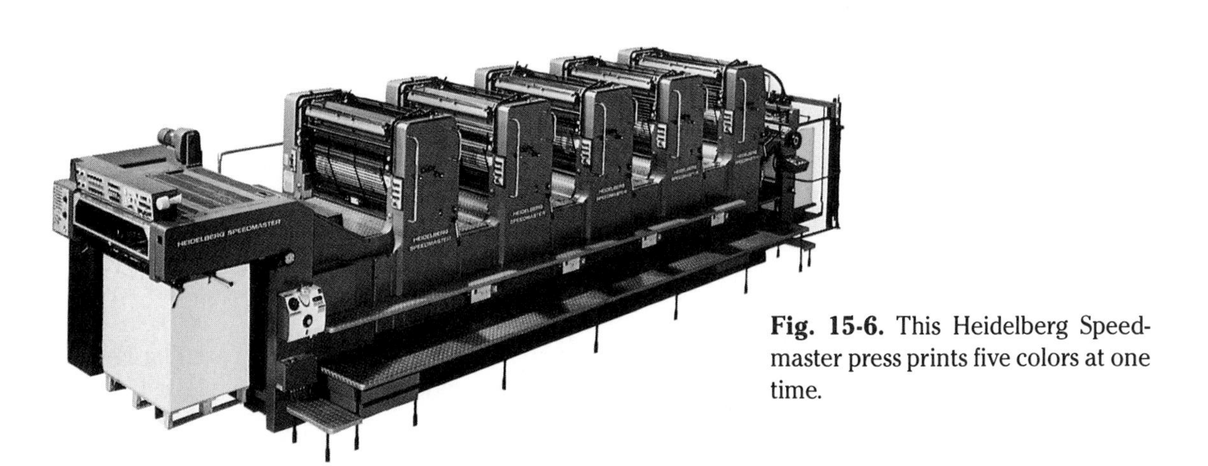

Fig. 15-6. This Heidelberg Speedmaster press prints five colors at one time.

RELIEF MESSAGE TRANSFER

Relief printing is the process of printing from a raised surface. Fig. 15-7. Ink is applied to the raised surface. This surface and the substrate are then pressed together to transfer the image.

Relief printing is the oldest method of printing. It was used by the Chinese as long ago as A.D. 200. The earliest known printed book was produced in China around A.D. 868 by relief printing, using images carved in wooden blocks.

Relief printing processes include letterpress and flexography. Letterpress uses metal type, while flexography uses a rubber or plastic plate.

Letterpress

A German metalsmith named Johannes Gutenberg invented movable metal type around 1450. This was a revolutionary invention. In fact, movable metal type is generally considered one of the most important inventions of all time. For the first time in history, books could be produced in large quantities. **Letterpress printing**, as it came to be known, was the primary means of graphic reproduction for five centuries!

In the last half of this century, however, use of letterpress printing has steadily declined. It is now used for less than 5 percent of all printing. However, letterpress is still used for finishing operations. Embossing, foil stamping, and die cutting (described later in this chapter) are all done on a letterpress.

Flexography

Flexography is a form of relief printing resembling letterpress. The message carrier looks similar to those used for letterpress. However, the carrier is made of rubber or plastic.

Flexography was first known as aniline printing. The inks used were made from aniline oil taken from the indigo plant. These inks dried very quickly. When cellophane was invented in the 1930s, aniline printing became the process of choice for printing on this new substrate. It was especially popular in the food packaging industry.

In the 1950s, however, aniline inks were believed to be toxic and unsuitable for food packaging. Although this turned out not to be true, by then the inks had a bad reputation. The industry therefore renamed the process flexography.

New and improved inks and presses and a wide variety of new plastic substrates have made flexography popular today. It now accounts for about 20 percent of all printing sales.

Flexographic Plates

Flexographic plates are of three basic types: rubber, liquid photopolymer, and sheet photo-

Fig. 15-7. This illustrates relief printing. Notice that the letters (image area) are raised.

polymer. A **photopolymer** is a light-sensitive plastic that looks like clear honey. Fig 15-8.

To make a rubber plate, a mold, or matrix, is made first by impressing the mold material with metal type and heating it. Liquid rubber is then poured into the mold. Heat and pressure cause the rubber to take the shape of the mold. A rubber stamp is an example of a simple flexographic plate.

To make plates from liquid photopolymer, the liquid is poured onto a thin sheet of clear film that covers a negative. This "sandwich" is then exposed. A developer hardens the exposed image areas. The unhardened, non-image areas are then washed away with a detergent. The image area remains as a raised, flexible, plastic surface.

Precast photopolymer plates are similar to liquid photopolymer plates. However, they begin as solid sheets of material. They are a little easier to work with but more costly than liquid photopolymers. The exposure and washing processes are roughly the same as for liquid photopolymer plates.

Flexographic Message Transfer

Flexographic presses use a continuous roll of substrate, known as a web. Narrow web flexographic presses are commonly used to print labels. Wide web presses print larger products, such as food containers.

With a flexographic press a very thin ink is used. Fig. 15-9. The fountain roller transfers this ink to the anilox roll, which has tiny grooves

SCIENCE FACTS

Polymers

A polymer is a large molecule made from a chain of smaller units called monomers. *Poly* means "many," and *mer* comes from the Greek word for "parts."

Plastics are polymers. There are basically two kinds of plastic: thermoplastics and thermosets. When thermoplastics are heated, the long chains of molecules slip past one another. The plastic softens and melts. In thermosets, however, the chains are linked. When a thermoset is heated, it does not melt, but it may soften or burn.

Fig. 15-9. This diagram shows the flexographic process. The flexographic plate is wrapped around the plate cylinder.

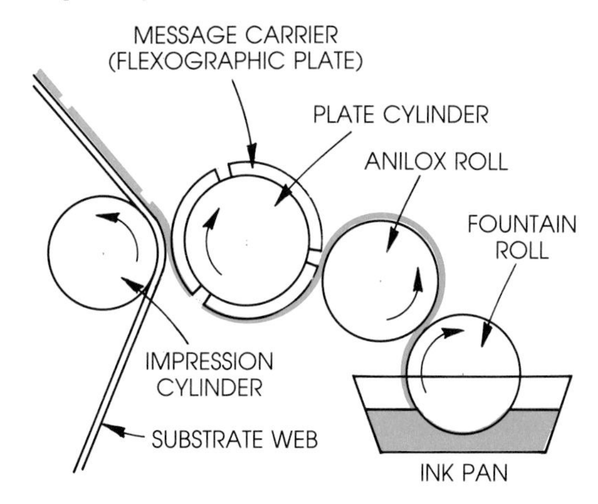

Fig. 15-8. Both of these plates are used for flexography. The one on the left is rubber and the one on the right is made from a photopolymer. Both types can produce millions of copies, making them good choices for certain kinds of high-volume printing.

machined on its surface. The anilox roll carries the ink to the plate. The plate is mounted on the plate cylinder with a two-sided sticky-back foam pad. Ink transfers from the plate to the substrate with the help of pressure from the impression roller. The unwind unit feeds the web into the press and the rewind unit takes it up again after printing.

Uses for Flexography

Because flexography's inks dry almost instantly on any substrate, including plastic films and foils, flexography is often used for packaging. Grocery and department stores are filled with packages and labels printed by the flexographic process. Fig. 15-10.

Flexography is not yet capable of printing fine detail, such as 133-line halftones, consistently. Therefore it is not used to print books, magazines, and other publications. Newspapers, however, are experimenting with flexography as a possible replacement for offset lithography.

GRAVURE

Gravure is the process of transferring ink from image areas below the surface of the plate. Fig. 15-11. In this way it is the opposite of relief printing. Gravure is a form of intaglio, a process used by artists as long ago as the fourteenth century. A chisel-like tool was used to carve lines into the end grain of wooden blocks. Ink was rubbed into the lines, and the surface of the wood was wiped clean. The image created by the chiseled lines was transferred to paper with pressure. Fig. 15-12. Copper plates soon replaced wooden plates. Although the illustrations were beautiful, the process was slow.

In 1879, Karl Klic, a Czech, invented a process for creating an image on a copper plate photographically. Klic later replaced flat plates with round cylinders. This greatly speeded up the process. Klic called his process "rotogravure,"a name now often shortened to simply gravure.

Fig. 15-10. This package was printed using the flexographic process.

Fig. 15-12. Gravure has been used by artists for centuries. This woodblock print was created by Albrecht Dürer (1471-1528).

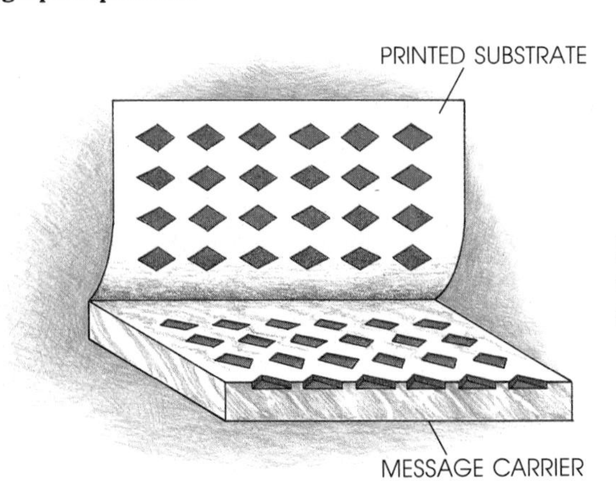

Fig. 15-11. This diagram illustrates the way gravure works. Notice how ink is captured by cells recessed in the surface of the plate.

Gravure Plates

A gravure plate, generally called a gravure cylinder, starts out as a perfectly smooth, copper-plated cylinder. The image is then etched or carved into its surface. In the past, a stencil-like mask was made by exposing a light-sensitive carbon tissue through a negative. This tissue was then attached to the cylinder and submerged in acid. The acid etched the image areas into the cylinder.

Gravure cylinders are now made with a computer-driven device known as a helioklischograph. The helioklischograph scans the copy and converts it to computer data. These data make a diamond-tipped stylus move against the rotating gravure cylinder the way a woodpecker pecks a tree. Thousands of tiny holes, called cells, make up the image area. Fig. 15-13.

Gravure Message Transfer

The gravure cylinder is mounted on a gravure press. Fig. 15-14. The cylinder rotates in an ink fountain and receives a film of very thin ink. A thin blade, known as the doctor blade, then removes all ink from the cylinder except that which collects in the ink cells. When pressure is applied to the substrate with an impression roller, the image is transferred.

Fig. 15-13. This diagram shows the gravure process. Data from the scanned artwork are fed to a computer.

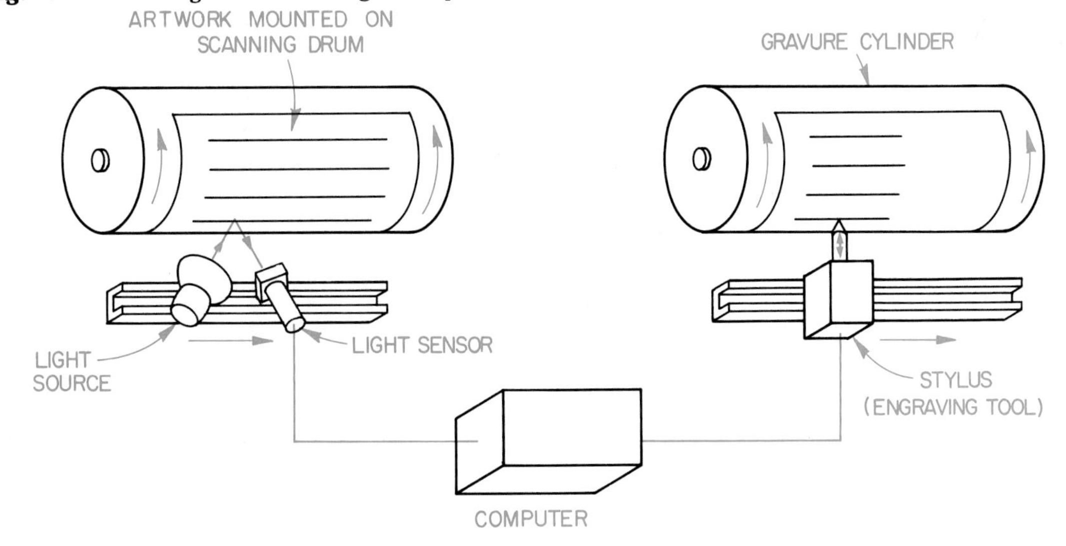

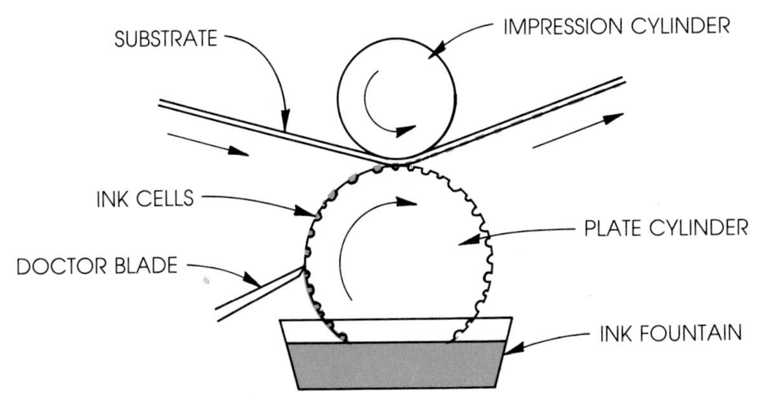

Fig. 15-14. The ink that remains in the ink cells is transferred to the substrate with pressure.

Uses for Gravure

Gravure accounts for about 20 percent of all printing sales. It is a costly process. However, gravure cylinders hold up for millions of impressions, making them ideal for long press runs. For this reason, gravure is used for printing United States paper money and postage stamps. More than twenty of the most popular magazines in the United States are also printed by gravure. Moreover, the color quality of gravure is better than any other graphic reproduction method. Therefore gravure is sometimes used for very high-quality work. Fig. 15-15.

Gravure inks dry almost instantly. For this reason gravure has a place alongside flexography in the packaging industry. Gravure has an edge when high-quality and/or fine detail is required.

A continuous image can be created on the cylinder with no top or bottom to it. Therefore, gravure is also used to print materials that have continuous patterns, such as rolls of wallpaper and vinyl flooring.

SCREEN PRINTING

If you cut a hole in a piece of paper, lay it down on another sheet of paper, and rub over it with a crayon, the crayon transfers the hole shape to the bottom sheet. The paper with the hole in it could be used again and again to "print" the same shape. After several uses, though, the paper might fall apart. However, it would last longer if you attached it to a screen like that on your front door and then spray-painted the hole shape.

Basically, this is what screen printing is all about. Instead of a door screen, a fine fabric screen is used. With a fine screen, the pattern of the screen doesn't transfer to the substrate. Ink is generally used, rather than crayons or spray paint.

Screen printing, then, is simply transferring ink through a stencil held in place by a screen. Fig. 15-16. A **stencil** is a thin sheet with holes cut through it in the shapes of letters, designs, and so on. The stencil allows ink to pass through the image areas (holes) while keeping the ink from the non-image areas.

In Asia, people printed from stencils as early as the tenth century, A.D. At first, human hairs were used to hold the stencils in place. Later, silk was used, and the process has long been known as "silk screen" printing. Silk, however, is costly and weak compared to modern synthetic fabrics. Silk is no longer used in the industry, and so the process has come to be known simply as screen printing.

Screen Printing Stencils

Stencils for screen printing come in a variety of materials. Some common types of stencils are described in the following paragraphs. Fig. 15-17.

Fig. 15-15. Magazines, newspapers, catalogs, and stamps are some of the many items that can be produced using the gravure process.

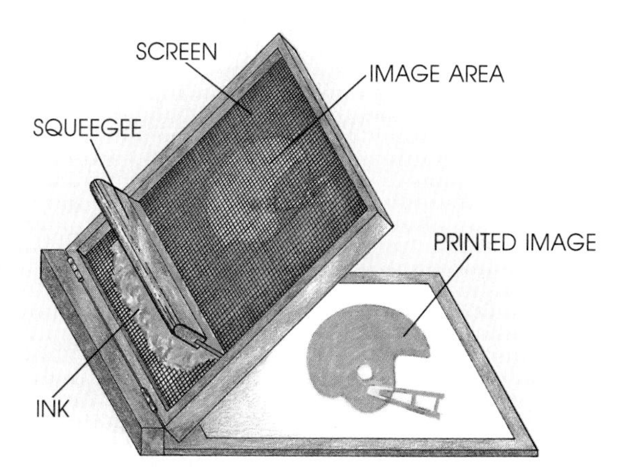

Fig. 15-16. The stencil is held in place by the screen. The squeegee drives the ink through.

Fig. 15-17. Hand-cut (top), photo-indirect (center), and photo-direct (bottom) stencils are all used for screen printing.

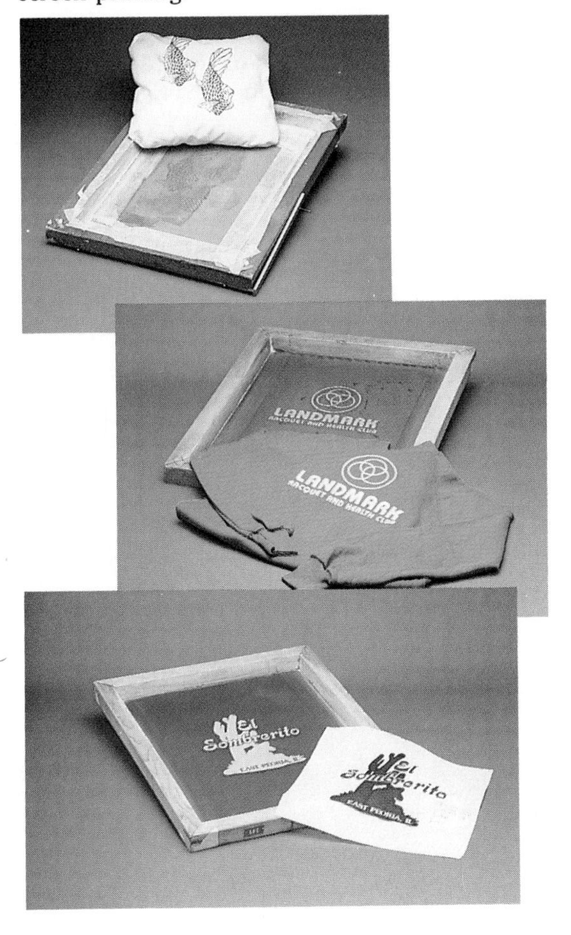

Thermofax Stencils

After the paper stencil, perhaps the simplest type is one made with a thermofax machine. A thermofax machine is used in many offices to produce overhead projection transparencies. Screen and stencil are combined in a single sheet of material. The material is placed in contact with line copy and fed through a thermofax machine. The resulting screen/stencil may then be mounted in a plastic frame and printed.

Hand-Cut Stencils

Hand-cut stencils consist of a clear sheet of plastic having a thin coating of a softer plastic. The soft plastic is cut and peeled away from the clear plastic, creating open image areas. The stencil is then attached to the screen. One type is attached with water. Another requires a lacquer fluid. Once the stencil is attached, the clear plastic, too, is peeled away, leaving the image areas open for the ink to pass through.

Photographic Stencils

Photographic stencils are made by exposing them through a clear positive image. Either a film positive or a line drawing on clear drawing film may be used. The stencil is then developed, which hardens the exposed (non-image) areas. The unhardened areas are then washed away. They form the image through which ink passes.

There are two common types of photographic stencils: photo-direct and photo-indirect. Photo-direct stencils are made from liquid emulsions coated directly on the screen and then exposed and processed. Photo-indirect stencils are made from sheets of material that are exposed, processed, and then attached to the screen.

Screen Fabrics

Synthetic fabrics such as polyester, nylon, or stainless steel are generally used for screen printing. Besides strength, the most important quality of screen printing fabrics is the openness of the weave. The smaller the spaces between threads in the fabric, the finer the detail printed. However, it is more difficult to print through a very small opening than through a coarse one.

The size of the spaces is called the mesh count. The most common method of measuring the mesh count is with a scale that ranges from 6XX to 25XX. A 6XX mesh is a very coarse screen and 25XX a fine one. For most work, 12XX to 16XX fabric is recommended.

Frames

Screen printing frames are generally made of wood or metal. Wooden frames are less costly and fairly durable. Metal frames are more stable and are used for high-quality work or where a longer-lasting frame is desired.

A common method of attaching the screen to a wooden frame is with cord that holds the fabric tightly in a groove. Another method is to stretch the fabric and glue it in place. Fig. 15-18.

Inks and Substrates

One of the advantages of screen printing is the wide variety of inks and substrates that may be used. There are inks for printing upon almost any surface, such as paper, metal, fabric, and a wide variety of plastics. Inks are also designed for outdoor use, such as for commercial signs.

Some screen printing doesn't use ink at all. Highway signs are sometimes made by screen printing an adhesive to which tiny glass beads are attached. These glass beads reflect the light from car headlights, making the sign more visible in the dark.

Transferring a Stencil Message

To print a stencil, ink is forced through the image areas with a squeegee, a firm but flexible rubber-like blade. For short runs, the frame may be hinged to a bench top or large board and the squeegee pulled across the screen by hand. After the image is printed, the substrate is placed

Fig. 15-18. Three methods are commonly used for mounting the screen on the wooden frame: rope and groove (left), adhesive (right), and staples (bottom).

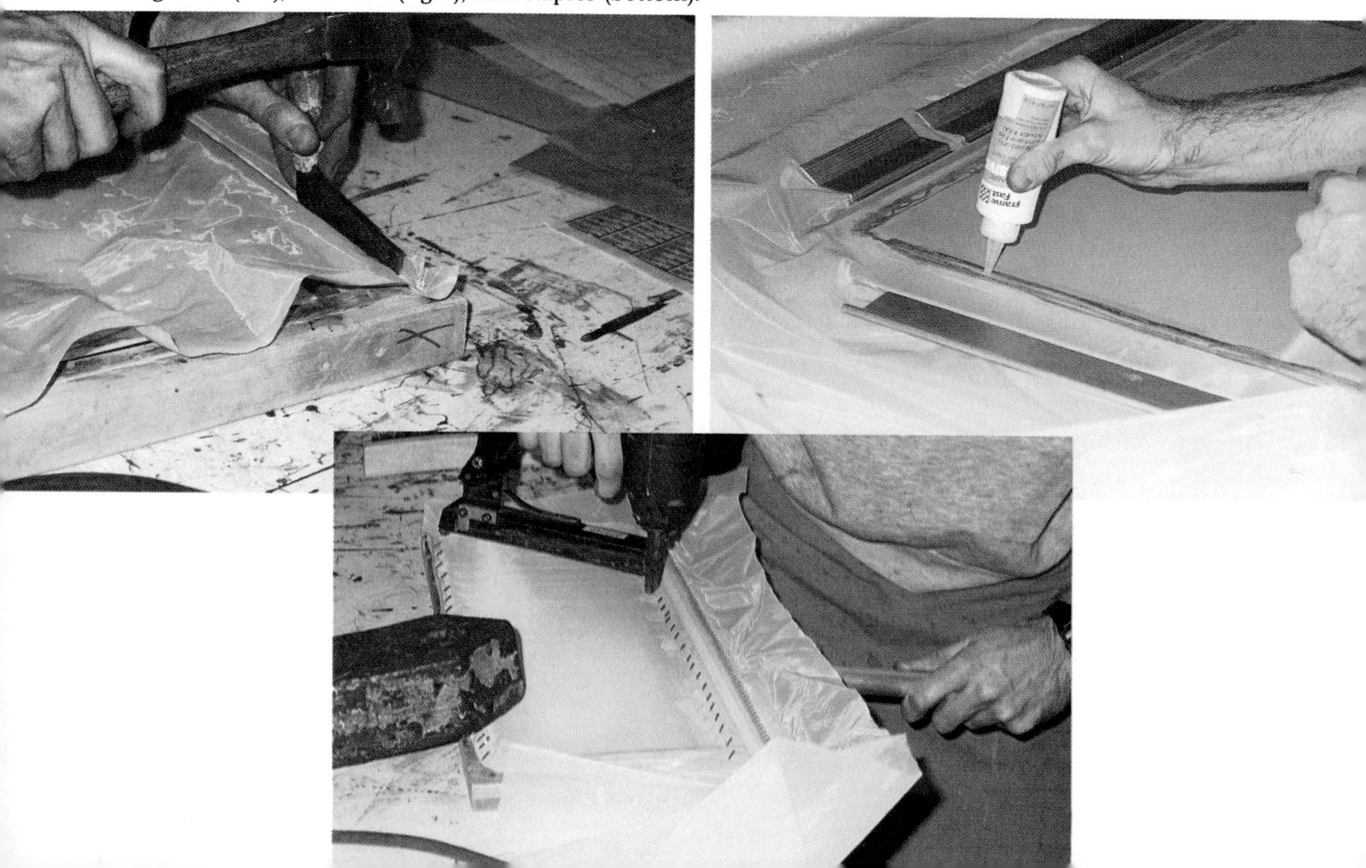

on a drying rack to air dry. For long runs, a screen printing press may be used. These presses do some or nearly all of the manual work. Fig. 15-19.

Fig. 15-19. Screen printing presses are used to print large quantities.

Uses for Screen Printing

Screen printing may be used to print on almost any two-dimensional surface. Screen printing substrates include paper, metal, glass, ceramic, fabric, and wood. Fig. 15-20.

The clothing industry relies upon screen-printed fabrics used for garments. Screen-printed T-shirts are sold in large numbers, as are screenprinted "iron-ons." Iron-ons are transferred from the paper they are printed upon to clothes with the use of a household iron or special press.

The ceramics industry uses screen printing to decorate glasses, drinking mugs, and other items. One method consists of screen printing tiny particles of crushed glass suspended in a liquid onto tableware. The items are then fired to fuse the glass to the piece. Another method is to screen print directly upon glasses and mugs with durable inks.

Screen printing is used to produce the conductive paths between components on plastic circuit boards in almost all electronic equipment. This may be done either by using a conductive ink or an acid resist. When an acid resist is used, it prevents acid from etching away a copper circuit board where conductive pathways are desired. Even the tiny electronic chips soldered to a circuit board are produced with the screen printing process.

Screen printing lends itself to printing small numbers of large images. A large wooden frame and stencil are far less expensive than the equipment needed for any other printing process. Therefore screen printing is commonly used for signmaking, large advertising posters, and even billboard advertisements.

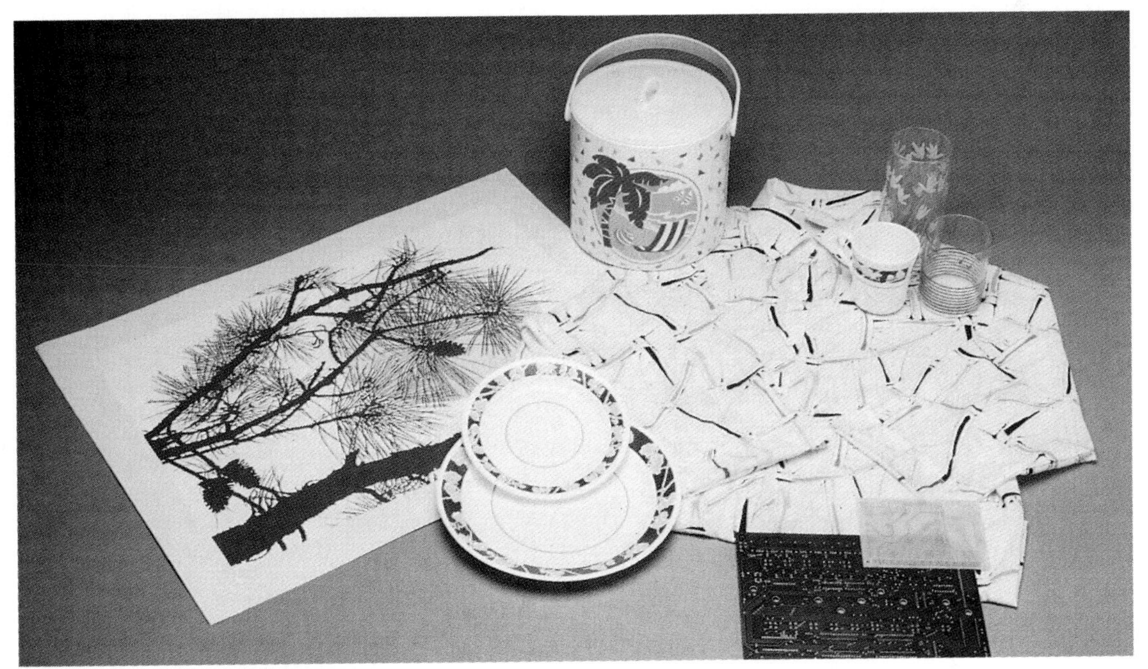

Fig. 15-20. These products have all been produced using screen printing.

ELECTROSTATIC MESSAGE TRANSFER

Electrostatic printing relies upon a charge of static electricity to transfer the message from the plate to the substrate. Fig. 15-21. A photocopy machine is the simplest form of electrostatic printing.

Light reflected from original copy creates a charged image area on a cylindrical drum. This charge is then transferred to the substrate (usually paper). Tiny grains of toner (usually black) are attracted to the image area by the charge. The toner is fused to the substrate with heat.

Photocopiers are very popular in offices for short-run copying. High-speed copiers, capable

In this photocopier, the original document moves from the document handler (1) to the platen (2). There it is exposed by lamps and mirrors through a lens (3) focusing the image onto the photoreceptor belt (4). Magnetic rollers (5) brush the photoreceptor belt with dry ink, which clings to the image area.

The copy paper moves from the main or auxiliary tray (6) to the belt, where the dry ink is transferred to it (7). The copy then goes between two rollers (8), where heat and pressure fuse the image to the paper.

Single copies go to the receiving tray (9). Multipage documents go to the sorter (10), which collates them. Sheets that are to be copied on both sides return by conveyor (11) to the auxiliary tray to repeat the process.

Users key in their instructions at the control console (12). Adjustment and testing are done at the maintenance module (13).

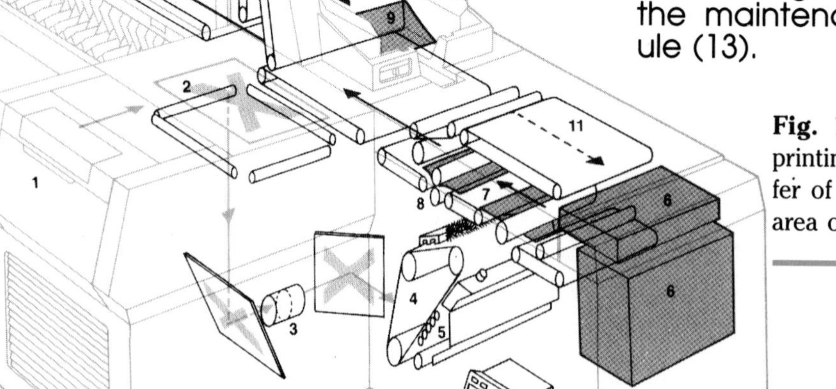

Fig. 15-21. Electrostatic printing involves the transfer of toner to the image area on the substrate.

of thousands of copies per hour, are also available. Fig. 15-22. Electrostatic printers have also been combined with microcomputers to create the desktop publishing systems described in Chapter 13.

Electrostatic Plates

The plate for electrostatic printing is the drum that carries the electrostatic charge. There are two basic ways of creating this charge on the drum. The first uses light reflected from the copy. This type is a common fixture in offices and quick-copy shops.

The second type, the laser printer, depends upon information relayed from a computer. Input devices for laser printers include computer keyboards for text and scanning devices for artwork. Fig. 15-23. The information is directed to the laser, which scans across the drum and creates the charged image area. As with an office copier, the charge is transferred to the substrate and fused to it with heat. Electrostatic laser printers are available in both inexpensive and costly, high-speed models.

SCIENCE FACTS

Static Electricity

You've probably had this happen to you often: you walk across a carpet, reach for a doorknob, and get an electric shock. What causes the shock? It's static electricity.

Electricity is the flow of electrons. Electricity may travel in a current, as in the wires that bring power to appliances. It may also remain in place, or "static."

What happens when you walk across that carpet? You pick up electrons from the carpet, thus building up a negative charge in your body. When your fingers come close to the doorknob, this static electric charge suddenly jumps to the doorknob. In a darkened room, you can actually see a spark.

Static electricity can be annoying and even dangerous. (Lightning is a discharge of static electricity.) Still, there are useful applications for static electricity, and electrostatic printing is one of them.

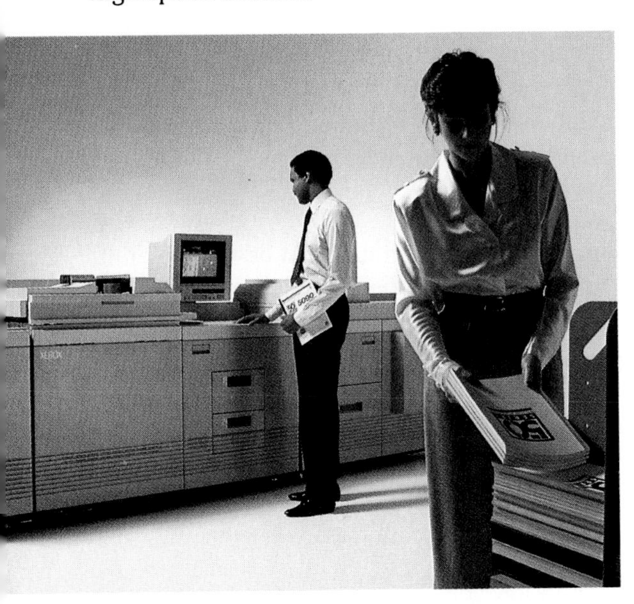

Fig. 15-22. This high-speed electrostatic copier makes 135 copies per minute. It will also bind or stitch sheets together.

Fig. 15-23. This Apple LaserWriter printer receives input from a computer.

INK JET MESSAGE TRANSFER

As its name implies, **ink jet printing** uses tiny spray guns to send very thin ink onto the substrate. Fig. 15-24. During ink jet printing the image carrier and the substrate do not touch. This is unlike most of the other graphic reproduction processes. Because the image carrier and the substrate do not touch, or impact, the process is sometimes called non-impact printing.

Because there is no contact during image transfer, the ink jet method can print on a wide range of substrates, including raised or delicate surfaces. Fig. 15-25.

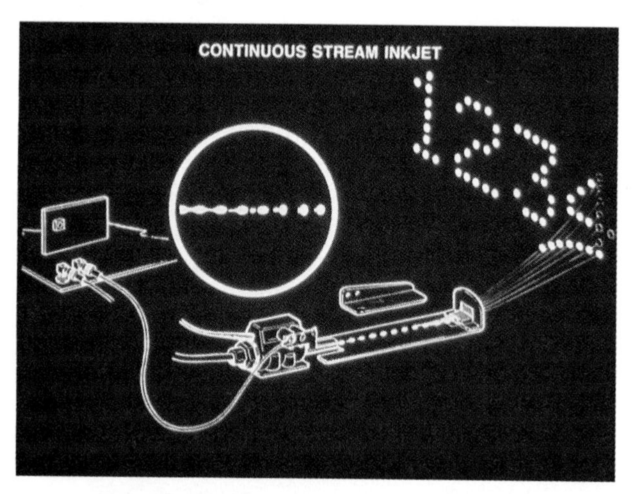

Fig. 15-24. The top diagram shows a continuous stream ink jet. Below is shown how each character is constructed in successive vertical scans.

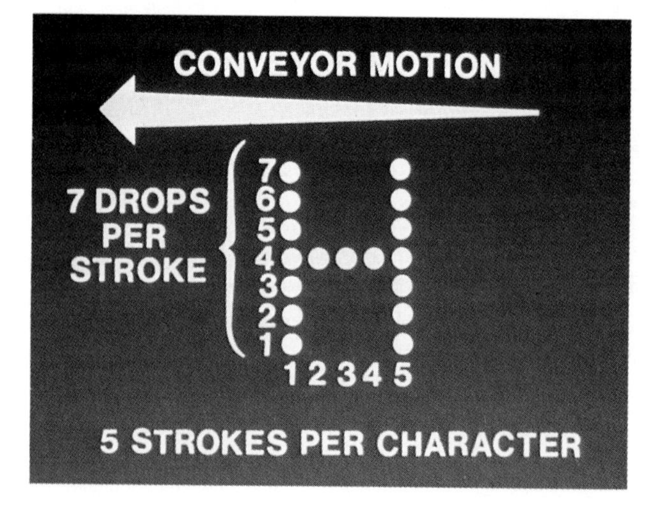

The message carrier for ink jet printing is actually a computer. Digital data control the tiny nozzles that spray the droplets of ink onto the substrate. The ink is also directed by an electrostatic field as it travels through the air to the paper. Ink jet printers include advanced, high-speed types or rather simple table-top models. Fig. 15-26.

LOW-COST DUPLICATION

Offices sometimes use ditto and/or mimeograph processes for short-run printing. The ditto master has a carbon image area. The image area is coated with a thin layer of alcohol. The carbon image is transferred to paper with pressure. As the carbon is used up, the copies get lighter. Ditto copying is therefore good for only about 100 copies.

In the mimeograph process, a stencil is cut, usually on a typewriter. Ink is forced through this stencil onto the paper. Mimeograph stencils can produce slightly better quality and longer runs than the ditto process. However, mimeograph is still used only for office copying where quality isn't as important and long runs aren't necessary.

Fig. 15-25. Ink jet printing can be used to print on almost any surface — even the yolk of an egg!

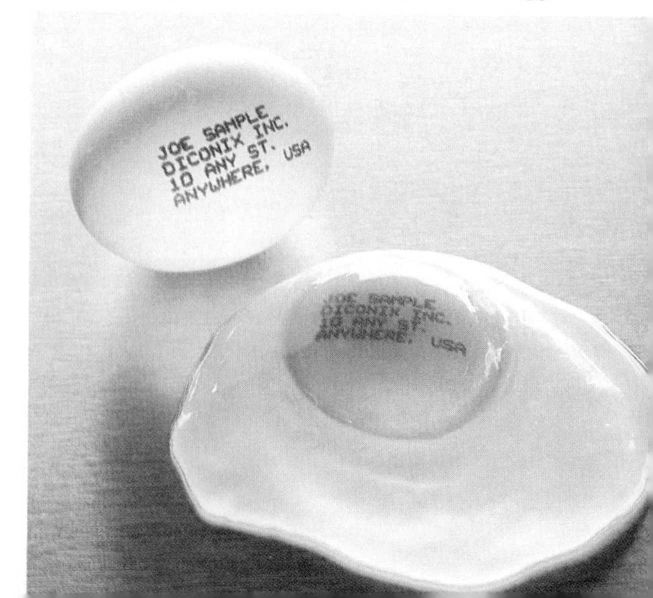

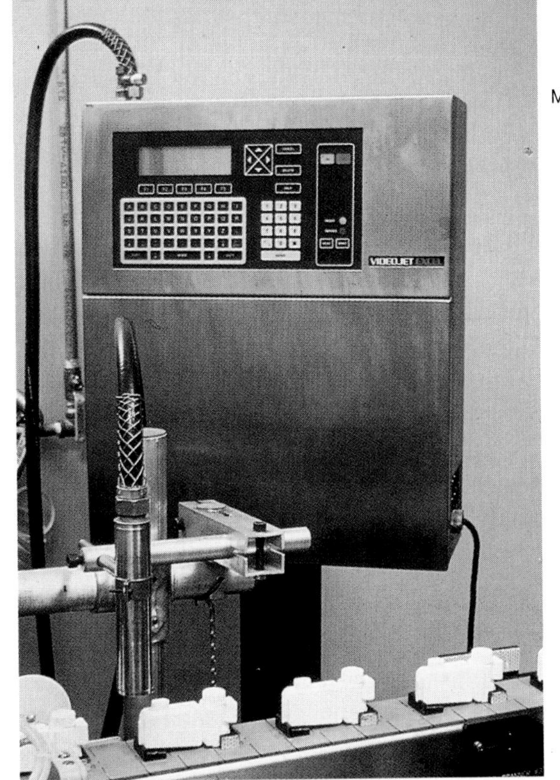

Fig. 15-26. This videojet printer is printing codes on cartridges used for medical tests.

PRODUCT CONVERSION

Product conversion refers to the wide range of operations performed on products after printing. Some examples include:

- trimming a press sheet to its final size
- folding and assembling a newspaper
- trimming, folding, assembling, and binding the pages of a book
- embossing, foil stamping, and die cutting a label
- forming a soft drink can
- cutting out, folding, and gluing a container

A few of these processes are discussed in the following.

Embossing, Foil Stamping, and Die Cutting

Embossing creates a raised image area on a substrate. Embossing is done by smashing the substrate between embossing dies. Fig. 15-27. The dies are three-dimensional molds that press their image into the paper under great pressure.

Foil stamping is similar to embossing. Colored metal foils and heat are used. The metal foil

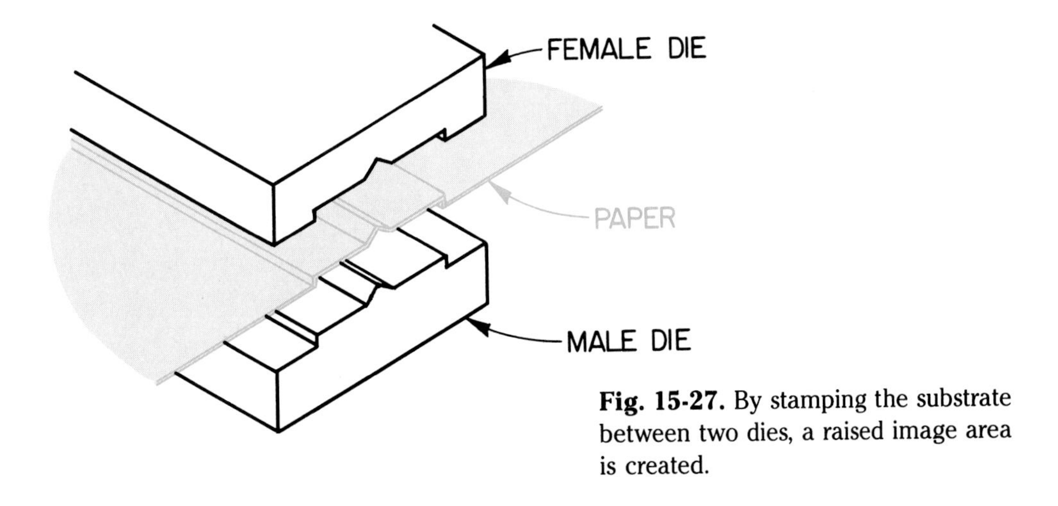

Fig. 15-27. By stamping the substrate between two dies, a raised image area is created.

image is fused to the substrate with heat and pressure.

Die cutting is to paper what cookie cutting is to cookie dough. Dies for die cutting consist of sharp knives mounted on a wooden base. Fig. 15-28. Part of the substrate is cut out by stamping it with the cutting die.

Each of these processes requires a lot of pressure between the die and substrate. Therefore letterpresses are often used because they provide the required pressure. High-quality greeting cards are made using embossing, foil stamping, and die cutting.

Thermography

Thermography, which means "writing with heat," produces a raised image area without embossing. The printed product is coated with a finely ground plastic powder before the ink has dried. The powder sticks to the image areas but can be removed from the non-image areas. The substrate is then carried under a heating device that fuses the powder into a smooth, glossy, raised image. Thermography is commonly used on wedding announcements and business cards.

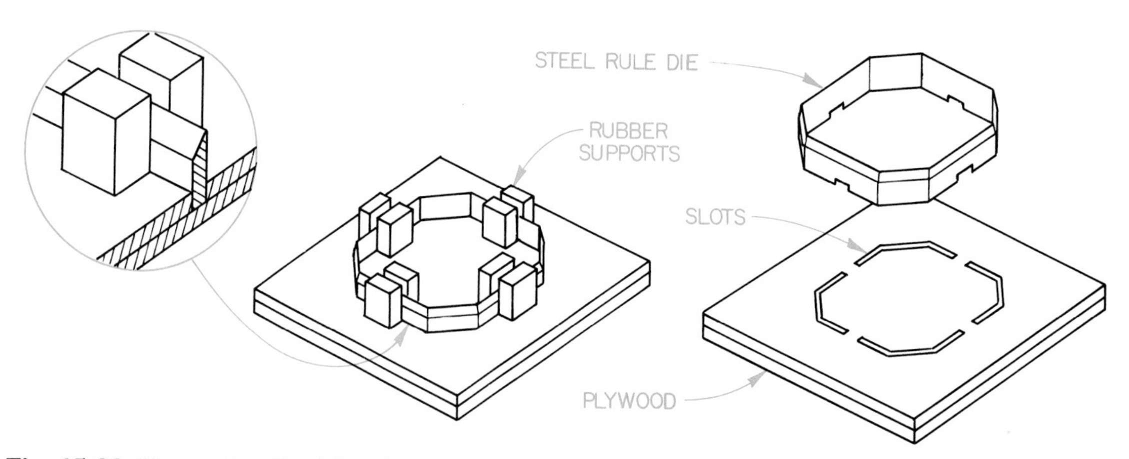

Fig. 15-28. The cutting dies (above) are held in place by rubber supports, which collapse when the die is smashed against the substrate. The die then cuts out the shape of the final product. Below are several examples of die-cut notecards.

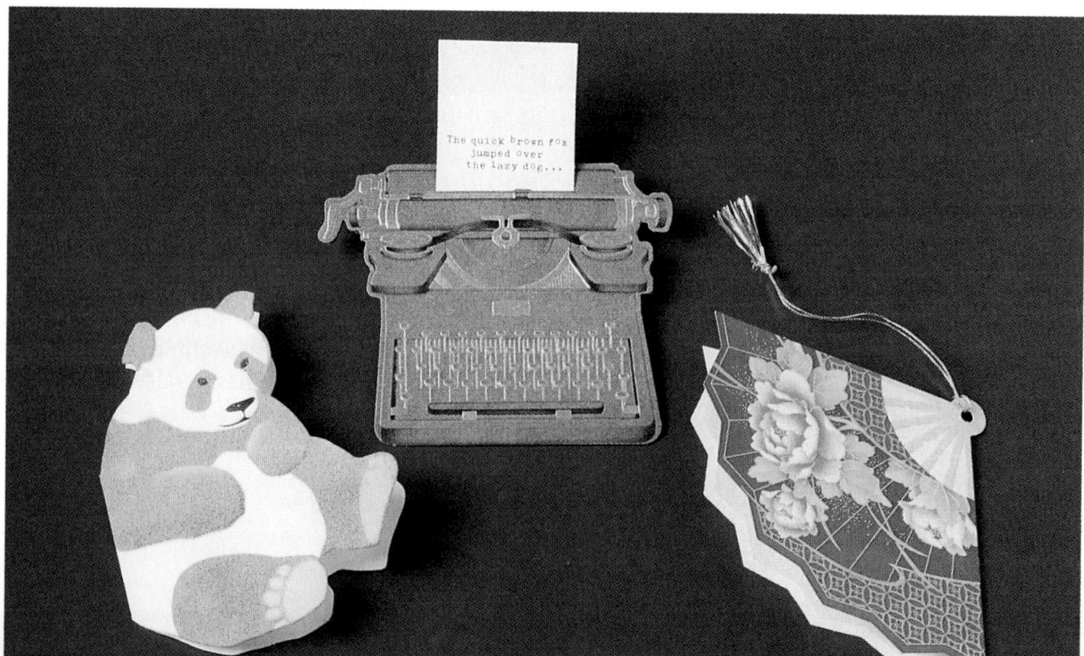

Assembling and Binding

Assembling is the process of bringing together printed sheets by collating, gathering, or inserting. Collating means assembling sheets in a special order. Gathering means assembling folded pages, known as signatures, one inside the other, as is done with magazines and books. Inserting means placing a single sheet inside a signature. For example, sometimes order forms or special advertisements are printed on heavy paper and inserted in a magazine. Fig. 15-29.

Binding is fastening assembled pages together permanently. Among the common binding methods are side stitching, saddle stitching, perfect binding, and case binding. Fig. 15-30.

Side stitching consists of stapling through the side of a collated product. Saddle stitching drives a wire staple through the back of signatures gathered together. These two techniques are best suited to products having few pages.

Perfect binding, also called adhesive binding, is the process of coating the back of assembled sheets or signatures with a hot glue. Generally, a soft cover is added while the glue is hot. The cover adds strength and a more finished appearance. Magazines and paperback books are often perfect bound.

Case binding creates a hardbound book. Signatures are assembled, sewn or glued together, and then given a rigid cover.

Folding, Drilling, and Cutting

Many kinds of folds are used to put a printed sheet in its final form. Folding machines range from small tabletop to large floor models.

A paper drill is basically a drill press with a hollow bit used to drill holes through a stack of paper. Drilling might be done on paper that will be used in a three-ring binder or on something hung on display in a store.

Paper is often cut to press sheet sizes from large basic sheet sizes before being printed. Press sheets are then generally trimmed to their final size after printing. Commercial paper cutters can cut large stacks of paper in a single pass.

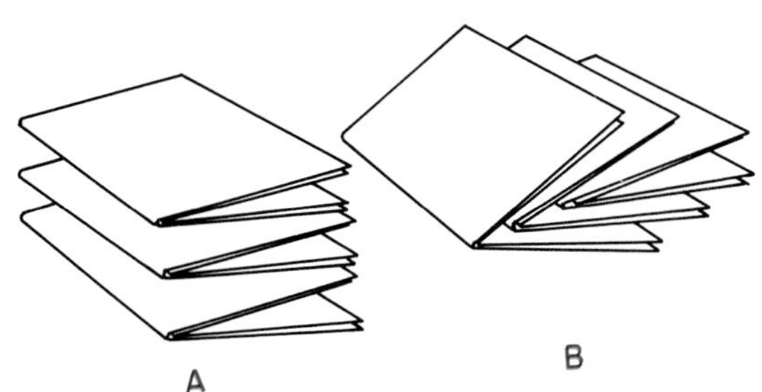

Fig. 15-29. Signatures may be gathered (A) or inserted (B).

Fig. 15-30. The common methods of binding are (A) side stitching, (B) saddle stitching, (C) perfect binding, and (D) case binding.

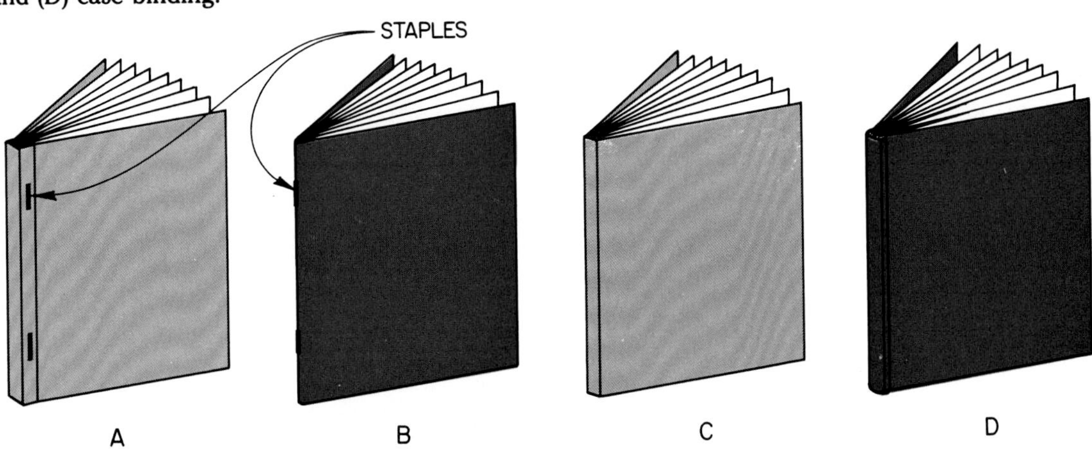

HEALTH & SAFETY

THE GOVERNMENT ACTS ON SAFETY

Twenty-five million Americans are exposed to chemical hazards each day. Because of this, Congress passed the Hazard Communication Act. The law requires employers to inform workers of possible hazards and take the necessary precautions.

Under this law, employers must clearly label any hazardous material. Information must be kept on file about all hazardous chemicals. In addition, employees must be trained to handle these materials safely.

The labels include the following information:

- name of the chemical
- the word "Caution," "Warning," or "Danger"
- a brief description of the major hazards, such as "extremely flammable"
- what to do to avoid the hazard, such as "wash thoroughly after handling"
- first aid instructions in the event of exposure to the chemical
- how to put out a fire caused by the chemical
- information for a physician on treatment
- directions for handling and storage

There are a number of chemicals used in the printing industry that fall under the Hazard Communication Act. Those used to develop film and the solvents used in the pressroom are good examples.

You should be aware of any possible hazards and be prepared to take precautions against harmful effects. Practice safe work habits, read labels carefully, and know basic first aid procedures. When in doubt, ask questions about potential hazards.

CHAPTER 15 REVIEW

Review Questions

1. Name five different types of offset plates.
2. Name and describe the six different operations that take place on an offset lithographic press.
3. Describe two different types of photographic screen stencils.
4. What is a photopolymer?
5. What is the purpose of the anilox roller in the flexographic printing process?
6. What is a gravure cell?
7. What is the purpose of the doctor blade in the gravure printing process?
8. How is the image transferred on an electrostatic printer?
9. Which finishing operations are often performed with letterpress equipment?
10. How are books assembled and bound?

Activities

1. Draw an image with a crayon on an offset plate supplied by your instructor. Sponge water over the plate. Next, roll ink over the plate. Try to transfer the image to a sheet of paper.
2. Using materials supplied by your instructor, stretch fabric in a screen printing frame.
3. Research photopolymers and write a report on your findings.
4. Find examples of side stitching, saddle stitching, perfect binding, and case binding. Make a display for the class.
5. Take a trip to a grocery or department store and try to find examples of each of the different message transfer processes. Report on your findings to the class.

Careers

Paper Sales Representative

When you want to buy a bicycle, you find a bicycle shop and talk to a salesperson. You find out which bike is best for you. When printers need to buy paper they go through the same process.

The paper industry produces about 700 million tons of paper a year—about $75 billion worth! Graphic communication industries change most of that into printed products, packaging materials, and publications. Someone has to know about the different types of papers available so customers select the right one for the job. Paper sales representatives do just that. As paper experts, they provide printers with the answers they need. They not only have to know all about paper but must be able to communicate well.

In 1986, the average salary for paper sales representatives was $46,000! It is a career for people who like working with people and knowing a lot about a certain product. The paper industry is one of the top ten in the United States, so there is a great future in paper sales!

Education

A broad understanding of graphic production systems is needed for a successful career in paper sales. A college or university that has a program in graphic communications, including such courses as paper science, printing management, or graphic communications technology, can provide this background. For a booklet listing all the schools that offer such programs, write the Graphic Arts Technical Foundation, 4615 Forbes Avenue, Pittsburgh, PA 15213. Ask for a copy of the publication *Technical Schools, Colleges, and Universities Offering Courses in Graphic Communications.*

Other Careers in Graphic Communications

Graphic designer
Layout artist
Printing management
Printing production
Graphic arts research scientist
Quality control supervisor
Customer relations
Printing sales
Job cost estimating
Desktop publishing

For More Information

To learn more about careers in graphic communications, contact:

Graphic Arts Technical Foundation
4615 Forbes Avenue
Pittsburgh, PA 15213

Gravure Association of America
90 Fifth Avenue
New York, NY 10011

Screen Printing Association International
10015 Main Street
Fairfax, VA 22031

Flexographic Technical Association
900 Marconi Avenue
Ronkonkoma, NY 11779

Correlations

Language Arts

1. One of the first steps in creating a graphic message is to write the copy (words). Write the copy that could be used to advertise an upcoming school function, such as a dance, sporting event, or other activity. Limit the length so that it will fit on an 8½ ″ × 11″ page. When you're finished, exchange papers with your classmates. Check the copy for errors in spelling, punctuation, and grammar. Is the message clear? Does it give all necessary information?
2. Look up the definitions for the following terms: calendering, colophon, duotone, hypothiosulphate, intaglio, logotype, posterization, ream, surprint. How does each relate to graphic production?

Science

Why do the exposed areas of lith film turn black when developed? Take a trip to the library and find out exactly what happens when photographic films are exposed to light and why they turn black in the developer. You may want to make your own photographic material as a science experiment.

Math

1. Photographs very often have to be enlarged or reduced to fit in a given space in a design. When you change one dimension, the other dimension is also changed proportionately. For example, let's say a photo measures 2″ wide by 4″ long. Your design calls for it to be 4″ wide. If you enlarge it that much, it will also become 8″ long. Calculate the percentage of enlargement or reduction necessary for the following and solve for the missing dimension.

Original Photo	Space on Layout
A. 8″ × 10″	2″ × ?
B. 5″ × 7″	7.5″ × ?
C. 4″ × 5″	5″ × ?
D. 11″ × 14″	2.75″ × ?
E. 2.5″ × 3.48″	5.75″ × ?

Social Studies

Printing has had a great impact on our lives. Create a list of ways in which printing has changed the society in which we live. You may wish to review the discussion "Categories of Impact" in Chapter 3.

Basic Activities

Basic Activity #1: Paste-Up/Mechanical

Most publications are not printed one page at a time. Instead, many pages are printed side by side on a large sheet of paper. The sheet is then folded and trimmed to produce a "signature." These signatures are then assembled with other signatures to create magazines and books.

With this in mind, find examples of the graphic elements listed below and paste them up to create a 16-page signature, as shown in Fig. V-1.

Materials and Equipment

11″ × 17″ sheet of white paper
non-reproducing blue pencil
wax or rubber cement
saddle stitcher or stapler

Procedure

1. Fold the 11″ × 17″ sheet into a 16-page right-angle book fold as shown in parts a through d of Fig. V-1.
2. Open up the sheet and draw non-reproducing blue lines along the folds to clearly indicate the outer edge of each page.
3. Number each page with a non-reproducing pencil *exactly* as it is numbered in Fig. V-1.
4. Cut the following items out of old magazines and carefully and accurately paste them on the sheet:

   All pages (except cover): page numbers in the bottom right or bottom left corner.
   Page 1 (cover): a title for the booklet that includes your name (you may use transfer lettering for this) and a color illustration of your choice.
   Page 2: a table of contents listing what appears on each page. This may be typeset, word processed on a microcomputer, or typewritten.
   Page 3: an illustration of formal balance.
   Page 4: an illustration of informal balance.
   Page 5: an illustration that shows emphasis.
   Page 6: a paragraph of Roman style text.
   Page 7: a paragraph of sans serif style text.
   Pages 8 & 9: a large black and white halftone that extends across both pages.
   Page 10: a line illustration.
   Page 11: a computer-generated graphic.
   Page 12: a continuous-tone image that is not a photograph.
   Page 13: an illustration with a screen tint.
   Page 14: a paragraph of script style type.
   Page 15: a paragraph of novelty style type.
   Page 16: a statement that identifies the class in which this booklet was created along with the name of the teacher, school, and date.
5. After the paste-up is complete, fold the sheet back up again and carefully trim off the fold edges at top and front, removing as little paper as possible.
6. Saddle stitch or staple the booklet together through the back fold.

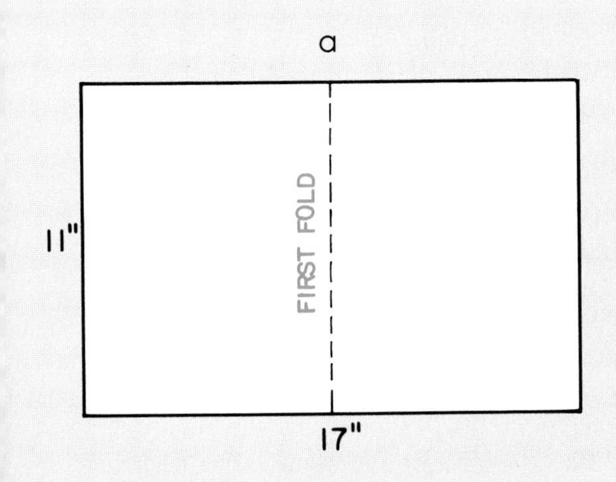
a
11"
FIRST FOLD
17"

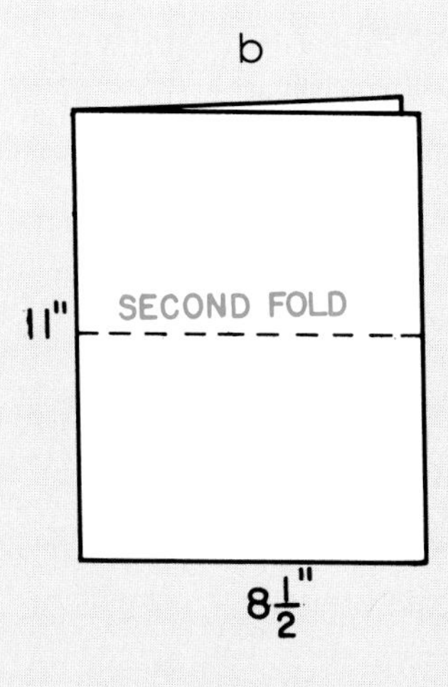
b
11"
SECOND FOLD
8 1/2"

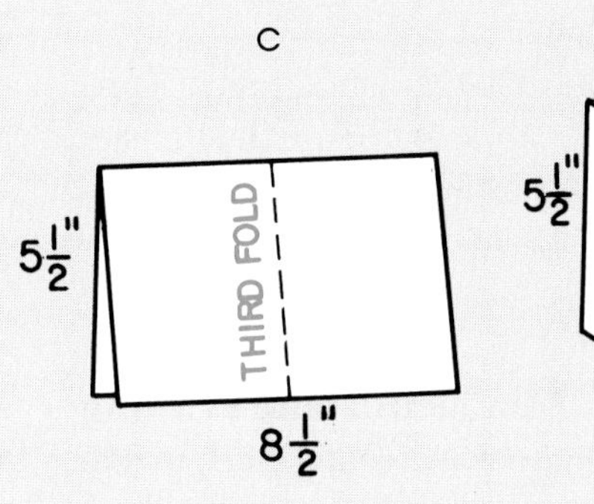
c
5 1/2"
THIRD FOLD
8 1/2"

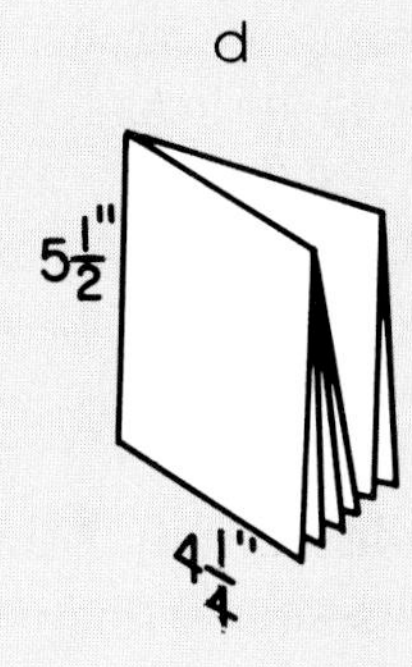
d
5 1/2"
4 1/4"

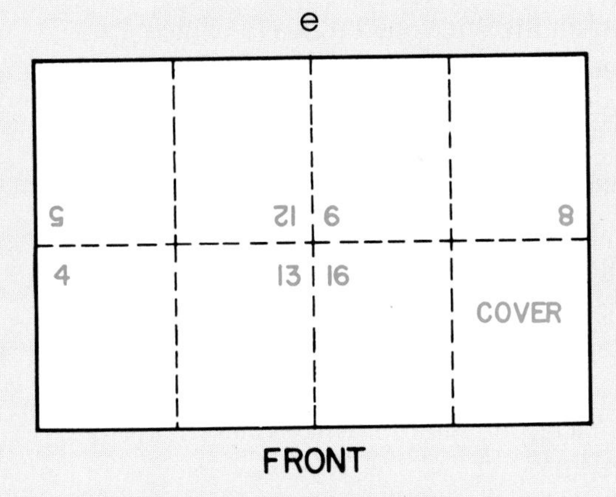
e
5
12
9
8
4
13
16
COVER
FRONT

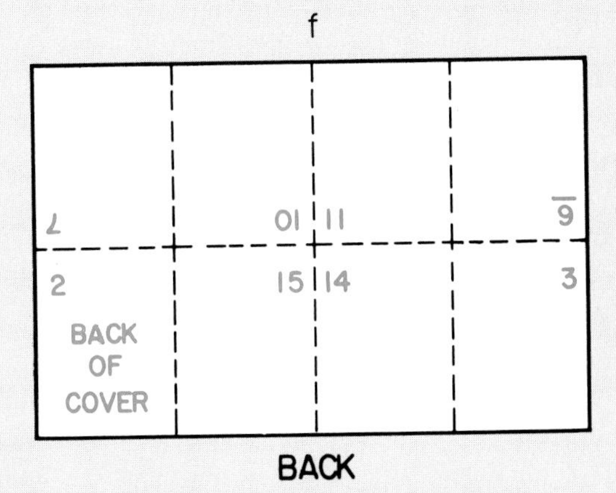
f
7
10
11
6
2
15
14
3
BACK OF COVER
BACK

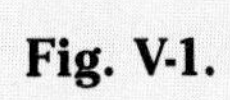

Fig. V-1.

Basic Activity #2: Line Negative

Using a page from Basic Activity #1 or line copy supplied by your instructor, you will make a line negative in the darkroom.

Materials and Equipment

line copy
process camera
lith developer
stop bath
fixer
12-step gray scale
heavy white paper

Procedure

1. Mount the line copy on heavy white paper and place the gray scale beside it. Put this on the center of the copy board on the process camera.
2. Make an exposure of the line copy on a sheet of orthochromatic (lith) film for the time indicated by your instructor.
3. Develop the film in lith developer until a solid Step 3 or Step 4 is reached on the gray scale.
4. Stop, fix, wash, and dry the line negative.
5. Scratch your name into the emulsion (wrong-reading) side.

Basic Activity #3: Computer Graphic Iron-On

Iron-on transfers are often used to decorate T-shirts and sweatshirts. In this activity, you will create a T-shirt iron-on using a computer graphics system.

Materials and Equipment

Apple II® or IBM PC® microcomputer system (or equivalent)
Print Shop® or Print Master® software (or equivalent)
dot matrix printer
Underware® brand ribbon for the dot matrix printer
T-shirt made of at least 50% synthetic fiber

Procedure

1. Boot the software and follow the menu directions to produce the design for the T-shirt.
2. Print the design on a dot matrix printer using an Underware® ribbon. This ribbon uses a special ink that may be ironed onto a 50% synthetic fabric.
3. Iron the image onto the T-shirt. Practice first on some scrap fabric to be sure the transfer process is working well.

Basic Activity #4: Electrostatic Printing

Design and produce a greeting card, invitation, or announcement using an electrostatic printing device (photocopier).

Materials and Equipment

several sheets of 8½″ × 11″ white paper
pencil
transfer letters
border tapes
clip art
electrostatic printing device (photocopier)

Procedure

1. Fold one sheet of paper into a "French fold," as shown in parts a through c of Fig. V-2.
2. Keeping the French fold, transfer letters, border tapes, and clip art in mind, produce a series of thumbnail sketches on scratch paper for a greeting card, invitation, or announcement.
3. Select the thumbnail sketch you like best and make a rough layout in pencil on the folded sheet of paper.
4. Using transfer letters, border tapes, and clip art, make a paste-up of the card. See part d of Fig. V-2 for positioning.
5. Reproduce the card using a photocopy machine.
6. Fold the card into its final form.

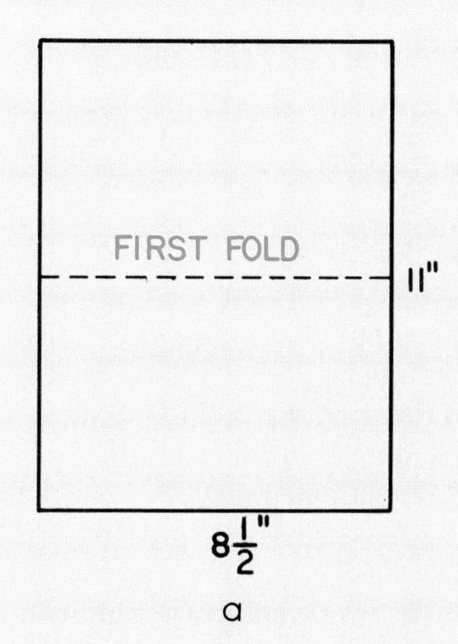

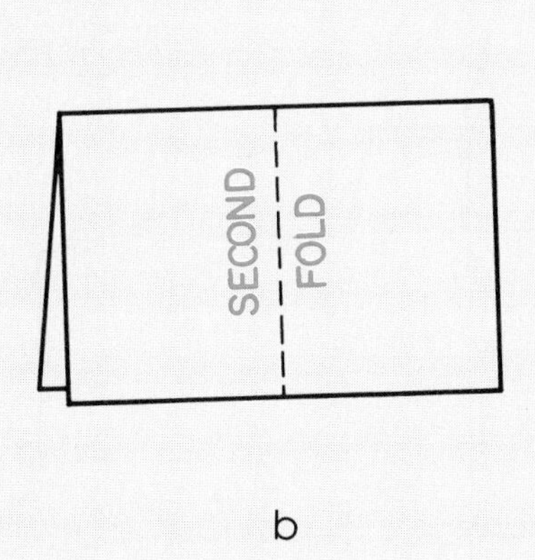

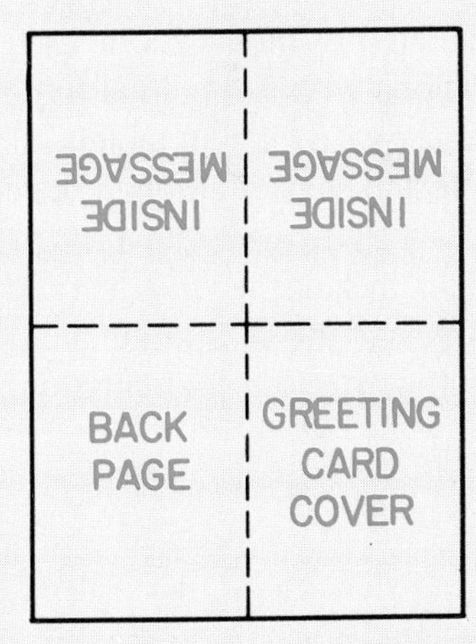

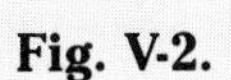
Fig. V-2.

Intermediate Activities

Intermediate Activity #1: Computer Graphics

Using a computer graphics system, design an 8½″ × 11″ eye-catching poster.

Materials and Equipment

computer graphics hardware, such as Apple Macintosh®, Apple IIe®, Apple GS®, IBM PC® (or equivalent)
dot matrix or laser printer
software, such as MacPaint®, Blazing Paddles®, PC Paint® (or equivalent)

Procedure

1. Using the computer graphics system, create three or more 2⅛″ × 2¾″ boxes.
2. Sketch possible locations for text and illustrations within the small boxes, using the mouse or digitizing tablet.
3. Select the best thumbnail sketch and, using it as a guide, generate a rough on the computer. Include the actual text and a reasonable sketch of the illustration.
4. Output this to a dot-matrix or laser printer.

Intermediate Activity #2: Package Prototype

Although we are told we should not "judge a book by its cover," we often buy things because of the package they are in. Design and produce a prototype of a package for a small bottle of perfume, cologne, or after-shave lotion.

Materials and Equipment

poster board or the equivalent
PMS color guide
PMS felt-tip marking pens
typesetter or transfer lettering
scissors
rubber cement

Procedure

1. Design the shape of the box by sketching layouts on scrap paper and cutting and folding them into final form. Be sure to include gluing tabs in the proper locations. See Fig. V-3 for an example of one type of carton.
2. After you have created a shape you like with enough gluing tabs, trace the outline of this pattern on a clean sheet of white paper.

Fig. V-3.

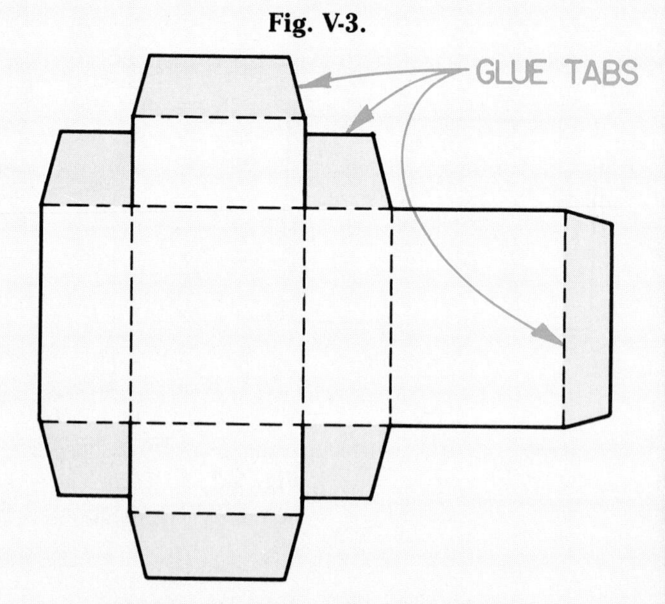

3. Make several photocopies of the outline.
4. Design the package decoration by sketching ideas on the photocopies.
5. When you settle upon a design you like, transfer the package and decoration layout to the posterboard.
6. Use PMS marking pens and typeset or transfer lettering to produce the final artwork for the package.
7. Score, fold, and glue the package prototype into its final form.

Intermediate Activity #3: Diffusion Transfer Halftones

Halftones are a method of reproducing photographs in printed materials. Make a diffusion transfer halftone of a black and white photograph.

Materials and Equipment

black and white photograph
diffusion transfer processor and activator
diffusion transfer negative and receiver material
diffusion transfer halftone screen
process camera
flash lamp

Procedure

1. Place a black and white photograph on the copy board of the process camera.
2. Adjust the process camera for the proper percentage of enlargement or reduction.
3. Place a sheet of diffusion transfer negative material on the camera back. Cover with a diffusion transfer halftone screen.
4. Make a "main" exposure for the time indicated by your instructor.
5. Open the camera back and make a "flash" exposure for the time indicated by your instructor.
6. Place a sheet of receiver material in contact with the negative material. Then process in a diffusion transfer processor.
7. Wait 30 seconds and peel the two sheets apart to reveal the halftone.

Intermediate Activity #4: Flexographic Stamp

Flexography is used to create labels as well as package designs. Produce a "rubber" stamp using flexographic plate material.

Materials and Equipment

process camera
lith film
PRINTIGHT® flexographic plate material
platemaker

Procedure

1. Locate or make the artwork for a small stamp.
2. Make a line negative of your artwork, reducing it so it will work as a stamp (about 4″ square or less).
3. Remove the protective cover film on the PRINTIGHT®.
4. Expose the film through the line negative on a platemaker. Ask your instructor for the correct exposure time.
5. Wash the unexposed plate material away by spraying it with water at 68°F (20°C) for several minutes.

6. Dry the plate with a hair dryer or let it air dry.
7. Re-expose the plate in the platemaker for 5 minutes.
8. Mount the flexographic plate on a flexographic press, a letterpress, or a small scrap of wood.
9. Print the flexographic stamp on a suitable sheet of paper.

Advanced Activities

Advanced Activity #1: Screen Printed Posterizations

Working from a black and white photograph, screen print a two- or three-color posterization.

Materials and Equipment

black and white photograph
gray scale
process camera
lith film
photo-indirect stencil material and developer
platemaker
screen printing frame and squeegee
screen printing ink
paper

Procedure

1. Place a black and white photograph on the copy board of the process camera. Place a 12-step gray scale next to it.
2. Make a line exposure on lith film for the normal line exposure time.
3. Develop the exposed film in lith developer until a solid Step 3 on the gray scale is reached.
4. On a second sheet of lith film, make a second exposure for three times the normal line exposure.
5. Develop this in lith developer until a solid Step 8 on the gray scale is reached.
6. Make contact positives of these negatives. An alternative method is to put them back on the process camera copy board with a white sheet of paper behind them and make normal line exposures. This process will create film positives.
7. Use these contact or film positives to expose photoindirect stencils for each positive.
8. Process the stencils.
9. Paste the stencils to a screen and mask the screen.
10. First, print the stencil with the most open area in a lighter color. Be sure to register the paper carefully with registration tabs.
11. Print the second color in register with the first.

Advanced Activity #2: Desktop Publishing

Create a promotional brochure using a desktop publishing system.

Materials and Equipment

computer hardware, such as Apple Macintosh®, IBM PC®, Apple IIe® (or equivalent system)
desktop publishing software
offset lithography equipment and supplies

Procedure

1. Identify an organization or group in your school that is in need of a promotional brochure. Ask your teacher which groups might need your help.
2. Work with the group to produce the necessary text, artwork, and photographs for the brochure.
3. Working closely with the group, prepare a rough layout for the brochure.
4. Keyboard all text using Microsoft Word® software (or equivalent).
5. Generate computer graphics using the proper software.
6. Using PageMaker® (or equivalent) software, merge all text and graphics to create and output camera-ready copy for the brochure. Leave windows for halftones that will be added later.
7. Following your teacher's instructions, digitize each of the photographs you intend to use or make diffusion transfer halftones of each.
8. Import the scanned halftones or paste-up the diffusion transfer halftones.
9. Reproduce the brochure using offset lithographic methods. Have your teacher show you the procedures.

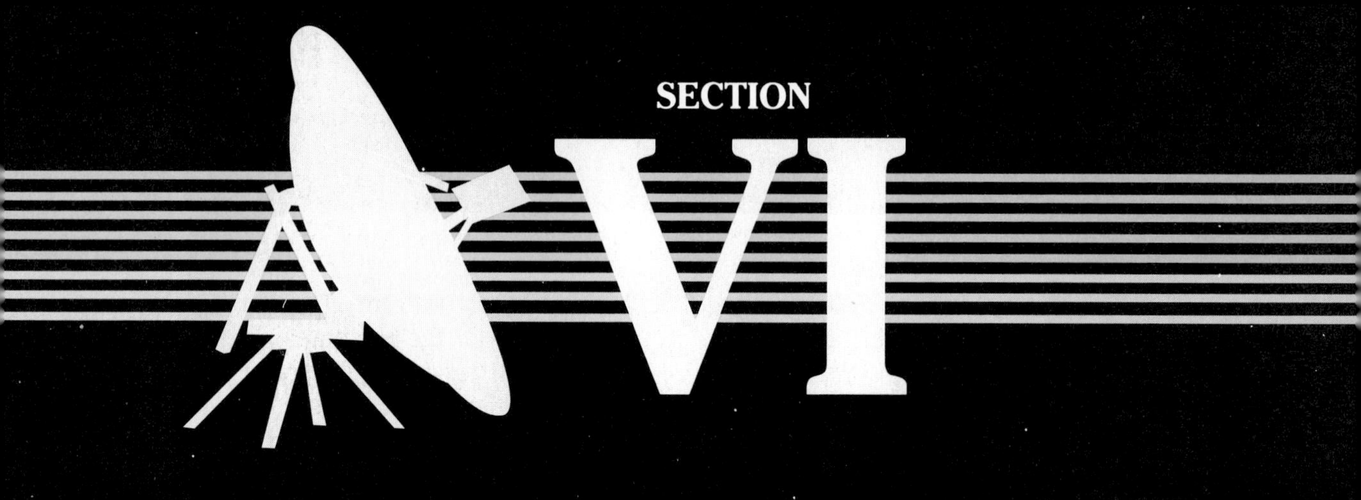

Audio and Video Systems

Thomas Edison, the brilliant inventor who gave us the light bulb, also invented the phonograph and the microphone. Even he found it hard to imagine how these inventions and others in the communications field would affect future lives.

In 1880 Edison said, "The phonograph . . . is not of any commercial value." He was equally gloomy about sound in movies. In 1913 he insisted, "The talking motion picture will not supplant the regular silent motion picture . . . There is such a tremendous investment in pantomime pictures that it would be absurd to disturb it." In 1922 he predicted, "The radio craze . . . will die out in time."

It is hard to believe that someone of Edison's brilliance could be so wrong. All three inventions have endured to enrich the modern world. What would our lives be like today without radio, records, or "talking motion pictures?"

Today audio and video technologies surround us, yet few people understand these technologies or how the equipment works. Of course, you don't have to know how things work in order to use them. However, knowing how things work will help you to buy products wisely and use them correctly.

In this section, you'll learn about the different types of audio and video communication systems that you come in contact with every day. You'll have a better idea of why and how they work. You may even enjoy using them more as a result.

CHAPTER 16

Principles of Audio and Video Communication

How much do you know about your television set? Have you any idea how the picture and sound find their way into your TV? What about the telephone? How does your voice become a message traveling down the wire? How is it that someone hears your voice at the other end?

Both TV and the telephone belong to the category of audio and video communication systems. (*Audio* is something we hear, and *video* is something we see.) A listing of audio and video systems would include the telegraph, telephone, record player, radio, and television. In this chapter, you will read about the scientific principles upon which audio and video communications are based.

Terms to Learn

alternating current
amplitude modulation
atmospheric transmission channels
electrical circuit
electromagnetism
frequency bands
frequency modulation
induction
physical transmission channels
radio waves
receiver
transmission channel
transmitter

As you read and study this chapter, you will find answers to questions such as:

- How do electricity and magnetism form the basis for audio and video communication?
- How is an electronic message sent and received?
- What are radio waves, and how are they used to transmit audio and video from one place to another?

HOW ELECTRONIC COMMUNICATIONS WORK

As you've seen in earlier chapters, communication models help us to understand abstract ideas. A basic electronic communication model consists simply of a transmitter, transmission channel, and receiver. Fig. 16-1. The **transmitter** sends the message. The **transmission channel** carries it. The **receiver** receives it. During a telephone conversation, for example, your voice is sent by the mouthpiece on your own phone (transmitter) through a wire (transmission channel) to the earpiece of another telephone (receiver).

The model tells us how electronic communications in general work. To really understand the different systems, you need to learn some basics about energy conversion, electricity, and magnetism.

Energy Conversion

All electronic communications begin with energy in one form or another. Audio communications, for example, start out as mechanical energy—sound waves made up of moving air. Video communications start as light energy. These forms of energy must be changed to electrical energy for use in a communication system.

In audio systems, sound waves (mechanical energy) are changed into electrical energy. In video systems, video cameras convert light to electrical energy. The electrical energy in the form of a signal is then transmitted.

Sometimes a signal is combined with another, stronger signal that "carries" it through the air for long distances. In radio communication, for example, "radio" waves carry the sound signal from the source to the receiver. Often, the signal must be adjusted in additional ways to cut down on noise (interference) or to give it strength.

These signals are sent through either the atmosphere or a wire or special fiber. They are then received by an antenna or other device.

At the receiving end, the process is reversed. The signal is decoded and changed back into an energy form that is usable. Fig. 16-1. In the case of radio, for example, the signal is changed back to its original form—sound.

Fig. 16-1. Energy, such as sound, is converted into an electrical signal, amplified, transmitted either by physical or atmospheric channels, decoded, and then converted back to sound.

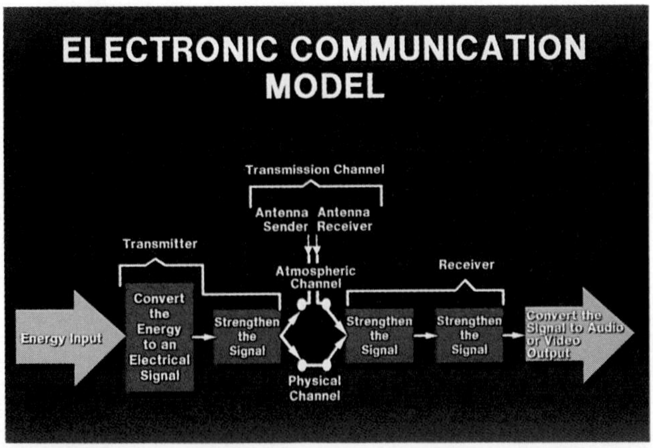

SCIENCE FACTS

How Your Ears Work

Sound—voices, a car engine, or a rock band—is a series of waves in the air. Inside your ear is a narrow passage that guides these waves to your eardrum. Your eardrum is a thin membrane that vibrates in tune with the sound waves. These vibrations are channeled to a system of nerves that send signals to your brain. Your brain understands these signals as the sounds you hear.

A wide range of sound waves exists in nature. Humans can hear waves produced at a rate of 40 to 15,000 per second. Sound waves travel through the air at about 770 miles per hour.

Electricity is the flow of free electrons. As you learned in Chapter 10, electrons are very tiny particles that circle the nucleus of an atom. Fig. 16-2. The nucleus of the atom is positively charged. The electrons are negatively charged. Positive and negative charges are attracted to one another. This attraction helps hold atoms together. However, the atoms of some materials will give up some electrons under certain conditions. As these electrons escape, energy in the form of electric current is released. All electronic communications are based on the movement of electrons.

Electricity is usually channeled through materials that allow it to pass easily. These are called conductors. Copper wire is an example of a good conductor. This is why copper wire is often used in communication systems. The pressure or force that moves the current through the wire is called voltage.

An **electrical circuit** is the path electricity follows from its source, through a conductor, to a receiving device. For example, a flashlight circuit contains a battery (source), wire (conductor), and lightbulb (receiving device). Fig. 16-3.

Electromagnetism

Some forms of communication, like the telephone, send messages by means of current traveling through a wire. Others, like the radio, must depend upon electromagnetic waves. These waves make communication without a connecting wire possible.

What are electromagnetic waves? You've no doubt seen the effects of a magnet. You can't really see the magnetism itself. We describe magnetism as a "field" or force. It is this magnetic field that allows a magnet to pick up a paper clip. Electricity may also be used to create a magnetic field. This form of magnetism is called **electromagnetism**. Any time electricity passes through a conductor, an invisible electromagnetic field is created.

For example, when electricity flows through copper wire, a magnetic field develops around the wire. You can see this by moving a magnetic compass in a plane perpendicular to a wire carrying current traveling in one direction. Fig. 16-4.

Fig. 16-2. When electrons escape an atom, they leave in the form of electricity.

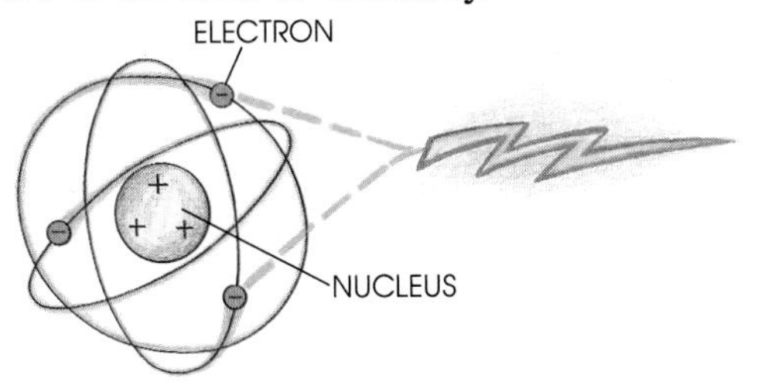

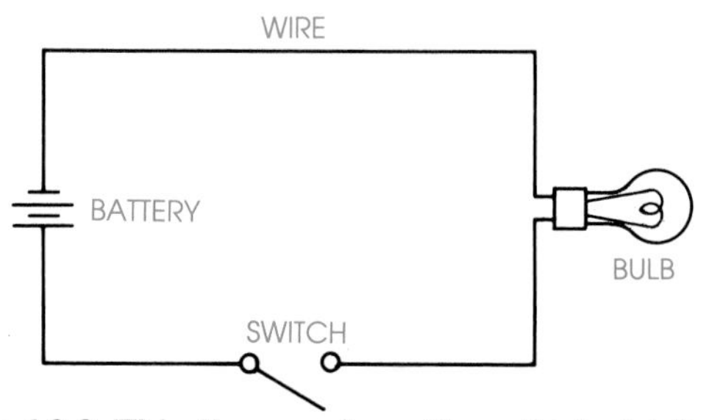

Fig. 16-3. This diagram shows the path electricity takes in a simple flashlight.

Fig. 16-4. As electricity flows through the wire, the compass needle follows the lines of magnetic force.

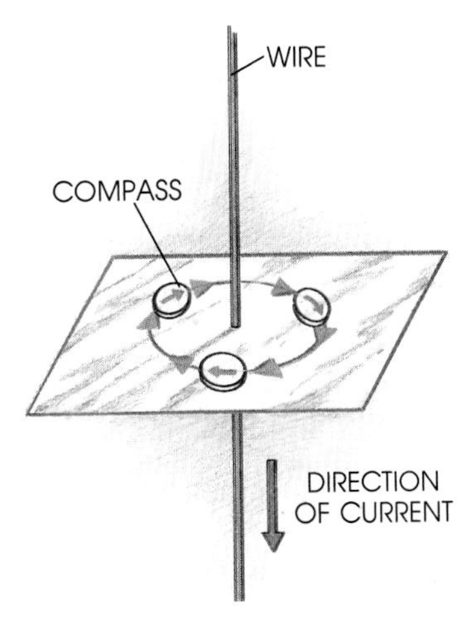

TECHNOLINKS

A MESSAGE THAT CHANGED THE WORLD

Can you read this?

•-- •••• •- -
•••• •- - ••••
--• •• -••
•-- ••• •• ••- --• •••• - -••-•

In 1844 the above message was sent from Washington, D.C., to Baltimore, Maryland, by telegraph, and nothing has been quite the same since.

The earliest practical telegraph was invented by an Englishman, Sir Charles Wheatstone, in 1837. But the model that caught on was invented by Samuel F. B. Morse, an American. Morse also developed Morse code. Morse code is a system of signals that made sending messages on the telegraph possible. The message shown above is in Morse code. It marked the beginning of teiecommunication in America. See if you can read it using the code symbols shown here.

A	•-	P	•••••
B	-•••	Q	••-•
C	•• •	R	• ••
D	-••	S	•••
E	•	T	-
F	•-•	U	••-
G	--•	V	•••-
H	••••	W	•--
I	••	X	•-••
J	-•-•	Y	•• ••
K	-•-	Z	••• •
L	—	Period	••--••
M	--	Comma	•-•-
N	-•	Question Mark	- ••-•
O	• •		

The first telegraph was little more than a battery, a sending device known as a "key," a wire, and a receiver. The key was a switch that opened and closed a circuit. When the circuit was closed, a pulse of current flowed through the line and activated the "sounder" on the receiving end. The sounder made a clicking noise. In Morse code, short clicks represented dots, and longer clicks were dashes. The letter "A" was dot-dash, "B" was dash-dot-dot-dot, and so forth. Using this code, Morse sent the message shown at the beginning of this article. It said, "What hath God wrought?"

The telegraph system rapidly became a major means of mass communication. By 1856, the Western Union Telegraph Company linked many parts of the United States. In 1866, after ten years of trying, a telegraph cable was successfully laid across the Atlantic Ocean, connecting America and Europe.

The telegraph was also the basis for other communication technologies. The teletype, for example, changed the code into typewritten letters. It became popular with newspapers. You will learn about some of these technologies in the next chapter.

The compass will always point *around* the wire, never toward it or away from it. The compass is following the magnetic lines of force.

If you wrap the wire around an iron core, such as a nail, the electromagnetic field gets much stronger. Fig. 16-5. As long as electricity flows through the wire, the magnetic field continues. When the current is turned off, the field collapses. You can see this at work if you move an electromagnet near a paper clip. When the current is on, the magnet attracts the clip. Turn it off and the clip falls away.

Induction

While electricity can create a magnetic field, the reverse can also happen. A magnetic field can create electricity. If you move a wire across a magnet, you can create an electric current in the wire. This is called **induction**. We say that a current has been "induced" in the wire by the magnet. This also happens if a magnet is moved close to a wire. Fig. 16-6.

Alternating Current

When a wire passes through a magnetic field, the electrons (current) flow in one direction. This is called direct current. If the wire is coiled and then rotated in the magnetic field, the electrons flow first in one direction and then in the opposite direction. The current produced is called **alternating current**.

Each change in direction is called a cycle. It is possible to create alternating current that reverses itself thousands and even billions of times per second. The electromagnetic waves resulting from such rapid cycles travel outward at great distances and can be used for radio and TV signals. The waves move at the speed of light.

Radio Waves

For radio and TV signals, a transmitting antenna carrying alternating current releases electromagnetic waves into the atmosphere. Although these electromagnetic waves are used for many types of broadcasting, they are commonly called **radio waves**. These waves may travel hundreds of miles. A receiving antenna picks up the waves. A weaker electric current, just like the one that started the waves, is induced in the receiving antenna.

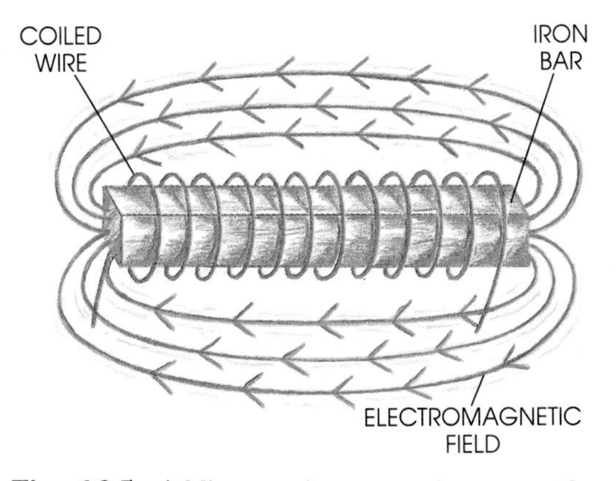

Fig. 16-5. Adding an iron core increases the strength of the magnetic field.

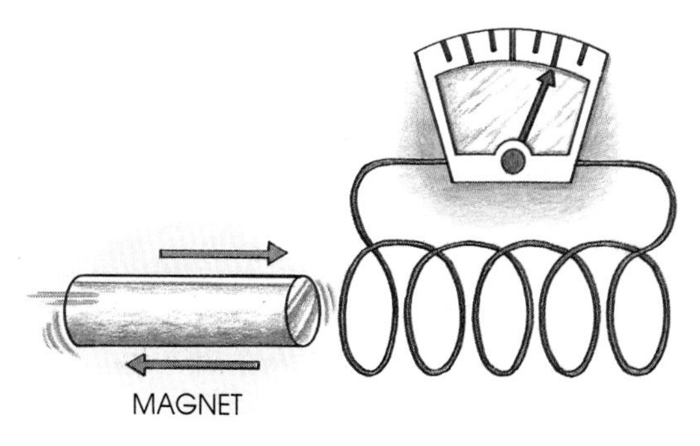

Fig. 16-6. As the magnet is moved close to the wire, current begins to flow.

Amplitude and Frequency

All wave forms—sound waves, radio waves, water waves—have both amplitude and frequency. Fig. 16-7. The strength of a wave is its amplitude. It is measured from its midpoint to its peak. A wave's length is measured from a point on the first wave to the same point on the next. The number of waves leaving a source each second is the frequency.

The basic unit of measure for radio wave frequency is one cycle per second, or one hertz (Hz). A kilohertz (kHz) is 1,000 cycles per second. A megahertz (MHz) is one million cycles per second. A gigahertz (GHz) is one billion cycles per second. The atmosphere is filled with radio waves of all different frequencies. To avoid confusion, the Federal Communications Commission (FCC) assigns each station its own frequency.

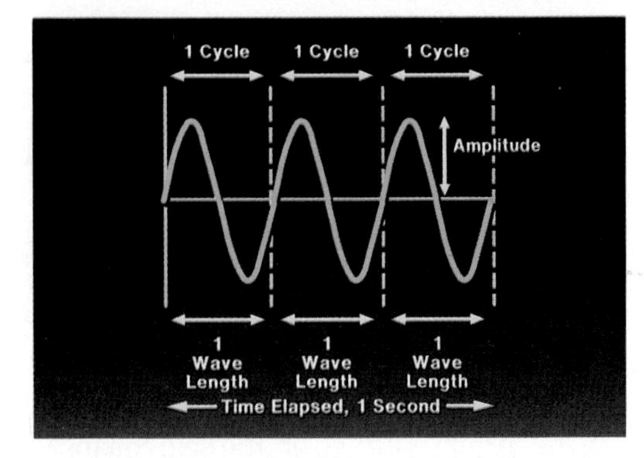

Fig. 16-7. Wave frequency refers to the number of cycles per second (hertz). The diagram shows a wave with a frequency of 3 hertz. The strength (height) of the wave is its amplitude.

Frequency Bands

Radio waves range in frequency from about 30 to 300 billion hertz (30 Hz to 300 GHz). All of these frequencies can carry information. To keep track of them, they are divided into ten **frequency bands**. Fig. 16-8. You have probably heard of VHF (very high frequency) and UHF (ultrahigh frequency). These are the frequency bands used for the channels on your television set.

Certain frequencies are used for specific types of audio or video communication. Fig. 16-9. There are two reasons for this. First, some frequencies are affected more by such things as weather and changes in the ionosphere. (The ionosphere is a layer of electrically charged particles from 60-200 miles above the surface of the earth.) During the day, medium frequency (MF) bands are good only for fairly short distances of several hundred miles or less. In the evening, however, the ionosphere loses some of its electrical charge. Then communications in the MF range may travel halfway around the world!

Secondly, the government regulates broadcast communication. As mentioned before, the FCC assigns frequencies. When you listen to AM radio, for example, you tune your dial to somewhere between 540 and 1600 kHz. The FM stations fall between 88.1 and 107.9 MHz. Channels 2-6 on your television set are assigned 54-88 MHz, while channels 7-13 are limited to the 174-216 MHz frequencies.

Modulation

When left alone, radio waves have a constant amplitude and frequency. They would sound like noise if you heard them on a radio. In order to send messages by radio waves, people change them. Altering radio waves so that they carry messages, or "intelligence," is known as modulation.

Radio-Frequency Bands		
30 — 300 Hz	Extremely low frequencies	ELF
300 — 3 kHz	Voice frequency	VF
3 — 30 kHz	Very low frequencies	VLF
30 — 300 kHz	Low frequencies	LF
300 — 3000 kHz	Medium frequencies	MF
3 — 30 MHz	High frequencies	HF
30 — 300 MHz	Very high frequencies	VHF
300 — 3000 MHz	Ultrahigh frequencies	UHF
3 — 30 GHz	Super high frequencies	SHF
30 — 300 GHz	Extremely high frequencies	EHF

Fig. 16-8. Frequency bands range from extremely low — 30 hertz — to extremely high — 300 billion hertz.

For example, in radio broadcasting, the sound vibrations to be transmitted (Fig. 16-10,A) are impressed on the radio, or carrier, waves (Fig. 16-10,B). First, the sound waves are changed (in a microphone, for example) to electrical signals. Then the low frequency sound signal is combined with the high frequency carrier signal.

In Fig. 16-10(C) the amplitude, or strength, of the carrier wave has been changed. The wave has received **amplitude modulation**. This signal can be received by an AM (amplitude modulation) radio. Frequency may also be modulated as in Fig. 16-10(D). During **frequency modulation** the waves are crowded together or spread farther apart. These signals are received by FM (frequency modulation) radios. For television, AM is used for the picture signal and FM for the sound signal.

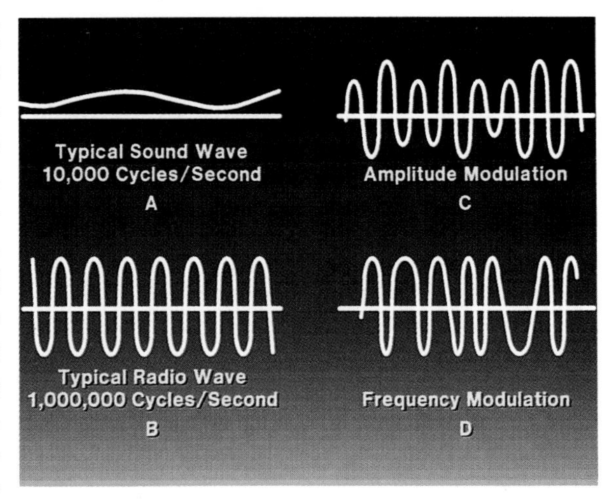

Fig. 16-10. A typical sound wave (A) is combined with a typical radio wave (B). The carrier (radio) wave may receive amplitude modulation (C) or frequency modulation (D).

Fig. 16-9. Can you find radio or television frequencies that you use on this chart?

Frequency	Description
30 Hz — 18 kHz	Range audible to human ear
10.02 — 13.6 kHz	OMEGA: US Navy submarine communication
100 kHz	LORAN-C: Long Range Navigation for ships and airplanes
285 — 325 kHz	Radiobeacons used for navigation by ships and airplanes; 2000 mile range
540 — 1600 kHz	AM radio (worldwide)
1.665 — 1.770 MHz	Older portable telephones
2.182 MHz	International marine emergencies
5.950 — 6.200 MHz 7.100 — 7.300 MHz 9.500 — 9.900 MHz 11.650 — 12.050 MHz	Shortwave radio bands (national and international)
26.965 — 27.405 MHz	Citizen's Band (CB) radio; has traditionally been the "trucker's" channel; channel 23 is for radio controlled garage door openers; intended for local communication
30 — 49 MHz	State police or highway patrol communication
46.610 — 46.970 MHz	Portable telephones and walkie-talkies (up to ½ mile)
54.00 — 88.00 MHz	VHF television; channels 2 — 6
88.1 — 107.9 MHz	FM broadcast radio; 200 channels
108 — 135 MHz	Aircraft operations and navigation
136 — 138 MHz	Weather satellites broadcasting weather maps
144 — 148 MHz	Most popular amateur radio (ham) band
156.050 — 157.425 MHz	Local marine communication
162 MHz	NOAA (national weather service)
162 — 174 MHz	Federal agencies (US Customs, FBI, FCC, NASA)
174 — 216 MHz	VHF television; channels 7 — 13
470 — 890 MHz	UHF television; channels 14 — 83
1 GHz plus	Microwaves; radar

Transmission Channels

When you make a telephone call to a friend down the street, your voice travels as an electrical signal along a wire to your friend's phone. The telephone wire is the channel used to transmit your voice. If your friend takes a trip to Hong Kong, you can still place a call to her or him by telephone, but your call will be sent through the air and relayed by a satellite. Audio or video communication may be transmitted in either of these two ways.

Atmospheric transmission channels use electromagnetic waves to carry information through the atmosphere. The waves originate from an antenna and may be relayed by a satellite. Radios, TV, and cellular phones all depend on this type of channel.

Physical transmission channels use a wire or some other connection between sender and receiver. Ordinary telephones and cable TV use this type of transmission channel. More about both types of channels will be discussed in Chapter 17.

PUTTING IT ALL TOGETHER

Now, let's go back to the communication systems model (Fig. 16-1) discussed on p. 348. How does what you have just learned fit into this model?

As you know, all audio and video systems can be described using the systems model. Let's use the radio as an example. Fig. 16-11. A disc jockey speaks, creating sound. The sound is changed into electrical energy and transmitted through the air in the form of radio waves. These waves reach the antenna of a radio. The waves' frequency is induced in the antenna. Inside the radio receiver they are changed back into sound.

In Chapter 17 you will learn how telephone, radio, television, record player, and audio and video tape systems work. Even though the systems are all different in some ways, they have the communication model in common. If you understand the model, you have a basic understanding of the systems themselves. Chapter 17 discusses the equipment that each system uses to transmit, channel, and receive the message.

Fig. 16-11. What other examples can you think of that will fit into this systems model?

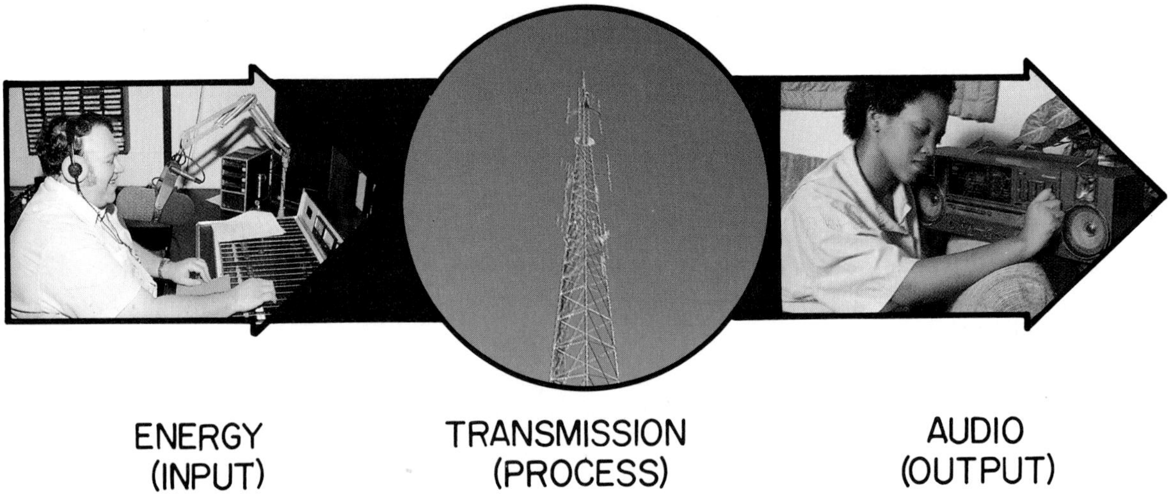

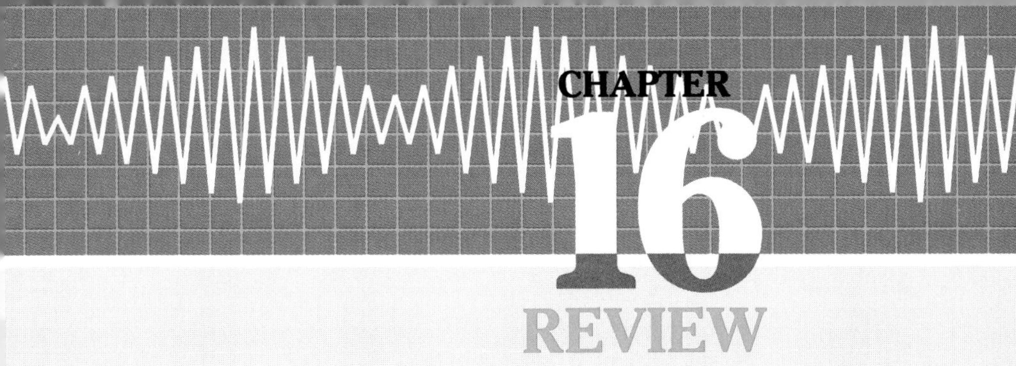

CHAPTER 16 REVIEW

Review Questions

1. What are the three general parts of an electronic communications model?
2. Describe electrical current and circuits.
3. Describe electromagnetism.
4. What is induction? How is it important to communication?
5. What is the difference between alternating and direct current?
6. What is the connection between alternating current and electromagnetic waves?
7. Define frequency and amplitude.
8. What does modulation mean? Describe the types of modulation.
9. What are the two types of transmission channels? How do they differ?
10. Explain how radio waves carry messages through the atmosphere.

Activities

1. Build a simple circuit with a flashlight bulb, copper wire, and a battery. Using Morse code, send a message to a friend across the room with this device.
2. Make an electromagnet using a battery, copper wire, and an iron nail. Demonstrate how the magnetic field is affected by the flow of electric current.
3. List by frequency the local AM radio stations you receive. Draw a line representing the AM radio frequency spectrum (540 kHz to 1600 kHz) and plot each of the local AM radio stations on this line.
4. Run a coil of copper wire through a sheet of paper. Sprinkle iron filings on the paper near the wire. Connect each end of the wire to the terminals of a battery. What happens to the filings? Draw a diagram showing the magnetic field.
5. Research and write a report on how the development of the telegraph affected our country's growth.

Audio and Video Equipment

Did you know that a TV picture is made of many dots of light? Did you know communication satellites are powered by the sun?

In Chapter 16 you learned about the scientific basis for audio and video communication. In this chapter you will discover how the equipment itself works.

Terms to Learn

amplifier
audio consoles
cartridge
diaphragm
frequency response
microwaves
mixing
multiplexing
parabolic reflectors
pickup tube
recording heads
signal plate
switcher
target
video formats

As you read and study this chapter, you will find answers to questions such as:

- In a telephone, how are sound waves converted to electrical signals and then back to sound waves?
- How does a radio work?
- How do satellites aid in radio transmission?
- How does a television camera convert what it "sees" into electrical signals?
- How are audio and video recorders alike?

THE TELEPHONE

On February 14, 1876, two Americans, Elisha Gray and Alexander Graham Bell, filed for separate patents. Each had developed a design for a telephone. After a long legal battle, Bell was awarded the patent rights. People had already learned the advantages of the telegraph. Thus, this new method of telecommunication, the telephone, was welcomed. Fig. 17-1.

In the early years, all phones had to be linked directly to one another with iron wire. By 1900, however, the signal was routed to intermediate points first. This made wiring simpler. Today's telephones are vastly different! Features such as automatic redialing, message recording, and memory storage of often-used numbers are all available.

Telephone Transmission

A telephone transmitter consists of the mouthpiece and the "body" of the phone where the dial is located. In the mouthpiece is a microphone that picks up the sound of your voice.

As you learned in the last chapter, sound waves must be changed to electrical signals before they are sent over long distances. That's what a microphone does. However, the microphone that picks up your voice in a telephone is very different from the microphone that a singer uses on stage.

The most important variation in microphones is their frequency response. **Frequency response** refers to the range of sound frequencies a microphone can reproduce well. The microphone in a telephone can reproduce only a small range. It is called a carbon microphone. Fig. 17-2.

The carbon microphone is made of a small cup filled with carbon granules. A small amount of electrical current constantly flows through the granules. Next to the cup containing carbon granules is a flexible piece of metal called a **diaphragm**. The diaphragm vibrates when sound waves strike it. The vibrating diaphragm presses against the carbon granules. This causes more electrical current to flow. When pressure is released, less current flows. This creates a changing electrical signal. The changing signal represents the sound as it is transmitted.

As you know, all telephones have their own number. When you dial a telephone number on a rotary phone, an electric switch inside the phone connects, then breaks, a circuit. It connects and breaks once for "1" twice for "2," and so on. Push-button phones send signals of different frequencies to represent each number dialed. Any device that can send those frequencies can dial a number. This is how many computer systems "dial" the telephone.

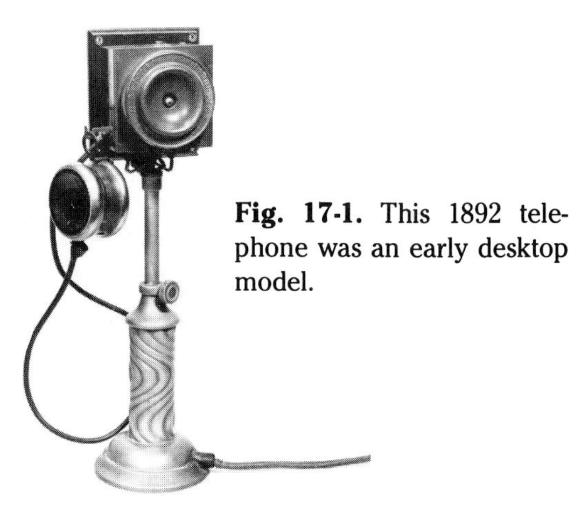

Fig. 17-1. This 1892 telephone was an early desktop model.

Fig. 17-2. When the diaphragm vibrates against the carbon granules in this microphone, a changing voltage is created.

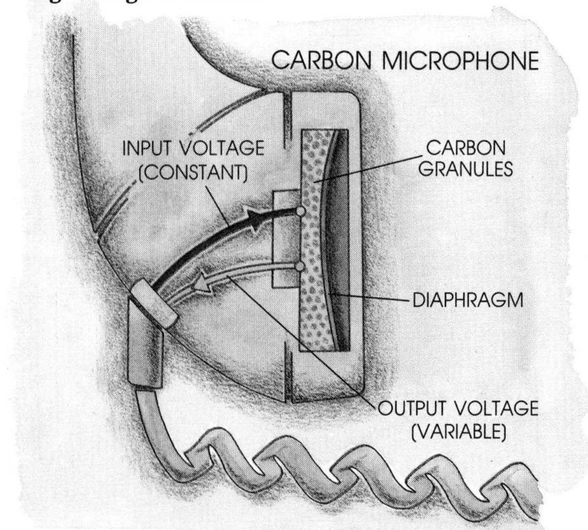

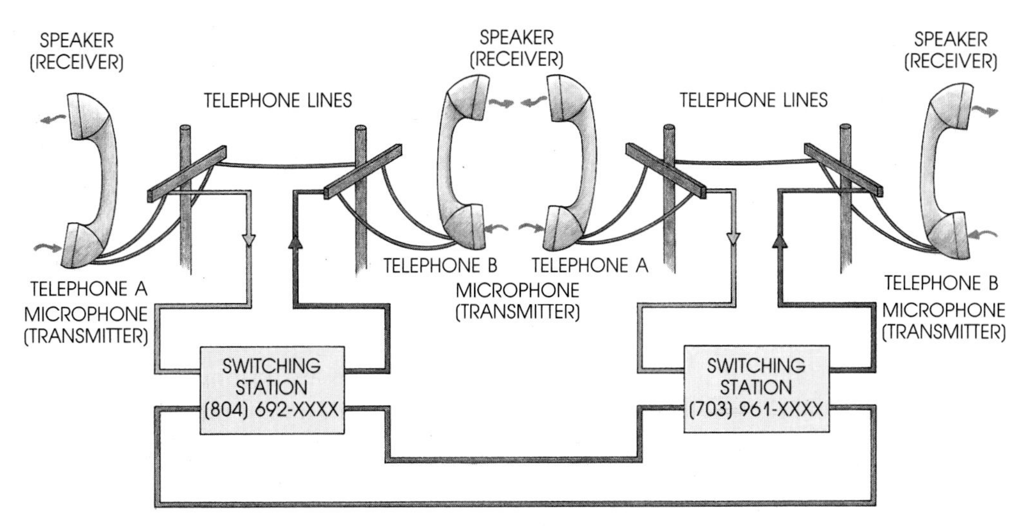

Fig. 17-3. Each switching station has an area code (the first three digits) and an exchange (the next three digits). The last four digits identify the individual phone.

Transmission Channels

Most telephone calls depend on physical transmission channels. These consist of wires, fibers, and cables.

The signal travels through a wire leading from your phone to a cable connected to your building. This cable leads to the switching station that supplies power for your local system. Fig. 17-3. The station routes your call to another switching station. The second station is indicated by the first three numbers in the number dialed. (This becomes the first six numbers if it is long distance). The last four numbers indicate the specific telephone connected to that station. All of this switching happens automatically.

When your phone is "on the hook," the circuit between it and the switching station is open. When someone calls you, the station sends a low-voltage current along the circuit that causes your phone to ring. When you pick up the receiver, the circuit is closed, which tells the station that your phone has been answered. The station then switches the call to your phone.

Copper Wire Channels

Twisted-pair wire is commonly used for local telephone transmission. This consists of two thin, insulated, copper wires twisted around each other. Twisted-pair wires may also be bundled together to form large cables that stretch all across the United States. Central switching points may be connected to each other with cables of copper wire. Cables run from these central locations to the telephones in individual homes.

Optical Fiber Channels

Optical fiber, a thin flexible glass tube, is now being used for some long distance transmission. (See Chapter 10, "Fiber Optics.") The signal travels as pulses of light. Fiber optic cables can carry many more messages than copper wire cables. Fig. 17-4. There is also less distortion of the signal. Fiber optic cables are replacing copper wire as the preferred channel.

Fig. 17-4. Many optical fibers may be run through a single cable. A core of steel gives the cable strength.

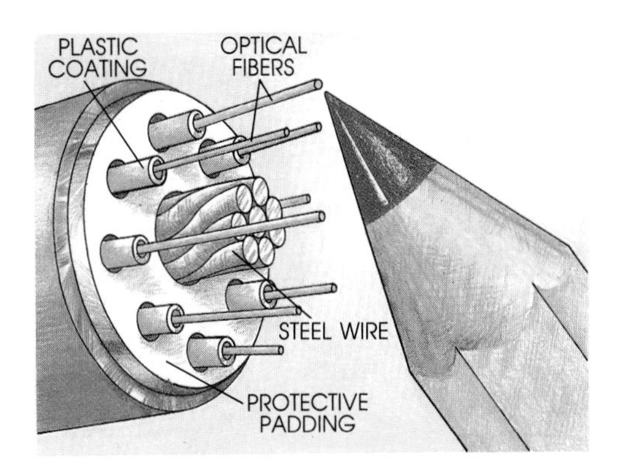

The light source used in optical fiber is a laser. (Lasers are also discussed in detail in Chapter 10.) The laser is generated from a microchip, or laser chip. Fig. 17-5. When excited by an electrical current, these chips give off laser light that is modulated to carry information. The modulated signal is then focused into the fiber, which transmits it. At the receiving end, the signal is demodulated and changed back into the sound of a voice.

Multiplexing

If only one signal at a time could be sent over a physical transmission channel, very little communication could ever take place. Imagine if you had to wait until all lines were clear before you could make a phone call. You'd probably never get to make it. **Multiplexing** allows two or more signals to be sent over a transmission channel at the same time.

There are two different ways of doing this. The first method, frequency division multiplexing (FDM), divides the channel into two or more frequency bands. Each message is sent using a carrier signal with a different frequency. This is much like radio broadcasting, but multiplexing takes place inside a cable or wire.

A second common method is time division multiplexing (TDM). TDM is usually used in digital transmission. Bits of data are sent at set time intervals. For example, if three computers sent data at once, the first might send during the first time slot, the second during the next time slot, and then the third during the third time slot. This process is repeated as long as messages are being sent. In actual practice, the data are sent so fast there seems to be no delay in communication.

Atmospheric Transmission Channels

Telephones are no longer limited to physical transmission. Portable phones, for instance, are actually low-powered radio transmitters/receivers. They send and receive messages electromagnetically.

Two types of transmission channels may be used for one call. A telephone call begun on a wire system may be changed to a microwave signal and sent to an atmospheric receiving station. **Microwaves** are electromagnetic waves that are shorter than radio waves but longer than infrared waves. They are more focused than radio waves.

Reception

Receivers work much like transmitters in reverse. A typical telephone receiver is located in the earpiece. Fig. 17-6. It consists of a coil of wire wound around an iron core. Together they form an electromagnet. Connected to the iron is a flexible metal diaphragm. When the electrical signal enters the receiver, it travels through the coil. This magnetizes the iron, which pulls on the diaphragm. The diaphragm vibrates and reproduces the sound.

Fig. 17-5. A laser beam created by a microchip is modulated to carry information. The beam then travels through the fiber optic cable.

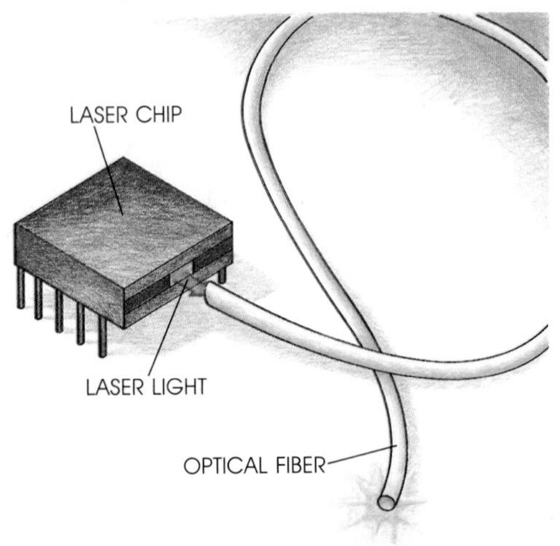

Fig. 17-6. The electromagnet inside a receiver causes the diaphragm to vibrate. This vibration reproduces the sound.

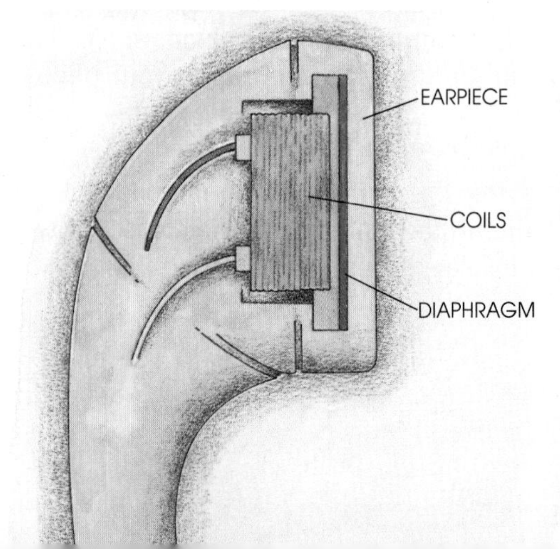

User's Guide to Technology

CHOOSING TELEPHONES AND PHONE SERVICES

It's a lot tougher to choose a telephone and telephone service options than it used to be! Not long ago, you had only a few simple decisions to make. Look at the choices you have today:

- *Regular desk and wall telephones* still exist, but they are getting scarcer every day. They come in many sizes, shapes, and colors.
- *Cordless telephones* use battery power and a small antenna to send signals to another antenna attached to the base unit, which rests on a table or wall. The base unit is plugged into the standard telephone service. Cordless phones allow you to walk and talk at the same time, but don't get too far from the "base station, " or the conversation will get a little fuzzy.
- *Cellular telephones* are like tiny radio transmitters. They transmit voice signals to a local "cell site" antenna, generally located within 25 miles of the phone. The cell site transmits the phone call to a "Mobile Telephone Switching Office" (MTSO). The MTSO relays the call to its destination. As the number of MTSOs increases, cellular telephones will become more widespread.
- *Videophones* allow you to see the person with whom you are talking. For those who already have a computer connected to the Internet, a low-cost videophone can be created by connecting a simple black and white (or color) video camera to the computer. A set-up such as this, along with the proper "video conferencing" software turns an ordinary computer into a videophone.

After you've decided on the telephone you want to buy, there are many services from which to choose. Some are available only to those with touch-tone telephones, so keep that in mind. As with the telephone itself, it's best to choose only those services you really need. Although costs for individual services may be small—perhaps only $2.50 a month—they can quickly add up.

- Call waiting. If you're on the phone, you get a signal that tells you another caller is trying to reach you.
- Call forwarding. If you're going to be away from home, your calls are automatically forwarded to a number where you can be reached.
- Threeway calling. Do you want to talk to two people in different places at the same time? This feature allows you to do so.
- TDD (Telecommunications Device for the Deaf). The TDD has a keyboard and monitor, enabling hearing-impaired people to communicate with other TDD users. If one party does not have a TDD, people at relay services can forward the messages to regular phones.
- Multi-ring service. Several phone numbers can use one line. A different ring sounds for each number.
- Call identification. With the aid of a video display device, you can learn the phone number of incoming calls.

If you have questions about your phone needs, call your telephone company representative. Someone will be glad to help you.

THE RADIO

A radio can send and receive signals without a connecting wire. In fact, a radio used to be called a "wireless." The invention of the radio made possible rapid communication over very long distances.

No one person was responsible for inventing the radio as we know it. An Italian, Guglielmo Marconi, built a device that in 1897 sent and received signals over a distance of four miles. Improvements on Marconi's device followed rapidly. One such improvement was the vacuum tube. Vacuum tubes greatly strengthened a weak signal so that it traveled farther and more people received it. Soon everyone wanted a radio. Fig. 17-7.

The radio industry took off in 1920 when a station in Pittsburgh produced the first program. This program broadcast the returns of the presidential election of that year. Then entertainers became interested in doing programs. Soon, listening to the radio became a national pastime.

Radio Transmission

The equipment discussed in this section transmits the radio signal. As you read, keep in mind that the various devices are linked together. In general, the process works like this.

Microphones change sound energy into an electrical signal. In a control room, audio engineers alter or combine this signal with other signals using an audio console. The completed signal is then strengthened and sent to the transmitter. At the transmitter, electromagnetic carrier waves are created and modulated with the audio signal. The combined signal is strengthened and sent to the antenna, which releases it into the atmosphere.

Microphones

Several types of microphones may be used for radio broadcasting as well as other forms of communication. They include dynamic, ribbon, crystal, and condenser microphones. Condenser microphones are currently the most widely used.

A condenser microphone has a metal diaphragm mounted near a fixed metal plate. The metal plate is electrically charged. Fig. 17-8. As sound waves strike the diaphragm, it vibrates. The vibration causes a change in voltage in the plate. The changing voltage represents the signal.

Condenser microphones have excellent frequency response. They are also well suited to miniaturization and are found on many cassette recorders.

Other microphones work in similar ways. Most involve a diaphragm of some kind and a changing voltage. Someone who moves around a great deal, such as a news reporter on location, may

Fig. 17-7. This early radio was called a crystal set. It used a piezoelectric crystal. Piezoelectric crystals produce a small voltage when pressure is applied to them.

Fig. 17-8. A condenser microphone uses a fixed metal plate to create the changing voltage for the signal.

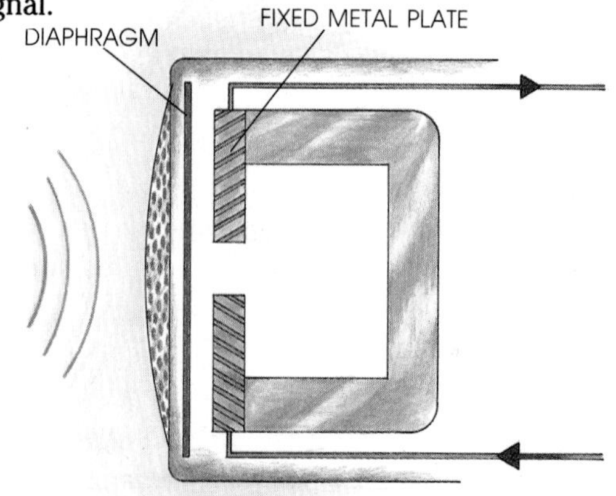

use a wireless microphone. The wireless microphone, like the portable telephone, contains a tiny battery-powered transmitter.

Control Room

The sound engineers at a radio station work in a soundproof audio control room. This room usually contains a patch board, a sound monitoring system, and an audio console. Fig. 17-9. The patch board is used to connect different input and output devices. The sound monitoring system allows engineers to monitor the different audio information being used or created.

Audio consoles, or mixers, come in many shapes and sizes, but they all do the same thing. They allow an engineer to control the volume and quality of incoming audio. Controls on the audio console make the following possible:

- Incoming volume is raised or lowered with sliding switches that allow delicate adjustments.
- Sounds can be balanced. Sound from a singer can be made louder and those from the band softer, for example.
- The amount of high (treble), midrange, and low (bass) frequencies can be equalized.
- "Echo" and "reverb" controls can add an echo-like effect.
- A number of sound inputs, such as those from several singers and instruments, can be combined into one signal and controlled with one switch.
- The overall strength of the signal output to a recording device can be monitored on a meter.

Most radio programs combine prerecorded and live sounds. Bringing together live sounds, prerecorded music, special audio effects, voices recorded over existing audio, and so on, is called **mixing**. Mixing is a challenging task for the audio engineer.

Recording live music, for example, presents unique problems. Ideally, the band sets up in a recording studio next to a control room where the engineer can monitor and record the music. All of the singers and instruments are assigned their own microphones. Volume and equalizer levels are set for each of the instruments.

The final recording is created by building up layers of sound, one over the other. For example, the instruments may be recorded in one session and the vocal track added later. The singers hear the instrumental playback as they record their vocals. This makes it easier on the audio engineer, since it makes mixing simpler.

Special sound effects are often added this way. They may be created in the studio or input from special-effects recordings.

Amplifiers

The sound signal is sent from the control room to an amplifier. An amplifier is a device used to strengthen an electrical signal. It contains transistors or other components that can control and increase the level of an audio signal without changing its waveform. Several different kinds of amplifiers are used during both transmission and reception.

Fig. 17-9. This audio engineer is at the audio console, also called a mixing console. The console allows her to adjust or change the sound in many different ways.

Transmitter

At the transmitter the electromagnetic waves used to carry the audio signal are created. Fig. 17-10. This is done by channeling direct current to a device called an oscillator. The oscillator changes the direct current to alternating current having a constant frequency and amplitude. The alternating current creates the carrier wave.

The audio signal is combined with the carrier wave. The signal then may be amplified again before being sent to the transmitting antenna.

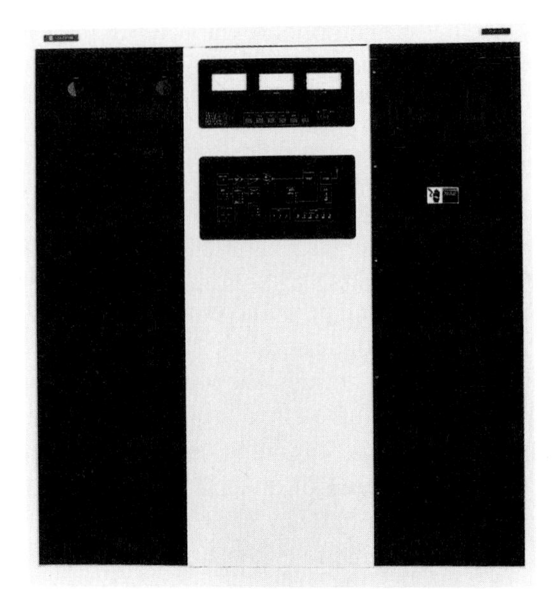

Fig. 17-10. This digital solid-state AM transmitter contains power amplifiers and a modulator.

Antennas

Antennas are used either to transmit or receive electromagnetic signals. Transmitting antennas may be towers or parabolic reflectors. Fig. 17-11. Radio waves leaving towers scatter in all directions. **Parabolic reflectors** ("dish" antennas) can transmit radio waves in straight lines toward a target.

Fig. 17-11. Antennas are used both to transmit and receive signals. They may be towers (left) or parabolic reflectors (right).

Transmission Channels

As you already know, radio signals are sent using atmospheric transmission channels. Atmospheric channels require no cable between the sender and receiver. Electromagnetic waves are launched into the atmosphere by the transmitting antenna. Three different types of waves may be generated: direct, ground, and sky. Fig. 17-12.

Direct waves travel in a straight line from point to point. Microwave transmitters are an example of direct wave systems. The microwaves may be sent directly to a receiving dish or relayed from a satellite. Fig. 17-13. Microwave dishes on towers are generally spaced about ten miles apart, depending upon the terrain.

Ground waves follow the curvature of the earth. They can travel several hundred miles before they weaken.

Sky waves radiate toward space. Lower-frequency sky waves are reflected back toward earth by the ionosphere. Shortwave radio broadcasting depends upon sky waves. When conditions are good, it is possible for sky waves to travel all the way around the world.

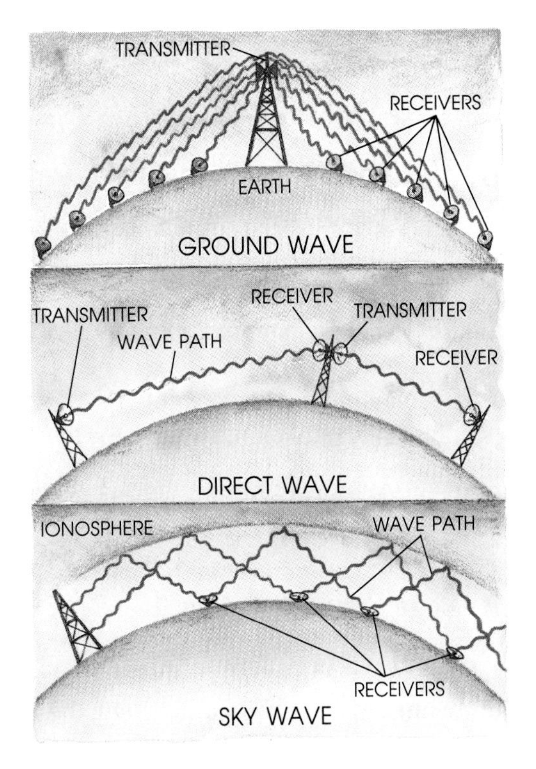

Fig. 17-12. Ground waves follow the curvature of the earth. Direct waves travel in a straight path. Sky waves are reflected back by the ionosphere.

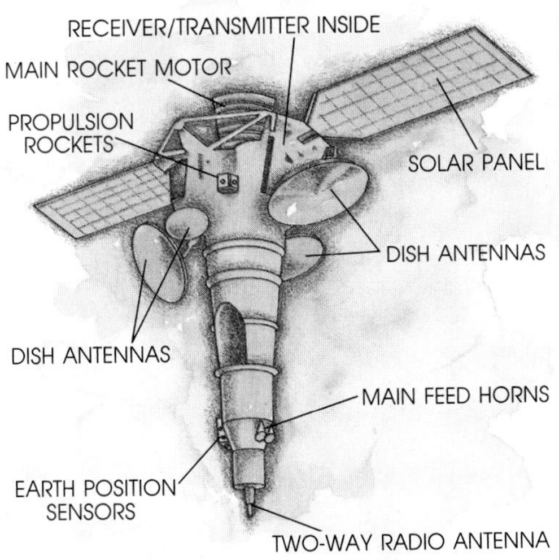

Fig. 17-13. This satellite collects radio signals with one set of antennas, amplifies them, and then transmits them with a different set of antennas. The solar panels collect energy from the sun, which is then used to operate the satellite.

Reception

A typical radio receiver includes an antenna, RF amplifier, mixer, IF amplifier, detector, AF amplifier, and speaker.

A receiving antenna is similar to a transmitting antenna. However, it collects, rather than transmits, electromagnetic radiation. Generally speaking, any length of wire can act as an antenna. Receiving antennas come in several shapes. Fig. 17-14.

Dipole antennas are lightweight and expensive. Their length is often one-fourth the length of the wave received. The arrow-shaped Yagi antenna can be pointed in the direction from which a signal is coming.

The radio waves induced in the antenna are channeled to three different amplifiers. The RF amplifier first selects the specific frequency to which the radio dial is tuned. For example, if you are tuned to AM radio station 1540, the RF amplifier selects only the radio waves that travel at this frequency.

Next, the RF amplifier strengthens the signal and passes it along to the mixer. The mixer converts the incoming signal to an intermediate frequency (IF), such as 455 kHz. Whether you tuned to AM station 540 or 1600, the mixer always changes the signal to 455 kHz. This is done because the signal must be amplified again. It is easier for a receiver to amplify one standard intermediate frequency than many different ones. The IF signal is then amplified again by an IF amplifier.

At this point, the signal is still in the form of a modulated carrier wave. It must be demodulated. This is done by the detector. The resulting signal is once again strengthened, this time by an audio frequency (AF) amplifier. It is then sent to the speaker.

Speakers

The speaker changes the signal back into sound. The most common is the moving coil dynamic speaker. There are two types: permanent magnet (PM) and electrodynamic.

The PM speaker consists of a permanent magnet and a voice coil mounted behind a speaker cone. Fig. 17-15. The electrical signal enters the voice coil. This produces a changing magnetic field that causes the voice coil to vibrate. Since the voice coil is connected to the flexible paper speaker cone, the cone vibrates as well. This vibration produces the sound waves that we hear.

Fig. 17-14. These basic antenna designs are used for home radio and television reception.

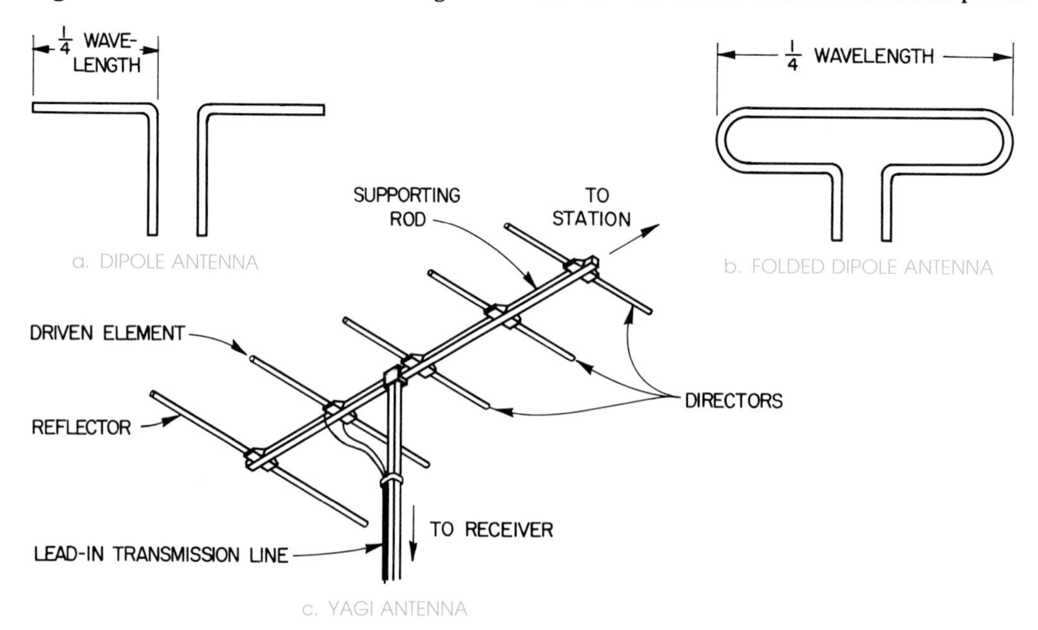

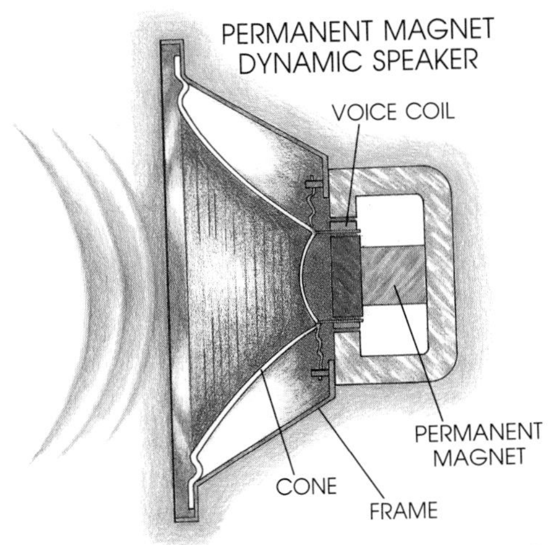

Fig. 17-15. The vibrations created in this speaker cone make the sounds we hear.

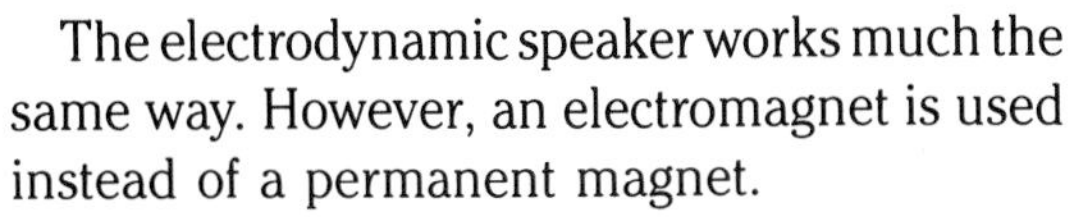

Fig. 17-16. These are only two of the designs in which speakers are made.

The electrodynamic speaker works much the same way. However, an electromagnet is used instead of a permanent magnet.

In both, the speaker cone is important. Treated paper of different kinds is used. Softer blotter-like cones reproduce bass sounds. Harder papers do a better job of reproducing the higher tones. In general, the larger the cone, the better the bass tones and the greater the speaker's power. Flared speaker cones do a better job with higher tones than straight-sided cones. Fig. 17-16. The higher the tone being reproduced, the smaller the cone area needed around the voice coil.

Not all speakers use a cone to generate sound waves. The "horn" type uses its long tapered shape to capture a column of air set in motion by a moving voice coil. The horn itself does not vibrate the way the cone does.

All of these factors are taken into account when speakers are designed. As a result, there are many different sizes, shapes, and styles. For example, some speakers have cones made of two or more different materials. Other cones are corrugated or have both a cone and a horn in one unit. Fig. 17-17.

Fig. 17-17. A woofer produces bass sounds, a tweeter treble sounds, and a midrange speaker those sounds in between.

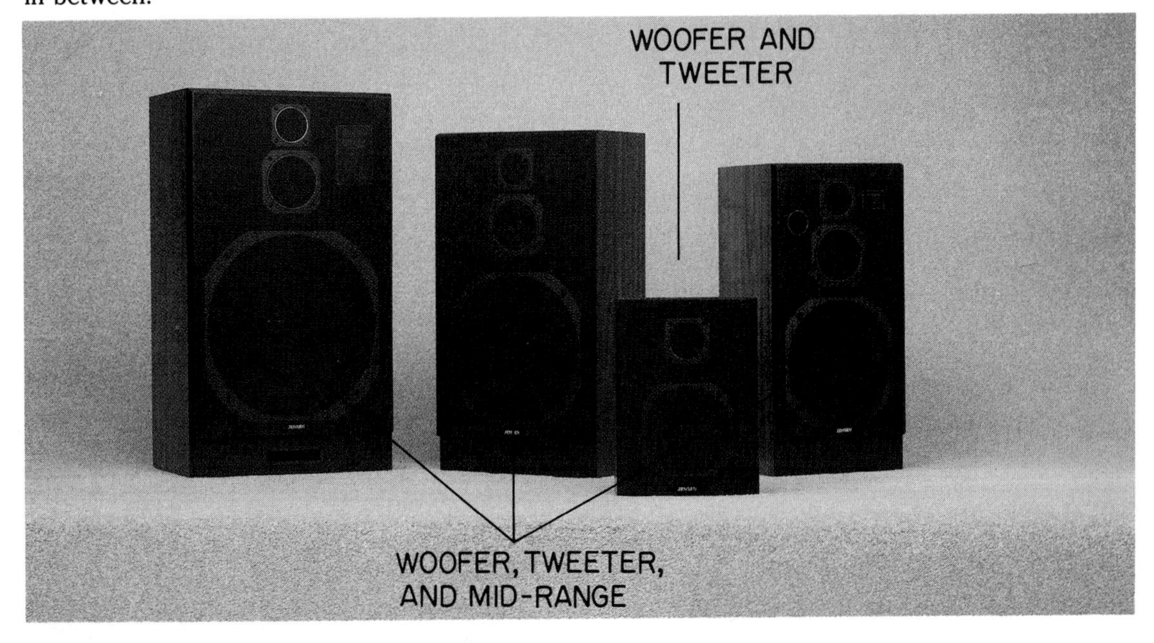

The cabinet or housing in which the speaker is mounted also plays a large part in sound quality. The size, shape, and material used all have an effect. For example, different types of speakers may be mounted in one cabinet to take advantage of their different tonal ranges.

Headphones are like tiny speakers held in place on your ears. Since the sound they generate is fed directly into the ear, they provide a richer listening experience. The speakers inside headphones are basically the same as those used for telephone receivers. Fig. 17-18.

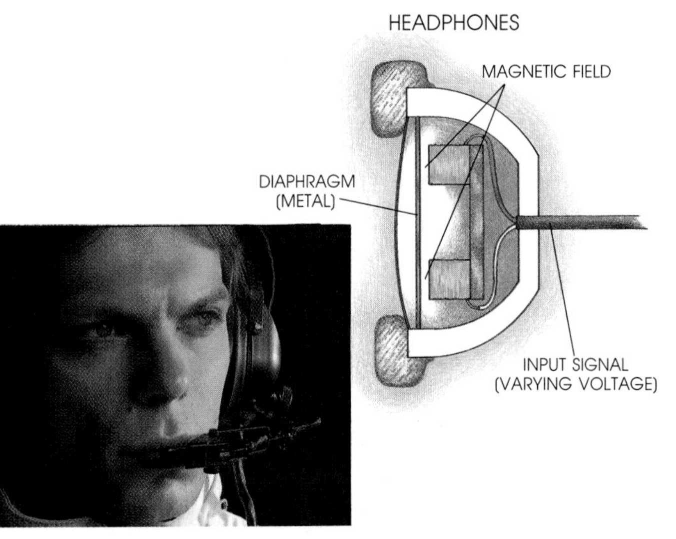

Fig. 17-18. Since headphone speakers rest directly on your ears, there is no loss in sound quality.

TELEVISION

No one person can be said to have invented television. However, in 1929 Vladimir Zworykin, an American physicist, first demonstrated a practical video transmitting/receiving system. Ten years later, the National Broadcasting Company began commercial television broadcasts. Fig. 17-19. Color broadcasts were introduced in 1953. In 1965 the *Early Bird* satellite relayed TV broadcasts between the United States and Europe. Today, there are more than 200 million TV sets in the United States alone.

In some ways, radio and television transmission are similar. Your TV set is like a radio with a picture attached. Television signals are transmitted in the VHF (very high frequency) and UHF (ultra high frequency) ranges. Each TV channel is assigned a different frequency by the FCC. The channel selector in the TV set (receiver) tunes the set to the right frequency, just as the dial does on a radio.

Television Transmission

For television, unlike radio, both audio and video signals must be created. Color video adds another level of complexity.

Television Cameras

Television cameras convert what they "see" to electrical signals that are then transmitted.

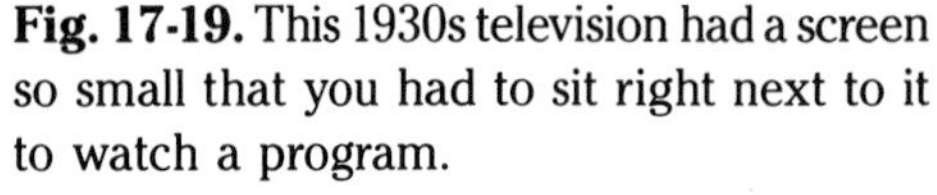

Fig. 17-19. This 1930s television had a screen so small that you had to sit right next to it to watch a program.

HEALTH & SAFETY

HOW LOUD IS TOO LOUD?

With so many new electronic devices available that reproduce excellent sound, it is important to remember that too much sound can be a health hazard. Intense noise over a long period can produce deafness. Continuous or even periodic noise can make people tired or irritable. High noise levels can affect a person's heart rate and blood vessels, alter the natural rhythm of brain waves, and create stress.

The intensity of sound is measured in decibels (dB). The softest sound humans can hear has been assigned an arbitrary value of 0 dB. Except for thunder or erupting volcanoes, nothing in nature is louder than 100 dB. A level of 75 dB will damage a person's hearing if it continues over a period of time. Following are some common noise levels. Numbers are decibels.

Quiet room	40
Normal conversation	60
Rush hour traffic	92
Jackhammer	105
Rock and roll band, near loudspeaker	110
Jet airplane engine, 100 ft. away	140

Young people who listen to rock music with the volume turned way up have been tested and found to have damaged hearing. While being surrounded by intense sound can be a pleasurable experience, it can also be dangerous. When using audiovisual equipment, such as stereos and tape players, practice a little control for the sake of your ears.

The following description is for a black and white camera.

The lens of the camera gathers light and directs it to the **pickup tube** inside the camera. Fig. 17-20. (The most common pickup tube is the *vidicon*.) The pickup tube has a glass faceplate. In back of the faceplate is a coating called the **signal plate**. Behind that is another coating called the **target**.

The light passes through the transparent faceplate and signal plate and strikes the target. The target is covered with a material that conducts electricity when exposed to light.

Light causes the negatively charged electrons in the target to move toward the signal plate. The number of electrons emitted is proportional to the amount of light. As the electrons leave the target, positively charged areas remain on the back of the target that correspond to the original image. How positive the areas are depends on how much light from the image reached them.

At the other end of the pickup tube, an electron gun generates a beam that scans across the target surface. It scans from the image's left to its right and from top to bottom. The scanning pattern is made up of 525 horizontal sweeps, 30 times each second. This creates the 30 frames per second characteristic of video.

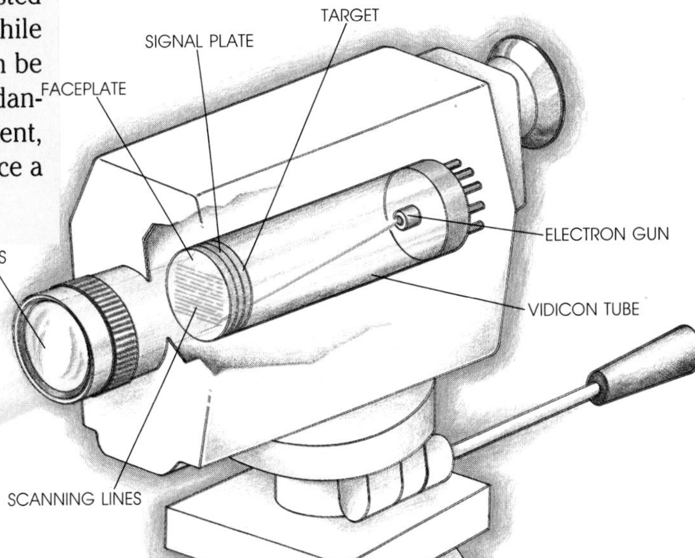

Fig. 17-20. A television camera uses an electron gun to create the signal.

As the beam scans, the negatively charged electrons are drawn to the positive areas. The lightest (most positive) areas on the target attract the most electrons. The darkest (least positive) areas attract very few. The electrons pass on through the target and strike the signal plate. When they hit the signal plate, an electrical current is created. The voltage of this current changes continually, depending on the number of electrons contacting it. This then is the electrical signal to be transmitted.

Charge-coupled devices (CCDs) are now used in some cameras instead of the traditional pickup tube. A CCD is a special type of microchip containing a very fine grid of light-sensitive capacitors. Capacitors store electrons. The output voltage from these capacitors varies depending upon the amount of light that strikes them. Fig. 17-21. The pattern of varying voltages creates the video signal. The CCDs are small, durable, lightweight, and adaptable to a wide range of light levels. They are thus a likely replacement for the pickup tube.

Color video. Color video is based upon the additive color system. (See Chapter 10.) Red, green, and blue are the primary additive colors. When all three are projected, one on top of the other, white results. By using red, green, and blue in various intensities and combinations, any shade or color may be produced.

Color video cameras have three pickup tubes: one each for red, green, and blue (RGB). Fig. 17-22A. Light passing through the lens strikes three mirrors or a prism. Each mirror image is directed to a different pickup tube. In front of each tube is a filter that allows only one color to pass. The tubes then process the images just like the black and white pickup tube already described. Fig. 17-22B. Information about the brightness and strength of the colors is combined with the other information.

The three signals are then combined to form one composite signal. An electrical pulse is added to the signal every time the electron beam scans across the target. Later, this pulse synchronizes TV set reception.

Fig. 17-21. Charge-coupled devices (CCDs) are used in place of pickup tubes in video cameras.

Fig. 17-22A. The three pickup tubes inside a color video camera process the three primary colors of light.

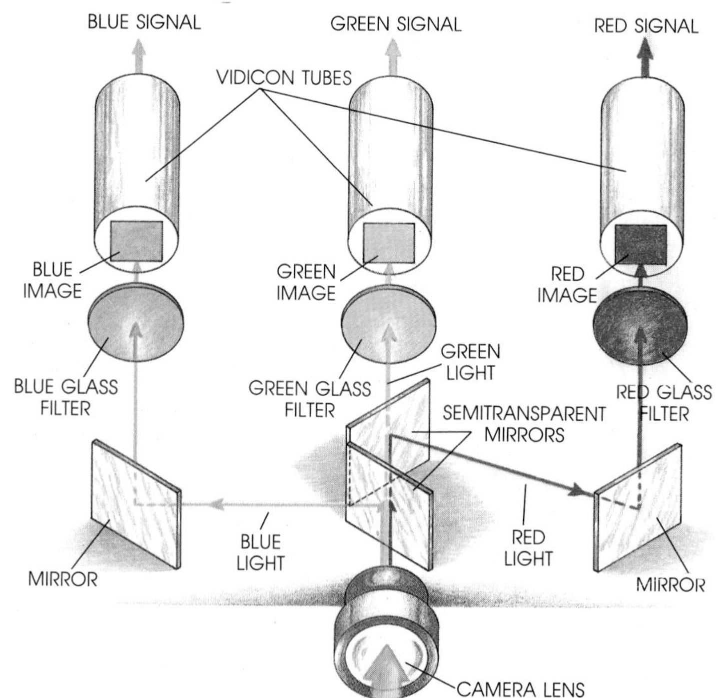

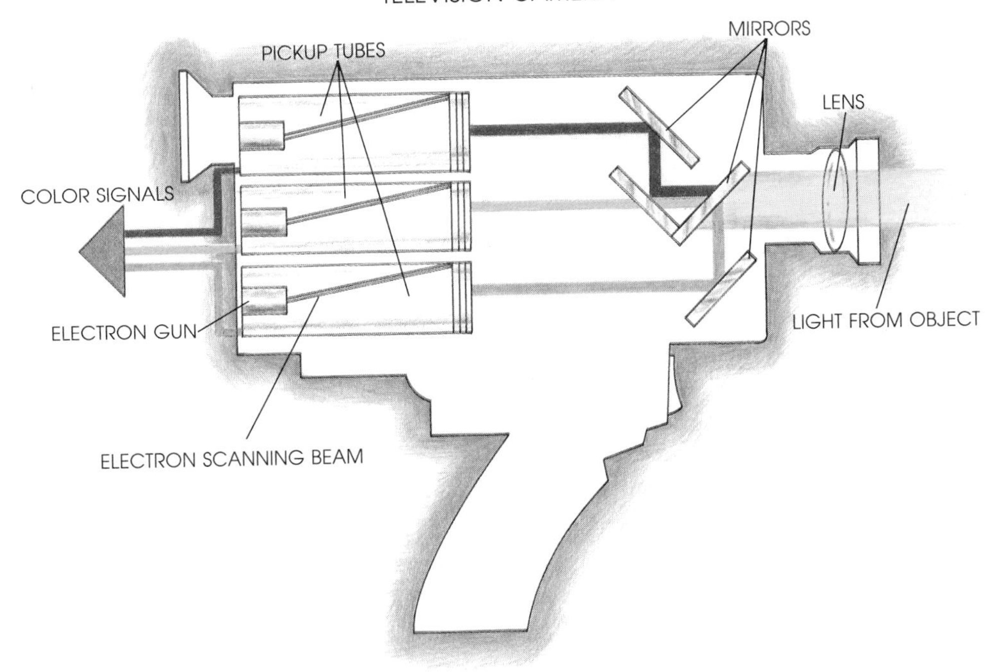

Fig. 17-22B. The color image is created by electron guns just as a black and white image would be.

Microphones

The audio portion of a TV signal is created in the same way as a radio signal. A highly sensitive microphone converts the sound waves into an electrical signal. The same microphones used for radio are used for television.

One type of microphone, the electret condenser, can be made as small as a tie tack. These are often used in TV productions, when the microphone should not be noticed.

Control Room

The audio control equipment for television is similar to that for radio and includes an audio console. Sometimes the audio and video control centers are combined; often, however, they are separate rooms. In the video control room, engineers operate the video equipment while the program takes place. The video control room contains a video switcher, video monitors, and an editor.

Video Switcher. Even a simple video production may require at least two cameras. More often, many different cameras are used. A television sporting event, such as a football game, may need *dozens* of cameras set up all over the stadium. The video **switcher** receives input from each camera. Fig. 17-23. The switcher

Fig. 17-23. The director sits at the video switcher and can see the images transmitted from each camera. Switching from one to the other can be done easily.

allows the director, who is in charge of the production, to choose which image to record by switching from one camera to another. The switcher enables the director to do the following:

- "Cut" from one camera to another or to a different video source, such as film or videotape.
- Go to a "black," or blank, screen.
- Fade or dissolve the image. In a fade, the image gradually appears on (or disappears into) a black screen.
- Show text over a video background. Text, also called "titles" or "credits," is generally produced by a computer called a character generator. Fig. 17-24. Letters of all different sizes and colors can be made. This is called keying.
- Superimpose one scene over another. "Dream" scenes use this technique. One camera shows the sleeping person and another the dream.
- Preview an image before it is recorded.
- "Wipe" an image right off the screen while replacing it with another. This can be done from side to side, top to bottom, or with a number of different patterns.
- Split the screen so that two or more images can be seen at the same time. Split screens can be done in a variety of sizes, shapes, and designs.
- Make one image appear in a hole in another image. TV news programs use this to show the announcer's face in one corner of the picture.

Video Monitors. A control room has several video monitors. These are TV sets that show scenes from different cameras or other sources. The director watches these and chooses which to record. The picture being recorded is shown on a master monitor.

Editor. Making changes in prerecorded programs is called editing. During editing the director selects from various recorded shots and combines them. Editing is done with the help of an electronic editor. Fig. 17-25.

The electronic editor controls a videotape playback machine and a videotape recorder. The original raw footage tape is placed in the play-

Fig. 17-24. Titles and lists of people who worked on a program are produced on the character generator.

Fig. 17-25. The electronic editor allows the technician to electronically "splice" two or more scenes together.

back machine, and a blank tape is placed in the recorder. The director selects parts of the original tape and records them onto the blank tape in any order desired.

Television programs are generally created in advance and stored on videotape. Later they are transmitted in much the same way a radio signal is transmitted.

Remote Telecasts

Remote telecasts — those done away from the studio — may be taped with hand-held cameras. A remote truck or van is parked nearby. Fig. 17-26. Inside the truck is a control room and equipment needed to create TV signals. A dish-shaped transmitting antenna may be used to transmit these signals to a relay antenna or satellite. A remote truck and cameras are used for on-the-scene newscasts and sports events.

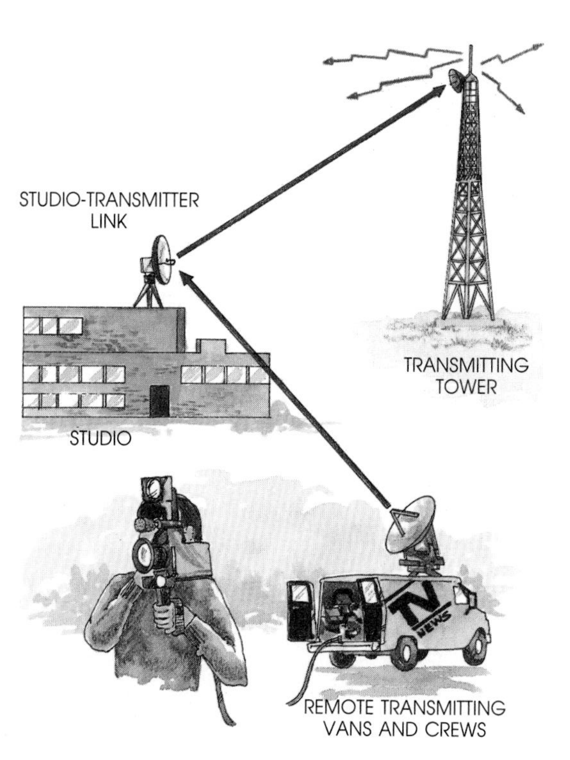

Fig. 17-26. Remote telecasts are made from trucks. The signal is transmitted back to the studio.

Antenna

The audio and video signals created in the control room pass through an amplifier and are then sent to the transmitter. An oscillator creates the carrier waves. The video signal is used to modulate the amplitude of a carrier wave. The audio signal is used to modulate the frequency of another carrier wave. Then the two waves are combined into one and amplified again. The combined signal is sent to the transmitting antenna.

Buildings and other large obstacles can interfere with TV signals. Thus they can travel only a fairly short distance (about 75 miles) without help. The range of a TV signal is approximately the area visible from the transmitting tower. This is called "line of sight" transmission. If the signal must travel farther, it is relayed by more than one tower or station. Sometimes satellites are used as relay stations. Each relay station has a receiving antenna and a transmitter. The station picks up the signal, amplifies it, and transmits it to the next station.

Transmission Channels

Most television signals travel through the atmosphere using electromagnetic waves, as do radio signals. Television signals use both the VHF and UHF frequency bands.

In recent years, cable TV has become popular. Since the cable is a physical transmission channel, it is not much affected by atmospheric conditions. Cable TV signals are stronger than those received by an antenna. They pick up less noise, or outside interference.

While a simple copper wire works well for the transmission of short-distance telephone conversations, it is not adequate for television transmission. The television signal has a higher frequency than a twisted-pair wire can carry efficiently. Coaxial cable is designed to carry video signals. Fig. 17-27. A coaxial cable consists of a number of copper wires surrounded by plastic insulators. The wires and insulators are held together inside a hollow cylinder. The electromagnetic signal travels between the wire and the walls of the cylinder. Because of its design, coaxial cable can carry a much larger number of signals than twisted pair wire.

Reception

The receiver for a TV signal is of course your TV set. The signal is induced in an antenna on the set or outside your house. The signals are amplified and sent to the tuner. By adjusting the channel selector on your set, you tell the tuner which frequency to look for. The tuner picks out that signal and sends it to a mixer where it is changed to an intermediate frequency. Another amplifier then strengthens it.

Next, the signal is sent to detectors that separate the carrier wave from the audio/video signal. At the audio detector, the audio signal is picked out and sent to the speaker. A television speaker is much the same as that for an FM radio. Most television sound is low-quality monaural (one channel), but there is a trend toward high-quality stereo sound.

At the video detector, the video signal is separated out. The color video portion of the TV signal is further separated into two signals for color and a brightness signal. A decoder (oscillator) changes these into the red, green, and blue signals of the original image. They are then sent to the picture tube.

Cathode Ray Tube

The picture tube is a cathode ray tube. Fig. 17-28. This tube is very similar to that in a computer monitor. (See Chapter 5.) The flat end of the tube is covered with phosphor salts. At the other end is an electron gun. Electrons shot from the gun excite the phosphor salts and make them glow. These glowing dots create the image you see.

A color picture tube has three electron guns that sweep across the flat surface. One gun is used for each of the three primary colors. The guns are designed to make 525 sweeps per image across the surface of the picture tube. The surface of the color CRT is covered by groups of red, green and blue phosphors. Each group is known as a "picture element," or pixel for short.

The signal broadcast by the television station controls the output of the three electron guns. If all three of the RGB phosphors in a pixel are excited in the proper proportions by the electron guns, the pixel displays white. If only the red phosphor is excited by the red electron gun, the pixel displays red. By changing the combination and intensity of the phosphors excited in a pixel, any color can be displayed on the screen.

The synchronization pulse added when the signal was created in the video camera now comes into play. This signal assures that scanning is an exact match to that done by the TV camera. If the picture isn't "in sync," you can adjust it with the horizontal and vertical controls on the TV set.

Fig. 17-27. Inside the coaxial cable (left) are wires surrounded by insulators. Many of these wires and insulators are combined (right) to make one cable.

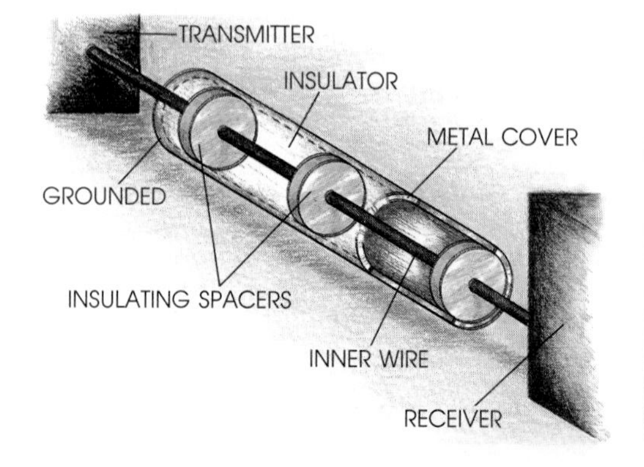

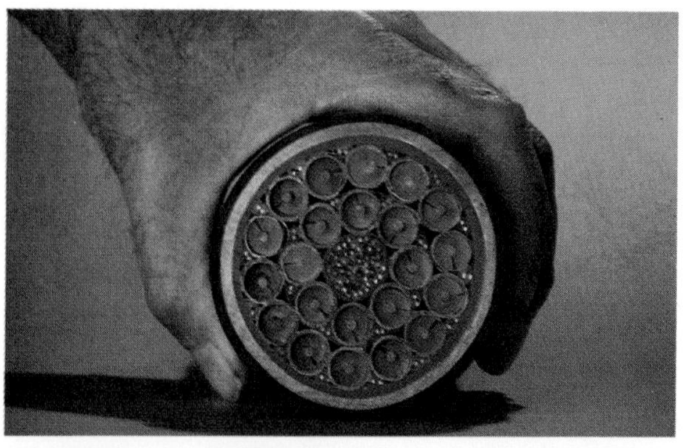

SCIENCE FACTS

How TV Fools Your Eye

A television picture is made of tiny points of light. Your eye receives the TV message in the same way it receives other light. (See Chapter 10.)

The success of a video image depends upon the fact that it can move so fast your brain can't keep up with it. A video program is really a series of rapidly-changing "still" images. Each scene is replaced about 30 times a second. Before your brain is through processing one image, another appears. Your brain interprets these rapid changes as a continuously moving image.

THE RECORD PLAYER

Thomas Edison's invention of the phonograph in 1877 astounded the world. Fig. 17-29. For the first time, recorded sounds could be played back in people's homes.

To make a record, Edison wrapped a metal cylinder with tinfoil. He placed the cylinder on an axle and rotated it. As the cylinder rotated, someone spoke into a mouthpiece. Inside the mouthpiece was a metal diaphragm that vibrated with the sound. A needle attached to the diaphragm made dents in the foil. The dents corresponded to the sound waves. To play the "record," the process was reversed. Another needle was placed against the cylinder. Attached to this needle was another diaphragm. As the cylinder was rotated, the diaphragm vibrated, producing sounds like the original.

Today's records are made using microphones that change sound waves into an electrical signal. The quality of sound produced is much better, but the basic principle is similar.

Record players, like video and audio tape recorders, differ from the other audio and video systems we have studied so far because they are not "live." The message is always recorded in advance and later transmitted.

Fig. 17-28. Each of three electron guns inside the cathode ray tube scans for one of the primary colors of light.

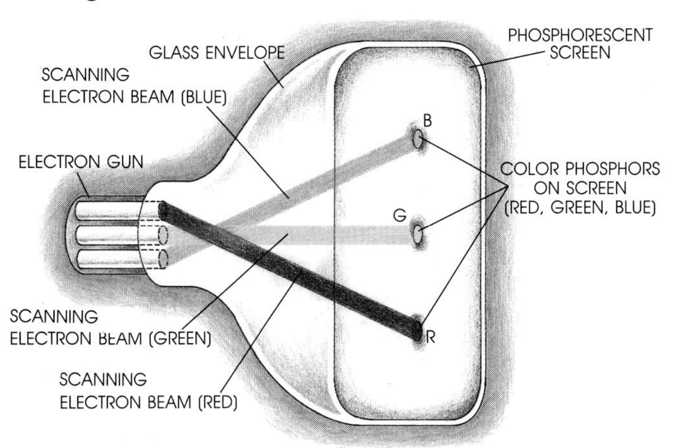

Fig. 17-29. Although deaf in later years, Thomas A. Edison had an ear for music. He supervised many recordings. By placing his ear against the phonograph speaker he could instantly detect a wrong note.

Recording the Message

The sound is recorded in a recording studio on tape. (See "Audio Tape Recorders and Players," page 377, for an explanation of how magnetic tape works.) A sound technician records sounds on different sections, or tracks, on the tape. For example, two singers' voices may be recorded on two different tracks using separate microphones.

As many as 48 tracks might be made. The sound technician can alter and combine the different tracks using a mixing board. Fig. 17-30. For example, sounds of different instruments may be made to fade in and out. Then the tracks are combined into two main tracks. The two tracks will eventually be sent to the two speakers on a stereo record player. A final master tape is then made.

Records are made with a needle that changes the magnetic impulse on the master tape into vibrations. These vibrations drive a stylus that cuts a groove into a chemically treated aluminum disc called a lacquer. Then a mold of the lacquer is made. This is done by coating the lacquer with silver and covering it with a nickel solution. When the solution hardens, it can be peeled away and used to mold vinyl records. Molding is done on a press. Hot vinyl is placed in the mold. The press is closed and pressure applied. This produces the record, which is a vinyl duplicate of the lacquer master.

Playing the Message

When you play a record on your turntable, the needle in the tone arm rests in the groove. As the turntable spins, the needle vibrates up and down and back and forth, traveling, or "tracking," the groove. This vibration is changed by the cartridge into an audio signal. The **cartridge** is the small rectangular object at the end of the tone arm. The signal is amplified and used to drive the speaker(s). The most common cartridges include crystal or ceramic, moving coil, induced magnet, and moving magnet. The moving magnet type is the most popular. Fig. 17-31.

In a moving magnet cartridge, the vibrations of the needle cause a small magnet to move freely next to a wire coil. This movement induces a current in the coil. The current reproduces the signal which originally created the record's groove.

Fig. 17-30. The mixing board allows the sound technician to alter and combine sounds.

Fig. 17-31. The movement of the magnet next to the coils induces a current in the coils. The current reproduces the signal.

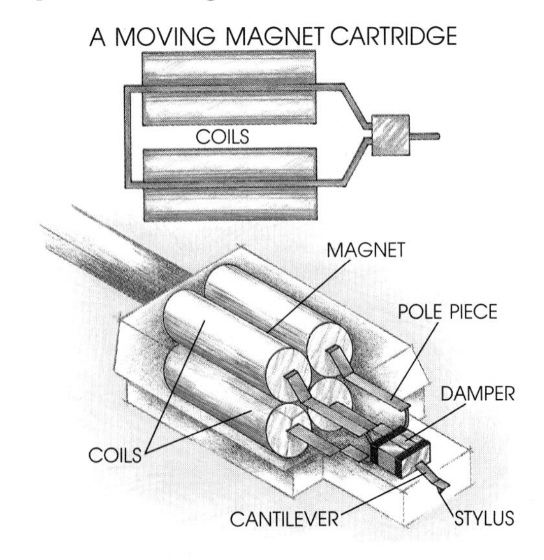

AUDIO TAPE RECORDERS AND PLAYERS

Valdemar Poulsen, a Dane, invented a machine in the 1880s that recorded sounds on wire. In the 1930s, magnetic tape was first used by German engineers. Today, of course, tape recorders and players are widely used. Tape decks and portable tape players are common items in almost every home.

Recording the Message

There are several different types of audio tape recorders: reel to reel, cassette, and digital. Each uses magnetism to store the signal.

Microphones, like those discussed earlier, are used to transform sound waves into electrical signals. The signal also may come directly from another audio device, such as a radio, tape recorder, or phonograph. The signal is amplified and directed to one or more **recording heads**, which are actually electromagnets. Fig. 17-32. The changing voltage of the signal creates a changing magnetic field in the recording head.

Magnetic recording tape consists of a plastic, such as polyester, coated with metal oxides. These metal oxides, usually iron or chromium, can be magnetized. When the magnetic tape passes over the recording head, the oxides are magnetized in a pattern that represents the audio signal. This is similar to the way magnetic storage tape for computers works. (See Chapter 5.) An erase head can demagnetize tape that is to be reused.

Audio recording tapes come in a range of widths and qualities. For ease of handling, home stereo systems generally use one-quarter-inch tape in a cassette. Professional audio recorders often use one-half-inch, one-inch, or even two-inch tape mounted on a reel. The sound quality depends on the type of oxides used and the speed at which the tape is recorded. The faster the tape moves, the better the quality. This is because, at higher speeds, the signal is spread out over more tape.

Fig. 17-32. (a) The electrical signal sent to the coil of the recording head causes a changing magnetic field. This magnetic field arranges the metal oxides coated on the recording tape. (b) An erase head demagnetizes (erases) the tape.

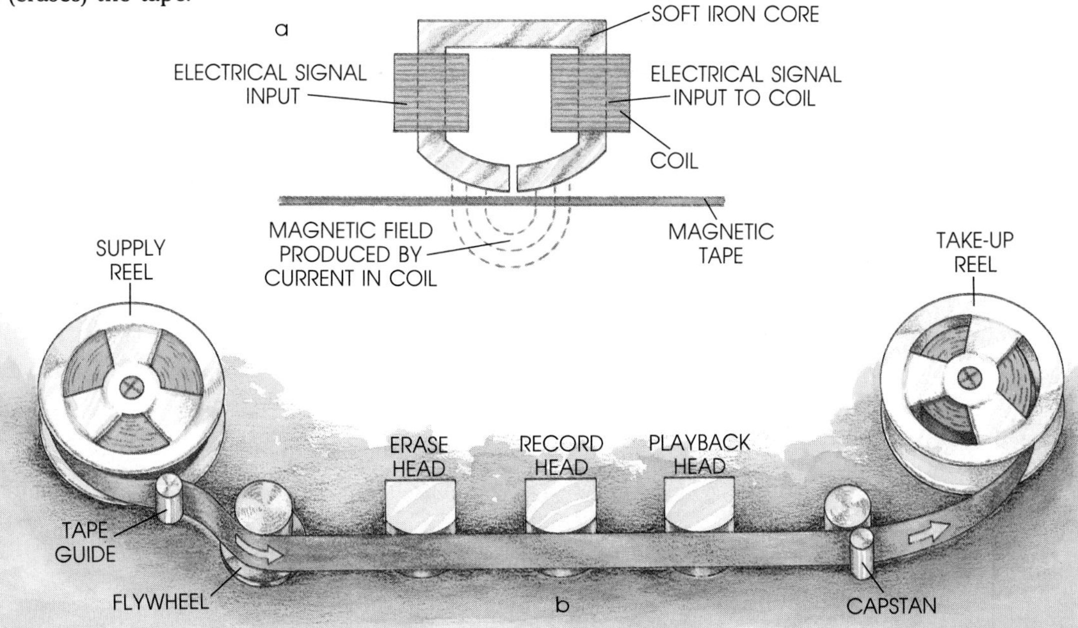

Tape recorders store the signal on the tape in tracks. Tracks are thin bands that run the length of the tape. Fig. 17-33. A full-track recorder uses the entire width of the tape to record the signal. A stereo recorder, which records two different audio signals at the same time, requires at least two tracks. Home stereo equipment uses four tracks. The second and fourth tracks are played when the tape is turned over.

Playing the Message

The playback head works in just the opposite way from the recording head. It, too, is an electromagnet. When the tape is passed over the playback head, the magnetized oxides induce a small current in the coil. This current recreates the electrical signal that was recorded. It is then amplified and directed to speakers. The speakers, like those described earlier, reproduce the original sound.

Digital Recording

The electrical signal we have described is an analog signal. An analog signal occurs in a continuous form that varies with the audio input. Digital audio tape (DAT) recording converts the signal created by microphones into bits of digital information. Fig. 17-34. (Digital systems are described in detail in Chapter 4.) The playback heads convert the signal back into an analog form. This analog signal is changed to audible music by the speakers.

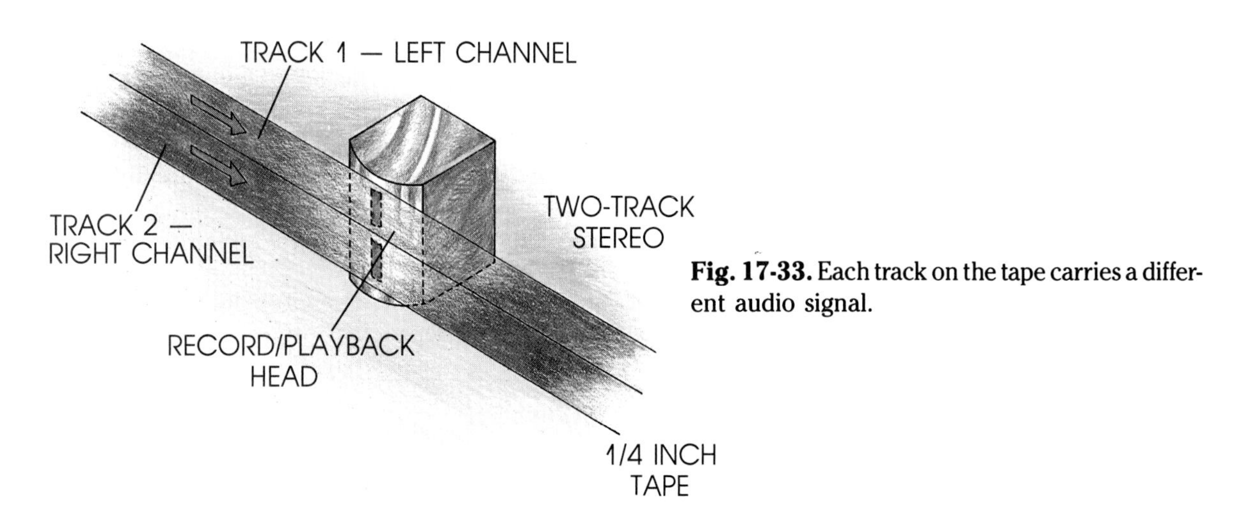

Fig. 17-33. Each track on the tape carries a different audio signal.

Fig. 17-34. Digital audio tape must be played on a digital tape deck, such as this one.

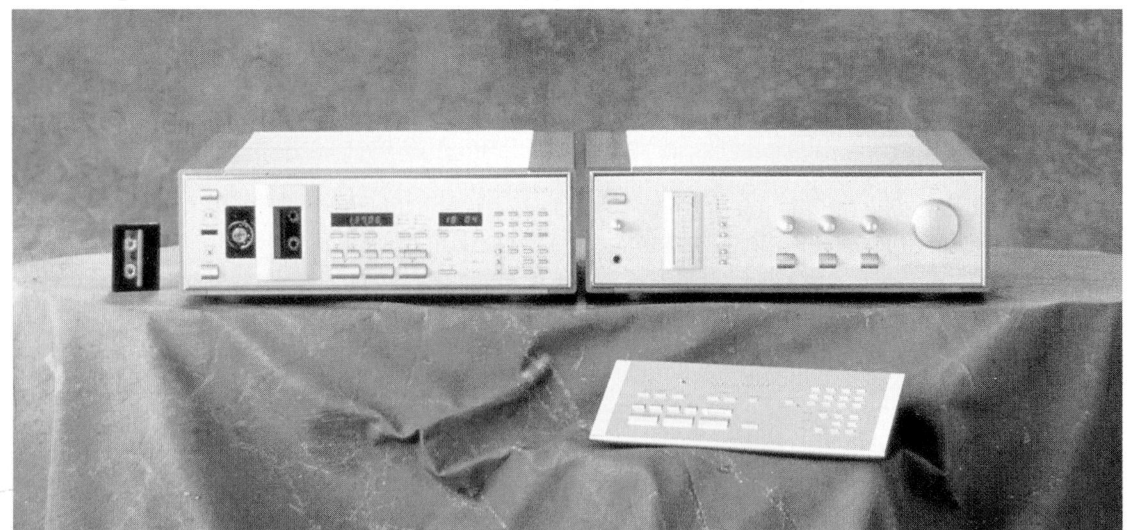

Digital systems produce truer sound than analog recording methods. When digital tapes are duplicated for sale, there is no loss in sound quality.

Compact disks, or CDs, are optical storage disks that store an audio signal digitally. The sound is recorded using a laser. The laser burns pits into the disk's surface. These represent the audio signal in digital form. Another laser inside the player "reads" reflected light from these pits that is changed back into sound. Optical storage disks are described in more detail in Chapter 5.

Since there are no grooves to wear out, CDs produce the highest quality audio available today. They hold more minutes of sound on one side than an ordinary record.

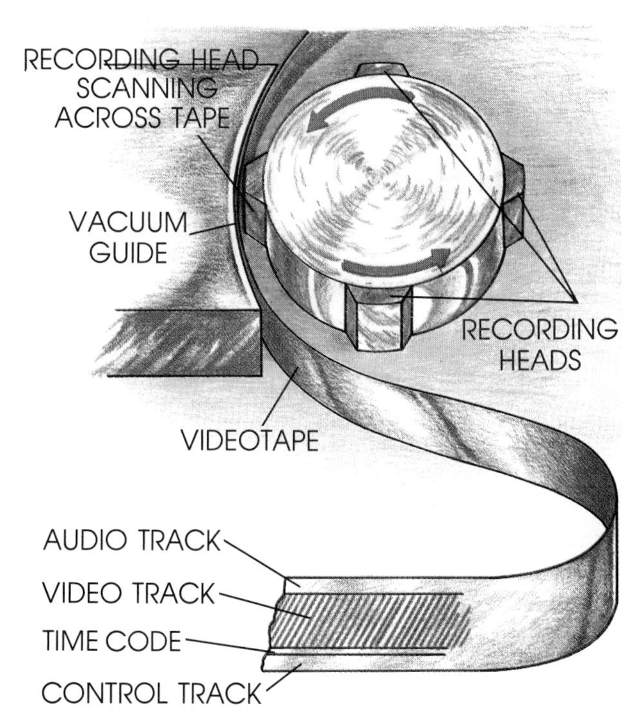

Fig. 17-35. The huge amounts of information contained in video/audio signals require a lot of tape area. In the system shown, the signal is recorded vertically.

VIDEO RECORDERS AND PLAYERS

Recorders that could record both sound and pictures on magnetic tape were developed in the 1950s. By the 1970s cassette-type recorders were being produced. They were inexpensive and quickly became popular for home use.

Since video and audio signals are basically the same, it is not hard to understand why video recorders (VTRs or VCRs) are very much like audio tape recorders. Electromagnetic recording heads magnetize oxides on the tape from a signal input by microphones and video cameras. The video playback heads change this back into a signal that drives the TV receiver. Since video recorders must handle both audio and video, however, they are more complex than audio recorders.

Because of the huge amounts of information contained in a video/audio signal, more tape area is required. One solution would be to run the tape at higher speeds. Yet, doing so would require hundreds of miles of tape for a one-hour recording! To solve this problem, VTRs use two or four recording heads that cover more tape area.

One type of system uses video record/playback heads mounted at right angles to the tape on a disk that spins at a high speed. Fig. 17-35. The video signal is recorded vertically along the tape as the heads scan it. Scanning allows more tape surface to be used for recording.

In addition to video recording heads, VTRs also have audio heads to record and play back one or two audio tracks. A pulse recorded on a control track is used to synchronize recording and playback.

Video Formats

There are a number of different video formats. **Video formats** refer to the type of videotape used for each system. They include 2″, 1″, ¾″, ½″, and 8 mm tape. One and two inch machines are primarily used for professional studio recording and editing. The ¾″ format has become the standard in the TV news industry. The ½″ videocassette recorders are popular for home use. Fig. 17-36. There are two different ½″ formats: Beta and VHS. The VHS format is more widely used than Beta.

The 8 mm format is strictly for home use. Its small size allows the camera and audio recorder to be combined into one unit. Fig. 17-37. These "camcorders" have also become very popular with consumers.

Fig. 17-36. Half-inch format VCRs have become very popular.

Fig. 17-37. A camcorder allows users to make their own movies and view them on a TV set.

CHAPTER 17 REVIEW

Review Questions

1. Explain how a carbon microphone works. How is it different from a condenser microphone?
2. What is multiplexing? How does it aid telephone communication?
3. Name the seven parts of a radio receiver and tell briefly what each part does.
4. How does a radio speaker change a signal back into sound?
5. Explain how a TV pickup tube works.
6. Describe the process of making a record.
7. How does magnetic tape record a sound signal?
8. How does a compact disk differ from a record?
9. Describe the recording heads on audio or video recorders.
10. Why is more tape area required for video recording?

Activities

1. Find out how an electric doorbell works and describe the process to the class. How does the doorbell fit the communication systems model?
2. Using any appropriate materials, construct an antenna that improves the reception of a radio receiver. Describe what did and did not work well.
3. Using a 35mm camera, photograph the picture on a television monitor. Use five different shutter speeds. Keep a record of which speeds you used for which pictures. Have the film developed. Explain the results to the class.
4. Construct a model of an early audio communication device, such as a radio or telegraph. Include with it a report that explains how the device works.
5. Obtain a portable video camera and tape a 60-second news "spot" showing an event taking place in your community or school. Get as much action in the scene as you can.

CHAPTER 18

Applications of Audio and Video Systems

It seems that the more communication devices we have, the more ways we can think of to use them. Although some simply replace devices already being used, others—such as the VCR—have inspired new activities and industries.

This chapter discusses some of the uses for the telephone, radio, television, phonograph, and audio and video tape players. Some you already know about, others may be a surprise. Also included in this chapter is a section on how the equipment and methods you have studied are applied to the actual production of a television program.

Terms to Learn

boom microphones
camera angles
CBs
closed-circuit television
conference call
director
electronic listening devices
hams
producer
radar
radio astronomy
script

As you read and study this chapter, you will find answers to questions such as:

- How have telephones replaced business conference rooms?
- How does the space program use radio?
- What has television done to improve medicine?
- What goes on behind the scenes as a TV program is made?

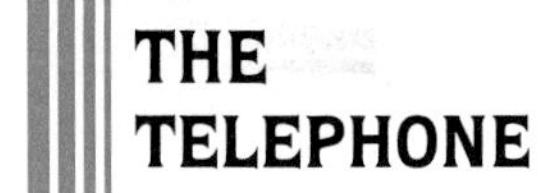

THE TELEPHONE

Use of the telephone for everyday communication has increased to such an extent that there is a constant demand for new communication channels. At least 71 countries in the world have more than 100,000 telephones in service.

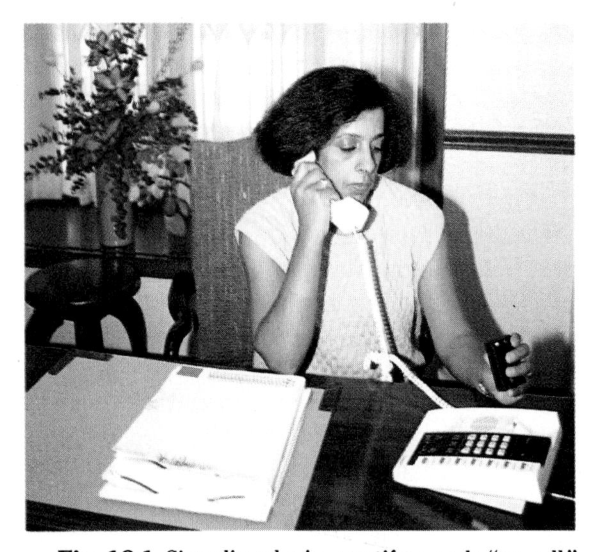

Fig. 18-1. Signaling devices notify people "on call," such as doctors, that they must report in.

Personal Communication

At present, the telephone remains the device of choice for personal communication. It is the way people stay in touch. They exchange the news of the day, make appointments, and enrich relationships by means of a phone call.

Answering machines and answering services have turned the telephone into a kind of private receptionist. Calls may be screened, answered later, or not answered at all. The person remains connected to the outside world without having to sacrifice privacy. With the addition of an electronic signaling device ("beeper"), doctors and other people who must stay in close contact with their offices can always be summoned to the phone. Fig. 18-1.

Communities with emergency "911" phone numbers now link telephones to systems designed to send aid to people in trouble. For example, if someone is injured but able just to dial the 911 number, the operator can locate the telephone and find out where the person is. An ambulance or police car is then dispatched. Fig. 18-2.

The introduction of cellular telephones—phones in cars—has met the needs of busy people stuck in traffic or driving long distances. A new pocket-size telephone may even exceed the popularity of the car phone. Both depend on radio batteries and atmospheric transmission.

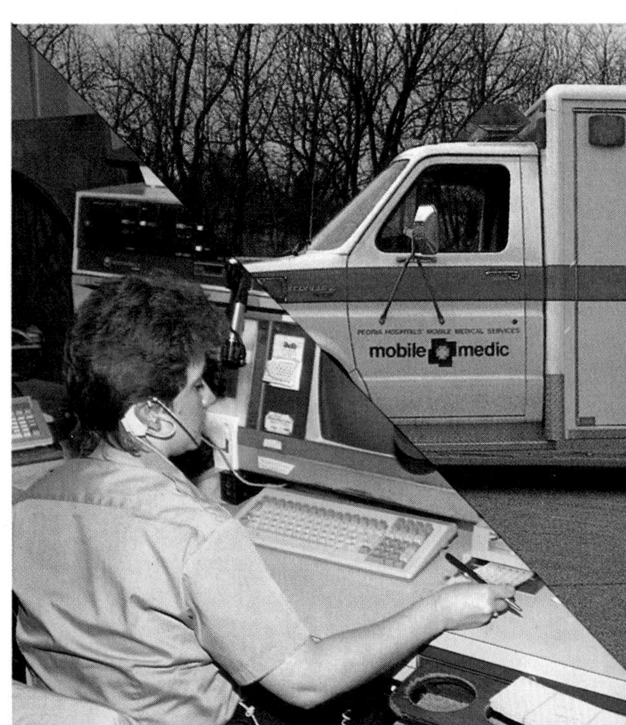

Fig. 18-2. An operator answering a 911 call can dispatch an ambulance at once to aid the victims.

Business

For businesses, telephone technology has made such things as conference calls possible. In a **conference call**, several people may talk using different phones but the same connection. A meeting among a number of people may take place without any of them having to leave their desks. The phone system itself has become the meeting place.

Some businesses use private telephone lines for special purposes. Radio and TV stations, for example, are connected by means of private lines into large networks. With the addition of fax machines (see p. 118), photos, drawings, and other information can be sent over these telephone lines and reproduced.

Telephones are also used by businesses for advertising and sales. Salespeople call potential customers at home, telling them about a special offer and hoping to persuade them to buy. Some organizations use telephone surveys to find out how customers feel about a particular product or service. Toll-free numbers have enabled businesses to give customers better service. At no charge, consumers may call to order an item, learn about a service or product, or voice a complaint.

Telephone's link to the computer has provided an efficient way to transmit data. With the use of a modem (see p. 113), information may be transmitted from one computer to another via telephone lines. The telephone has made long-distance computer networks possible, enabling people to access information rapidly and without leaving their offices or homes.

RADIO

The radio is primarily an entertainment and information medium. However, it also has a number of special uses.

Entertainment and Information

Before the advent of television, radio stations offered a variety of programs. Comedies, such as *Burns and Allen* and mysteries, such as *The Shadow* kept listeners enthralled much as TV programs do today. Over the years, however, radio has changed. Today, music programs, news, sports, and talk shows seem to be what listeners want to hear.

Many such programs are designed for people who lead busy lives. For example, some stations offer nothing but news. The same facts are repeated several times during an hour. Listeners who can't wait for an hourly "newsbreak" can learn what's happening right away. Those driving cars may tune in to frequent traffic reports that help them avoid long delays.

Radio talk shows have become extremely popular. Although they have existed for years in one form or another, their numbers have increased in the last ten years. Some are information-based. A specialist in a field such as medicine, psychology, or gardening hosts the program and listeners call in with questions. Other programs seem to revolve around the personality of the host. Fig. 18-3. The host asks what people think about something, such as foreign affairs, and listeners call in with their opinions. This type of program seems to provide people such as those who work late at night with a way to stay in touch and enlarge the scope of their

Fig. 18-3. Larry King, an extremely popular radio show host, inspires listeners to call in and talk.

lives. Many are devoted fans who call in so often that the host knows them by name.

Radio has also been used for special interest groups to contact those who share their concerns. For example, political organizations use radio to tell potential supporters about their positions.

Radio is used by businesses for advertising. Radio programs are interrupted regularly with commercials that sell products or services.

Public radio stations are nonprofit organizations that do not broadcast commercials. They depend on donors to fund their operation. Public radio programs tend to be of more cultural interest than those on commercial radio.

Two-Way Communication

Broadcast radio is one-way communication, but radio has long been used for two-way communication as well. In recent years, **CBs**—citizens' band radios—have been popular. Originally they were used by people in wilderness areas or on boats. Then long-distance truck drivers installed them in their trucks. Fig. 18-4. The radios allowed them to talk with family members or friends who were within range and who also had a CB radio. The radios were also handy in an emergency if a repair crew had to be summoned. Soon, however, everybody wanted a CB. Instead of using call letters to identify themselves, people gave themselves special nicknames, such as "Silly Sal from Sunny Cal." CBs became a new social outlet—a way for people to make and keep friends.

Amateur radio operators, called **hams**, use shortwave radio signals to contact other ham operators all over the world. Unlike signals used for CBs, shortwave radio signals travel long distances. Ham operators are licensed by the government.

Two-way radios are also used for safety and defense. Every police car has one, as do ambulances, fire trucks, and many military vehicles. A guard patrolling an industrial area at night can call for help on a small two-way radio called a walkie-talkie. Nursery monitors, for use within the home, let parents in another room know if children are awake or in trouble.

Some two-way radios are used for commercial and business purposes. Every taxi has a radio. A dispatcher contacts the driver and tells him or her about a customer that must be picked up. Paging systems may also be two-way radios. The person being summoned speaks into the signaling device, which contains a receiver. Airplane pilots receive takeoff and landing instructions by means of radios, and construction workers building skyscrapers use radios to communicate with workers on the ground.

In 1932 it was discovered that the stars emit radio waves. Soon afterward **radio astronomy** was born. Giant dish antennas receive the radio waves, and the patterns of the waves tell astronomers more about our universe. Fig. 18-5. This has led to the question of whether there are beings on other planets who might be trying to contact us using radio transmission. Today some scientists are trying to contact those on other planets by sending out a steady radio signal.

Fig. 18-4. CB radios allow those who may not have access to a phone to stay in touch.

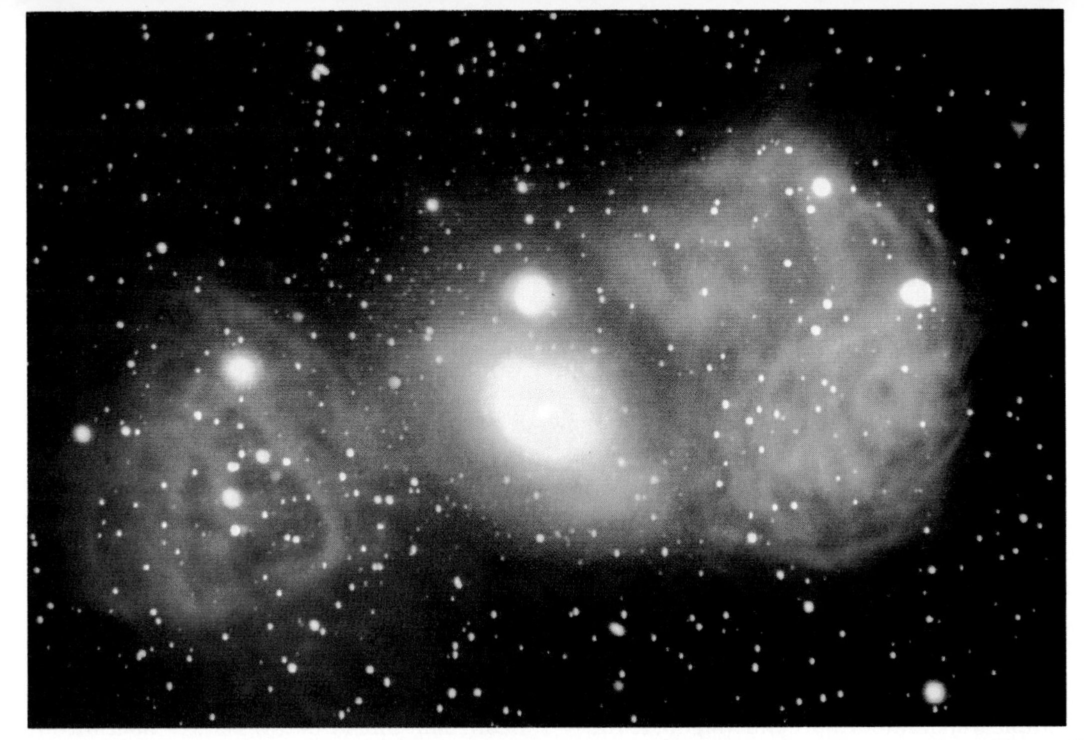

Fig. 18-5. This composite picture of galaxy NGC 1316 shows both radio and optical emissions. The radio emissions are the two large red lobes. The optical emissions are the blue-white area. By studying such radio and optical data, astronomers have learned that NGC 1316 has consumed its neighboring galaxies. The next victim may be the small blue-white galaxy just to the "north" in this photo.

SCIENCE FACTS

Radio Telescopes

Stars and other objects in space produce many different wavelengths of electromagnetic radiation. However, most of them are either reflected or absorbed by Earth's atmosphere. Only optical and radio wavelengths penetrate to ground level.

Scientists have long used optical telescopes to learn about the heavens. Since the 1930s they have also used radio telescopes to gain information. Most radio telescopes in use today consist of an antenna in a dish-shaped reflector. The reflector collects the signals and focuses them onto the antenna, which converts them to electrical signals. The signals are then amplified and sent to an output device.

The radio signals from space are very weak. If all the energy received by all the radio telescopes in the world during the past 50 years could be combined, it still wouldn't be enough to power a flashlight. Because the signals are so weak, many of them must be collected in order to produce an image. One way to do this is to build a very large dish. In Arecibo, Puerto Rico, there is a dish 1000 feet in diameter. A series of smaller dishes can also be electronically linked to create the same effect as a very large dish. The VLA (Very Large Array) near Socorro, New Mexico, consists of twenty-seven dishes arranged in a Y shape. Each dish is only 85 feet in diameter, but the combined effect is the same as that of a single dish 17 *miles* across.

Other Uses

A few years ago the recently built American embassy in Moscow was declared unusable because it was discovered that the Soviets had "bugged" it. Fig. 18-6. In other words, **electronic listening devices** had been installed so that the Soviet government could monitor what was being said. Such listening devices, which are really tiny one-way radios, are often used when political or industrial spies want to gather information.

Remote control devices also depend on radio transmission. Garage door openers are one example. Another is the remote control on your TV set. The control turns the set on or off, changes channels, and adjusts volume. Remote controls are simple transmitters that send a signal to the TV receiver. Information such as channel number and sound level is carried by this signal.

Many ships navigate by means of radio signals beamed from shore. A similar method is used by astronauts in guiding spacecraft. The equipment on a spacecraft is also controlled by means of radio signals. Not only does NASA communicate with astronauts by radio, computerized information can be sent to the space vehicles themselves.

Another important use of radio transmission is radar. **Radar** stands for Radio Direction And Ranging. Radar uses reflected radio signals to detect objects that are too far away to be seen in other ways. Narrowly focused microwaves are sent out in short bursts from a transmitter and antenna. If a solid object gets in the way of the beam, the energy is reflected back to a receiver. The object appears on the radar screen. Fig. 18-7. Aircraft use radar for navigation and safety. A beam sent ahead of the plane tells it whether another plane or some other object is in the way. Anti-aircraft missiles can be detected with radar, as can tropical storms. The police use radar to detect speeding cars.

Fig. 18-6. The brick U.S. Embassy office building in Moscow has never been used because of the bugging devices it contains.

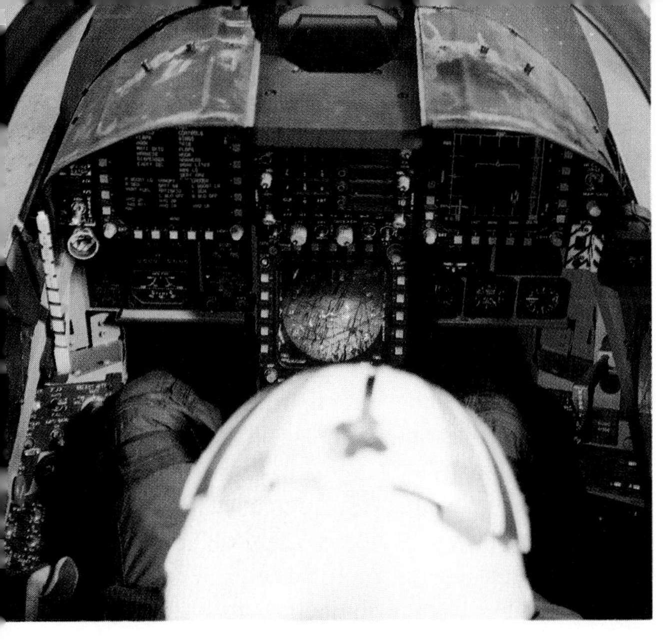

Fig. 18-7. The small screen directly over the pilot's head is for radar. The reflected radar signal shows up on the screen.

AUDIO AND VIDEO TAPE PLAYERS

Audio tape is used primarily for music recordings, although books and plays have been produced on tape as well. Most commercial radio programs are first recorded on audio tape and aired at a later time. Businesses use audio tape for dictaphone machines. A letter is dictated on tape and the secretary plays it back and types the letter.

Videotape has a wider range of uses. Virtually anything viewed by a video camera may be stored on videotape. Closed-circuit security systems record what the camera sees on tape. In case of a robbery, the culprit's face may appear on the recording. As mentioned earlier, businesses use taped messages for meetings and job training. Like radio programs, many TV programs are taped in advance and broadcast later.

Perhaps the most popular use for videotape is with home VCRs. People who are too busy to watch a TV program as it is broadcast or who want to keep a copy of it simply record it on tape. They can then watch it at a more convenient time or store it in a tape library. Hollywood film producers release taped versions of major motion pictures for rent or sale to VCR owners. Videotape sales have done much to help film producers regain some of the huge investment needed to make motion pictures.

PHONOGRAPH

The phonograph has been a popular entertainment device for many years. It is used primarily to play music, but plays, comedy routines, and books for the blind have all been put on records. The phonograph as we know it may soon be replaced by the compact disc player (see p. 379). Compact discs are smaller and their sound quality so much better that they have proved immensely popular. The phonograph of only a few years ago may in the very near future be considered an antique.

TELEVISION

Although the most common use of television is for entertainment and information broadcasting, it has other uses as well. Most of these involve closed-circuit TV.

THE DIGITAL MULTIMEDIA HOME ENTERTAINMENT SYSTEM

The shift from analog to digital format is having a profound effect on the audio and video broadcast industries. While most of the world now receives analog audio and video signals with their radios and television sets, all that will change dramatically in the next decade or two.

Currently, the home entertainment system includes some or all of the following:

- a large format television set
- a stereo videocassette recorder (VCR)
- an audio/video amplifier that feeds the television's audio signal to large speakers around the room to create "surround sound"
- a stereo sound system that includes a radio receiver, tape player, and a compact disc (CD) audio deck
- a videodisc or compact disc interactive (CD-I) player that feeds video to the television set

The home entertainment system is going digital — and the view from the couch has never looked (or sounded) better! Virtually all of the technology is already available, but options are still limited and expensive. That will change. The digital home entertainment system will be less expensive, have much higher quality, be interactive, and offer far more options than now imaginable.

The heart of the digital home entertainment system will be the digital television. It is like morphing a computer and a television set. The television signal will be broadcast digitally and received digitally, either over optical fiber, coaxial cable, or via wireless transmission (with a satellite dish). It will display on your computer screen, and you will control it with sophisticated computer software; so the options will be incredible!

You will be able to edit the video/audio almost any way you wish. You might alter the image size and colors; create as many split screens as you like; search among the hundreds of television channels with key words (for example, type "horror" to find a horror movie); create audio effects; create your own instant replays; take digital "snapshots" of any image on the screen; and so on.

Like the television signal, the music source you listen to will be transmitted digitally to your entertainment system. It will, in turn, output this signal digitally to speakers placed around the room, probably via a wireless transmission system.

Since the signal is digital, the music will be CD quality. This service is currently available from the DSS (digital satellite system) companies, but it will expand greatly in the future. You will pay for this music as people now pay for cable television. With commercial-free CD-quality "radio" signals coming to your home entertainment system, you won't be purchasing CDs in the future. Rather, you will simply receive your favorite music with a small satellite dish or download it from the Internet. Once it's in your computer, you can edit it any way you wish. You could even edit your voice into the original sound track to create your own version of the latest hits!

Likewise, there won't be any need for a VCR or video game systems. You will simply download movies and games from the "digital jukebox" located somewhere else on the planet. It won't matter where it is. Digital signals don't degrade over distance the way analog signals do.

While all of this is technically feasible right now, the trick is providing affordable cable or wireless technology to every home. Telephones "work" because most people have one or more in their homes. The same is true for the digital home entertainment system of the future.

Closed-Circuit TV

In **closed-circuit television** the signal travels to only certain TV receivers. Hospitals make frequent use of closed-circuit TV to monitor the condition of many patients at one time. A nurse watches the monitors and responds in case of trouble. Hospitals also use TV cameras in operating rooms. Students get a closeup view of an operation that would not be possible otherwise. Several surgical procedures can now be done using tiny incisions. A flexible fiber optic tube is inserted through the incision, and doctors can see inside the patient by watching the TV monitor. Fig. 18-8. Long-handled instruments are then inserted and the procedure carried out.

Space satellites are often equipped with TV cameras to detect weather patterns. Movements of clouds across a given area may signal a weather change. The information is relayed to weather forecasters who then try to predict what it means. Other satellites contain cameras used for spying. When a plant for manufacturing poison gas was built in Libya a few years ago, a camera on a space satellite got a picture of it, and other countries were forewarned. TV has also been used for space exploration. By means of TV cameras, scientists have seen close-ups of the moon and planets. Fig. 18-9.

Business and industry use television as a teaching aid. Taped programs played for new employees tell them about the company or instruct them in how to do a job. Sales meetings can be held nationwide by means of prerecorded programs. Workers in plants using computer-controlled robots can monitor the robots' work using TV cameras.

Engineers in nuclear plants watch what's happening inside the reactor on closed-circuit TV. In this way the highly radioactive areas can be observed without danger to workers.

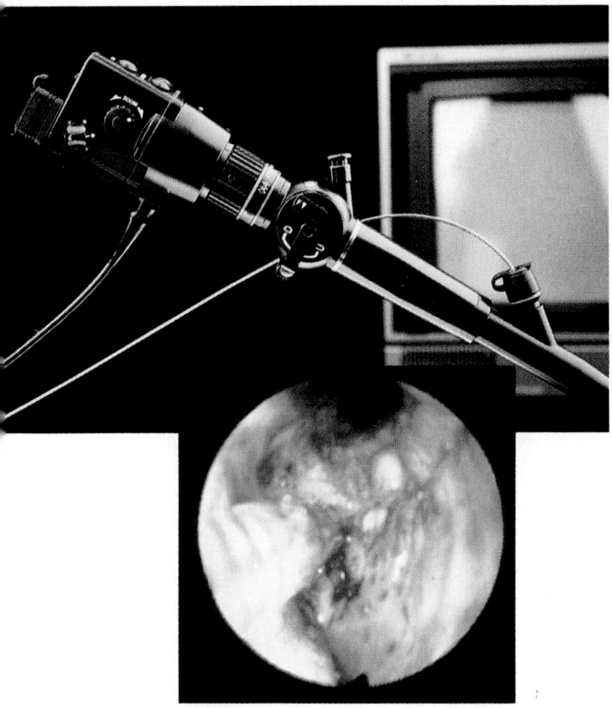

Fig. 18-8. Using the video camera inside this scope (top), surgeons can spot such things as cancerous growths (bottom).

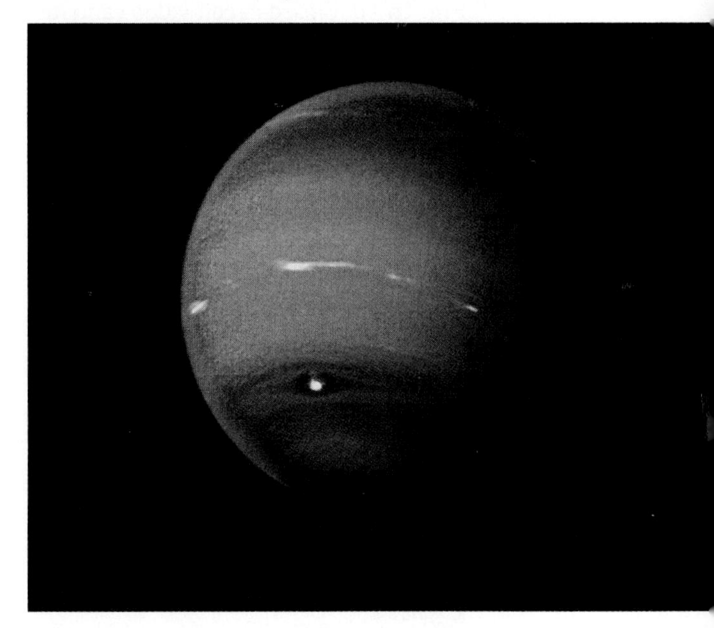

Fig. 18-9. Video cameras carried on *Voyager II* have sent back views of the other planets, such as this one of Neptune.

Closed-circuit TV is also useful for security purposes. Cameras in banks or stores watch for shoplifters and other thieves. Fig. 18-10. In prisons, TV cameras allow one guard to monitor an entire bank of cells. Some private individuals install security systems in their homes. The cameras pick up the movements of any intruders.

TV has also been helpful to archaeologists working underwater. Video cameras can operate at depths beyond the range of divers. The cameras are also better than the human eye for viewing conditions in muddy water. Archaeologists bringing up the remains of sunken ships use cameras to record the scene before it's touched. Then when the ship is retrieved in bits and pieces, they can use the recordings to recreate the scene exactly.

Commercial and Public Television

Most television broadcasting is done by the four major networks—ABC, CBS, NBC, and Fox. These networks serve over 730 local commercial stations. In addition to the major networks, public stations, which are funded by viewers and other contributors, provide programming. In recent years, cable networks as well have offered feature films and other programs to subscribers for a fee. News, advertising, educational programs, and entertainment can all be found on commercial and public TV.

Information. TV is a powerful news medium. In the morning, at noon, in the evening, and again at night, all the networks and many private stations broadcast the day's events. Other

Fig. 18-10. Closed-circuit video cameras in this bank constantly record the scene. If the bank is robbed, pictures of the robbers will be on tape.

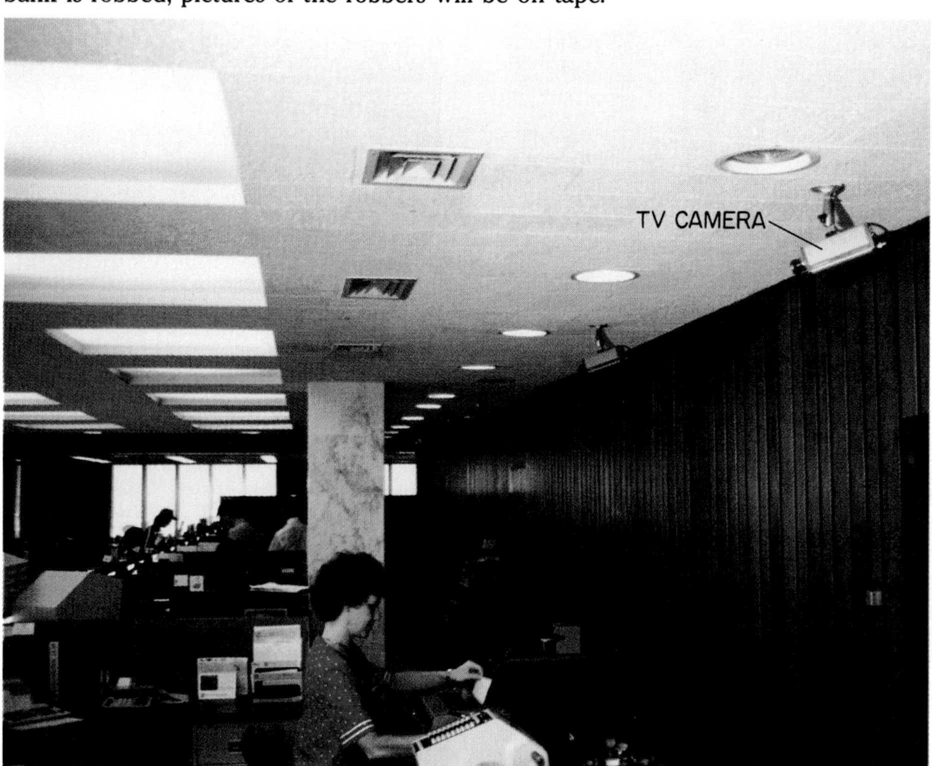

news-related programs, such as documentaries, government hearings, and spacecraft launches keep viewers in touch with what's happening.

Special interest programs also have a place on TV. The Olympic Games and other sporting events, travel programs, and shows that teach us how to do something are all popular.

Advertising. By the age of eighteen, the average young person sees a total of 360,000 TV commercials. Viewers are urged to buy everything from dog food to underwear by means of clever, insistent advertising. Political candidates who appreciate television's power hire special consultants to help them use TV to win votes.

Education. Both commercial and public television can educate, although public TV devotes many more hours of programming to it. Shows on cooking, home repair, painting, and exercise are all popular. Fig. 18-11. In recent years talk shows, such as the *Phil Donahue Show*, have become a kind of open forum during which individuals or groups air their views.

Entertainment. Most television broadcasting is done to entertain. In addition to comedy and drama, variety shows, soap operas, children's programs, and movies all reach us via our TV sets. Its influence on our lives is considerable.

Just what is the process through which a TV show is created? The following section will give you a general idea of what takes place from the time a show is conceived until it is broadcast on the air.

Television Production

When you watch a TV program, you see only a few of the many people who create the show. While the actors are the ones whom we tend to praise (or blame) for the show, many others are involved before and during the production.

Television production begins with planning. People in the programming department of a network or station decide what ideas will be produced as programs. Networks and stations produce many programs themselves. Independent companies produce others. In either case, once an idea is approved, a **producer** then becomes responsible for the program.

The producer obtains the script, oversees set and lighting plans, selects the performers, and is involved in working out every detail of the production. In addition, the producer usually controls the money that is spent and keeps the production within its budget.

Fig. 18-11. Many TV programs show us how to do something, such as stay in shape.

Writing the Script

Video production usually begins with a script of some kind. A **script** is a line-by-line description of the program. Fig. 18-12. It includes the content of the show. For example, a script for a commercial would list the different characters and the lines to be read. The script also includes information for production specialists, such as camera operators.

The actual format for the script varies with the type of show. Television programming requires scripts for dramas, commercials, news, sports, and talk shows. Each of these would use a different script format.

The Director

The producer hires a **director** to oversee the actual making of the program. After the script has been approved by the producer, the producer and director select the performers and other workers.

Next, the director decides how the production will look when finished. Any scene in a script may be produced in many different ways. It is up to the director to decide exactly how the scene will be done. Fig. 18-13.

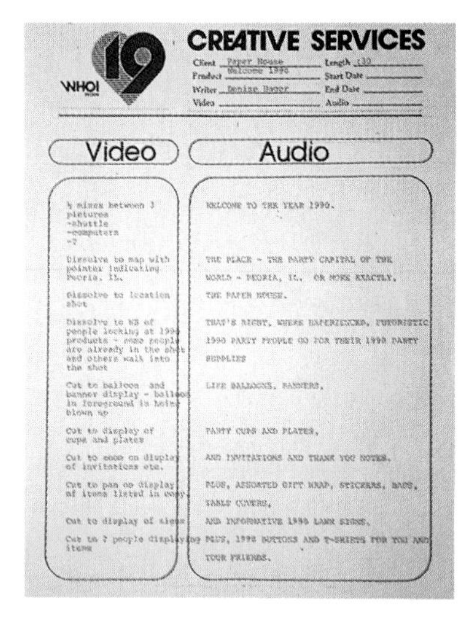

CREATIVE SERVICES

WHOI 19

Client Paper House Length :30
Product Welcome 1990 Start Date
Writer Denise Bauer End Date
Video Audio

Video	Audio
[illegible] between 3 pictures -shuttle -computers -?	WELCOME TO THE YEAR 1990.
Dissolve to map with pointer indicating Peoria, IL.	THE PLACE - THE PARTY CAPITAL OF THE WORLD - PEORIA, IL. OR MORE EXACTLY,
Dissolve to location shot	THE PAPER HOUSE.
Dissolve to MS of people looking at 1990 products - some people are already in the shot and others walk into the shot	THAT'S RIGHT, WHERE EXPERIENCED, FUTURISTIC 1990 PARTY PEOPLE GO FOR THEIR 1990 PARTY SUPPLIES
Cut to balloon and banner display - balloon in foreground is being blown up	LIKE BALLOONS, BANNERS,
Cut to display of cups and plates	PARTY CUPS AND PLATES,
Cut to [illegible] on display of invitations etc.	AND INVITATIONS AND THANK YOU NOTES.
Cut to pan on display of items listed in copy	PLUS, ASSORTED GIFT WRAP, STICKERS, BAGS, TABLE COVERS,
Cut to display of signs	AND INFORMATIVE 1990 LAWN SIGNS.
Cut to ? people displaying items	PLUS, 1990 BUTTONS AND T-SHIRTS FOR YOU AND YOUR FRIENDS.

Fig. 18-12. A video script may contain technical notes as well as the "words" actors or announcers will say.

Production Specialists

Producing a program calls for many abilities. Costumes must be created. Sets must be designed. Special props, such as furniture, must be obtained. Cameras, lights, microphones, and other equipment must be selected and operated. If you watch the credits at the end of a TV show, you get an accurate idea of just how many people are needed.

Rehearsals

Rehearsal, or practice, is necessary to the success of any production. Recording artists, for example, often practice long hours to perfect their music. Video programs can require even more rehearsal time, because there are so many more people involved. Taking charge of rehearsals is also a task of the director.

Production

When a program is taped, and most programs are, a television or recording studio is used. The studio contains all the equipment needed. Video recording studios require overhead lighting grids, backdrops, set storage, and enough space for cameras to move around. Studios are also designed with the best sound quality in mind. Complete sets are built for prime time television programs and movies. The large production companies in Hollywood have entire "lots" that are covered with different sets.

Fig. 18-13. This director is seated in the control room while the show is being taped. She communicates with the camera operators and other technicians by means of her headset.

Cameras. Cameras used to shoot TV programs are usually large and heavy. They are mounted on wheels for easy movement. All the shots taken by a camera operator have been decided on in advance by the director.

As you learned in Chapters 11 and 12, there are many ways to photograph a subject. The same is true when videotaping a TV show. Each time you change the point of view of the camera, you change the effect of the photograph. As with a regular camera, field of view can also be changed.

To make it easy to adjust the lens, video cameras have a zoom feature. The zoom lens allows the operator to change to any desired field of view with the push of a button. Fig. 18-14.

Fig. 18-14. The shots below were taken with a zoom lens. The operator can go from far away to very close up with the touch of a button.

(a) Extreme long shot.

(b) Long shot.

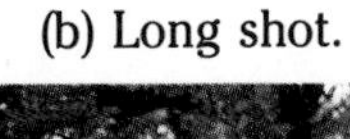

(c) Medium shot.

(d) Close-up.

(e) Extreme close-up.

Different effects may also be created by moving the camera around the subject and shooting from different angles, called **camera angles**. The "normal" camera angle is at the eye level of the subject. Fig. 18-15. High angle shots, with the camera positioned above eye level, make the subject look smaller. High angle shots help "set the scene." A low angle, on the other hand, makes the subject look larger. A low angle shot of a bad guy makes him look twice as tough to beat.

A tilted angle creates an eerie feeling. A subjective angle is used to make it appear as if we are viewing the scene through one of the characters' eyes.

Camera movement adds interest and variety to a scene. Common moves are shown in Fig. 18-16.

Lighting. As you know, without light, most cameras would be useless. For this reason, lighting is essential to video production.

Light may be used to add depth to a scene. "Spot" lighting is commonly used to draw attention to a particular part of a scene. Lighting can also be used to create different moods by varying its brightness and color.

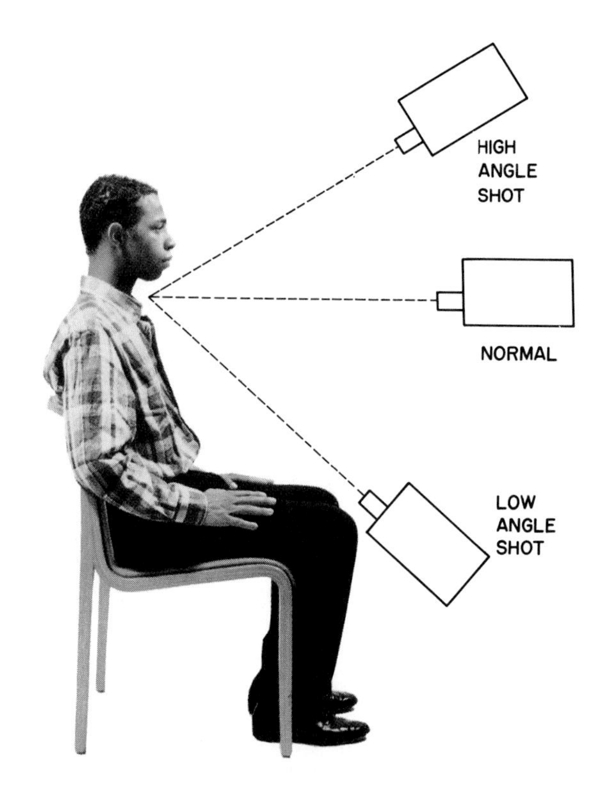

Fig. 18-15. A normal camera angle is at the eye level of the subject. A high-angle shot is above eye level, and a low-angle shot is below eye level.

Fig. 18-16. The following camera movements are common. *Pan.* The camera moves horizontally, while the pedestal remains in the same place. *Tilt.* The camera tilts up and down on the pedestal. *Dolly.* The camera and pedestal are moved away from or toward the subject. *Truck.* The camera and pedestal are moved from side to side. *Arc.* The camera and pedestal are moved in a semicircle around the subject. *Pedestal.* The camera is raised or lowered. *Crane.* The camera and operator are both carried on a crane.

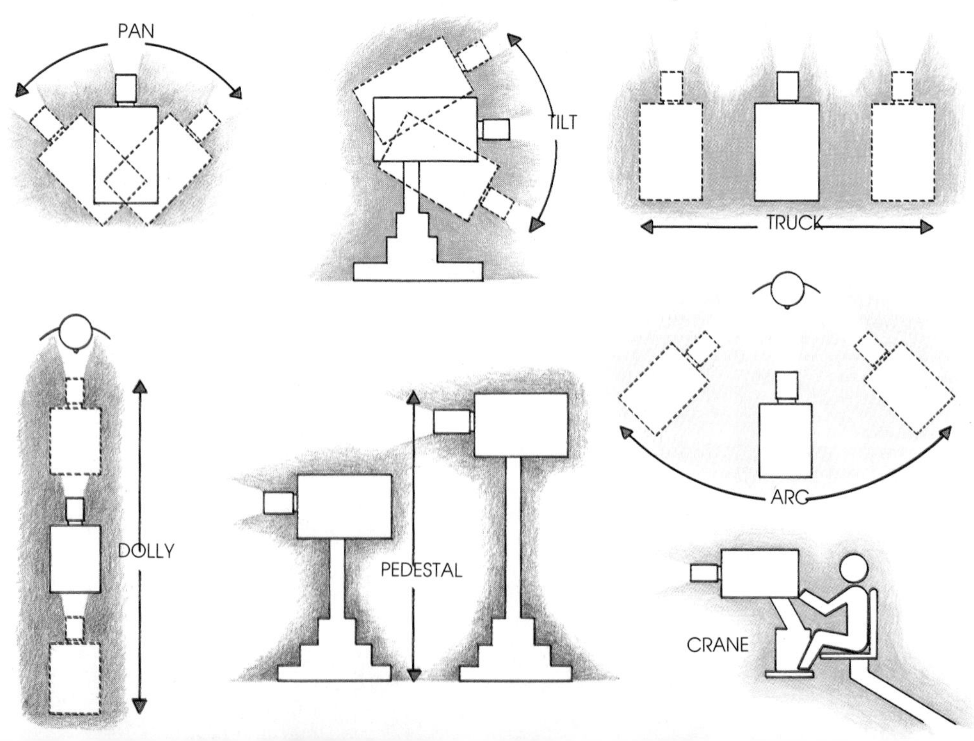

There are four types of lighting commonly used in video production: key, fill, back, and background lighting. Fig. 18-17. The key light is the main source of light for the subject. It is often positioned above and to the side of the subject.

Soft fill lights are used to get rid of shadows created by the key light. They are generally located to the side of the subject opposite the key light.

The back light is placed behind the subject. It lights the edges of the subject, such as hair and shoulders. This helps to create a three-dimensional appearance.

Finally, background lighting is used to light the background of the set. This is done to highlight certain areas or remove unwanted shadows.

Studio lighting is generally mounted overhead on a grid made of pipes. The grid allows the lights to be placed wherever they are needed and keeps them up out of the camera's view. Lights may also be mounted on floorstands. Floorstands are particularly useful outside the studio. They are also used for backlighting and other low angle lighting needs.

Sound. The lavalier microphone is commonly used on the set of video productions. It may be attached to an actor's shirt or lapel and can't be easily seen.

Hand-held microphones are excellent for singers, announcers, or reporters in the field. Floor stand microphones are often used to record music played by a band or orchestra.

For video productions, the microphone is usually held out of the camera's view. **Boom microphones** are mounted on a pole or cart which may be driven around the set. Fig. 18-18.

Microphones may also be mounted on a set of headphones. These are positioned very close to the speaker's mouth. They pick up only that person's voice and little other noise in the area.

Taping the Show

Most TV shows are first taped and then broadcast at a later time. The director sits in the control room along with the audio and video engineers and other special production workers. By means of a headset containing a microphone and earphones the director can communicate with camera operators and other workers on the set. The director watches the monitors that show what each camera is viewing. The decision of which view to record is the director's.

Later the tape may be edited. Background music may be added. New scenes may be inserted at different points. Then the completed tape is stored until it is put on the air.

Fig. 18-17. This lighting arrangement is commonly used for many live TV programs.

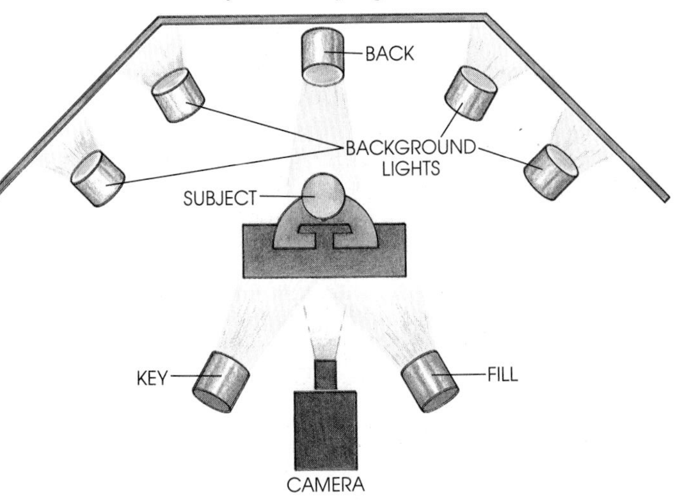

Fig. 18-18. The boom microphone shown here is mounted on a cart so that it may be moved from place to place.

PAINTING *YANKEE DOODLE DANDY* IN RED, WHITE, AND BLUE

When Jimmy Cagney first danced across movie screens in *Yankee Doodle Dandy*, he did it in black and white. Today, with the aid of computers, that and other old black and white movies are being colorized. When Cagney sings "It's a Grand Old Flag," the grand old flag has been painted in its true colors.

Two methods are used for colorization. Both require a black and white videotape of the film. Then a list of all the shots in each scene is fed into a computer. Teams of workers study the scenes and decide which colors to use for every item. Brightness and varying levels of gray on the videotape help determine colors and shades.

One system assigns colors to each pixel in the image. As the scene changes, pixels that carry over keep their colors. If something new comes into the picture, a technician must assign it a color.

The second system uses masks over areas with the same brightness values. Colors are added in these broad areas rather than item by item. The masks allow colors to blend, giving soft edges.

Colorization is an expensive and time-consuming process. It takes a company of about 200 people 24 hours to color six to ten minutes of film. An entire full-length movie requires about four months. Costs average about $3000 per minute or about $400,000 for a two-hour film.

Is it all worth it? Some people in the film industry say yes; others, no. Audience reaction, too, seems mixed. Try to see a colorized film in both versions and judge for yourself.

CHAPTER 18 REVIEW

Review Questions

1. What is a conference call?
2. How do ham radios differ from CBs?
3. Describe radio astronomy.
4. How does radar work?
5. How do hospitals use closed-circuit TV?
6. How has closed-circuit TV helped archaeologists?
7. Describe the jobs of a TV producer and director.
8. What is the function of a TV script?
9. Name the kinds of lighting used for a TV show.
10. Describe the process of taping a TV show.

Activities

1. Find out if your community uses a 911 emergency phone number. If so, read about it in your phone book or contact a phone representative to learn how the system works. Report your findings to the class.
2. Develop a 30-second radio commercial script for a new toothpaste.
3. If your local library loans out audio tapes, it may have copies of old radio programs. Obtain some tapes and listen to them. Then write a brief "review" of the programs. How do they compare to programs on television today?
4. Watch three television commercials. If possible, tape them on a VCR so you can view them more than once. How long does each last? How many edits (changes of view or scene) can you count in each? Describe the different camera angles, fields of view, and other techniques you can identify.
5. Cable television companies provide television facilities to the public. These are called public access channels. Contact your local cable companies and find out if there is an access channel in your area. Then call the channel and find out about how you can produce a program for them or work as a volunteer on the technical crew. Share what you've learned with the class.

Careers

Radio Announcer (Disc Jockey)

Many job opportunities exist in radio announcing, including those of disc jockey, sports commentator, news reporter, and talk show host.

Disc jockeys do more than just play the music that you hear. They must present the music, commercials, and other programming in a way that holds your interest. They lend their own personalities to the show. They conduct interviews, stage programs in public places, and reply to telephone call-ins.

Sports announcers are of two basic types: play-by-play announcers and sports analysts. Both have to be very knowledgeable about the sports they cover. The play-by-play announcer concentrates on explaining what is happening in the game. The analyst's job is to offer interesting details about such things as the players' performance.

Radio talk show hosts provide listeners with an opportunity to "speak their minds." Talk shows attract a large audience, partly because one never knows for sure what people will say when they call in. Talk show hosts have to field questions and comments and keep the show on track. A quick wit and the ability to respond to questions and comments on-the-spot are important in this job.

Education

A two- or four-year college degree in communication studies provides a solid background for work in radio. College programs generally include work experience as a part of the degree requirements. Through internships or "co-op" work programs, students gain valuable experience in radio broadcasting. Many colleges also have their own radio stations staffed by students.

Another good way to train for this field is to volunteer to work at a local college or public station. Many colleges do not have enough students to fill all the program slots, and so they rely on volunteer help. Although volunteer work is done for free, the experience gained from it is priceless.

Other Careers in Broadcasting

Radio programming specialist
Television announcer

For More Information

To learn more about careers in broadcasting, contact:

National Association of Broadcasters
1771 N Street N.W.
Washington, D.C. 20036

Correlations

 *Language Arts*

The history of audio and video communication systems is rich with great inventors and their inventions. In addition to those you have read about in this section, other inventors have contributed to our wealth of audio and video technology. Using the resources in your library, research one of these inventors. Prepare a written report that tells about the circumstances surrounding his or her invention.

 *Science*

As you know, color television is based upon three primary additive colors: red, green, and blue. Using colored cellophane, make your own red, green, and blue filters. Set up three light sources (slide projectors work well for this) and attach one of the three filters to each. Aim the light from all three sources to a single spot on a white wall or screen. What do you see? Next, experiment using only two of the sources and filters. What are the results? Now, try varying the intensity of the light sources. (Most slide projectors have two brightness settings, or you can add thin sheets of light gray cellophane to your filters.) What happens when the brightness of the colors is changed? Report your experience to the class.

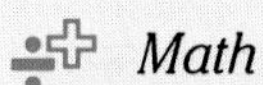

There is a mathematical formula that describes the relationship of current, voltage, and resistance. The formula, known as Ohm's Law, is

$$\text{Current} = \frac{\text{Voltage}}{\text{Resistance}}$$

Current is measured in amperes (amps), resistance is measured in ohms, and voltage is measured in volts. Given Ohm's Law, how is current related to voltage? If you double the amount of voltage in an electrical circuit, how much does the current change? How much does the resistance change? It may help you to substitute some numbers into the Ohm's Law formula. If you double the resistance, how does the voltage change? How about the current?

 *Social Studies*

Few countries in the world enjoy the same levels of technology that ours does. Take out a world map and select a country. Then try to learn more about the use of television or the telephone there. What is the country's population? How many people own TVs or telephones? Are all cities and villages linked by phone? How many TV stations are there? How many hours a day do they operate? What is the country's economy like in general? How do you think increased use of TV or the telephone might affect that country? Report your findings to the class.

Basic Activities

Basic Activity #1: Building an Electroscope

You can't see electricity, but you *can* see the effects of electricity. As you know, some materials (conductors) allow electricity to move freely through them. "Insulators" are materials, like plastic, that do not conduct electricity. An electroscope is a simple device that will allow you to see the difference between conductors and insulators.

Materials and Equipment

small clear plastic pill bottle
cork
copper wire
monofilament fishing line
plastic rod
wood or cardboard for base
tape
aluminum foil
wire hook
wool cloth

Procedure

1. Assemble the electroscope as shown in Fig. VI-1. Choose a day with low humidity to try the experiment.
2. Attach the copper wire to the hook holding the foil. Tape the wire in place on the cardboard or wooden base. Be sure some wire extends beyond the wood.
3. Charge the plastic rod with static electricity by rubbing it with a piece of wool cloth.
4. Touch the end of the copper wire with the plastic rod. Observe what happens to the foil.
5. Repeat this experiment using the fishing line instead of copper wire. Observe what happens to the aluminum foil.

Fig. VI-1.

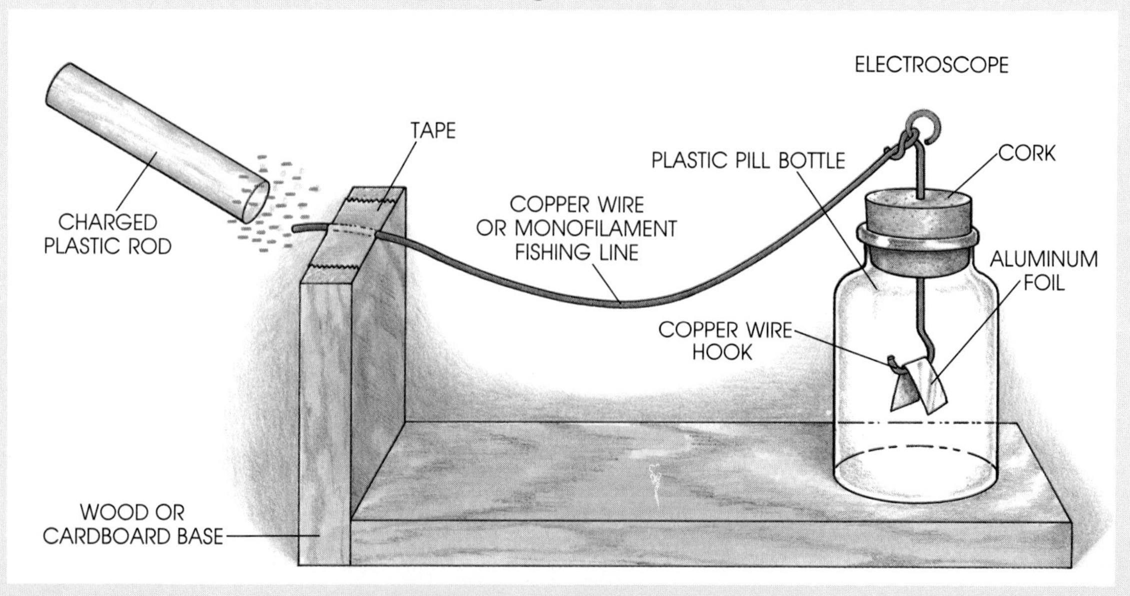

Basic Activity #2: Headphone Microphone

Most headphones work opposite to the way a microphone works, but this exercise demonstrates how much the two are alike.

Materials and Equipment

audio tape recorder
headphone set
blank audio tape

Procedure

1. Plug the headphone set into the "microphone in" (or "aux in") spot on the audio tape recorder. Be sure the plug fits the recorder. Sometimes an adapter is needed.
2. Set the tape recorder in the "record" mode.
3. Talk or sing into the earpiece of the headphone set.
4. Rewind the tape and play it back. Was any sound recorded on the tape? Why or why not?

Basic Activity #3: Communication Satellite

More than 200 communication satellites orbit the earth. They are used to relay radio, television, telephone, and computer data messages all over the globe.

Materials and Equipment

foam core board
cardboard
aluminum foil
glue
tape
soda straws

Procedure

1. Find several different pictures of communication satellites (in addition to the ones in the textbook).
2. Design a model of a communication satellite, using the various pictures as a guide.
3. Construct the model using the materials noted above, and any others you feel you need.
4. Communication satellites work in similar ways, but they don't all look alike. Why do you think this is so?

Basic Activity #4: "Gel" Filters

Lighting for video productions sometimes requires the lights to be colored for special effects. This is done by mounting colored filters, known as "gels," in front of the lights. In this activity, you will make a set of gels for use in videotaping.

Materials and Equipment

cellophane in several different colors
medium weight wire
tape or glue

Procedure

1. Bend a wire frame for each gel. It is a good idea to design this frame so that it may be easily mounted on the lights used during videotaping. Experiment with different ways of doing this.
2. Remember that the lights get hot, so you will have to mount the gels so that they don't warp, burn, or melt. Experiment to see how far away the gels will need to be.
3. Once you get a frame design that works, cut cellophane to fit it and tape or glue the cellophane in place on the frame.
4. Repeat this process for several different colors of cellophane. Later, you can experiment to see what different effects they create when used for videotaping.

Intermediate Activities

Intermediate Activity #1: Induction

As you learned in this section, current may be caused to flow in a wire when it is drawn over a magnet or when a magnet is moved inside a wire coil.

Materials and Equipment

soda straw (3" section)
30 feet of 30-gauge insulated wire
tape
iron nail
magnet
meter that measures microamperes

Procedure

1. Scrape the insulation off the ends of the wire. Make a coil by wrapping the wire around the soda straw as in Fig. VI-2.
2. Insert the nail inside the straw.
3. Connect the ends of the wire to the meter.
4. Move the magnet back and forth along the coil. Watch the meter to see the effect.
5. Remove the nail and move the magnet again. What happens to the meter?

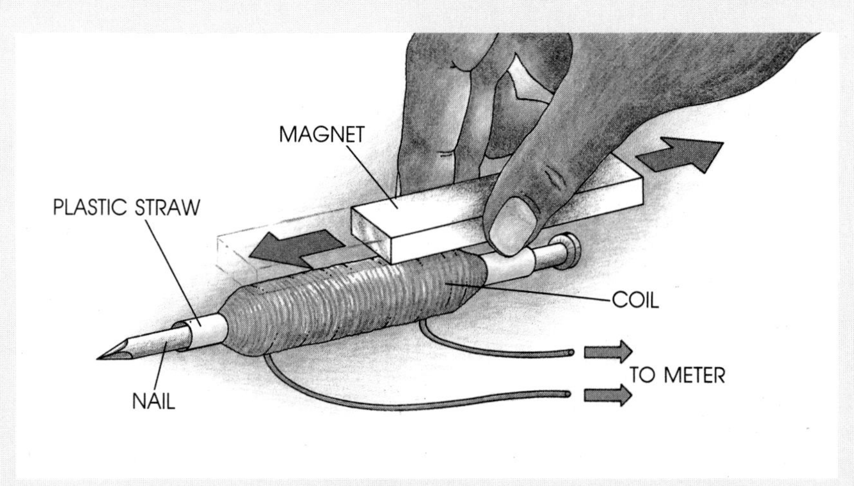

Fig. VI-2.

Intermediate Activity #2: Sound Effects

For radio or television programs, prerecorded audio effects are commonly used. A barking dog, a door opening and closing, running water, and rain are all typical sound effects. Suppose you were an audio engineer for one of your favorite programs. What sounds would you have on hand?

Materials and Equipment

portable audio cassette recorder with microphone
blank audio cassette

Procedure

1. Record your own voice to check the audio levels on the recorder. Practice adjusting the controls and listening to the results in order to make a good quality recording.
2. Record as many different sounds as you can. Keep a log of each sound and its location on the tape. You should be able to find it easily when you need it.
3. Play the tape for the class. Which sounds were easily recognized? Which were not?

Intermediate Activity #3: Microphone/ Light Stand

A boom microphone is mounted on a pole so that it may be held out of view during videotaping. These mikes may be held by hand, or they may be mounted on a floor-stand. Similarly, lights used in videotaping may be mounted on a floor-stand for ease of use. In this activity, design and build a stand that could be used for mounting either a boom microphone or a light.

Materials and Equipment
2″ × 2″ lumber
one- or two-gallon can
concrete
several 12d nails

Procedure
1. Cut the 2″ × 2″ lumber to about a 5-foot length.
2. Drive the nails through the piece of lumber several inches from one end. Place this end in the can.
3. Mix the concrete and pour it into the can around the pole. Support the pole in a vertical position while the concrete sets up overnight.
4. Design a means of attaching a light or microphone to this pole. You may want to attach another 2″ × 2″ pole to act as an "arm," for example.

Intermediate Activity #4: Network Simulation

Telephone networks are a series of switching stations. You can make your own switching network to demonstrate this concept.

Materials and Equipment
insulated copper wire
one or more flashlight batteries
several flashlight bulbs
2-foot square of cardboard or thin plywood
several multiple pole switches
staples and/or tape

Procedure
1. Design a network connecting several switches to each other. Plan to connect a flashlight bulb to each of the poles of the switches. Be sure the entire network can receive power from the flashlight batteries. An example is shown in Fig. VI-3.

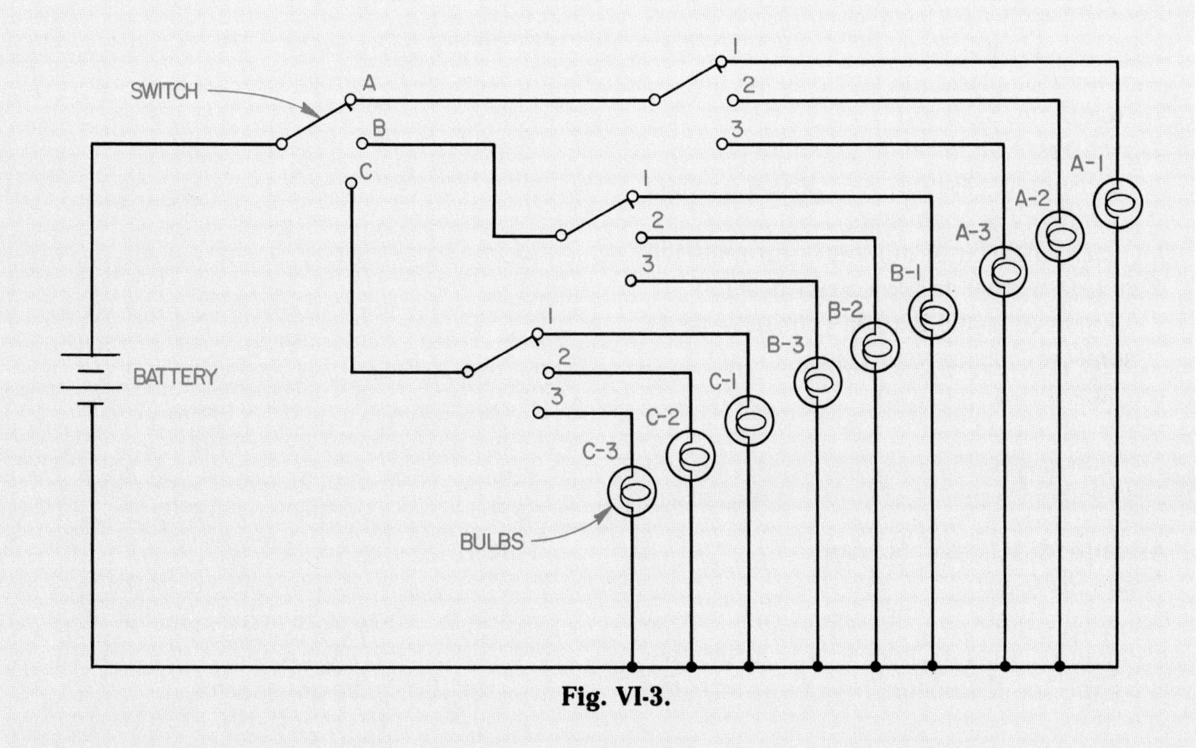

Fig. VI-3.

2. Construct the network on the sheet of plywood or cardboard. Hold the wire and switches in place with tape or small staples.
3. Number each of the poles on each of the switches. Number the light bulbs with the same number as is on the pole of the switch to which it is connected.
4. "Dial" a light bulb's number by turning each of the switches to the correct position.
5. Demonstrate and explain your network to someone outside your class.

Advanced Activities

Advanced Activity #1: Audio Advertisement

Plan and produce a 30-second radio advertisement for either a school activity, an elective class you think students may want to know about, or a community service or facility.

Materials and Equipment

audio cassette recorder with microphone
blank audio cassette tape

Procedure

1. Think about different ideas that might work well to attract attention to your message. Remember, you can only use sound.
2. Develop a script complete with sound effects, background music, spoken lines, etc.
3. Rehearse your script to be sure it exactly fits the 30-second time allotted. Change the script as needed to fit the time.
4. Tape the radio ad. You will probably need several tries to get it right. Don't be discouraged. This is just how it is done in the real world of radio advertising.
5. Play your commercial for the class.

Advanced Activity #2: Video Advertisement

Making a TV advertisement will give you an idea of how much work goes into any video production. Plan and make a 30-second TV commercial for one of the following: going out for a school sport, opposing drugs, improving the school environment (removing litter, planting trees, etc.), a school election, or another subject approved by your teacher.

Materials and Equipment

storyboard material (paper, pencil, colored pencils/marking pens, etc.)
video recording and playback equipment
blank videotape
video editing equipment (optional)

Procedure

1. Think about different ideas that could be used. Choose the one that will make best use of visual qualities.
2. Develop a storyboard that illustrates each shot or setup.
3. Develop a script that follows the storyboard. The script should include all spoken lines, background music, special effects, etc.
4. Edit the script and storyboard so they will fit into the 30-second time allotment. Design a set, including lighting and microphone arrangements. This set design should be put on paper so that it can be referred to during the video production.
6. Mark the script with video and audio instructions to be followed during taping.
7. Record all special effects, background music, etc.
8. Rehearse the script until you are able to run through it smoothly.
9. When you are comfortable with all aspects of the commercial, videotape the production. If you have video editing equipment, you may shoot different scenes and then edit them together later. If not, plan on a series of short "takes" that will "add up" to your commercial, or one long take.
10. Show your production to the rest of the class.

Appendix A: U.S. Common Weights and Measures

Linear Measures

1 mile = 1760 yards = 5280 feet
1 yard = 3 feet = 36 inches
1 foot = 12 inches
1 mil = 0.001 inch
1 fathom = 2 yards = 6 feet
1 rod = 5.5 yards = 16.5 feet
1 hand = 4 inches
1 span = 9 inches

Area Measures

1 square mile = 640 acres = 6400 square chains
1 acre = 10 square chains = 4840 square yards = 43,560 square feet
1 square chain = 16 square rods = 484 square yards = 4356 square feet
1 square rod = 30.25 square yards = 272.25 square feet = 625 square links
1 square yard = 9 square feet
1 square foot = 144 square inches

Diameter Measures

1 circular inch = area of circle having a diameter of 1 inch = 0.7854 square inch
1 circular inch = 1,000,000 circular mils
1 square inch = 1.2732 circular inches

Cubic Measures

1 cubic yard = 27 cubic feet
1 cubic foot = 1728 cubic inches

Liquid Measures

1 U.S. gallon = 0.1337 cubic foot = 231 cubic inches = 4 quarts = 8 pints
1 quart = 2 pints = 8 gills
1 pint = 4 gills

Appendix B: Metric Conversion Factors

Linear Measures

1 kilometer = 0.6214 mile
1 meter = 39.37 inches = 3.2808 feet = 1.0936 yards
1 centimeter = 0.3937 inch
1 millimeter = 0.03937 inch

1 mile = 1.609 kilometers
1 yard = 0.9144 meter
1 foot = 0.3048 meter = 304.8 millimeter
1 inch = 2.54 centimeters = 25.4 millimeters

Area Measures

1 square kilometer = 0.3861 square mile = 247.1 acres
1 hectare = 2.471 acres = 107,639 square feet
1 are = 0.0247 acre = 1076.4 square feet
1 square meter = 10.764 square feet
1 square centimeter = 0.155 square inch
1 square millimeter = 0.00155 square inch

1 square mile = 2.5899 square kilometers
1 acre = 0.4047 hectare = 40.47 ares
1 square yard = 0.836 square meter
1 square foot = 0.0929 square meter
1 square inch = 6.452 square centimeters

Cubic Measures

1 cubic meter = 35.315 cubic feet = 1.308 cubic yards
1 cubic meter = 264.2 U.S. gallons
1 cubic centimeter = 0.061 cubic inch
1 liter = 0.0353 cubic foot = 61.023 cubic inches
1 liter = 0.2642 U.S. gallons = 1.0567 U.S. quarts

1 cubic yard = 0.7646 cubic meter
1 cubic foot = 0.02832 cubic meter = 28.317 liters
1 cubic inch = 16.38706 cubic centimeters
1 U.S. gallon = 3.785 liters
1 U.S. quart = 0.946 liter

Weight Measures

1 metric ton = 0.9842 ton (long) = 2204.6 pounds
1 kilogram = 2.2046 pounds = 35.274 ounces
1 gram = 0.03527 ounce
1 gram = 15.432 grains

1 long ton = 1.016 metric ton = 1016 kilograms
1 pound = 0.4536 kilogram = 453.6 grams
1 ounce = 28.35 grams
1 grain = 0.0648 gram
1 calorie (kilogram calorie) = 3.968 Btu

GLOSSARY*

A

abacus — a calculator that works by sliding beads along its wires. (74)

additive primary colors — the colors of light from which all other colors can be created; red, green, and blue. (216)

aligned dimensioning — dimensions placed so they are read parallel to either the bottom or right side of the drawing. (146)

alternating current — electrons flowing first in one direction and then in the opposite direction. (351)

American National Standards Institute — organization that sets all drafting standards. (142)

American Sign Language (ASL) — a communication system developed for those who cannot hear. (21)

amplifier — a device used to strengthen an electrical signal or control the level of an audio signal without changing its waveform. (363)

amplitude — a measurement of the intensity (strength) of a wave form. Sometimes referred to as a wave's height. (352)

amplitude modulation — the process of changing the amplitude, or strength, of a carrier wave; AM. (353)

analog computers — computers that work with data that vary in a continuous way, such as temperature, pressure, or speed. (74)

analog signal — a constantly changing voltage of electricity used to send a message. (112)

animal communication — ways in which animals relay information to one another. (22)

answering machines — devices that answer phone calls and record messages. (384)

antennas — devices used either to transmit or receive electromagnetic signals. (364)

antihalation layer — a film layer that absorbs light and prevents it from reflecting back through the film emulsion. (242)

aperture — a camera lens opening that is created by the diaphragm. (257)

applications software — computer programs that handle many different tasks. (85)

artificially intelligent programs — computer programs that have the ability to store user input, process it, and react differently in the future based upon this input. (85)

ASCII — American Standard Code for Information Interchange; computer code system. (80)

assembling — the process of bringing together printed sheets by collating, gathering, or inserting. (331)

ASSEMBLY — a general-purpose computer programming language. (85)

assess — evaluate. (48)

atmospheric transmission channels — electromagnetic waves used to carry information through the atmosphere. (354)

audio — something we hear. (347)

audio and video communication — relaying information by means of radios, televisions, telephones, record players, and tape players. (24)

audio console — mixer; a device that allows an audio engineer to control the volume and quality of incoming signals. (363)

automatic film processing — a mechanical setup that automatically moves film through the developing process. (305)

*The numbers in parentheses after each definition are page numbers on which more information can be found.

auxiliary views — views in a multiview drawing used to describe an oblique feature. They are drawn at an angle, so that the oblique feature can be seen in its true size and shape. (167)

axonometric projection — a drawing method similar to orthographic projection except that the object being viewed is tilted or rotated so three sides are seen in the frontal plane or view. (168)

B

back light — a light placed behind a subject that helps to create a three-dimensional appearance in a photograph. (397)

balance — an impression of steadiness created by placing all the elements of a design in a certain way. (284)

base layer — layer beneath the emulsion layer in a piece of film; consists of a thin sheet of clear plastic. (242)

BASIC — a general purpose programming language. (85)

baud — bits per second. (114)

Bell, Alexander Graham — inventor of the telephone. (358)

bilateral tolerancing — tolerancing in which variations are allowed in both directions. (150)

binary system — a counting system based on multiples of 2. (79)

binding — fastening assembled pages together permanently after printing. (331)

bit — a tiny amount of computer data. (80)

block diagrams — drawings that show the flow of electrical power in a system. (174)

blueprint — a building plan with white lines on a deep blue background. (157)

boom microphone — a microphone mounted on a pole or cart that may be driven around a TV set. (397)

bulk load — the practice of buying 100-foot rolls of film and winding it onto individual small rolls. (244)

burning-in — the process of darkening a too-light area in a photograph during enlarging. (264)

bus — system by which bits of data constantly travel from one component to another inside a microcomputer. (83)

byte — a chunk of eight bits of computer data. (80)

C

C — a general-purpose computer programming language. (85)

cable release — a tool that attaches to the shutter release of a camera. It helps prevent jiggling the camera when using slow shutter speeds. (242)

CAD — computer-aided drafting or computer-aided drafting and design. The drawing is displayed on a computer monitor as the drafter creates it. (120)

CAM — computer-aided manufacturing. (120)

camcorders — camera and audio recorders combined into one unit. (380)

camera angle — a camera position. The "normal" camera angle is at the eye level of the subject. (396)

carbon microphone — a microphone using a small cup filled with carbon granules. (358)

cartridge — the small rectangular object at the end of the tone arm on a phonograph; picks up vibrations from the record groove. (376)

case binding — the process of creating a hard bound book; signatures are assembled, sewn or glued together, and then given a rigid cover. (331)

cathode ray tube — a large glass vacuum tube that is flat on one end and coated with phosphorescent salt. The salts glow and create the image seen on a computer monitor or TV screen. (99)

CBs — citizens' band radios. (386)

CD-ROM disk — an optical computer storage disk. CD-ROM stands for "compact disk, read-only memory." Data cannot be erased from, changed, or added to a CD-ROM disk. (105)
cellular telephones — phones in cars that use radio batteries and atmospheric transmission. (384)
central processing unit (CPU) — a microprocessor; it processes all input and provides control for the computer. (81)
chain dimensioning — a drafting technique in which every dimension on a drawing begins with another dimension. (151)
character generator — a computer that produces text, titles or credits for video. (372)
charge-coupled device — a special type of microchip containing a very fine grid of light-sensitive capacitors that creates a video signal; CCD. (370)
CIM — computer-integrated manufacturing. (121)
cladding — a covering material used on optical fibers. (226)
clearance fits — fits needed where freedom of motion around moving parts is required. (151)
clip art — artwork (often bound in a large book) that can be clipped out for use in graphic designs. (293)
closed-circuit television — a TV transmission that travels to only certain receivers. (391)
coaxial cable — a cable designed to carry video signals; consists of a number of copper wires surrounded by plastic insulators. (374)
collating — the process of assembling printed sheets in a special order. (331)
color balance — a process done to color films to balance them for the light source being used. (246)
color blind — a condition caused when cone cells in a person's eyes are absent or not fully formed. (221)
color correction filters — filters used with color film to correct for the type of lighting used. (240)
color scanners — computers that create quality color separations. (307)
color sensitivity — a rating that describes how sensitive different films are to different colors. (243)
color separation — a process used to create the illusion of a range of colors on the printed page. (306)
color system — a series of colors that have each been given a number for identification. (291)
color transparencies — photographic slides. (246)
Colossus — one of the first true computers built by the British in 1943; designed to decipher coded messages used by the Germans in World War II. (76)
communication — the sharing of information, thoughts, and ideas. (15)
communication concepts — principles of designing, encoding, storing, retrieving, transmitting, receiving, and decoding. (20)
communication technology — using knowledge, tools, and skills to communicate. (17)
communications system model — a description of a communications system; message, sender, channel, and receiver. (19)
compact disks — optical storage disks that store an audio signal digitally; CDs. (379)
compass — a drafting tool used to draw circles and arcs. (157)
composition — [1] the way in which all the elements in a photograph or design are arranged. (252) [2] The process of converting words and illustrations into a form that will be ready for printing. (292)
compound lens — a lens formed when concave and convex lenses are combined. (219)
comprehensive layout — a full-color layout that shows how a product will look after it has been printed. (291)
computer — a machine that makes calculations and processes information very quickly. (71)
computer-aided design (CAD) — using a computer to make drawings. (119)

computer-aided manufacturing (CAM) — the use of computers to control manufacturing processes. (120)

computer control systems — computer systems that collect input, process data, and then output controlling signals to other devices. (23)

computer-integrated manufacturing (CIM) — the complete manufacturing process overseen by computers. (121)

computerization — the reliance of more and more systems upon the power of computers to make them work. (33)

computer systems analysts — the people who figure out how a business can make the most of computers. (128)

computer virus — a computer program hidden within another program. (57)

concave — thinner in the center and thicker at the edges; a term used to describe lenses. (219)

condenser microphone — the most widely used type of microphone. Uses a metal diaphragm mounted near a fixed metal plate. (362)

conductors — materials that allow electricity to pass through them easily. (349)

cones — color-sensitive cells of the eye. Some record red, some green, and some blue light waves. (221)

conference call — a phone call in which several people may talk using different phones but the same connection. (385)

contact print — a sheet of small prints that allows photographers to preview the pictures; a proof sheet. (263)

content — words and illustrations in a graphic design. (283)

continuous lights — lights that remain turned on. (239)

continuous-tone images — illustrations composed of varying shades of gray. Black and white photographs are a form of continuous-tone images. (294)

contour drawings — drawings used to show the elevation, or height, of land forms. (180)

contrast — the amount of difference between light and dark areas in a photograph. (244)

contrast filters — filters used to enhance the difference between light and dark areas in a photograph. (264)

control — one machine directing another. (23)

control key — a key that lets you call upon special features of a computer and software. (91)

control room — the part of a recording studio where the control equipment, director, technical director, and engineers are located. (371)

conventions — graphic symbols and presentations used for drafting in a specialized field. Conventions provide information about a part without having to draw it as it would appear or write notes about its makeup. (172)

convex — thicker in the center and thinner at the edges; a term used to describe lenses. (218)

copy — words set into type. (292)

copyboard — the surface of a process camera on which the mechanical or camera-ready copy is placed. (300)

copyfitting — the process of making text fit the available space in a layout. (293)

cornea — a clear, curved tissue located at the front of the eye. It lets in the light rays and helps focus them on the back of the eye. (220)

cultural — having to do with the skills and arts developed during a given period in time. (49)

cursor — a short blank line that blinks and indicates position on a computer screen. (91)

curves — drafting tools used to draw curved lines. (156)

D

Da Vinci, Leonardo — a great fifteenth-century artist, architect, and engineer who used technical drawings. (139)

dampening — the process of moistening an offset plate with fountain solution. (316)

data communication systems — computer systems. (24)

data security — the protection of computer data from theft or tampering. (57)

data transmission channel — a means of carrying a message between an input device and an output device. There are two types of data transmission channels: physical and atmospheric. (113)

data — information. (32)

database — a long list of short computer records. (118)

depth of field — the amount of a camera's field of view, from front to back, that remains in focus. (257)

design elements — line, shape, form, space, color, texture, and shades of dark or light. (286)

desktop publishing — computer-aided pub lishing. (294)

detail drawing — a drawing that gives all the information needed to make a part or a section of construction. (173)

developer — a chemical that causes the silver crystals on exposed film to turn black. (247)

developing tank — a light-tight container that holds one or more developing reels of film. (248)

development — research done with the idea of solving a specific problem; results in a product or method. Applied research. (25)

diagrammatic drawings — drawings that represent the location of and distances between structural members in civil engineering projects. (180)

diagrams — electrical drawings. (174)

diaphragm — a flexible piece of metal that vibrates when sound waves strike it. (358)

diazo prints — reproduction prints having dark lines on a white background; sometimes called whiteprints. (157)

dictaphone machine — a machine that allows a letter to be dictated on tape and played back later for typing. (389)

Dictionary of Occupational Titles — a government publication describing specific careers. (27)

die cutting — a process by which part of a substrate is cut out by stamping it with a cutting die. (330)

digital — measured in precise units. (74)

digital camera — a camera that stores images in digital format rather than on film. The images can be downloaded to a computer. (236)

digital computers — computers that require data to be in precise amounts, such as binary numbers. (74)

digital editing — the process of changing information in computer files. (56)

digitization — the shift from analog to digital systems. (37)

digitizing tablet — an electronic drawing board. An electronic pen, connected with a wire to the tablet, allows the user to create an image on a computer screen. (93)

dimension lines — lines drawn between extension lines on a technical drawing. The dimension is written in a break in the line. (147)

dimensions — measurements. (145)

dimetric drawings — technical drawings that show an object using two equal axes. (169)

dipole antennas — lightweight, expensive antennas that have a horizontal shape. (366)

direct current — electrons (current) flowing in one direction. (351)

director — a person who oversees the actual making of an audio or video program. (394)

dividers — a drafting tool used to transfer measures from one location to another or to check measurements made on a drawing against the final product. (157)

dodging — the process of lightening a dark area in a photograph to improve detail. (264)

dot matrix printer — a printer, the print head of which is made up of a series of thin blunt-ended metal pins called wires. (100)

drafter — a worker who makes technical drawings. (176)

drafting film — medium on which technical drawings are made. (153)

drafting machine — a drafting tool with two perpendicular blades attached to a rotating head used to draw horizontal, vertical, and angled lines. (154)

drawing to scale — drawing objects or features larger or smaller than true size but in the correct proportions. (152)

drum plotter — a plotter in which both the paper and pen move; used for larger drawings. (188)

E

E-mail — messages sent through computer networks; electronic mail. (40)

economic — having to do with the economy. (49)

Edison, Thomas — the inventor of the phonograph. (375)

editing — making changes in prerecorded programs. (372)

electret condenser — a microphone that can be made as small as a tie tack. (371)

electrical circuit — the path electricity follows from its source, through a conductor, to a receiving device. (349)

electricity — the flow of free electrons. (349)

electromagnetic waves — waves created by a magnetic field that make communication without a connecting wire possible. (349)

electronic banking — banking done by means of automatic teller machines (ATMs). (123)

electronic communication model — a systems model consisting of a transmitter, transmission channel, and receiver. (348)

electronic editor — a device that allows a director to select parts of an original tape and record them onto a blank tape in any order desired. (372)

electronic flash — light that illuminates a subject only at the exact instant the picture is taken. (239)

electronic listening devices — hidden devices that monitor conversations. (388)

electronic pagination system — a system that combines computer-created text with color scanning. A single operator can compose an entire full-color page electronically. (308)

electronic publishing systems — microcomputers and laser printers combined into publishing systems. (294)

electrons — tiny, negatively-charged particles that make up part of an atom. (99)

electrostatic printing — a printing method that relies upon a charge of static electricity to transfer the message from a plate to the substrate. (326)

elevation — a drawing that shows how a building looks from the front, rear, or sides. (177)

embossing — a process that creates a raised image on a substrate. (329)

emphasis — making one element in a design stand out. (285)

emulsion layer — a film layer that contains tiny light-sensitive silver halide crystals. (242)

emulsion side — the dull side of a line negative on which the image appears backwards. (302)

ENIAC — Electronic Numerical Integrator and Computer; an early computer. (76)

enlarger — device used to "blow up" (enlarge) film negatives. (248)

entrepreneur — a person who organizes and manages a new business. (78)

environmental — relating to our physical environment. (49)

erasable optical disk — a storage disk that can be erased and rewritten over and over. (106)

ethical — relating to matters of right and wrong. (49)

exposure control — a camera adjustment regulating the right amount of light. (257)

exposure — the amount of light reaching film in a camera. (301)

extension lines — lines used to show the beginning and end of a dimension. (147)

extension tubes — hollow tubes that fit between the camera body and the lens and are used for close-focus photography. (238)

exterior elevations — drawings that show how a building looks from the outside. (177)

F

f-stops — camera settings, each of which lets in half or twice as much light as the next. (257)

facsimile transmission (fax) — sending data, such as text or photographs, from place to place via telephone lines. (118)

fade — a video technique in which the image gradually appears on (or disappears into) a black screen. (372)

fax machines — devices that send drawings and other information over telephone lines. (118)

FCC — Federal Communications Commission. (58)

feedback — something that happens as a result of the outputs of a technological system. (18)

feeding unit — a device that carries paper or another substrate into an offset press. (316)

field of view — what a camera sees. (237)

fill lights — lights used to get rid of shadows created by a key light. (397)

film animation — the process of creating color images on clear cellulose film to make a cartoon. (267)

film assembly — grouping together negatives and preparing them for printing. (281)

film conversion — photographing graphic components such as mechanicals or continuous-tone art to produce film negatives or film positives. (300)

film positive — a film image that is black in the image areas and clear in the non-image areas; opposite of a negative. (306)

film recorders — recorders designed to output to light-sensitive films. (103)

film winders — automatic devices inside a camera that advance the film after each exposure. (240)

filters — devices used to screen out certain wavelengths of light. (240)

finish — the appearance of a paper's surface. (290)

first angle projection — a method of laying out a multiview drawing in which the front view is on top, the left side view directly to the right, and the top view on the bottom. Used in some European countries. (166)

fisheye lens — a wide-angle lens. (238)

fixer — a chemical that makes an exposed film image permanent and removes any unexposed silver crystals. (247)

flash exposure — an exposure made in a process camera to reinforce the dots in the shadow areas of a halftone. (305)

flat — a series of negatives taped to a masking sheet. (309)

flatbed plotter — a plotter in which the paper (or other substrate) remains stationary and the pen moves. (188)

flexography — a form of relief printing; the carrier is made of rubber or plastic. Also called aniline printing. (318)

flood lights — continuous lighting that fills an area. (239)

floor plans — drawings that show the inside of a building viewed from above. (176)

floppy disks — magnetic storage medium for computers. (104)

focal length — the distance between the center of the lens of a camera and the film, when the lens is focused to infinity. (237)

focal point — the point where the rays meet when light is concentrated. (218)

foil stamping — a process by which a colored metal foil image is fused to a substrate with heat and pressure. (329)

fonts — an assortment of sizes, thicknesses, and other variations of a given typeface. (288)

formal balance — a type of balance achieved when a line drawn through the center of a design would create two halves that are symmetrical. (284)

FORTRAN — a general-purpose programming language. (85)

foundation plans — drawings used to show how the weight of a building or other structure will be spread out and supported. (176)

fovea — the center of the macula of the eye; the point of clearest vision. (220)

frequency — the number of waves (of light, sound, etc.) that pass through a given point in one second. (212)

frequency bands — certain groups of radio frequencies used for specific types of audio or video communication. (352)

frequency modulation — the process of crowding carrier waves together or spreading them farther apart to change their frequency; FM. (353)

frequency response — the range of sound frequencies a microphone can reproduce well. (358)

full section — a sectional view drawing in which the object is shown cut in half. (168)

function keys — computer keys marked on the keyboard as "F1," "F2," "F3," etc. Each performs a special function when struck. (91)

G

gate — a computer logic circuit. (80)

general design engineering drawings — drawings used to show the arrangement of structure and how its members are assembled. (178)

geodesy — the practice of locating the exact positions of features along the earth's surface. (180)

geostationary orbit — the orbit an object makes when it travels at the same speed as the earth rotates; it remains above the same part of the earth at all times. (43)

grain — a pattern that appears in the silver crystals when film is developed. (244)

graphic designer — an artist who creates designs using graphic materials. (284)

graphic message — words and illustrations in printed materials. (295)

graphic production — printing. (24)

graphics — text and illustrations in printed matter. (283)

gravure — the process of transferring ink from image areas below the surface of a printing plate. (320)

gray scale — a strip of special paper divided into steps that serves as a guide in determining how long a photo negative should be left in the developer. (302)

ground waves — radio waves that follow the curvature of the earth. (365)

H

half-section — a sectional view in a technical drawing in which only a quarter of the object is removed. (168)

halftone photography — photography that changes continuous-tone images to patterns of dots. (304)

hams — amateur radio operators. (386)

hard disks — rigid "platters" permanently located inside a computer drive and used for information storage. (105)

hardware — parts of a computer system, other than software. (79)

harmony — the effect achieved when the various elements of a message design work well together. (286)

Harvard Mark I — the first full-scale computer; built by IBM. (76)

hertz — one cycle per second; unit of measurement used to measure radio waves. A kilohertz (kHz) is 1,000 cycles per second. A megahertz (MHz) is one million cycles per second. A gigahertz (GHz) is one billion cycles per second. (352)

highlights — light reflecting off the lightest areas of a photograph. (304)

holography — the use of lasers to record realistic three-dimensional images. (225)

hue — the name of a color, such as red, green, or blue. (216)

human-to-human communication — one person talking to another. (21)

human-to-machine communication — humans communicating with machines. (23)

hybrid computers — computers that combine digital and analog operations. (74)

I

incident-light meters — meters that measure the amount of light falling on a subject. (239)

induction — the process of creating electricity by means of a magnetic field. (351)

industrial photographer — a photographer who makes pictures of products and processes, visual materials for presentations and displays, and photographs for brochures, catalogs, and other promotional materials for industry. (270)

informal balance — balance achieved in a design by means of an arrangement of elements that may look different but that have equal weight to the eye. (284)

infrared films — films sensitive to nearly all visible light, plus invisible infrared wavelengths. (243)

ink jet printer — a printer, the print head of which is made up of tiny nozzles that spray ink onto the paper. (101)

inking — spreading a thin film of ink over the image area on an offset printing plate. (316)

input — [1] information, materials, energy, financial resources, and human effort put into a technical system (18); [2] information put into a computer. (90)

instrumentation — using a machine to collect data. (23)

insulators — materials, like plastic, that do not conduct electricity. (401)

integration — combining or overlapping of communication systems. (37)

intensity — the brightness or dullness of a hue. (216)

interactive television — TVs and computers hooked up together. (38)

interface — connect; refers to a computer network. (114)

interference fits — fits indicated on drawings for parts that fit together without use of fasteners, welding, or adhesive. (151)

interior elevation — a drawing that shows construction or design features of the inside of a building. (177)

Internet — a worldwide computer network used to send and receive audio and video data. (41)

ionosphere — a layer of electrically charged particles from 60-200 miles above the surface of the earth. (352)

iris — the colored part of the eye. It looks like a flat doughnut, and its opening can be made larger or smaller. (220)

ISO — International Standards Organization. (142)

isometric drawings — drawings in which the object is tilted so its edges become axes that form equal angles. (168)

J

Jobs, Steve — co-founder of Apple Computer Company. (78)

joystick — a computer control that when rotated moves an object around the screen. (92)

K

kelvin — the unit of measurement for the temperature of light. (217)

key light — the main source of studio light for a photographic subject. (397)

keyboard — a group of keys like those of a typewriter used for inputting data into a computer. (90)

L

lacquer — a treated aluminum disc used to create records. (376)

laser — (Light Amplification by Stimulated Emission of Radiation) a narrow beam of parallel light waves created using a crystal, a gas, chemicals, dyes, or semiconductors. (224)

laser printer — a printer that uses a computer-driven laser beam to leave an electrostatically charged image on a drum. The charged image area is transferred to the paper. (102)

layers — parts of a computer (CAD) drawing similar to the transparent overlays used in manual drawings. (190)

lead substitutes — plastic drawing medium designed for surfaces where lead wear and smearing is a problem, such as polyester drawing films. (154)

leader lines — straight lines used to help dimension or describe geometric shapes in technical drawings. (147)

lens — [1] a clear, bean-shaped tissue in the eye that can become thicker or thinner to adjust for viewing close up or far away. It helps to focus the light rays on the back of the eye. (220) [2] A piece of transparent material that is used to focus light. (218)

letterpress printing — printing with movable metal type. (318)

light — a stream of escaping photons (atomic particles) that sometimes also behaves like a system of waves. (212)

light integrator — a timer that controls film exposure. (301)

light meter — a device that measures the amount of light present for photography. (239)

light sensitivity — the speed with which film reacts to light; also, film "speed." (243)

limits and fits — specifications that tell how parts will fit together for proper assembly. (151)

line art — illustrations made up of solid, dark (usually black) lines and shapes drawn on a white surface. (293)

line of sight transmission — the range of a TV signal; approximately the area visible from the transmitting tower. (373)

line photography — the process of converting line copy to a line negative. (302)

local area networks — computer networks that send information over short distances; LANs. (114)

location drawings — simple drawings of a structure and the locations of its parts. (179)

logic diagrams — electrical drawings that show the operation of a circuit made up of logic gates. (174)

long distance network — computer network in which data is sent over telephone lines or by satellites. (115)

M

machine communication — machines communicating with one another. (23)

machine-to-human communication — machines communicating with humans. (23)

macro lenses — lenses that focus on very small objects at very close range. (238)

macula — an oval area in the retina of the eye responsible for color and detail vision. (220)

magnetic media — computer storage media that rely upon oxides coated onto a plastic base; includes floppy disks, hard disks, and tape. (104)

magnetic tape — computer storage medium wound on a reel; usually used for back-up copies of data. (105)

Maiman, Theodore H. — American who built the first ruby laser in 1960. (225)

mainframe computers — computers that have a large capacity but are also bulky. A single mainframe computer can serve many users. (78)

Marconi, Guglielmo — the inventor of the radio. (362)

mass media — media that reach many people; television, radio, newspapers, magazines, and books. (49) mass — the amount of matter in an object. (20)

mechanical — a paste-up; used for graphic production. (295) message transfer — printing. (314)

microchips — integrated circuits on a base of silicon; the basic building blocks of a computer. (77)

microfiche — a sheet of microfilm with several rows of small drawings on it. (158)

microfilming — a storage method in which drawings are reproduced at a greatly reduced scale and only the film is stored. (158)

microphones — devices that change sound energy into an electrical signal. (362)

microprocessor — a circuit that processes all input and provides control for a computer. (81)

microwaves — electromagnetic waves that are shorter and more focused than radio waves. (360)

miniaturization — the process of making things smaller. (36)

mixing — bringing together live and prerecorded sounds to make an audio transmission. (363)

modem — a device that changes digital data output from a computer into a series of analog tones, or vice versa. (113)

modulation — the alteration of radio waves in order to send messages using them. (352)

mouse — a hand-held device that is moved around on a surface and in turn moves images on a computer monitor. (95)

moving coil dynamic speaker — the most common type of speaker used; uses a voice coil connected to a flexible paper speaker cone. (366)

multimedia — the combining of several media, such as text, sound, animation, and video. (96)

multi-mode optical fibers — optical fibers that have larger cores and can accept light from a variety of sources. (226)

multiplexing — a system that allows two or more signals to be sent over a transmission channel at the same time. (360)

multiview or working drawings — a type of drawing that shows an object from several different views, or angles; as a result, it describes the object completely. Multiview drawings are used in manufacturing an object. (164)

N

nanosecond — one billionth of a second. (81)

negative — exposed film; the light areas of the scene appear dark and the dark areas appear light. (222)

networking — linking computers to one another. (113)

notes — information placed on a drawing to improve the dimensioning process or describe something about the entire drawing. Local notes are used to describe local features. (149)

novelty typefaces — typefaces that do not fit neatly into any of the other major styles. (289)

nucleus — the center of an atom; the nucleus is positively charged. (212)

numeric keypad — computer number keys that look like the keys of a simple calculator. (90)

O

object beam — a beam that shines on an object and is reflected onto a photographic plate to create a hologram. (225)

oblique drawing — a drawing in which the irregular side of the object (front, top, or side) is seen face on. The other two sides are drawn along two axes and a look of depth is created. (169)

offset lithography — printing an image by first transferring, or offsetting, it onto a rubber blanket. (314)

Ohm's Law — current equals voltage divided by resistance. (401)

one-point perspective drawing — a perspective drawing that uses one vanishing point. (172)

on-line database — a package of information that may be stored and retrieved by computer. (40)

on-line services — computer services, such as CompuServe® and America Online®, that allow users to send and receive e-mail, purchase goods, access information, and download computer programs. (116)

optical computer — a computer that uses laser beams instead of electric wires to process information. (36)

optical fibers — transparent fibers of glass; used to send messages. (226)

optical storage disks — medium used to store computer data; include CD-ROM's, WORM's, and erasable optical disks. A laser is used to write and read the data. (105)

optic systems — technical systems that use light to transmit and record an image or other type of information. (211)

orthochromatic films — films sensitive to nearly all visible light except red. (301)

orthographic projection — a method of making a multiview drawing in which the views are drawn at right angles, or perpendicular, to one another. (164)

oscillator — a device that changes direct current to alternating current having a constant frequency and amplitude. (364)

outputs — the result of the process that takes place in a technological system. (18)

P

panchromatic films — films sensitive to nearly all visible light, especially blue light. (243)

paper drill — a drill press with a hollow bit used to drill holes through a stack of paper. (331)

parabolic reflectors — "dish" antennas that can transmit radio waves in straight lines toward a target. (364)

parallax — the optical effect caused by one eye seeing things from a slightly different angle than the other eye. (157)

parallel port — a bridge inside a computer that allows bits of data to travel side by side. (83)

parallel straightedges — drafting tool used for drawing parallel horizontal lines. (154)

parametric design — the design of families of similar parts that can be applied to different situations. (190)

PASCAL — a computer programming language used for science. (85)

Pascal, Blaise — Frenchman who built fifty different experimental calculators. (74)

paste up — copy and line art for graphic production pasted in place on a white "board." (295)

patch board — a device used to connect different input and output devices for audio transmission. (363)

perfect binding — the process of coating the back of assembled sheets or signatures with hot glue to hold them together. (331)

peripherals — computer hardware other than the basic components. (79)

perspective drawings — technical drawings that resemble the way something would look in real life. The object appears three-dimensional, and receding parallel lines appear to come together in the distance. (170)

photogrammetry — the use of aerial photography as the source for surveying information. (182)

photographic papers — papers used to make photographic prints. (245)

photographic stencils — stencils made by exposing them through a clear positive image. (323)

photon — an extremely tiny atomic particle. Escaping photons make up what we see as light. (212)

photopolymer — a light-sensitive plastic used to make printing plates. (319)

physical transmission channel — a wire or some other connection that carries a message between sender and receiver. (354)

pica — a type measurement equal to 12 points (or 1/6 of an inch). (289)

pickup tube — a tube inside a television camera that converts what it sees to electrical signals that are then transmitted; a vidicon tube. (369)

pictorial drawings — technical drawings that create the illusion of three dimensions by changing an object's proportions. (164)

pixels — glowing dots that make up the image on a computer monitor or TV screen. (100)

planes of projection — planes used in drawing that correspond to the six sides of a box: front, right side, rear, left side, top, and bottom. (165)

plotter — a device similar to a printer except that it actually draws, or "plots," the images on paper with a pen. Plotters are the standard output device for computer-aided design (CAD). (102)

point — a type measurement equal to (1/72) of an inch. (289)

polarized light — light, the travel of which is limited to a single plane. (215)

political — having to do with the government. (49)

polymer — a large molecule made from a chain of smaller units called monomers; plastics are polymers. (319)

port — a bridge between a computer and the device to which it is connected. (83)

potentiometer — an electrical device that varies the amount of voltage passing through it. (92)

Poulsen, Valdemar — the first person who recorded sounds on wire magnetic tape. (377)

principles of design — design rules regarding rhythm, balance, proportion, variety, emphasis, and harmony. (284)

printing on demand — documents stored in computer files and printed as needed. (42)

problem solving method — the way in which researchers go about discovery and invention. (26)

process — what happens to the input in a technological system; includes technical processes as well as the concepts and principles upon which the technology is based. (18)

process camera — a camera used to create film negatives for printing. (300)

producer — a person responsible for an audio or video program. (393)

product conversion — the wide range of operations performed on products after printing. (329)

profiles — drawings that show the rise and fall of land forms. (180)

programmable — a term describing a microchip that can carry a set of instructions. (78)

programming language — a code that allows communication with a computer. (85)

projection lines — lines that extend from all corners and edges on the front view to the other views of a multiview drawing. (167)

projection printing — making prints by projecting light through a negative onto a sheet of photographic paper. (260)

PROM — a ROM chip that may be programmed. (82)

proportion — the size relationship of one part of a design to another. (285)

public radio stations — nonprofit organizations that do not broadcast commercials. They depend on donors to fund their operations. (386)

puck — a plastic lens with two crosshairs that slides around a digitizing tablet when pushed. Used to make CAD drawings. (93)

pupil — the opening in the iris of the eye. It controls the amount of light allowed into the eye. (220)

Q

qwerty keyboard — a keyboard with the letters Q, W, E, R, T, and Y all in a row near the top left corner. (90)

R

radar — Radio Direction and Ranging. Radar uses reflected radio signals to detect objects that are too far away to be seen in other ways. (388)

radio — a device that can send and receive signals without a connecting wire. (362)

radio astronomy — a form of astronomy using giant dish antennas that receive radio waves. The patterns of the waves tell astronomers more about the universe. (386)

radio waves — electromagnetic waves used for many types of broadcasting. (351)

random-access memory (RAM) — computer memory that stores information temporarily. (82)

rangefinder camera — a camera that uses a separate lens for the viewfinder. (231)

read-only memory (ROM) — computer memory with permanent storage of information that cannot be added to or changed. (82)

receiver — a device that receives a message. (366)

recording heads — electromagnets that change sound vibrations into electrical signals; used in tape recorders. (377)

reducer — chemical used to lighten a dark film negative. (247)

reference beam — a beam that is reflected from a mirror and falls on a photographic plate at an angle in order to create a hologram. (225)

reflected-light meters — meters that measure the amount of light reflected off a photographic subject. (239)

refracted — bent. (215)

registration — holding paper in precisely the same spot each time as it goes through a press so that colors and images align properly when one sheet is printed upon two or more times. (316)

rehearsal — practice for an audio or video program. (394)

relief printing — the process of printing from a raised surface; the oldest printing method. (318)

remote telecasts — programs done away from the TV studio. (373)

research — the search for new knowledge. (25)

resolution — the degree of sharpness of an image on a TV or computer monitor. (100)

retina — a light-sensitive coating at the back of the eye. (220)

reverse — a light color of type on a dark background, often a halftone. (310)

RGB — red, green, and blue; color system used for TV transmission. (370)

rhythm — repetition in a design. (284)

rods — cells in the eye that register amounts of light. The rods are color blind. (221)

Roman typefaces — typefaces made with a serif on most letters. The width of the strokes used to form each letter varies. (288)

rotogravure — a form of intaglio printing done with round cylinders. (320)

rough layout — a layout that contains all the information a printer needs to produce a final product. (291)

rule — a line used in graphic design. (289)

rule of thirds — a rule for placing the subject within a photo or design composition. The rule of thirds divides the viewing area into thirds both vertically and horizontally. (252)

S

saddle stitching — driving a wire staple through the back of printed signatures gathered together; a form of binding. (331)

safelights — lights with coatings that filter out certain colors of light; the remaining light does not expose photographic paper. (260)

sampling — electronically capturing a piece of music. (56)

sans — "without." (288)

scales — drafting tools used to make proportional measurements. (156)

scanner — a device that sends a "picture" of text or artwork to a computer. (98)

schematic diagrams — drawings that illustrate how the different electrical and electronic parts in a device are connected to each other as well as to the power sources. (175)

screen printing — transferring ink through a stencil held in place by a screen. (322)

screen tint — an area of uniform dots that creates the illusion of a lighter shade in a printed area. (309)

script typefaces — typefaces that look like handwriting. (288)

script — a line-by-line description of an audio or video program. (394)

sectional view — a drawing of an object that looks as though the object has been cut open; used when inner details are complex. (167)

semaphore — a telecommunication system that uses two flags held in different positions to represent letters in the alphabet. (22)

semiconductor — a substance that conducts electricity better than an insulator but not as well as a conductor. (77)

Senefelder, Aloys — the inventor of lithography. (314)

sensors — devices that detect such things as light, pressure, temperature, or sound. (121)

serial port — a bridge through which computer data can travel one bit at a time. (83)

shop drawings — drawings made for parts or steel members that are made in the shop rather than at a construction site. (178)

shutter — a curtain at the front of a camera that opens and closes the lens opening. (230)

side stitching — stapling through the side of a collated product; a binding method. (331)

signal plate — a coating inside a TV pickup tube that converts an image area to an electrical signal. (369)

signatures — folded pages, one inside the other, that are put together into magazines and books. (331)

silicon — the second most common element in the earth's crust; found in clay, sand, and many rocks. An ingredient in many products, including glass, silicones, and microchips. (77)

single lens reflex camera — a camera with which both viewing and picture taking are done through a single lens. A mirror inside the camera reflects the image to a five-sided prism. The prism reflects the image to the viewfinder. (233)

single-mode optical fibers — optical fibers requiring special lasers as their light source. (226)

sketches — quick, freehand drawings that remain fairly rough and unfinished. (162)

skylight filter — a clear filter that helps to screen out ultraviolet light that can produce a haze on a photograph. (240)

sky waves — radio waves that radiate toward space. (365)

social — having to do with the ways in which communities of people live. (49)

software — the operating system (set of instructions) that tells a computer what to do with the data it receives. (84)

sound — a series of waves in the air that makes an eardrum vibrate. (348)

speaker — a device that changes an audio signal back into sound. (366)

special effects filters — filters used to create a wide range of interesting photographic images, such as sunbursts or mirror images. (240)

speed of light — 186,000 miles per second. (214)

spot lights — lights that direct a narrow beam of light at a small area. (239)

spot retouching — painting over a spot on a photograph with a gray spotting solution. (264)

spreadsheet — computer software that keeps track of budgets and other financial matters. (118)

square serif typefaces — typefaces made with square serifs. (288)

squeegee — a firm but flexible rubber-like blade. (324)

standards — drafting rules that say what symbols, and so on, should be used for different things. (142)

star network — *see* switching network. (114)

stencil — a thin sheet with holes cut through it in the shapes of letters, designs, and so on. (322)

step test — a test made to save time and photographic paper by exposing a small strip first. (262)

stick lead holder — a drafting tool similar to a pencil but it needs no sharpening and keeps a constant line width. (153)

stop bath — chemical solution of 18 percent acetic acid that halts the film developing process by neutralizing the developer. (247)

storage media — media on which computer data is stored; includes programming cards, magnetic tape, floppy disks, hard disks, and optical disks. (39)

stress analysis — tests that determine whether a structure will be strong enough; may be done using a model of the structure created on a computer. (190)

stress — tension created inside a material by forces acting on it from the outside. (191)

stripping — the process of attaching film negatives to a masking sheet. (308)

strobe lighting — *see* electronic flash. (239)

substrate — any material on which printing is done. (290)

subtractive colors — colors of light created by mixing additive primary colors. (217)

surface model — a computer model that resembles a wireframe model with a skin over it. (192)

surprint — the overlaying of one printed image (generally text) on top of another (usually a halftone). (310)

survey — the process of determining and locating forms and boundaries of a tract of land by measurement. (180)

switcher — a device that receives input from video cameras and allows the director to choose which image to record. (371)

switching network — a network in which each computer is linked to a central switching unit. (114)

symbol library — a collection of predrawn symbols for CAD. (189)

synchronization pulse — a signal that assures that TV receiver scanning is an exact match to that done by the TV camera. (370)

synthesizer — a device that creates sound from computer data. (103)

T

T-square — a straightedge drafting tool, with a crosspiece at one end. (154)

target — a coating inside a TV pickup tube that helps create the video signal. (369)

technical communication systems — communication systems that depend upon specific tools and equipment. (24)

technical design — drafting, mechanical drawing, and engineering design. (24)

technical drawing pen — a pen with nibs in a variety of standard inch and metric widths. (154)

technical illustrations — technical drawings that resemble artwork. (164)

technology — using knowledge, tools, and skills to solve problems. (15)

technology assessment — the process of studying the effect of a new technical device. (48)

tele — "over a distance." (124)

telecommunication — communicating over distance. (22)

telephoto lenses — lenses that make faraway objects appear closer and enlarge subjects. (238)

templates — patterns used to draw standard shapes and figures. (155)

text typefaces — typefaces styled after the work of scribes who copied books by hand during the Middle Ages. They are very detailed. (288)

thermal printer — a printer that uses heat to create images on the paper. (102)

thermofax machine — a device used to produce overhead projection transparencies. (323)

thermography — a method that produces a raised image on a substrate without embossing. (330)

third angle projection — a method of laying out a multiview drawing in which the top view is on top, the front view is below it, and the right side view is on the right; used in most industrialized countries. (166)

thumbnail sketches — small, quick sketches drawn to try out ideas and show how something might look. (291)

title block — block of notes that gives the name of a drawing and other important information. (143)

tolerance — the amount a measure can vary from drawing specifications without harming the product. (150)

toner — chemical used to change the color of photographic paper. (247)

topographic mapping — mapping the surface features in an area or region of the earth to show natural features, such as hills and lakes, as well as structures and boundaries. (180)

touchpad — pads similar to digitizing tablets but cheaper and less accurate. A two-layer sandwich of materials conducts electricity and creates the image on a computer screen. (94)

touch screens — screens similar to digitizing tablets. They sense where a finger has touched them and send this information to a computer. (96)

trackball — a computer device that works the same as a mouse, except that the ball is spun by hand rather than by moving it around on a surface. (96)

tracks — thin bands that run the length of an audio or video tape. (378)

transistor — a device used to amplify (increase) or control electric current. (77)

transmission channel — device that carries a message. (348)

transmitter — a device that creates electromagnetic waves used to carry a signal. (364)

triangles — drafting tools used to draw vertical and angled lines. (155)

trimetric drawings — drawings that show an object using three axes. (169)

tripod — a three-legged stand that holds a camera while the photograph is being taken. (242)

true solid model — a computer model that behaves as a real-life solid object based on information about physical properties, such as density, volume, and weight; used for stress analysis. (192)

tuner — a device inside an audio or video receiver that locates the correct frequency. (374)

tungsten light — light from ordinary lightbulbs. (246)

twin lens reflex camera — a camera with two lenses, one for viewing the subject and the other for focusing the light onto the film. The term "reflex" means there is a mirror inside the camera that reflects the image of the subject up to the photographer's eye. (232)

twisted-pair wire — two thin, insulated, copper wires twisted around each other; used for local telephone transmission. (359)

two-point perspective drawing — a perspective drawing in which the vertical lines and axis are parallel and do not meet at a vanishing point. All other lines are drawn to one of two vanishing points. (172)

two-way radios — radios that allow both sending and receiving; used for safety and defense. (386)

typeface — a certain design of type. (288)

typesetting — the process of turning written text into the final typefaces. (292)

typographer's rule — a rule with six different scales on it, including those for points and picas. (289)

typography — the use of type. (288)

U

umbrella reflector — a device used to spread light over a wide area for photography. (240)

unidirectional dimensioning — dimensions placed so that they are all read in line with the bottom of a technical drawing. (146)

unilateral tolerancing — tolerancing in which variation is allowed in only one direction. (150)

UNIVAC I — computer brought out by the Sperry Rand Corporation in 1951. (77)

universal product code — black stripes or bars printed on product packages that when read by a scanner indicate price and inventory information. (121)

universal systems model — a description of technical systems; input, process, output and feedback. (18)

user friendly — term that describes a computer or program that is easy to use. (96)

V

vacuum back — the back of a process camera that holds the film in place. (300)

value — the darkness or lightness of a hue; depends on the amount of light reflected. (216)

variety — difference; adds interest and excitement to a graphic design. (285)

vellum — a drawing medium similar to tracing paper but more translucent. (153)

video format — the type of videotape used for a recording system. (380)

video monitors — TV sets in a control room that show scenes from different cameras or other sources. (372)

video — relating to the use, transmission, or reception of visual images. (347)

videodisk — optical disk that closely resembles a CD-ROM. The laser that writes on the disk is driven by a videotape. (106)

view camera — a large camera the lens of which is attached with a collapsible bellows. (232)

visible spectrum — colors that can be seen by the human eye; red, orange, yellow, green, blue, indigo, and violet. (216)

voice input devices — devices that change the spoken word into data a computer can work with. (98)

W

walkie-talkie — a small two-way radio. (386)

wall section — a drawing that shows how a wall is to be constructed and what type of materials are to be used. (178)

water wash — a water bath that removes all fixer and excess chemicals that might otherwise ruin processed film or photographic paper after it dries. (247)

wave length — a measurement made from a point on one wave to the same point on the next. (352)

wireframe model — a computer model that appears as if it were made of bent wires. (192)

wireless microphone — a microphone containing a tiny battery-powered transmitter. (363)

wiring diagrams — drawings that show how electrical products are wired together. (175)

word processing — computer software that handles writing tasks. (118)

World Wide Web — a network of servers that store files for distribution over the Internet. People with Internet access and client software can browse (look at) and download files on the WWW. (116)

WORM disk — an optical storage computer disk. WORM stands for "write once, read many." (106)

Wozniak, Steve — co-founder of Apple Computer Company. (78)

X

xerography — a form of photocopying using an electrostatic charge and heat processing. (157)

Y

yagi antenna — an arrow-shaped antenna that can be pointed in the direction from which a signal is coming. (366)

Z

zoom lenses — camera lenses that allow the photographer to change the focal length as needed. A subject that is far away "zooms" into focus. (238)

Zworykin, Vladimir — the person who first demonstrated a practical video transmitting/receiving system. (368)

CREDIT LIST

AP/Wide World Photos, 388
AP/Wide World Photos/Vince Bucci, 51
AT&T Archives, 34, 226
Alvin & Company, 154, 156, 158
American Heart Association, CPR Learning System/Actronics, Inc., 125
Apple Computer, Inc., 35, 36, 78, 98, 102, 134
Apple Computer, Inc./John Greenleigh Photography, 82, 96, 100
Arnold & Brown, 13, 32, 36, 39, 52, 106, 113, 123, 125, 152, 209, 215, 217, 219, 222, 231, 232, 235, 236, 239-241, 244, 245, 248, 252, 257, 258, 261, 265, 292, 296, 312, 325, 330, 361, 363, 380, 390, 395, 397
Art Resource/Sotheby Parke-Bernet, 314
Autodesk, Inc., 194

Roger B. Bean, 53, 55, 63, 91, 130, 219, 230, 238, 250, 253, 254, 282, 294, 307, 315, 327, 345, 346, 393
Bell & Howell Document Management Products Company, 158
The Bettman Archive, 25, 26, 34, 35, 50, 75, 77, 213, 320, 350, 358, 368, 375

CH Products, 92
Connecting Point Computer Center/ Arnold & Brown, 39
Chevrolet Motor Division/David Kimble, 138
Chevrolet Motor Division, General Motors Corporation, David Kimble, artist, 138
City of Peoria Police Department/ Roger B. Bean, 384
City of Peoria Police Department, 384
CS&A, 18, 20, 26, 40, 62, 79, 116, 121, 214, 216, 217, 243, 348, 352-353
Curt Campbell, 110
Canon U.S.A., Inc., 327
Matt Coker/Arnold & Brown, 281
Control Data, 192, 193
Corning Glass Works, 34, 126
Coventry Creative Graphics, 10-13, 287, 350
Crossfield Electronics, Ltd., 56, 98, 294, 298, 303
Custom Design and Prototype, 175
Custom Video, Dallas, Texas/Ann Garvin, 35

Dahle U.S.A., 158
Design Associates, 284
Diebold, Inc., 119, 123
Dolphin Research Center, Grassy Key, FL/ Laurel Canty, 11
Dow Corning Corporation, 116
Dycam, Inc., Chatsworth, CA, 236

Eastman Kodak Company, 328
Epson America, Inc., 23, 101

David Falconer/Frazier Photolibrary, 69
Marc Featherly, 74
Flexographic Technical Association, 319, 320
David R. Frazier Photolibrary, 46
Fujitsu America, 191

Bob Gangloff, 19, 22, 33, 38, 48, 55, 57, 83, 86, 92, 94, 95, 97, 99, 101, 102, 104, 105, 108, 112, 114, 115, 137, 144, 171, 180, 214-216, 220-222, 224-226, 230-234, 242, 246, 247, 259, 266, 284-286, 290, 294, 304, 306, 314, 315, 320, 323, 349, 351, 358-360, 362, 364, 365, 367-371, 374-379, 396, 397, 402, 405
Gannett Company, Inc., 44
Ann Garvin, 35, 48, 65, 77, 344, 354, 392
Georgia Port Authority, 182
Gerber Scientific Instrument Company, 310
Graphic Associates, 53
Gravure Association of America, 322

Harris Corporation, 364
Heidelberg, Inc., 317
Hell Graphics, 307
Hedrich-Blessing/Pentair, 280
Jeffery R. Henline, OD./Arnold & Brown, 223
Hewlett-Packard Company, 2, 36, 39, 70, 101, 382
Charles Hofer, 384
Honeywell, Inc., 23
Houston Instruments, 103, 188
Hughes Aircraft Company, 15, 35, 225, 389

Intel Corporation, 30, 34, 77
International Business Machines Corporation, 34, 36, 39, 72, 75, 76, 84, 97, 101, 187
International Jensen, Inc., 367
InterVision Systems, Inc., Raleigh, NC, 12

Jupiter Transportation Company/Ann Garvin, 386

Knight-Ridder, Inc., 27, 228
Knight-Ridder, Inc./Paul Barton Photography, 27
Knight-Ridder, Inc./Chuck Mason Photography, 27
Knoxville Clinic/Roger B. Bean, 384

Interior Design by Benoit Design & Nancy Norall DESIGN

A

D

E

F

G

H

I

J

K

L

M

N

O

P

Q

R

S

T